BASIC ABSTRACT ALGEBRA

Second edition

Basic abstract algebra

Second edition

P. B. BHATTACHARYA
Formerly, University of Delhi

S. K. JAIN
Ohio University

S. R. NAGPAUL
St. Stephen's College, Delhi

Published by the Press Syndicate of the University of Cambridge
The Pitt Building, Trumpington Street, Cambridge CB2 1RP
40 West 20th Street, New York, NY 10011-4211, USA
10 Stamford Road, Oakleigh, Melbourne 3166, Australia

First edition published 1986
Second edition published 1994
Reprinted 1995

Library of Congress Cataloging-in-Publication Data is available.

A catalogue record for this book is available from the British Library.

ISBN 0-521-46081-6 hardback
ISBN 0-521-46629-6 paperback

Transferred to digital printing 2001

For
PARVESH JAIN
To whom we owe more than we can possibly expres

Contents

Part II Groups

Part III Rings and modules

Preface to the second edition

The following are the main features of the second edition.

More than 150 new problems and examples have been added. The new problems include several that relate abstract concepts to concrete situations. Among others, we present applications of G-sets, the division algorithm and greatest common divisors in a given euclidean domain. In particular, we should mention the combinatorial applications of the Burnside theorem to real-life problems. A proof for the constructibility of a regular n-gon has been included in Chapter 18.

We have included a recent elegant and elementary proof, due to Osofsky, of the celebrated Noether–Lasker theorem.

Chapter 22 on tensor products with an introduction to categories and functors is a new addition to Part IV. This chapter provides basic results on tensor products that are useful and important in present-day mathematics.

We are pleased to thank all of the professors and students in the many universities who used this textbook during the past seven years and contributed their useful feedback. In particular, we would like to thank Sergio R. Lopez-Permouth for his help during the time when the revised edition was being prepared. Finally, we would like to acknowledge the staff of Cambridge University Press for their help in bringing out this second edition so efficiently.

P. B. Bhattacharya
S. K. Jain
S. R. Nagpaul

Preface to the first edition

This book is intended for seniors and beginning graduate students. It is self-contained and covers the topics usually taught at this level.

The book is divided into five parts (see diagram). Part I (Chapters 1–3) is a prerequisite for the rest of the book. It contains an informal introduction to sets, number systems, matrices, and determinants. Results proved in Chapter 1 include the Schröder–Bernstein theorem and the cardinality of the set of real numbers. In Chapter 2, starting from the well-ordering principle of natural numbers, some important algebraic properties of integers have been proved. Chapter 3 deals with matrices and determinants. It is expected that students would already be familiar with most of the material in Part I before reaching their senior year. Therefore, it can be completed rapidly, skipped altogether, or simply referred to as necessary.

Part II (Chapters 4–8) deals with groups. Chapters 4 and 5 provide a foundation in the basic concepts in groups, including G-sets and their applications. Normal series, solvable groups, and the Jordan–Hölder theorem are given in Chapter 6. The simplicity of the alternating group A_n and the nonsolvability of S_n, $n > 4$, are proved in Chapter 7. Chapter 8 contains the theorem on the decomposition of a finitely generated abelian group as a direct sum of cyclic groups, and the Sylow theorems. The invariants of a finite abelian group and the structure of groups of orders p^2, pq, where p, q are primes, are given as applications.

Part III (Chapters 9–14) deals with rings and modules. Chapters 9–11 cover the basic concepts of rings, illustrated by numerous examples, including prime ideals, maximal ideals, UFD, PID, and so forth. Chapter 12 deals with the ring of fractions of a commutative ring with respect to a multiplicative set. Chapter 13 contains a systematic development of

TABLE OF INTERDEPENDENCE OF CHAPTERS

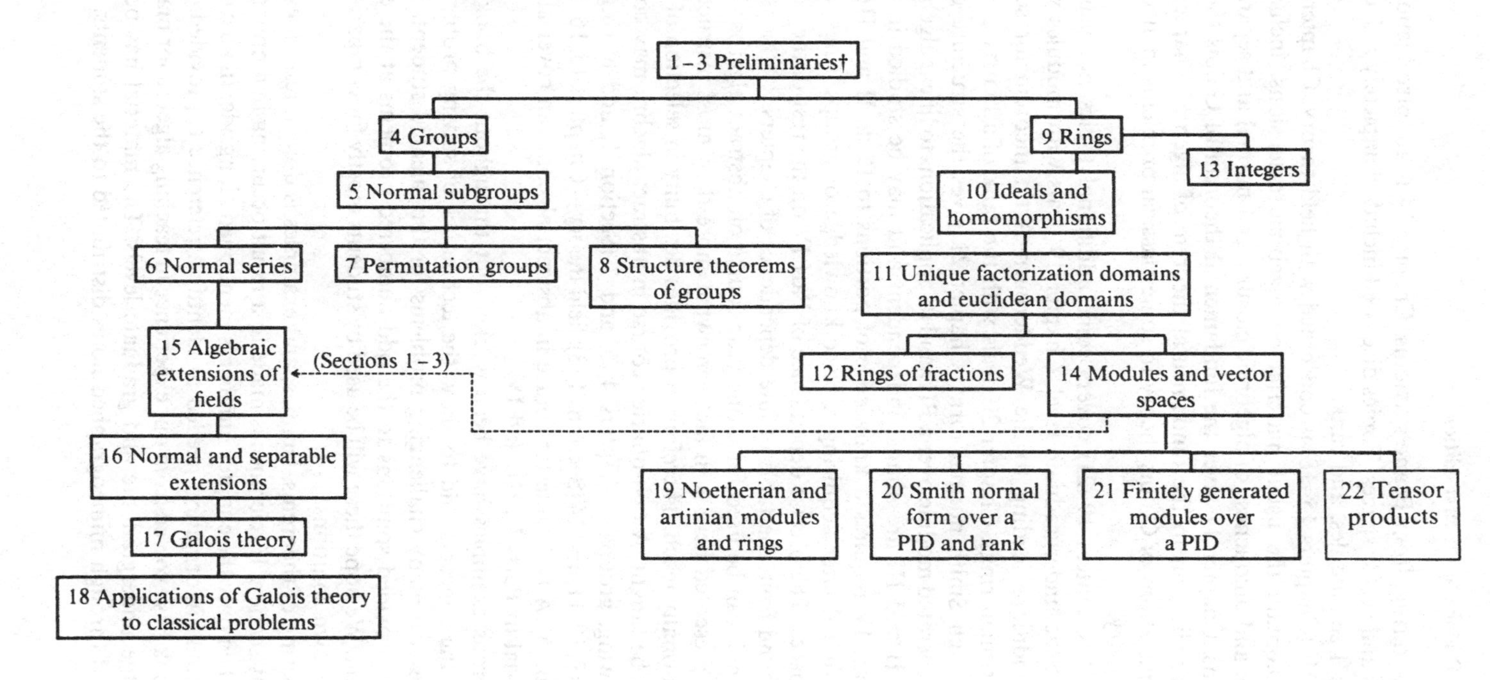

† To be read as and when needed.

integers, starting from Peano's axioms. Chapter 14 is an introduction to modules and vector spaces. Topics discussed include completely reducible modules, free modules, and rank.

Part IV (Chapters 15–18) is concerned with field theory. Chapters 15 and 16 contain the usual material on algebraic extensions, including existence and uniqueness of algebraic closure, and normal and separable extensions. Chapter 17 gives the fundamental theorem of Galois theory and its application to the fundamental theorem of algebra. Chapter 18 gives applications of Galois theory to some classical problems in algebra and geometry.

Part V (Chapters 19–21) covers some additional topics not usually taught at the undergraduate level. Chapter 19 deals with modules with chain conditions leading to the Wedderburn–Artin theorem for semi-simple artinian rings. Chapter 20 deals with the rank of a matrix over a PID through Smith normal form. Chapter 21 gives the structure of a finitely generated module over a PID and its applications to linear algebra.

Parts II and III are almost independent and may be studied in any order. Part IV requires a knowledge of portions to Parts II and III. It can be studied after acquiring a basic knowledge of groups, rings, and vector spaces. The precise dependence of Part IV on the rest of the book can be found from the table of interdependence of chapters.

The book can be used for a one-year course on abstract algerba. The material presented here is in fact somewhat more than most instructors would normally teach. Therefore, it provides flexibility in selection of the topics to be taught. A two-quarter course in abstract algebra may cover the following: groups – Chapters 4, 5, and 7 (Section 1) and 8; rings – Chapters 9, 10, 11, and 14 (Sections 1–3); field theory – Chapters 15, 16, and 18 (Section 5). A two-semester course in abstract algebra can cover all of the material in Parts II, III, and IV.

Numerous examples have been worked out throughout the book to illustrate the concepts and to show the techniques of solving problems. There are also many challenging problems for the talented student. We have also provided solutions to the odd-numbered problems at the end of the book. We hope these will be used by students mostly for comparison with their own solutions.

Numbering of theorems, lemmas, and examples is done afresh in each chapter by section. If reference is made to a result occurring in a previous chapter, then only the chapter number is mentioned alongside. In all cases the additional information needed to identify a reference is provided.

The book has evolved from our experience in teaching algebra for many years at the undergraduate and graduate levels. The material has been class tested through mimeographed notes distributed to the students.

We acknowledge our indebtedness to numerous authors whose books have influenced our writing. In particular, we mention P. M. Cohn's *Algebra*, Vols. 1, 2, John Wiley, New York, 1974, 1977, and S. Lang's *Algebra*, Addison-Wesley, Reading, MA, 1965.

During the preparation of this book we received valuable help from several colleagues and graduate students. We express our gratitude to all of them. We also express our gratefulness to Ohio University for providing us the facilities to work together in the congenial environment of its beautiful campus.

It is our pleasant duty to express our gratitude to Professor Donald O. Norris, Chairman, Department of Mathematics, Ohio University, whose encouragement and unstinted support enabled us to complete our project. We also thank Mrs. Stephanie Goldsberry for the splendid job of typing.

P. B. Bhattacharya
S. K. Jain
S. R. Nagpaul

Glossary of symbols

$\forall$	for all
$\exists$	there exists
$\in$	is an element of
$\notin$	is not an element of
$\{x \in A \mid P(x)\}$	set of all $x \in A$ satisfying condition $P(x)$
$\mathscr{P}(x)$	power set of x
$(X_i)_{i\in\Lambda}$	family indexed by set Λ
$\bigcup\limits_{i\in\Lambda} X_i$	union of $(X_i)_{i\in\Lambda}$
$\bigcap\limits_{i\in\Lambda} X_i$	intersection of $(X_i)_{i\in\Lambda}$
$X \times Y$	Cartesian product of X and Y
$\varnothing$	empty set
$\subset$ or $\subseteq$	is a subset of
$\subsetneqq$	is a proper subset of
$\supset$ or $\supseteq$	contains
$\supsetneqq$	properly contains
$\Rightarrow$	implies
$\Leftrightarrow$	if and only if
iff	if and only if
$f: X \to Y$	f is a map of X into Y
$f(x)$	image of $x \in X$ under $f: X \to Y$
$f: x \mapsto y$	$y = f(x)$ where $f: X \to Y$, $x \in X$, $y \in Y$
$\circ$	composition
ϕ	Euler's function

(a,b)	in number theory, the greatest common divisor of a and b; in rings and modules, the ideal or submodule generated by a and b
$a \mid b$	a divides b
$a \nmid b$	a does not divide b
δ_{ij}	Kronecker delta
det	determinant
$\operatorname{sgn} \sigma$	± 1, according as the permutation σ is even or odd
e_{ij}	square matrix with 1 in (i, j) position, 0 elsewhere
$\mathbf{n}$	the set of positive integers $\{1,2,3,\ldots,n\}$
$\mathbf{N}$	set of all natural numbers
$\mathbf{Z}$	set of all integers
$\mathbf{Q}$	set of all rational numbers
$\mathbf{R}$	set of all real numbers
$\mathbf{C}$	set of all complex numbers
$\mathfrak{c}$	the cardinal of the continuum (cardinality of the reals)
$Z/(n)$ or Z_n	integers modulo n
$\|X\|$ or card X	cardinality of X
$\|G\|$	order of group G
$[S]$	subgroup generated by S
C_n	cyclic group of order n
S_n	as a group, the symmetric group of degree n; as a ring, the ring of $n \times n$ matrices over S
A_n	alternating group of degree n
D_n	dihedral group of degree n
$\mathrm{GL}(m, F)$	group of invertible $m \times m$ matrices over F
$Z(G)$	center of G
$\lhd$	is a normal subgroup of
A/B	quotient group (ring, module) of A modulo B
$[L:K]$	in groups, the index of a subgroup K in a group L; in vector spaces, the dimension of a vector space L over K; in fields, the degree of extension of L over K
$N(S)$ $(N_H(S))$	normalizer of S (in H)
$C(S)$ $(C_H(S))$	conjugate class of S (with respect to H)
$\prod_{i\in\Lambda} X_i$	product of $(X_i)_{i\in\Lambda}$
$\oplus \sum_{i\in\Lambda} X_i$	direct sum of $(X_i)_{i\in\Lambda}$
$\operatorname{Im} f$	image of homomorphism f
$\operatorname{Ker} f$	kernel of homomorphism f
$\subsetneq$	is isomorphic into (embeddable)
$\simeq$	is isomorphic onto

R^{op}	opposite ring of R
(S)	ideal (submodule) generated by S
$(S)_r$	right ideal generated by S
$(S)_l$	left ideal generated by S
$\sum_{i\in\Lambda} X_i$	sum of right or left ideals (submodules) $(X_i)_{i\in\Lambda}$
$R[x]$	polynomial ring over R in one indeterminate x
$R[x_1,...,x_n]$	polynomial ring over R in n indeterminates, $x_1,...,x_n$
$R[[x]]$	formal power series ring
$R\langle x\rangle$	ring of formal Laurent series
$Z(p\infty)$	rationals between 0 and 1 of the form m/p^n, $m,n > 0$ under the binary operation "addition modulo 1"
$\mathrm{Hom}_R(X,Y)$	set of all R-homomorphisms of R-module X to R-module Y
$\mathrm{Hom}(X,Y)$	set of all homomorphisms of X to Y
$\mathrm{End}(X)$	endomorphisms of X
$\mathrm{Aut}(X)$	automorphisms of X
R_S	localization of a ring R at S
$F(\alpha)$	subfield generated by F and α
$F[S]$	subring generated by F and S
$F(S)$	subfield generated by F and S
$GF(q)$	Galois field (finite field) with q elements
$\overline{F}$	algebraic closure of F
E_H	fixed field of H
$G(E/F)$	Galois group of automorphisms of E over F
$\phi_n(x)$	cyclotomic polynomial of degree n
$M\otimes_R N$	tensor product of M_R and $_RN$
$\square$	end of the proof

PART I

Preliminaries

CHAPTER 1

Sets and mappings

1 Sets

The concept of set is fundamental in mathematics. It is not our purpose to present here an axiomatic account of set theory. Instead we shall assume an intuitive understanding of the terms "set" and "belongs to." Informally speaking, we say that a *set* is a collection of objects (or elements).

If S is a set and x is an element of the set S, we say *x belongs to S,* and we write $x \in S$. An element of a set S is also called a member of S. If x does not belong to S, we write $x \notin S$.

Let A and B be sets. We say that A and B are *equal,* written $A = B$, if they consist of the same elements; that is,

$$x \in A \Leftrightarrow x \in B.$$

(The symbol $\Leftrightarrow$ stands for "if and only if.") A set is thus determined by its elements.

A set with a finite number of elements can be exhibited by writing all of its elements between braces and inserting commas between elements. Thus, $\{1,2,3\}$ denotes the set whose elements are 1, 2, and 3. The order in which the elements are written makes no difference. Thus, $\{1,2,3\}$ and $\{2,1,3\}$ denote the same set. Also, repetition of an element has no effect. For example, $\{1,2,3,2\}$ is the same set as $\{1,2,3\}$. Given a set A and a statement $P(x)$, there is a unique set B whose elements are precisely those elements x of A for which $P(x)$ is true. In symbols, we write $B = \{x \in A | P(x)\}$. When the context is clear, we sometimes also write $B = \{x | P(x)\}$.

There are standard symbols for several sets that we frequently deal with. Some of these are given below. Others will be introduced subsequently as the occasion arises.

N denotes the set of all natural numbers 1,2,3,....
Z is the set of all integers $0,\pm 1,\pm 2,\ldots$.
Q is the set of all rational numbers – that is, fractions a/b, where a,b are integers and $b \neq 0$.
R is the set of all real numbers.
C is the set of all complex numbers $x + iy$, where x,y are real numbers and $i^2 = -1$.

For any positive integer n, the set $\{1,2,\ldots,n\}$ is denoted by **n**. A set S is called *finite* if it has no elements, or if its elements can be listed (counted, labeled) by natural numbers 1,2,3,... and the process of listing stops at a certain number, say n. The number n is called the cardinality of S, and we write $|S| = n$. A set whose elements cannot be listed by the natural numbers $1,2,\ldots,n$ for any n whatsoever is called an *infinite* set.

A set S is said to be *empty* if S has no elements; that is, the statement $x \in S$ is not true for any x. If S and T are both empty sets, then $S = T$, since the condition $x \in S \Leftrightarrow x \in T$ is satisfied because there is no element x in either S or T to which the condition may be applied. (In such a case we say that the condition is satisfied *vacuously*.) Because any two empty sets are equal, there is just one empty set, which is denoted by $\varnothing$. The empty set is also called the *null* set or the *void* set.

Definition. *Let A and B be sets. A is called a* subset *of B if every element of A is an element of B; that is, if*

$$a \in A \Rightarrow a \in B.$$

(The symbol $\Rightarrow$ stands for "implies.")

If A is a subset of B, we write $A \subset B$. (Some authors write $A \subseteq B$.) Further, if A is a subset of B, we also say that B contains (or includes) A, and we write $B \supset A$ (or $B \supseteq A$).

It follows immediately from the definition that A and B are equal if and only if $A \subset B$ and $B \subset A$. Thus, every set is a subset of itself. Moreover, the empty set $\varnothing$ is a subset of every set because the condition $x \in \varnothing \Rightarrow x \in A$ is satisfied vacuously.

If $S \subset A$, but $S \neq A$, then S is a *proper subset* of A written as $S \subsetneq A$.

Definition. *Let A and B be subsets of a set U. The* union *of A and B, written $A \cup B$, is the set*

$$A \cup B = \{x \in U | x \in A \text{ or } x \in B\}.$$

The intersection *of A and B, written $A \cap B$, is the set*

$$A \cap B = \{x \in U | x \in A \text{ and } x \in B\}.$$

The difference *of A and B, written $A - B$, is the set*

$$A - B = \{x \in U | x \in A \text{ and } x \notin B\}.$$

If $B \subset A$, then $A - B$ is called the complement *of B in A. A and B are said to be* disjoint *if $A \cap B$ is empty ($A \cap B = \varnothing$).*

For example, if $A = \{1,2,3\}$ and $B = \{3,4,5\}$, then $A \cup B = \{1,2,3,4,5\}$, $A \cap B = \{3\}$, $A - B = \{1,2\}$, and $B - A = \{4,5\}$.

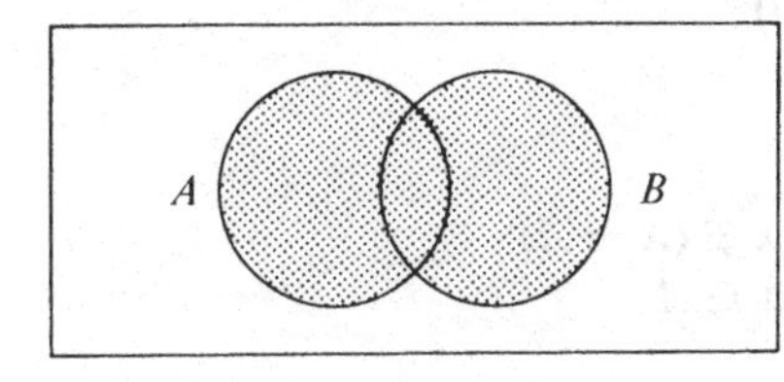

$A \cup B$

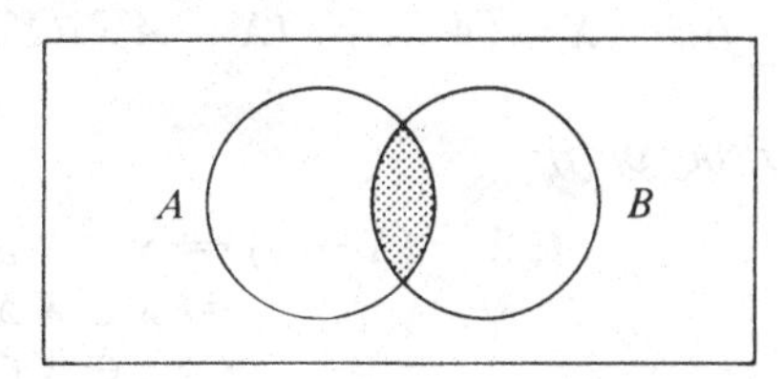

$A \cap B$

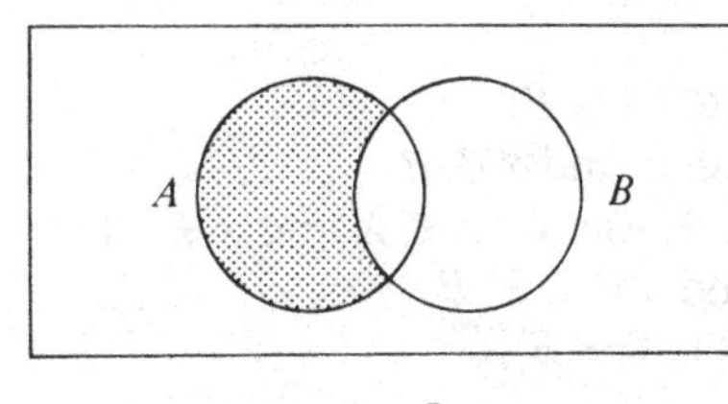

$A - B$

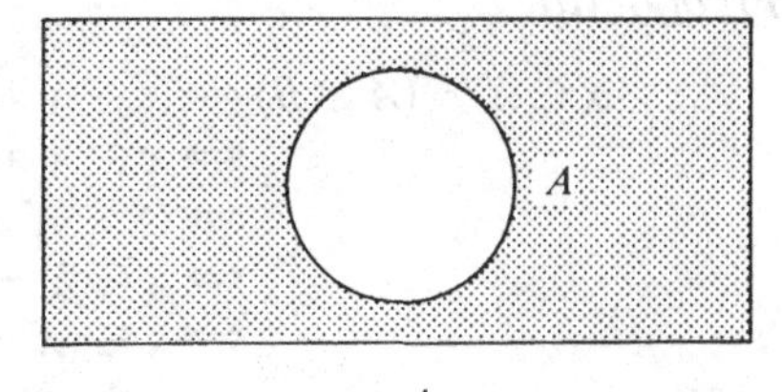

A'

The term *universal set* is sometimes used for a set U that contains all sets in a given context; that is, $X \subset U$ for every set X under consideration. The complement of X in U (namely, the set $U - X$) is then simply called the complement of X and is written X' without explicit reference to U.

It is sometimes helpful to illustrate union, intersection, difference, and complement by means of *Venn diagrams.* We draw circles to represent the given sets A and B and enclose them within a rectangle representing the universal set U. The shaded area in each diagram represents the set $A \cup B$, and so forth, as indicated.

1.1 **Theorem.** *Let A, B, and C be sets. Then*

(i) $A \cup A = A = A \cap A$.
(ii) $A \cup B = B \cup A$; $A \cap B = B \cap A$.

(iii) $(A \cup B) \cup C = A \cup (B \cup C)$, $(A \cap B) \cap C = A \cap (B \cap C)$.
(iv) $A \cup (B \cap C) = (A \cup B) \cap (A \cup C)$,
$A \cap (B \cup C) = (A \cap B) \cup (A \cap C)$.
(v) $A \cup (A \cap B) = A = A \cap (A \cup B)$.

Proof. Left as an exercise. □

1.2 Theorem (DeMorgan's rules). *Let A, B, and X be sets. Then*

(i) $X - (X - A) = X \cap A$.
(ii) $X - (A \cup B) = (X - A) \cap (X - B)$.
(iii) $X - (A \cap B) = (X - A) \cup (X - B)$.

Proof of (i):

$$\begin{aligned} x \in X - (X - A) &\Leftrightarrow x \in X \text{ and } x \notin (X - A) \\ &\Leftrightarrow x \in X \text{ and } x \in A \\ &\Leftrightarrow x \in X \cap A. \end{aligned}$$

Hence, $X - (X - A) = X \cap A$. □

Proof of (ii):

$$\begin{aligned} x \in X - (A \cup B) &\Leftrightarrow x \in X \text{ and } x \notin (A \cup B) \\ &\Leftrightarrow x \in X \text{ and } x \notin A \text{ and } x \notin B \\ &\Leftrightarrow (x \in X \text{ and } x \notin A) \text{ and } (x \in X \text{ and } x \notin B) \\ &\Leftrightarrow x \in X - A \text{ and } x \in X - B \\ &\Leftrightarrow x \in (X - A) \cap (X - B). \end{aligned}$$

Hence, $X - (A \cup B) = (X - A) \cap (X - B)$.
The last part is proved similarly. □

In view of the equality $(A \cup B) \cup C = A \cup (B \cup C)$ (Theorem 1.1), we can do away with the parentheses and simply write $A \cup B \cup C$ to denote unambiguously the union of the sets A, B, and C. Moreover, it is clear that the set $A \cup B \cup C$ consists of all those elements that belong to at least one of the sets A, B, and C. Likewise, $A \cap B \cap C$, the intersection of A, B, and C, is the set of those elements that belong to each of the sets A, B, and C. This suggests the following definition for the union and intersection of an arbitrary number of sets.

Definition. *Let S be a set whose elements are themselves sets. The* union *of all sets in S is defined to be the set*

$$\{x \mid x \in X \text{ for some } X \text{ in } S\}$$

and is denoted by $\bigcup_{X \in S} X$. *The* intersection *of all sets in S is defined to be the set*

$$\{x | x \in X \text{ for every } X \text{ in } S\}$$

and is denoted by $\bigcap_{X \in S} X$.

If S contains only a finite number of sets, say $X_1, \ldots, X_n$, their union is written $\bigcup_{i=1}^{n} X_i$ or $X_1 \cup \cdots \cup X_n$, and their intersection is written $\bigcap_{i=1}^{n} X_i$ or $X_1 \cap \cdots \cap X_n$.

Definition. *Let X be a set. The set of all subsets of X is called the* power set *of X and is denoted by* $\mathcal{P}(X)$. *That is,*

$$\mathcal{P}(X) = \{S | S \subset X\}.$$

Recall that the empty set $\varnothing$ and the set X itself are subsets of X and are therefore elements of $\mathcal{P}(X)$. For example, let $X = \{1,2\}$. Then the subsets of X are $\varnothing$, $\{1\}$, $\{2\}$, and X. Hence,

$$\mathcal{P}(X) = \{\varnothing, \{1\}, \{2\}, X\}.$$

If X is the empty set $\varnothing$, then $\mathcal{P}(X)$ has just one element: $\varnothing$.

It should be noted that a and $\{a\}$ are not the same. If $a \in X$, then $\{a\} \in \mathcal{P}(X)$.

1.3 Theorem. *Let X be a finite set having n elements. Then* $\mathcal{P}(X)$ *has* 2^n *subsets. Consequently,* $|\mathcal{P}(X)| = 2^{|X|}$.

Proof. Let us first consider those subsets of X that have r elements each, where $0 \leq r \leq n$. It is shown in high school algebra that the number of ways in which r elements can be selected out of n elements is

$$\binom{n}{r} = \frac{n!}{r!(n-r)!},$$

which is therefore the number of subsets of X having r elements each. Hence, the total number of subsets of X is $\sum_{r=0}^{n} \binom{n}{r}$. On putting $a = 1$ in the binomial expansion

$$(1 + a)^n = \sum_{r=0}^{n} \binom{n}{r} a^r,$$

we get

$$\sum_{r=0}^{n} \binom{n}{r} = (1 + 1)^n = 2^n.$$

This proves that X has exactly 2^n subsets. Hence, $\mathcal{P}(X)$ has 2^n elements. Since $n = |X|$, we get $|\mathcal{P}(X)| = 2^{|X|}$. □

Incidentally, the equality $|\mathcal{P}(X)| = 2^{|X|}$ explains why the set of all subsets of X is called the power set of X.

Definition. *Let X be a set. Let π be a set whose elements are nonempty subsets of X; that is, $\pi \subset \mathcal{P}(X)$ and $\varnothing \notin \pi$. If the elements of π are pairwise disjoint and their union is X, then π is called a* partition *of X, and the elements of π are called* blocks *of the partition π.*

For example, the set $\pi = \{\{1,2\},\{3\},\{4,5\}\}$ is a partition of the set $X = \{1,2,3,4,5\}$.

The next theorem follows immediately from the definition.

1.4 Theorem. *Let X be a set. Let π be a set whose elements are nonempty subsets of X. Then π is a partition of X if and only if each element of X belongs to exactly one element of π.*

Proof. Exercise. □

Note that if X is empty, it has only the partition $\pi = \varnothing$.

Problems

1. Prove Theorem 1.1.
2. Prove Theorem 1.2(iii).
3. If a set A has m elements and a set B has n elements, find the number of elements in $A \cup B$. Assume that $A \cap B$ has k elements.
4. After the registration of 100 freshmen, the following statistics were revealed: 60 were taking English, 44 were taking physics, 30 were taking French, 15 were taking physics and French, 6 were taking both English and physics but not French, 24 were taking English and French, and 10 were taking all three subjects.
   (a) Show that 54 were enrolled in only one of the three subjects.
   (b) Show that 35 were enrolled in at least two of them.
5. During quality control checking of a sample of 1000 TV sets, it was found that 100 sets had a defective picture tube, 75 sets had a defective sound system, 80 sets had a defective remote control system, 20 sets had a defective picture tube and a defective remote control, 30 sets had a defective picture tube and a defective sound system, 15 sets had a defective sound system and a defec-

tive remote control system, and 5 sets had all three defects. Use Venn diagrams to show that

(a) 195 sets had at least one defect.
(b) 805 sets had no defects.
(c) 55 sets had a defective picture tube only.
(d) 35 sets had a defective sound system only.
(e) 50 sets had a defective remote control only.

2 Relations

Definition. *Let a,b be elements of a set S. Then the set $\{\{a\},\{a,b\}\}$ is called an* ordered pair *and is denoted by (a,b); a is called the first component (or coordinate), and b is the second component (or coordinate).*

We now show that $(a,b) = (c,d)$ if and only if $a = c$ and $b = d$.

If $a = c$ and $b = d$, then trivially $(a,b) = (c,d)$. Conversely, let $(a,b) = (c,d)$. Then

$$\{\{a\},\{a,b\}\} = \{\{c\},\{c,d\}\}.$$

By definition of equality of sets, this implies

$$\{a\} = \{c\} \quad \text{or} \quad \{a\} = \{c,d\}.$$

If $\{a\} = \{c\}$, then we must have $\{a,b\} = \{c,d\}$. This yields $a = c$, $b = d$. If, on the other hand, $\{a\} = \{c,d\}$, then we must have $\{a,b\} = \{c\}$. So $a = c = d$ and $a = b = c$, which implies $a = c = b = d$.

Definition. *Let A,B be sets. The set of all ordered pairs (x,y), where $x \in A$ and $y \in B$, is called the* cartesian product *of A and B, in that order, and is denoted by $A \times B$. In symbols,*

$$A \times B = \{(x,y) | x \in A, \ y \in B\}.$$

For example, if $A = \{1,2\}$ and $B = \{a,b,c\}$ then $A \times B = \{(1,a),(1,b),(1,c),(2,a),(2,b),(2,c)\}$.

The term "cartesian" is borrowed from coordinate geometry, where a point in the plane is represented by an ordered pair of real numbers (x,y) called its cartesian coordinates. The cartesian product $\mathbf{R} \times \mathbf{R}$ is then the set of cartesian coordinates of all points in the plane.

Definition. *Let A and B be sets, and let R be a subset of $A \times B$. Then R is called a* relation *from A to B. If $(x,y) \in R$, then x is said to be in relation R to y, written $x \ R \ y$. A relation from A to A is called a relation on A (or in A).*

Strictly speaking, a relation is determined by three sets, A,B and a subset R of $A \times B$, although we call it simply the relation R. If R is a relation from A to B, and S is a relation from C to D, then R and S are *equal* if $A = C$, $B = D$, and, for all $x \in A$, $y \in B$, $x\ R\ y \Leftrightarrow x\ S\ y$.

Definition. *Let R be a relation in the set X. R is said to be*

(a) reflexive *if $x\ R\ x$ for all $x \in X$;*
(b) symmetric *if $x\ R\ y$ implies $y\ R\ x$ for all $x,y \in X$;*
(c) antisymmetric *if $x\ R\ y$ and $y\ R\ x$ imply $x = y$ for all $x,y \in X$;*
(d) transitive *if $x\ R\ y$ and $y\ R\ z$ imply $x\ R\ z$ for all $x,y,z \in X$.*

If R is reflexive, symmetric, and transitive, then R is called an equivalence relation *on X. If R is reflexive, antisymmetric, and transitive, then R is called a* partial order *on X.*

2.1 Examples

(a) Let X be the set of all lines in a plane. For $x,y \in X$ let $x \| y$ mean that x is parallel to y. Let us further agree that every line is parallel to itself. Then $\|$ is an equivalence relation on X. Similarly, congruence of triangles and similarity of triangles are equivalence relations.

(b) Let X be a set whose elements are themselves sets. Consider the relation $\subset$ determined by "set inclusion." For any sets $A,B,C \in X$ we see that

(i) $A \subset A$;
(ii) if $A \subset B$ and $B \subset A$, then $A = B$;
(iii) if $A \subset B$ and $B \subset C$, then $A \subset C$.

Hence, set inclusion is reflexive, antisymmetric, and transitive; therefore it is a partial order on X.

(c) The relation $\leq$ ("less than or equal to") on the set $\mathbf{R}$ of real numbers is reflexive, antisymmetric, and transitive; therefore, it is a partial order on $\mathbf{R}$.

(d) The relation *congruence modulo n* on $\mathbf{Z}$ is defined as follows. Let n be a fixed positive integer. For any $x,y \in \mathbf{Z}$, x is said to be *congruent* to y (modulo n), written

$$x \equiv y \pmod{n},$$

if n divides $x - y$. Now, for any x,y,z in $\mathbf{Z}$, it is true that

(i) n divides $x - x = 0$; hence, $x \equiv x \pmod{n}$;
(ii) if n divides $x - y$, then n divides $y - x$;
(iii) if n divides $x - y$ and also $y - z$, then n divides $x - z$.

This proves that congruence modulo n is an equivalence relation on **Z**.

Let X be a set, and let $\leq$ be a partial order on X. (The symbol $\leq$ here denotes an arbitrary partial order and does not necessarily have its usual meaning of "less than or equal to" in real numbers.) The set X together with the partial order $\leq$ is called a *partially ordered set* or, briefly, a *poset*. We refer to it as the poset $(X,\leq)$ or simply the poset X.

Let $(X,\leq)$ be a poset, and let $x,y \in X$. If $x \leq y$, then x is said to be *contained* in y. If $x \leq y$ and $x \neq y$, then x is said to be *properly contained* in y, written $x < y$. If $x < y$ and there is no element a in X such that $x < a < y$, then y is said to *cover* x.

A finite poset X can be represented by a diagram in the following manner. Represent each element in X by a small circle (or a point) in such a way that whenever $x < y$, then y is higher than x in the diagram. Further, join x and y by a straight segment whenever y covers x. As an illustration, we give below the diagrams for the following three posets:

(a) $\{1,2,3,4,5,6\}$ ordered by the usual relation of "less than or equal to";
(b) $\{1,2,3,4,5,6\}$ ordered by divisibility;
(c) $\mathscr{P}(\{1,2,3\})$ ordered by set inclusion.

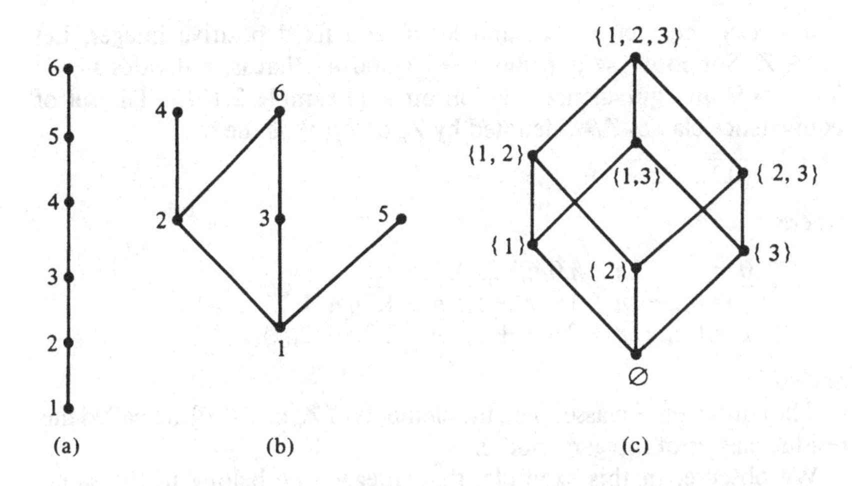

Let $(X, \leq)$ be a partially ordered set. Let S be a subset of X. An *upper bound* of S is an element $b \in X$ such that $x \leq b$ for all $x \in S$. A *least upper bound* (l.u.b.) of S is an element $m \in X$ such that (i) $x \leq m$ for all $x \in S$ and (ii) if $x \leq m'$ for all $x \in S$ then $m \leq m'$. A *lower bound* and a *greatest lower bound* (g.l.b.) are defined analogously.

It can be easily shown that a l.u.b. (g.l.b.) of S, if it exists, is unique.

Definition. A partially ordered set L is called a *lattice* if every pair of elements in L has a least upper bound and a greatest lower bound.

The least upper bound and the greatest lower bound of $\{x, y\}$ are written $x \vee y$ and $x \wedge y$, respectively (called *join* and *meet* of x, y, respectively).

The diagrams (a) and (c) above represent lattices.

A poset $(X, \leq)$ in which for all $x, y \in X$, either $x \leq y$ or $y \leq x$ is called a *chain* or a *totally ordered set*. Every chain is a lattice.

For any set S, the power set $\mathscr{P}(S)$, ordered by set inclusion, is a lattice. Here $A \vee B = A \cup B$ and $A \wedge B = A \cap B$ for all $A, B \in \mathscr{P}(S)$.

Definition. *Let E be an equivalence relation on a set X, and let $a \in X$. The set of all elements in X that are in relation E to a is called the* equivalence class *of a under E and is denoted by E(a). That is,*

$$E(a) = \{x \in X \mid x\, E\, a\}.$$

A subset C of X is called an equivalence *class of E (or an E-class) in X if $C = E(a)$ for some a in X. The set of all equivalence classes of E in X is called the* quotient set *of X by E and is written X/E. That is,*

$$X/E = \{C \subset X \mid C = E(a) \text{ for some } a \in X\}.$$

For example, let $X = \mathbf{Z}$, and let n be a fixed positive integer. Let $x, y \in \mathbf{Z}$. Suppose $x \equiv y$ means $x \equiv y \pmod{n}$; that is, n divides $x - y$. Then $\equiv$ is an equivalence relation on $\mathbf{Z}$ [Example 2.1(d)]. The set of equivalence classes $\mathbf{Z}/\equiv$, denoted by $\mathbf{Z}_n$ or $\mathbf{Z}/(n)$, is the set

$$\{\bar{0}, \bar{1}, \ldots, \overline{(n-1)}\},$$

where

$$\bar{0} = \{\ldots, -2n, -n, 0, n, 2n, \ldots\},$$
$$\bar{1} = \{\ldots, -2n+1, -n+1, 1, n+1, 2n+1, \ldots\},$$
$$\bar{2} = \{\ldots, -2n+2, -n+2, 2, n+2, 2n+2, \ldots\},$$

and so on.

The equivalence classes [i.e., the elements of $\mathbf{Z}_n$ or $\mathbf{Z}/(n)$] are called the residue classes of integers mod n.

We observe, in this example, that integers a, b belong to the same equivalence class if and only if they differ by a multiple of n, and any two equivalence classes are either the same or disjoint. The following theorem shows that this property holds for every equivalence relation.

2.2 Theorem. *Let E be an equivalence relation on a set X. Then X/E is a partition of X. Conversely, given a partition π of X, there is a unique equivalence relation E such that $X/E = \pi$.*

Proof. For any $a \in X$, $a \, E \, a$; hence, $a \in E(a)$. Suppose $a \in E(b)$. Then $a \, E \, b$; hence, for every $x \in E(a)$, $x \, E \, a$, $a \, E \, b$, and, therefore, $x \in E(b)$. Hence, $E(a) \subset E(b)$. By symmetry, $b \, E \, a$; hence, $E(b) \subset E(a)$. Therefore, $E(a) = E(b)$. This proves that each a in X belongs to exactly one equivalence class, namely, $E(a)$. Hence, by Theorem 1.4, X/E is a partition of X.

Conversely, let π be a given partition of X. Define the relation E as follows. For any $x,y \in X$, $x \, E \, y$ if and only if x,y are in the same block of π. Clearly, E is an equivalence relation on X, and the equivalence classes of E are the blocks of π. Hence, $X/E = \pi$. Suppose E' is also an equivalence relation on X such that $X/E' = \pi$. Then for all $x,y \in X$,

$$\begin{aligned} x \, E \, y &\Leftrightarrow x,y \text{ are in the same block of } \pi \\ &\Leftrightarrow x \, E' \, y. \end{aligned}$$

Hence, $E = E'$. Therefore E is the unique equivalence relation on X such that $X/E = \pi$. □

Problems

1. Let $A = [0,1]$, $B = \{0,3,5\}$. Draw the graph of $A \times B$ in the xy-plane. Is $A \times B = B \times A$?
2. If either A or $B = \emptyset$, what is $A \times B$? If $A \times B = \emptyset$, what can you say about A and B?
3. Suppose $X \subset A$ and $Y \subset B$. Show that $X \times Y \subset A \times B$. If for given sets X,Y,A,B, we have $X \times Y \subset A \times B$, does it necessarily follow that $X \subset A$ and $Y \subset B$?
4. Let U be the set of all people. Which of the following relations on U are equivalence relations?
   (a) is an uncle of
   (b) is a roommate of
   (c) is a friend of
5. Describe the equivalence relation corresponding to the following partition of $\mathbf{Z}$:

   $$\begin{aligned} \{\ldots,-8,-4,0,4,8,\ldots\} \cup \{\ldots,-7,-3,1,5,\ldots\} \cup \{\ldots,-6,-2,2,6,\ldots\} \\ \cup \{\ldots,-5,-1,3,7,\ldots\}. \end{aligned}$$

6. Show that the relation defined on $\mathbf{N} \times \mathbf{N}$ by

   $(a,b) \sim (c,d)$ iff $a + d = b + c$

   is an equivalence relation.
7. Let S be a nonempty set, and let R be a relation on S. Suppose the following properties hold:
   (i) For every $x \in S$, $x \, R \, x$.
   (ii) For every $x,y,z \in S$, $x \, R \, y$ and $y \, R \, z$ imply $z \, R \, x$.

Prove that R is an equivalence relation. Also, prove that any equivalence relation satisfies (i) and (ii).

8. Determine the fallacy in the following argument, which attempts to show that a symmetric and transitive relation R on a set X implies that the relation is also reflexive: "By symmetry $x\,R\,y$ implies $y\,R\,x$; by transitivity $x\,R\,y$ and $y\,R\,x$ imply $x\,R\,x$." [*Hint:* A relation R is a subset of $X \times X$. If $R = \emptyset$, then R is symmetric and transitive but not reflexive.]
9. Define a relation $\sim$ in the set of integers $\mathbf{Z}$ as follows: $a \sim b$ iff $a + b$ is an even integer. Is $\sim$ (a) reflexive, (b) symmetric, (c) transitive? If so, write down the quotient set $\mathbf{Z}/\sim$.

3 Mappings

Definition. *Let A and B be sets. A relation f from A to B is called a* mapping *(or a* map *or a* function*) from A to B if for each element x in A there is exactly one element y in B (called the* image *of x under f) such that x is in relation f to y.*

If f is a mapping from A to B, we write

$$f: A \to B \quad \text{or} \quad A \xrightarrow{f} B.$$

Let $f: A \to B$. The sets A and B are called, respectively, the *domain* and *codomain* of the mapping f. Let $x \in A$. If y is the image of x under f, we write $f(x) = y$ and say that f takes x to y; in symbols,

$$x \overset{f}{\mapsto} y$$

or simply $x \mapsto y$. Again, if $f(x) = y$, then x is called a *preimage* of y.

The *graph* of a mapping $f: A \to B$ is defined to be the set $G = \{(x,y) \in A \times B \mid f(x) = y\}$. As a matter of fact, G is (the relation) f itself interpreted as a subset of $A \times B$.

Two mappings $f: A \to B$ and $g: C \to D$ are said to be *equal,* written $f = g$, if $A = C$, $B = D$, and $f(x) = g(x)$ for all $x \in A$.

We emphasize that if the mappings f and g have different codomains, they are not equal even if their domains are equal and $f(x) = g(x)$ for every x in the domain.

A mapping $f: X \to X$ such that $f(x) = x$ for all $x \in X$ is called the *identity* mapping on X and is denoted by i_X (or I_X).

Let A be a nonempty subset of a set X. Then the mapping $f: A \to X$ such that $f(a) = a$ for each $a \in A$ is called an *inclusion* (or *insertion*) map of A into X.

Let $f: A \rightarrow B$. The subset of B consisting of every element that is the image of some element in A is called the *image* (or *range*) of the mapping f and is denoted by $\operatorname{Im} f$. That is,

$$\operatorname{Im} f = \{y \in B \mid y = f(x) \text{ for some } x \in A\}.$$

Let X and S be sets. The set of all possible mappings from X to S is denoted by S^X. That is,

$$S^X = \{f \mid f: X \rightarrow S\}.$$

Suppose X and S are finite sets having m and n elements, respectively. Consider any mapping $f: X \rightarrow S$. The image of any given element in X can be any one of the n elements in S. Therefore, the m elements in X can be assigned images in $n \times n \times \cdots \times n = n^m$ ways. This just means that there are exactly n^m distinct mappings from X to S. Thus, for finite sets X and S, we have the result that $|S^X| = |S|^{|X|}$.

For example, let $X = \{0,1\}$. Then there are $2^2 = 4$ distinct mappings from X to X:

$$\begin{aligned}
&f_1: X \rightarrow X \quad \text{with } 0 \mapsto 0, \quad 1 \mapsto 0,\\
&f_2: X \rightarrow X \quad \text{with } 0 \mapsto 0, \quad 1 \mapsto 1,\\
&f_3: X \rightarrow X \quad \text{with } 0 \mapsto 1, \quad 1 \mapsto 0,\\
&f_4: X \rightarrow X \quad \text{with } 0 \mapsto 1, \quad 1 \mapsto 1.
\end{aligned}$$

We observe that if X is empty, there is just one mapping $X \rightarrow S$, namely, the empty mapping in which there is no element to which an image is to be assigned. This might seem strange, but the definition of a mapping is satisfied vacuously. Note that this is true even if S is also empty.

On the contrary, if S is empty but X is not empty, then there can be no mapping from X to S.

Let $f: A \rightarrow B$ and $g: B \rightarrow C$ be mappings. Then the mapping $h: A \rightarrow C$ given by

$$h(x) = g(f(x)) \qquad \text{for all } x \in A$$

is called the *composite* of f followed by g and is denoted by $g \circ f$ (or more commonly by gf). Thus, $g \circ f(x) = g(f(x))$ for all $x \in A$.

The mappings f and g are called factors of the composite $h = g \circ f$.

Note that the notation for the composite mapping is such that the order in which the mappings act is from *right* to *left*. In $g \circ f$, f acts first.

The composite $g \circ f$ is defined whenever the domain of g contains the range of f.

The composite of mappings $f: A \rightarrow B$, $g: B \rightarrow C$ can be represented by the diagram

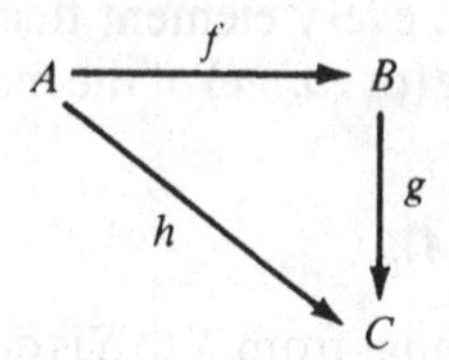

One can go from A to C directly,

$$A \xrightarrow{h} C$$

or along the path

$$A \xrightarrow{f} B \xrightarrow{g} C.$$

If the result is the same – that is, the image of any $x \in A$ is the same along either path – then we say that the *diagram commutes.* It is clear that the diagram commutes if and only if $h = g \circ f$.

In the sequel we denote the composite by gf (instead of $g \circ f$).

3.1 **Theorem.** *Let f: $A \to B$, g: $B \to C$, and h: $C \to D$. Then*

$$h(gf) = (hg)f.$$

Proof. Clearly, $h(gf)$ and $(hg)f$ have the same domain A and the same codomain D.

Let $x \in A$. Then, by definition of the composite,

$$h(gf)(x) = h(gf(x)) = h(g(f(x))),$$

and

$$(hg)f(x) = hg(f(x)) = h(g(f(x))).$$

Hence, $h(gf)(x) = (gf)f(x)$ for every x in A; therefore, $h(gf) = (hg)f$. □

Definition. *A mapping f: $A \to B$ is*

(a) injective *(or* one-to-one *or* 1-1*) if, for all $x_1, x_2 \in A$,*

$$x_1 \neq x_2 \Rightarrow f(x_1) \neq f(x_2)$$

or, equivalently, $f(x_1) = f(x_2) \Rightarrow x_1 = x_2$;

(b) surjective *(or* onto*) if, for every $y \in B$,*

$$y = f(x) \text{ for some } x \in A.$$

A mapping that is injective and surjective is said to be bijective. *If f: $A \to B$ is a bijective mapping, we may write f: $A \simeq B$. A bijective mapping*

$f: A \to A$ is called a permutation *of A. An injective (surjective, bijective) mapping is also called an* injection (surjection, bijection). *A bijective mapping is also called a* one-to-one correspondence.

The following theorem characterizes injective, surjective, and bijective mappings.

3.2 Theorem. *Let $f: A \to B$. Then*

(i) f is injective if and only if no element in B has more than one preimage.
(ii) f is surjective if and only if every element in B has a preimage.
(iii) f is bijective if and only if every element in B has exactly one preimage.

Proof. Obvious. □

3.3 Theorem. *An injective mapping from a finite set to itself is bijective.*

Proof. Let $f: A \to A$ be injective. Let $a \in A$. We shall find $b \in A$ such that $f(b) = a$, which proves that f is surjective and, hence, bijective. Now consider the effect of performing f repeatedly. Let us write f^2 for ff and, in general, $ff \cdots f$ (with n factors) as f^n. The elements $a = f^0(a), f(a), f^2(a), \ldots$ are all in a finite set A; hence, there must be repetitions. So assume that $f^r(a) = f^s(a)$, where $r > s$, say. If $s = 0$ then choose $b = f^{r-1}(a)$. Otherwise, because f is injective,

$$f^r(a) = f^s(a) \Rightarrow f(f^{r-1}(a)) = f(f^{s-1}(a)) \Rightarrow f^{r-1}(a) = f^{s-1}(a) \Rightarrow \cdots \Rightarrow f^{r-s}(a) = a \Rightarrow f(b) = a,$$

where $b = f^{r-s-1}(a)$. □

We wish to remark that Theorem 3.3 does not hold for infinite sets. For example, let $f: \mathbf{N} \to \mathbf{N}$, defined by $f(x) = x + 1$, be a mapping from the set of natural numbers to itself. Then f is injective but not surjective. Why?

Recall that i_X denotes the identity mapping on X.

Definition. *Let $f: A \to B$. A mapping $g: B \to A$ is called an* inverse *of f if*

$$fg = i_B \quad \textit{and} \quad gf = i_A.$$

If f has an inverse, we say f is invertible.

If f is invertible, then its inverse, written f^{-1}, is unique.

3.4 **Theorem.** *f is invertible if and only if f is bijective.*

Proof. Let $f: A \to B$ be invertible with g as its inverse. Then $f(x) = f(y) \Rightarrow gf(x) = gf(y) \Rightarrow x = y$. Therefore, f is 1-1. Further, if $z \in B$, then $fg(z) = z$, which implies that $g(z) \in A$ is a preimage of z under f. This proves f is onto. Hence, f is a bijection.

Conversely, let $f: A \to B$ be a bijection. Define $g: B \to A$ as follows: Let $b \in B$ and define $g(b) = a$, where a is the unique preimage of $b \in B$ under f. Obviously, g is the inverse of f. □

Definition. *Let $f: A \to B$. A mapping $g: B \to A$ is*

(a) *a* left inverse *of f if $gf = i_A$,*
(b) *a* right inverse *of f if $fg = i_B$.*

Clearly, the inverse of f, if it exists, is a left inverse as well as a right inverse of f.

3.5 **Theorem.** *A mapping $f: A \to B$ is*

(i) *injective if and only if f has a left inverse,*
(ii) *surjective if and only if f has a right inverse.*

Proof. Exercise. □

We shall frequently come across mappings that are determined in a natural way – *induced,* as we say – by another mapping or by some inherent property of the domain or codomain. In the next section we describe some of these induced mappings.

3.6 Examples of induced mappings

(a) Let $f: A \to B$. Let $S \subset A$ and $T \subset B$ such that $f(x) \in T$ for all $x \in S$. Then f induces the mapping $g: S \to T$ given by $g(x) = f(x)$ for all $x \in S$.

The mapping g is called a *restriction* of f.

(b) There is an important mapping determined by a subset of a set. Let $S \subset X$ and let $A = \{0,1\}$. Then S determines the mapping $f_S: X \to A$ given by

$$f_S(x) = \begin{cases} 1 & \text{if } x \in S, \\ 0 & \text{if } x \notin S. \end{cases}$$

The mapping f_S is called the *characteristic function* of S (as a subset of X).

Conversely, given a mapping $f\colon X \to A$, let

$$S = \{x \in X \mid f(x) = 1\}.$$

Then it is clear that the characteristic function of S is the given mapping f. Moreover, S is the unique subset of X with f as its characteristic function. This proves that the mapping $\psi\colon \mathcal{P}(X) \to A^X$, given by $S \mapsto f_S$, is bijective.

It is clear that we could have used the set $\mathbf{2} = \{1,2\}$ in place of A. Thus, there is a one-to-one correspondence between the set $\mathcal{P}(X)$ and $\mathbf{2}^X$. Because of this, the set $\mathcal{P}(X)$ is also written as $\mathbf{2}^X$.

(c) Let E be an equivalence relation on the set X. Then E induces the surjective mapping $p\colon X \to X/E$, with $x \mapsto E(x)$, where $E(x)$ is the equivalence class of x under E. The mapping p is called the *canonical* (or *natural*) mapping from X to the quotient set X/E.

(d) Given sets S and T, there are two natural mappings with domain $S \times T$–namely,

$$p\colon S \times T \to S \quad \text{with} \quad (x,y) \mapsto x \quad \text{for all } (x,y) \in S \times T,$$

and

$$q\colon S \times T \to T \quad \text{with} \quad (x,y) \mapsto y \quad \text{for all } (x,y) \in S \times T.$$

The mappings p and q are called the *projections* from $S \times T$ onto S and T, respectively.

(e) Given mappings $f\colon A \to S$ and $g\colon B \to T$, let us define

$$\phi\colon A \times B \to S \times T \quad \text{with} \quad (x,y) \mapsto (f(x),g(y)) \quad \text{for all} \quad (x,y) \in A \times B.$$

The mapping ϕ is called the *cartesian product* of f and g and is denoted by $f \times g$ or (f,g).

(f) Let $f\colon A \to B$. Then f induces a mapping

$$f_*\colon \mathcal{P}(A) \to \mathcal{P}(B)$$

given by

$$f_*(S) = \{y \in B \mid y = f(x) \text{ for some } x \in S\}$$

for every $S \subset A$. Note, in particular, that $f_*(A) = \operatorname{Im} f$, $f_*(\varnothing) = \varnothing$, and for every singleton $\{a\} \subset A$, $f_*(\{a\}) = \{f(a)\}$. It is usual to drop the subscript * and to denote the induced mapping by f also. That is,

$$f(S) = \{y \in B \mid y = f(x) \text{ for some } x \in S\}.$$

The context usually makes clear whether f denotes the mapping $A \to B$ or the induced mapping $\mathcal{P}(A) \to \mathcal{P}(B)$. The mapping $f\colon A \to B$ also induces a mapping in the reverse direction, namely,

$$f^*: \mathscr{P}(B) \to \mathscr{P}(A)$$

given by

$$f^*(T) = \{x \in A \mid f(x) \in T\} \qquad \text{for every } T \subset B.$$

Once again, the induced mapping f^* is commonly written f^{-1}, so that $f^{-1}(T) = \{x \in A \mid f(x) \in T\}$. But it should be carefully noted that f^{-1}, used in this sense, is not the inverse of f as defined earlier. In fact, the induced mapping $f^*: \mathscr{P}(B) \to \mathscr{P}(A)$ always exists whether f is invertible or not.

Let S be a nonempty set and n a positive integer. As usual, let $\mathbf{n}$ denote the set $\{1,...,n\}$. A mapping $f: \mathbf{n} \to S$ is called a *list* (or an *ordered set*) of elements of S. The element $f(i)$ is usually written f_i, and the list f is exhibited as $(f_1, f_2,..., f_n)$. The positive integer n is called the *length* of the list, and $f_1,..., f_n$ are called *members* or *elements* of the list. A list of length n is also called an *n-tuple*.

A mapping $f: \mathbf{N} \to S$ is called a *sequence* of elements of S and is written $(f_i)_{i=1,2,...}$ or $(f_1, f_2,...)$, where $f_i = f(i)$, as before.

More generally, a mapping $f: X \to S$ is called a *family* of elements of S and represented as $(f_i)_{i \in X}$. The set X is called the *index* set of the family f, and f is said to be *indexed* by X or X-indexed.

One may wonder why we introduced the new term "family," since, in general, there seems to be no difference between a family and a mapping. The difference lies in the point of view. When we refer to the mapping f: $X \to S$ as a family, we think of the elements of X as labels for locating elements of S. One may visualize a set of pigeonholes or post office boxes, each labeled with exactly one element of X. In each box we put exactly one element of S (with repetitions allowed; that is, we may put the same element in several boxes), f_i being the element in the box labeled i. Thus, we see that the elements of the index set X serve as addresses for the locations of the elements of S in the range of f.

Let $(A_i)_{i \in X}$ be an arbitrary family of sets indexed by X. The cartesian product of $(A_i)_{i \in X}$, denoted by $\Pi_{i \in X} A_i$, is defined as

$$\prod_{i \in X} A_i = \left\{ f: X \to \bigcup_{i \in X} A_i \,\middle|\, f(i) \in A_i \text{ for each } i \in X \right\}.$$

Note that each element of $\Pi_{i \in X} A_i$ is a family $(a_i)_{i \in X}$, where $a_i \in A_i$.

In the particular case when $A_i = A$ for each $i \in X$, the cartesian product $\Pi_{i \in X} A_i$ is in fact A^X.

The cartesian product of a finite family (that is, a list) of sets $A_1, A_2,..., A_n$ is written $\Pi_{i=1}^n A_i$ or $A_1 \times A_2 \times \cdots \times A_n$.

Problems

1. Prove Theorem 3.2.
2. Let $f: A \to B$ and $g: B \to C$ be mappings. Show that
   (a) gf is injective if both f and g are injective.
   (b) gf is surjective if both f and g are surjective.
   (c) gf is bijective if both f and g are bijective.
3. If fg is injective (or surjective), what can be said about f and g?
4. Prove Theorem 3.5.
5. Let $f: A \to B$ and $g: B \to A$ be mappings such that $gf = I_A$. If f is surjective or g is injective, show that f and g are bijective. Then show that $fg = I_B$.
6. Let $C[0,1]$ be the set of all continuous functions from the closed interval $[0,1]$ to $\mathbf{R}$, and let S be the set of all differentiable functions $f \in C[0,1]$ such that $f(0) = 0$ and f' is continuous. Show that the mapping $D: f \mapsto f'$ is a bijection from $C[0,1]$ to S.
7. Find the number of injective mappings from a set with m elements to a set with n elements if (a) $m = n$, (b) $m < n$, (c) $m > n$.
8. If f is a mapping from $\mathbf{N}$ to $\mathbf{N}$ defined by $f(x) = 2x + 1$, show that f has a left inverse but no right inverse. Also show that f has infinitely many left inverses.
9. Let S be a set with n elements. Find the number of distinct mappings from S to S. How many are bijective?

4 Binary operations

Definition. *A mapping*

$$*: S \times S \to S$$

is called a binary operation *on the set S.*

A binary operation on S thus assigns to each ordered pair of elements of S exactly one element of S. Binary operations are usually represented by symbols like $*, \cdot, +, \circ$, instead of letters f, g, and so on. Moreover, the image of (x,y) under a binary operation $*$ is written $x * y$ instead of $*(x,y)$.

Addition $(+)$ and multiplication $(\cdot)$ in the set $\mathbf{Z}$ of integers or, more generally, in the set $\mathbf{R}$ of real numbers are the most familiar examples of binary operations. We have come across a few others in this chapter, without having called them by that name. If A and B are subsets of X, then $A \cup B$, $A \cap B$, and $A - B$ are also subsets of X. Hence, union, intersection, and difference are all binary operations on the set $\mathcal{P}(X)$. Again, given mappings $f: X \to X$ and $g: X \to X$, their composite gf is also a mapping $X \to X$. Hence, the composition ($\circ$) of mappings is a binary operation on the set $S = X^X$ of all mappings from X to X.

More generally, for any positive integer n, a mapping $f: S^n \to S$, where $S^n = S \times S \times \cdots \times S$ (n factors), is called an *n-ary operation* on S. When $n = 1$, we get a *unary* operation, $u: S \to S$.

Definition. *A binary operation* $*: S \times S \to S$ *on the set* S *is*

(a) commutative *if*

$$x * y = y * x \qquad \text{for all } x,y \in S;$$

(b) associative *if*

$$x * (y * z) = (x * y) * z \qquad \text{for all } x,y,z \in S.$$

If $\circ$ *is another binary operation on* S, *then* $*$ *is*

(c) left-distributive *over* $\circ$ *if*

$$x * (y \circ z) = (x * y) \circ (x * z) \qquad \text{for all } x,y,z \in S;$$

(d) right-distributive *over* $\circ$ *if*

$$(y \circ z) * x = (y * x) \circ (z * x) \qquad \text{for all } x,y,z \in S.$$

If $*$ *is both left- and right-distributive over* $\circ$, *then* $*$ *is said to be* distributive over $\circ$.

4.1 Examples

(a) Addition (+) and multiplication ($\cdot$) in $\mathbf{Z}$ (or, more generally, in $\mathbf{R}$) are both commutative and associative. Moreover, $\cdot$ is distributive over + (but the converse is not true).

(b) Union and intersection in the set $\mathcal{P}(X)$ are both commutative and associative. Moreover, each is distributive over the other.

(c) Composition ($\circ$) of mappings is an associative binary operation on the set X^X of all mappings from X to X. If X has more than one element, then $\circ$ is not commutative.

Let $*$ be an associative binary operation on the set S. Then $a * (b * c) = (a * b) * c$ for all $a,b,c \in S$. Hence, we may omit the parentheses and simply write $a * b * c$ without ambiguity. One would intuitively expect that the result can be generalized to any finite number of elements. This is indeed true. To prove it formally, we first introduce some notation.

Let $*$ be a binary operation (not necessarily associative) on S. Let us call

$a * b$ the product of a,b in that order. Given a list $a_1,\dots,a_n \in S$, we can form a product of the elements $a_1,\dots,a_n$, in that order, in several ways by inserting various parentheses and repeatedly applying the binary operation $*$. But the parentheses must be suitably placed so as to get a meaningful product, that is, one without ambiguity. For example, if $*$ is not associative, the expression $a * b * c$ is ambiguous and, hence, not a meaningful product. The meaningful products of a,b,c in that order are $(a * b) * c$ and $a * (b * c)$. For four elements, a,b,c,d, there are five meaningful products: $((a * b) * c) * d$, $(a * (b * c)) * d$, $(a * b) * (c * d)$, $a * ((b * c) * d)$, $a * (b * (c * d))$.

In general, let $\phi(a_1,\dots,a_n)$ denote any meaningful product of $a_1,\dots,a_n$ in that order. Then, clearly, $\phi(a_1,\dots,a_n) = \phi_1(a_1,\dots,a_k) * \phi_2(a_{k+1},\dots,a_n)$ for some k, where $1 \le k \le n$ and $\phi_1(a_1,\dots,a_k)$ and $\phi_2(a_{k+1},\dots,a_n)$ are themselves meaningful products.

The *standard product* of elements $a_1,\dots,a_n$ in that order is defined inductively (see Theorem 1.7, Chapter 2) as follows:

$$\prod_{i=1}^{1} a_i = a_1,$$
$$\prod_{i=1}^{n} a_i = \left(\prod_{i=1}^{n-1} a_i\right) * a_n \qquad \text{for } n > 1.$$

For example, the standard product of a_1,a_2,a_3,a_4,a_5 in that order is $(((a_1 * a_2) * a_3) * a_4) * a_5$.

4.2 Theorem. *Let $*$ be an associative binary operation on the set S. Given a list $a_1,\dots,a_n \in S$, every meaningful product of $a_1,\dots,a_n$ in that order is equal to the standard product $\Pi_{i=1}^{n} a_i$.*

Proof. We prove the theorem by induction on n. For $n = 1$ the result is trivially true. Let $n > 1$ and suppose the result holds for products of m elements for every $m < n$. Let $\phi(a_1,\dots,a_n)$ be any meaningful product of $a_1,\dots,a_n$ in that order. Then $\phi(a_1,\dots,a_n) = \phi_1(a_1,\dots,a_k) * \phi_2(a_{k+1},\dots,a_n)$ for some k, where $1 \le k \le n$ and $\phi_1(a_1,\dots,a_k)$ and $\phi_2(a_{k+1},\dots,a_n)$ are meaningful products of $a_1,\dots,a_k$ and $a_{k+1},\dots,a_n$, respectively. By the induction hypothesis, we have

$$\phi_1(a_1,\dots,a_k) = \prod_{i=1}^{k} a_i,$$
$$\phi_2(a_{k+1},\dots,a_n) = \prod_{j=1}^{n-k} a_{k+j}.$$

Hence,

$$\phi(a_1,\dots,a_n) = \left(\prod_{i=1}^{k} a_i\right) * \left(\prod_{j=1}^{n-k} a_{k+j}\right).$$

If $k = n - 1$, then

$$\phi(a_1,\dots,a_n) = \left(\prod_{i=1}^{n-1} a_i\right) * a_n = \prod_{i=1}^{n} a_i.$$

If $k < n - 1$,

$$\begin{aligned}\phi(a_1,\dots,a_n) &= \left(\prod_{i=1}^{k} a_i\right) * \left(\left(\prod_{j=1}^{n-k-1} a_{k+j}\right) * a_n\right)\\ &= \left(\left(\prod_{i=1}^{k} a_i\right) * \left(\prod_{j=1}^{n-k-1} a_{k+j}\right)\right) * a_n.\end{aligned}$$

Now

$$\left(\prod_{i=1}^{k} a_i\right) * \left(\prod_{j=1}^{n-k-1} a_{k+j}\right)$$

is a meaningful product of $n - 1$ elements; therefore, by the induction hypothesis, it is equal to $\Pi_{i=1}^{n-1} a_i$. Therefore, once again

$$\phi(a_1,\dots,a_n) = \left(\prod_{i=1}^{n-1} a_i\right) * a_n = \prod_{i=1}^{n} a_i. \quad \square$$

Theorem 4.2 is known as the *generalized associative law.* As a result of the theorem, if $*$ is an associative binary operation on S, then given a list $a_1,\dots,a_n \in S$ there is a unique product of $a_1,\dots,a_n$ in that order that is written $a_1 * \cdots * a_n$ (without any parentheses).

Definition. *Let $*$ be an associative binary operation on a set S. Let $a \in S$ and $n \in \mathbf{N}$. We define inductively the* power of an element *a as follows:*

$$a^1 = a, \qquad a^{n+1} = (a^n) * a.$$

The validity of this definition follows from the Recursion Theorem (Theorem 1.7, Chapter 2).

4.3 Theorem. *Let $*$ be an associative binary operation on S. Let $a \in S$, and let m,n be positive integers. Then*

(i) $a^m * a^n = a^{m+n}$.
(ii) $(a^m)^n = a^{mn}$.

Proof. Follows from the definition of a^n and generalized associative law. □

Problems

1. Determine if the binary operation $\circ$ on $\mathbf{R}$ is associative (commutative) in each of the following cases:
   (a) $x \circ y = x^2 y$
   (b) $x \circ y = \min(x,y)$
   (c) $x \circ y = x^y + y^x$
   (d) $x \circ y = 1$
2. Let S be a nonempty set with associative binary operation $\circ$. Let $x,y,z \in S$. Suppose x commutes with y and z. Show that x commutes with $y \circ z$ also.
3. Let S be a nonempty set with associative binary operation $\circ$. Let $a \in S$. Show that the binary operation $\circ$ given by $x \circ y = x \circ a \circ y$ is also associative. If $\circ$ is commutative, is $\circ$ commutative?
4. Let S be a set with n elements. Find the number of binary operations on S.
5. How many binary operations in Problem 4 are commutative?
6. Suppose S is a nonempty set with associative binary operation $\circ$. Suppose $e, f \in S$ such that $x \circ e = x$ and $f \circ x = x$ for all $x \in S$. Prove that $e = f$. Furthermore, if $x \circ y = e = z \circ x$, show that $y = z$.

5 Cardinality of a set

Earlier in this chapter when we defined a finite set as having a finite number of elements, we relied on our intuitive notion of counting. We shall now make this notion precise and extend it to arbitrary sets.

Let A,B be sets. If there exists a bijective mapping $f: A \to B$, we say that A is *equipotent* to B and write $A \simeq B$. It is clear that equipotence is an equivalence relation (on any given collection of sets). This enables us to assign to every set S a *cardinal number* $|S|$ in accordance with the following definition.

Definition. *Let A and B be sets. If there exists a bijection $f: A \to B$, then A and B have the same* cardinal number *(or* cardinality*)* $|A| = |B|$.

If m and n are natural numbers, it can be easily proved that the sets $\mathbf{m} = \{1,\dots,m\}$ and $\mathbf{n} = \{1,\dots,n\}$ are equipotent if and only if $m = n$. We use this fact to define a finite set and its cardinality.

Definition. *Let S be a set. If S is equipotent to the set $\{1,...,n\}$ for some $n \in \mathbf{N}$, then S is called a* finite *set of* cardinal number *n. An empty set is a finite set of cardinal number* 0.

A set that is not finite is called an *infinite* set. The cardinal number of an infinite set is called a *transfinite* cardinal number. An infinite set S is said to be *countable* (or *denumerable*) if S is equipotent to the set $\mathbf{N}$. The cardinal number of $\mathbf{N}$ (and hence of every countable set) is denoted by $\aleph_0$. (Some authors call a set countable if it is finite or equipotent to $\mathbf{N}$. The letter $\aleph$, pronounced *aleph*, is the first letter of the Hebrew alphabet.)

We shall prove that $\mathbf{Z}$ and $\mathbf{Q}$ are both countable sets; hence, $|\mathbf{Z}| = |\mathbf{Q}| = \aleph_0$. But the set $\mathbf{R}$ of real numbers is uncountable. We shall prove that, in a sense to be explained later, the cardinal number $\mathfrak{c}$ of $\mathbf{R}$ is strictly greater than $\aleph_0$.

The usual ordering of natural numbers gives an ordering of the cardinal numbers of finite sets. We extend this ordering to arbitrary (finite or infinite) sets.

Definition. *Let A and B be sets. If there exists an injective mapping f: $A \to B$, then $|A|$ is said to be* less than or equal to *$|B|$ (written $|A| \leq |B|$).*

For cardinal numbers of finite sets, the relation $\leq$ coincides with the usual ordering of natural numbers. Moreover, it is clear that for cardinal numbers of arbitrary sets the relation $\leq$ is reflexive and transitive. For any set A the identity map $A \to A$ is injective; hence, $|A| \leq |A|$. If mappings f: $A \to B$ and g: $B \to C$ are both injective, then gf: $A \to C$ is also injective. Hence, $|A| \leq |B|$, $|B| \leq |C|$ imply $|A| \leq |C|$. The following theorem shows that $\leq$ is also antisymmetric and, therefore, a partial order on the set of cardinal numbers.

For any mapping $\phi: X \to X$, let $\phi^n, n \in \mathbf{N}$, have the usual meaning [i.e., $\phi^n = \phi \circ \phi \circ \cdots \circ \phi$ (n times)], and let ϕ^0 denote the identity mapping.

5.1 Theorem (Shröder-Bernstein). *Let A and B be sets, and let f: $A \to B$ and g: $B \to A$ be injective mappings. Then there exists a bijection ψ: $A \to B$.*

Proof. If f is not surjective, let $C = B - \operatorname{Im} f$.

Set

$$B^* = \{(fg)^n(c) | c \in C, n \geq 0\}$$

and

$$A^* = g(B^*) = \{g(b) | b \in B^*\}.$$

The restriction of g to B^* is obviously a bijection $B^* \to A^*$. Therefore, every a in A^* has a unique preimage $g^{-1}(a)$ in B^*. Consider the mapping $\psi: A \to B$ given by

$$\begin{aligned} \psi(a) &= g^{-1}(a) && \text{if } a \in A^*, \\ &= f(a) && \text{if } a \notin A^*. \end{aligned}$$

We claim that ψ is bijective. We first prove that $a \in A^*$ if and only if $f(a) \in B^*$. Let $a \in A$. If $a \in A^*$, then $a = g(b)$ for some $b = (fg)^n(c)$, where $c \in C$, $n \geq 0$. Hence, $f(a) = fg(b) = (fg)^{n+1}(c) \in B^*$. Conversely, if $f(a) \in B^*$, then $f(a) = (fg)^n(c)$ for some $c \in C$ and $n > 0$. Hence, $a = g(fg)^{n-1}(c) = g(b)$, where $b = (fg)^{n-1}(c) \in B^*$. Therefore, $a \in A^*$.

Let $a,a' \in A$ and $\psi(a) = \psi(a')$. If $\psi(a) = \psi(a') \in B^*$, then $a,a' \in A^*$; hence, $g^{-1}(a) = g^{-1}(a')$, which yields $a = a'$. If $\psi(a) \notin B^*$, then $a,a' \notin A^*$. Thus,

$$\psi(a) = \psi(a') \Rightarrow f(a) = f(a') \Rightarrow a = a'.$$

This proves ψ is injective. To prove ψ is surjective, let $b \in B$.

If $b \in B^*$, then $a = g(b) \in A^*$; hence, $\psi(a) = g^{-1}(a) = b$. If $b \notin B^*$, then $b \notin C$ because $C \subset B^*$. Hence, $b = f(a)$ for some $a \in A$. Since $f(a) \notin B^*$, $a \notin A^*$; hence, $\psi(a) = f(a) = b$. Therefore, ψ is surjective and, hence, bijective. □

We saw earlier that if X is a finite set with n elements, then $\mathcal{P}(X)$ has 2^n elements (Theorem 1.3). Hence, $|X| < |\mathcal{P}(X)|$. The following theorem shows that this result holds for any set. (As usual, $|A| < |B|$ means that $|A| \leq |B|$ and $|A| \neq |B|$.)

5.2 Theorem. *For any set X, $|X| < |\mathcal{P}(X)|$.*

Proof. The mapping $j: X \to \mathcal{P}(X)$ given by $j(x) = \{x\}$ for all $x \in X$ is obviously injective. Hence, $|X| \leq |\mathcal{P}(X)|$. To prove that $|X| \neq |\mathcal{P}(X)|$, we have to show that there is no surjective (and hence no bijective) mapping from X to $\mathcal{P}(X)$. Since we have already proved [Example 3.6(b)] that $\mathcal{P}(X)$ is equipotent to $\{0,1\}^X$, it will suffice to prove that there is no surjective mapping from X to $\{0,1\}^X$.

Consider any mapping $f: X \to \{0,1\}^X$. For any x in X, let f_x denote the image of x under the mapping f. Then f_x, being an element of $\{0,1\}^X$, is a mapping $f_x: X \to \{0,1\}$. Consider now the mapping $g: X \to \{0,1\}$ given by

$$\begin{aligned} g(x') &= 0 && \text{if } f_{x'}(x') = 1, \\ &= 1 && \text{if } f_{x'}(x') = 0, \end{aligned}$$

for all $x' \in X$. Thus, $g \in \{0,1\}^X$, but $g \notin \operatorname{Im} f$. Hence, the mapping $f: X \to$

$\{0,1\}^X$ is not surjective. This proves that $|X| \neq |\mathscr{P}(X)|$. Therefore $|X| < |\mathscr{P}(X)|$. □

For any set X let us denote the cardinal number of $\mathscr{P}(X)$ by $2^{|X|}$. This notation is in agreement with the equality $|\mathscr{P}(X)| = 2^{|X|}$ for finite sets (Theorem 1.3).

We shall now prove the important result that the set of real numbers is not countable.

5.3 Theorem. *The cardinal number of the set* **R** *of real numbers is*

$$\mathfrak{c} = 2^{\aleph_0}.$$

Hence, **R** *is not countable.*

Proof. Any mapping f: **N** → {0,1} determines an infinite decimal

$$x_f = 0 \cdot f(1) f(2) f(3) \cdots.$$

Then $f \mapsto x_f$ is a mapping $\phi: \{0,1\}^{\mathbf{N}} \to \mathbf{R}$, which is clearly injective. On the other hand, any real number x can be expressed in the binary scale in the form

$$x = \cdots x_5 x_3 x_1 \cdot x_2 x_4 x_6 \cdots,$$

where $x_n = 0$ or 1 for every $n \in \mathbf{N}$. Hence, x determines a mapping

$$g_x: \mathbf{N} \to \{0,1\} \quad \text{with} \quad n \mapsto x_n.$$

Thus, we have an injective mapping

$$\psi: \mathbf{R} \to \{0,1\}^{\mathbf{N}} \quad \text{given by} \quad x \mapsto g_x.$$

Therefore, by Theorem 5.1, **R** is equipotent to $\{0,1\}^{\mathbf{N}}$, hence,

$$\mathfrak{c} = |\mathbf{R}| = |\{0,1\}^{\mathbf{N}}| = 2^{\aleph_0}.$$

By Theorem 5.2, $|\mathbf{N}| = \aleph_0 < 2^{\aleph_0}$, so there is no bijection from **N** to **R**. Therefore, **R** is uncountable. □

We next prove some properties of countable sets. It follows immediately from the definition that a set S is countable if and only if $S = \{S_1, S_2, \ldots\}$, where the S_i are all distinct. If $S = \{S_1, S_2, \ldots\}$, but the S_i are not all distinct, we can delete all repetitions in the sequence $\{S_1, S_2, \ldots\}$ and be left with either a sequence or a finite list. Hence, S is either countable or finite. Put differently, if there is a surjective mapping $\mathbf{N} \to S$, then S is either countable or finite.

5.4 Theorem. *A subset of a countable set is countable or finite.*

Proof. Let S be a countable set, and let $T \subset S$. If T is empty, there is nothing to prove. Suppose T is nonempty and let $a \in T$. Because S is countable, there is a bijection $f\colon \mathbf{N} \to S$ that induces a surjective mapping $g\colon \mathbf{N} \to T$ given by

$$\begin{aligned} g(n) &= f(n) \qquad \text{if } f(n) \in T, \\ &= a \qquad \text{if } f(n) \notin T. \end{aligned}$$

Hence, T is either countable or finite. □

5.5 Theorem. *The union of a sequence of finite sets is countable or finite.*

Proof. Let $\{A,B,\ldots\}$ be a sequence of finite sets, and let S be their union. Let $A = \{a_1,a_2,\ldots,a_p\}$, $B = \{b_1,b_2,\ldots,b_q\}$, and so on. Then

$$S = \{a_1, a_2, \ldots, a_p, b_1, b_2, \ldots, b_q, \ldots\}.$$

Thus, there is a surjective mapping $\mathbf{N} \to S$, so S is countable or finite. □

5.6 Theorem. *Let A and B be countable sets. Then $A \cup B$ and $A \times B$ are countable. More generally, the union and cartesian product of a finite number of countable sets are countable.*

Proof. Let $A = \{a_1,a_2,\ldots\}$, $B = \{b_1,b_2,\ldots\}$. Then $A \cup B = \{a_1,b_1,a_2,b_2,\ldots\}$. Because $A \subset A \cup B$, $A \cup B$ is not finite. Hence, $A \cup B$ is countable.

The set $A \times B$ can be partitioned into finite subsets

$$S_p = \{(a_i,b_j) | i + j = p + 1\}, \qquad p = 1,2,\ldots.$$

Hence, by Theorem 5.5, $A \times B$ is countable.

The more general result follows by induction on the number of sets. □

5.7 Theorem. $\mathbf{Z}$ *and* $\mathbf{Q}$ *are countable sets.*

Proof. $\mathbf{Z} = \mathbf{N} \cup \mathbf{N}'$, where $\mathbf{N}' = \{0,-1,-2,\ldots\}$. Hence, by the previous theorem, $\mathbf{Z}$ is countable. Therefore, $\mathbf{Z} \times \mathbf{N}$ is countable. The mapping

$$\phi\colon \mathbf{Z} \times \mathbf{N} \to \mathbf{Q} \qquad \text{given by} \qquad \phi(a,b) = a/b$$

is surjective. Hence, $\mathbf{Q}$ is countable. □

CHAPTER 2

Integers, real numbers, and complex numbers

1 Integers

In this section we give some properties of integers that the reader will encounter throughout the book. We do not intend to give here an axiomatic development of this topic (we do this in Chapter 13); rather we assume our familiarity with addition, multiplication of integers, and the usual properties of these operations.

We first give some definitions and notation. An integer b is a *factor* or *divisor* of an integer a (or a is a *multiple* of b) if there exists an integer c such that $a = bc$. We say that b divides a and write $b|a$ ($b\nmid a$ means b does not divide a). A nonzero integer p is called a *prime* if $p \neq \pm 1$ and the only factors of p are $\pm p$ and ± 1. Let a be an integer. The absolute value (or modulus) $|a|$ of a is defined as follows:

$$|a| = \begin{cases} a & \text{if } a > 0, \\ 0 & \text{if } a = 0, \\ -a & \text{if } a < 0. \end{cases}$$

The set of integers is denoted by **Z**, and the set of positive integers (also called *natural numbers*) by **N**.

We now state the well-ordering property of positive integers, which is proved in Chapter 13.

If S is a nonempty subset of **N**, *then S contains a* **smallest element** (*also called* **least element**).

This property along with our assumed familiarity with addition and multiplication of integers provides a basis for proving a number of other important properties of **Z**.

We now derive some consequences of the well-ordering property of positive integers. To begin, we derive

1.1 First principle of induction

Let S be a subset of $\mathbf{N}$ *satisfying the following properties:*

(i) $1 \in S$.
(ii) *If* $a \in S$, *then* $a + 1 \in S$.

Then $S = \mathbf{N}$.

Proof. Let $T = \{x \in \mathbf{N} | x \notin S\}$. If $T = \varnothing$, we are done. So let $T \neq \varnothing$. Then by the well-ordering property of positive integers, T has a smallest element m. Clearly, $m \neq 1$ because $1 \in S$. Then, by (ii), $m - 1 \notin S$, which contradicts the minimality of m. Thus, $T = \varnothing$, and, hence, $S = \mathbf{N}$. □

Next we derive

1.2 Second principle of induction

Let S be a subset of $\mathbf{N}$ *such that*

(i) $1 \in S$, *and*
(ii) $n \in S$ *whenever* $m \in S$ *for all positive integers* $m < n$.

Then $S = \mathbf{N}$.

Proof. Let $T = \{x \in \mathbf{N} | x \notin S\}$. If $T = \varnothing$, we are done. Otherwise by the well-ordering property T contains a smallest element m. By (i), $m \neq 1$. Then for all positive integers $x < m$, $x \in S$. But then by (ii), $m \in S$, a contradiction. Therefore $T = \varnothing$. □

One of the important results that follow from the second principle of induction is

1.3 Theorem (fundamental theorem of arithmetic). *Every positive integer is either* 1 *or it can be written in one and only one way as a product of positive primes.*

Proof. The theorem is true if $n = 1$. So let $n > 1$. If n is prime, we are done. Else, let $n = n_1 n_2$, where $n_1, n_2 > 1$, so n_1, $n_2 < n$. Assume that the theorem is true for all positive integers $< n$. We shall prove that it is true also for n. Now, by the induction hypothesis each n_1, n_2 can be written uniquely as a product of positive primes. Thus, n can be written as a product of positive primes. We proceed to prove that the factorization so obtained is unique. Let

$$n = p_1 p_2 \cdots p_s = q_1 q_2 \cdots q_t,$$

where p_i and q_i are positive primes. If $p_1 = q_1$, then

$$\frac{n}{p_1} = p_2 \cdots p_s = q_2 \cdots q_t < n.$$

But, by the induction hypothesis, this implies $s - 1 = t - 1$, and by renumbering the subscripts (if necessary) $p_i = q_i$, $i = 2,\dots,s$. This proves the uniqueness of prime factorization if $p_1 = q_1$ (or if some $p_i = q_j$). So let $p_1 \neq q_1$. For definiteness, let $p_1 > q_1$. Then

$$\begin{aligned} n > (p_1 - q_1)p_2 \cdots p_s &= p_1 p_2 \cdots p_s - q_1 p_2 \cdots p_s \\ &= q_1 q_2 \cdots q_t - q_1 p_2 \cdots p_s = q_1(q_2 \cdots q_t - p_2 \cdots p_s). \end{aligned}$$

By the induction hypothesis either $q_1 = p_i$ for some $i = 2,\dots,s$ or $q_1 | (p_1 - q_1)$. If $q_1 | (p_1 - q_1)$, then $q_1 = p_1$, a contradiction. Thus, $q_1 = p_i$ for some $2 \leq i \leq s$, which reduces to the case considered in the beginning. The proof of the theorem then follows by the second principle of induction. □

Next we prove

1.4 Theorem (division algorithm). *Let $a,b \in \mathbf{Z}$, $b > 0$. Then there exist unique integers q and r such that*

$$a = bq + r, \qquad 0 \leq r < b,$$

where q is called the quotient, *and r the* remainder *of a modulo b.*

Proof. Let $S = \{a - bx | x \in \mathbf{Z}\}$. S has a positive element $a - b(-|a|)$. By the well-ordering property of positive integers, S contains a smallest positive integer, say, $m = a - bx$. If $m > b$, then $m - b = a - b(x + 1) \in S$, which contradicts the minimality of m. Thus, $m \leq b$.

If $m = b$, then $a - bx = b$ implies $a = b(x + 1)$; therefore, choose $q = x + 1$ and $r = 0$. If $m < b$, then choose $q = x$ and $r = m$. For uniqueness of q and r, let

$$a = bq + r = bq' + r'.$$

Then

$$b(q - q') = (r' - r).$$

Because $|r - r'| < b$, $q - q' = 0$. Thus, $q = q'$, which gives, in turn, $r = r'$. □

Definition. *Let $a,b \in \mathbf{Z}$. Then d is called a* greatest common divisor *(g.c.d.) of a and b if the following hold:*

(i) $d|a$ and $d|b$.
(ii) If $c \in \mathbf{Z}$, $c|a$, and $c|b$, then $c|d$.

Remark. d is unique to within the factors ± 1. The positive g.c.d. of a and b is usually denoted by (a,b).

1.5 Theorem. *Let a and b be two nonzero integers. Then*

(i) a,b have a positive g.c.d., say d;
(ii) there exist integers m,n such that $d = am + bn$.

Proof. Let $S = \{ax + by | x,y \in \mathbf{Z}\}$. Then obviously S contains positive integers. By the well-ordering property, let $d = am + bn$ be the smallest positive integer in S. We prove that d is the g.c.d. of a and b.

By the division algorithm $a = dq + r$, where $0 \leq r < d$. Then

$$r = a - dq = a - (am + bn)q = a(1 - mq) - bnq \in S,$$

which contradicts the minimality of d in S unless $r = 0$. Hence, $r = 0$ and $d|a$. Similarly, $d|b$. Further, let $c \in \mathbf{Z}$, $c|a$, and $c|b$. Then $d = am + bn$ implies $c|d$. Hence, $d = (a,b)$. □

1.6 Corollary. *Let p be prime and $a,b \in \mathbf{Z}$ such that $p|ab$. Then $p|a$ or $p|b$.*

Proof. Suppose $p \nmid a$. Then $(p,a) = 1$. This implies that there exist $u,v \in \mathbf{Z}$ such that $up + va = 1$. Then $upb + vab = b$; so $p|b$ since $p|ab$. □

Sometimes a mapping $f: \mathbf{N} \to S$ is described inductively; that is, by giving $f(1)$ and prescribing a rule that determines $f(n + 1)$ when $f(n)$ is known. For example, the standard product $\Pi_{i=1}^{n} a_i$ was defined in this manner in the last chapter (Section 4). Another example is the definition of the factorial function: $1! = 1$, $(n + 1)! = (n + 1)\cdot n!$ for all $n \in \mathbf{N}$. Now the question is whether this is a valid procedure for describing a mapping. In other words, does there exist a unique mapping $f: \mathbf{N} \to S$ satisfying the conditions stated earlier? Intuitively, the answer seems to be yes. We prove it formally in the following theorem.

1.7 Theorem (recursion theorem). *Let S be a set and $a \in S$. Given a mapping $\phi: \mathbf{N} \times S \to S$, there exists a unique mapping $f: \mathbf{N} \to S$ such that*

$$f(1) = a \quad \text{and} \quad f(n + 1) = \phi(n, f(n)) \qquad \text{for all } n \in \mathbf{N}.$$

Proof. We first show that if a mapping satisfying the given conditions exists, then it is unique. Let f, g: $\mathbf{N} \to S$ be two such mappings. Then $f(1) = a = g(1)$. Further, if $f(n) = g(n)$ for any $n \in \mathbf{N}$, then $f(n+1) = \phi(n, f(n)) = \phi(n, g(n)) = g(n+1)$. Hence, by the induction principle, $f(n) = g(n)$ for all $n \in \mathbf{N}$.

To prove the existence of f, let M be the set of all subsets X of $\mathbf{N} \times S$ satisfying the following conditions:

$$(1, a) \in X; \text{ and if } (n, x) \in X, \text{ then } (n+1, \phi(n, x)) \in X \text{ for all } n \in \mathbf{N}. \quad (*)$$

Clearly, the set $\mathbf{N} \times S$ itself satisfies (*); hence, M is nonempty. Let $R = \bigcap_{X \in M} \{X\}$ be the intersection of all sets in M. Then R satisfies (*); hence, $R \in M$. Let P be the set of positive integers n for which there exists a unique element x_n in S such that $(n, x_n) \in R$. If $1 \notin P$, there exists $b \in S$, $b \neq a$, such that $(1, b) \in R$. But then $R - \{(1, b)\}$ satisfies (*); hence, $R \subset R - \{(1, b)\}$, a contradiction. Therefore, $1 \in P$.

Now let $n \in \mathbf{N}$ and suppose $n \in P$, but $n + 1 \notin P$. Then there is a unique $x_n \in S$ such that $(n, x_n) \in R$; hence, by condition (*), $(n + 1, \phi(n, x_n)) \in R$. Since $n + 1 \notin P$, there exists $c \in S$, $c \neq \phi(n, x_n)$, such that $(n + 1, c) \in R$. Again, it is easily verified that $R - \{(n + 1, c)\}$ satisfies (*); hence $R \subset R - \{(n + 1, c)\}$, a contradiction. Therefore, if $n \in P$, then $n + 1 \in P$. Therefore, $P = \mathbf{N}$. So, for every $n \in \mathbf{N}$ there is a unique element $x_n \in S$ such that $(n, x_n) \in R$.

Now $(n, x_n) \in R \Rightarrow (n + 1, \phi(n, x_n)) \in R$; hence, by uniqueness of x_{n+1}, $\phi(n, x_n) = x_{n+1}$. Therefore the mapping f: $\mathbf{N} \to S$ given by $f(n) = x_n$ satisfies the required conditions, namely, $f(1) = a$ and $f(n + 1) = \phi(n, f(n))$ for all $n \in \mathbf{N}$. □

Problems

1. Prove by induction the following summation formulas:

   (a) $$1^2 + 3^2 + \cdots + (2n-1)^2 = \frac{n(4n^2-1)}{3}.$$

   (b) $$1^3 + 2^3 + \cdots + n^3 = \left[\frac{n(n+1)}{2}\right]^2.$$

2. Prove by induction that

$$\frac{d^n}{dx^n}(1+x)^{-1} = (-1)^n n! \frac{1}{(1+x)^{n+1}}.$$

3. Let $A_1, \ldots, A_n$ be subsets of a set. Prove by induction the following laws:

   (a) $$\left(\bigcap_{i=1}^{n} A_i\right)' = \bigcup_{i=1}^{n} A_i'.$$

(b) $\left(\bigcup_{i=1}^{n} A_i\right)' = \bigcap_{i=1}^{n} A_i'$.

4. If $a,b,c \in \mathbf{Z}$ such that $a|b$ and $a|c$, show that $a|mb + nc$, for $m,n \in \mathbf{Z}$.
5. Let $a,b \in \mathbf{Z}$. The *least common multiple* (l.c.m.) of a and b, denoted by $[a, b]$, is defined to be a positive integer d such that
   (i) $a|d$ and $b|d$,
   (ii) whenever $a|x$ and $b|x$, then $d|x$.

   Suppose $a = p_1^{e_1} \cdots p_k^{e_k}$ and $b = p_1^{f_1} \cdots p_k^{f_k}$ be factorizations of a and b as a product of primes, where $e_i, f_i \geq 0$. Prove that
   (a) $[a,b]$ is unique.
   (b) $[a,b] = p_1^{g_1} \cdots p_k^{g_k}$, where $g_i = \max(e_i, f_i)$.
   (c) $[a,b] = ab/(a,b)$, where (a,b) is the greatest common divisor of a and b.
6. Show that any integer $n > 1$ is either a prime or has a prime factor $\leq \sqrt{n}$.
7. Show that the polynomial
   $$x^n + a_1 x^{n-1} + \cdots + a_{n-1}x + a_n,$$
   where $a_i \in \mathbf{Z}$, has no rational root that is not integral. Also show that any integral root must be a divisor of a_n.
8. Prove Wilson's theorem: For any prime p, $(p-1)! + 1 \equiv 0 \pmod{p}$.
9. Prove Fermat's theorem: For any $a, p \in \mathbf{Z}$ with p prime, $a^p \equiv a \pmod{p}$.
10. Let n be a positive integer, and let $\phi(n)$ denote the number of positive integers less than and prime to n. Prove Euler's theorem: For any integer a relatively prime to n, $a^{\phi(n)} \equiv 1 \pmod{n}$. ($\phi$ is called *Euler's function*). Furthermore, note that for a prime p, $\phi(p) = p - 1$. Thus, Euler's theorem generalizes Fermat's theorem.

2 Rational, real, and complex numbers

We assume that the reader is familiar with rational and real numbers and the usual operations of addition and multiplication in them. Constructing the rational numbers from integers is given in Chapter 12. However, the construction of real numbers is not given because it is normally done in books on analysis (see, for example, Landau 1960).

A real number may be taken to be a (terminating or nonterminating) decimal. A terminating or a recurring decimal is a rational number that is, equivalently, a fraction of the form a/b, where $a,b \in \mathbf{Z}$ and $b \neq 0$.

The system of *complex numbers* is defined to be the set $\mathbf{C} = \mathbf{R} \times \mathbf{R}$

(where **R** is the set of real numbers) with addition and multiplication given by

$$(a,b) + (c,d) = (a + c, b + d),$$
$$(a,b) \cdot (c,d) = (ac - bd, ad + bc).$$

It is easily verified that addition and multiplication in **C** are both associative and commutative and, moreover, that multiplication is distributive over addition. For any $(a,b) \in \mathbf{C}$,

$$(a,b) + (0,0) = (a,b) = (a,b)(1,0),$$
$$(a,b) + (-a,-b) = (0,0).$$

Furthermore, if $(a,b) \neq (0,0)$,

$$(a,b)\left(\frac{a}{a^2 + b^2}, \frac{-b}{a^2 + b^2}\right) = (1,0).$$

Thus, addition and multiplication in **C** have the same properties as the corresponding operations in **R** (or **Q**), with (0,0) and (1,0) playing the roles of 0 and 1, respectively.

Consider complex numbers of the form $(x,0)$. Addition and multiplication of two such complex numbers give

$$(a,0) + (b,0) = (a + b,0),$$
$$(a,0) \cdot (b,0) = (ab,0).$$

This allows us to identify every complex number $(x,0)$ with the real number x and thereby treat **R** as a subset of **C**. Let us henceforth write x for $(x,0)$. Then for any $(a,b) \in \mathbf{C}$,

$$(a,b) = (a,0) + (0,1)(b,0)$$
$$= a + ib \qquad \text{where } i = (0,1).$$

Now $i^2 = (0,1) \cdot (0,1) = (-1,0) = -1$.

Thus, every complex number can be written as $a + ib$, where a and b are real and $i^2 = -1$. This is the common notation for complex numbers.

Let $z = a + ib \in \mathbf{C}$, where $a,b \in \mathbf{R}$. Then a and b are called the *real part* and the *imaginary part* of z, respectively. The complex number $\bar{z} = a - ib$ is called the *conjugate* of z. The nonnegative real number $|z| = +\sqrt{a^2 + b^2}$ is called the *modulus* of z.

3 Fields

Introductory algebra is concerned with the operations of addition and multiplication of real or complex numbers as well as the related opposite operations of subtraction and division. On the other hand, abstract algebra is concerned with deriving properties of an abstract system (such as the systems of integers, real numbers, or complex numbers) from its axioms in a formal, rigorous fashion. The purpose of this section is to introduce the richest algebraic structure, for which the systems of rational, real, and complex numbers are concrete examples.

Definition. *A system $(F, +, \cdot)$, where F is a nonempty set and $+, \cdot$ are binary operations on F, is called a* field *if the following axioms hold:*

(i) *The additive operation is associative:* $(x+y)+z = x+(y+z)$.
(ii) *F has a zero element 0 such that* $x+0 = x = 0+x$.
(iii) *Every element $x \in F$ has a negative $-x \in F$ such that* $x+(-x) = 0 = (-x)+x$.
(iv) *The additive operation is commutative:* $x+y = y+x$.
(v) *The multiplicative operation is associative:* $(xy)z = x(yz)$.
(vi) *F has $1 \neq 0$ such that* $x1 = x = 1x$.
(vii) *Every element $0 \neq x \in F$ has an inverse $x^{-1} \in F$ such that* $xx^{-1} = 1 = x^{-1}x$.
(viii) *The multiplicative operation is commutative:* $xy = yx$.
(ix) *The multiplicative operation distributes over the additive operation:* $x(y+z) = xy+xz$.

Examples of fields are the systems of rational numbers, real numbers, and complex numbers under the usual operations of addition and multiplication. The system of integers **Z** is not a field under the usual operations because each nonzero integer is not invertible in **Z**. **Z** is an example of another algebraic structure – "commutative ring" – discussed later in the book.

Note that although familiar statements such as

(a) $x0 = 0$
(b) $(-x)y = -(xy) = x(-y)$
(c) $x(y-z) = xy - xz$

are not included in the axioms, they can be easily deduced from them. For example, to prove (a) consider

$$\begin{aligned} x0 &= x(0+0) && \text{[axiom (ii)]} \\ &= x0 + x0 && \text{[axiom (ix)]} \end{aligned}$$

By axiom (iii), $-(x0) \in F$. Thus,

$$\begin{aligned} x0 + (-(x0)) &= (x0 + x0) + (-x0) \\ &= x0 + (x0 + (-x0)) \qquad \text{[axiom (i)];} \end{aligned}$$

so $0 = x0$.

Not every field is infinite. We close this section by giving an example of a finite field.

3.1 Example (integers modulo n).
Let n be a fixed positive integer, and let

$$\mathbf{Z}_n \text{ or } \mathbf{Z}/(n) = \{\bar{0},\bar{1},\ldots,\overline{(n-1)}\}$$

be the set of equivalence classes of integers mod n (Section 2, Chapter 1). We define addition and multiplication in $\mathbf{Z}_n$ as follows:

$$\bar{x}+\bar{y}=\overline{x+y}, \qquad \bar{x}\bar{y}=\overline{xy}, \tag{1}$$

$\bar{x}, \bar{y} \in \mathbf{Z}_n$. Since $\bar{x}=\overline{x+kn}$ for all $k \in \mathbf{Z}$, the question arises: If $\bar{x}=\bar{x}'$ and $\bar{y}=\bar{y}'$, is it true that $\bar{x}+\bar{y}=\bar{x}'+\bar{y}'$ and $\bar{x}\bar{y}=\bar{x}'\bar{y}'$?; in other words, are addition and multiplication in (1) well defined? We proceed to show that this is indeed the case.

Now

$$\begin{aligned}\bar{x}=\bar{x}' \text{ and } \bar{y}=\bar{y}' &\Rightarrow n \text{ divides } x-x' \text{ and } n \text{ divides } y-y'\\ &\Rightarrow n \text{ divides } (x-x')+(y-y')\\ &\Rightarrow n \text{ divides } (x+y)-(x'+y')\\ &\Rightarrow \overline{(x+y)}=\overline{(x'+y')}\\ &\Rightarrow \bar{x}+\bar{y}=\bar{x}'+\bar{y}', \qquad \text{by (1).}\end{aligned}$$

This proves that addition in $\mathbf{Z}_n$ is well defined.

Similarly,

$$\begin{aligned}\bar{x}=\bar{x}' \text{ and } \bar{y}=\bar{y}' &\Rightarrow n \text{ divides } (x-x')(y-y')\\ &\Rightarrow n \text{ divides } (xy+x'y')-xy'-x'y\\ &= (xy-x'y')+x'(y'-y)+y'(x'-x)\\ &\Rightarrow n \text{ divides } xy-x'y'\\ &\Rightarrow \overline{xy}=\overline{x'y'}=\bar{x}\bar{y}=\bar{x}'\bar{y}',\end{aligned}$$

proving that multiplication is also well defined.

That addition satisfies field axioms (i)–(iv) for $(\mathbf{Z}_n,+,\cdot)$ is trivial. The zero element is $\bar{0}$, and the negative of $\bar{x}$ is $\overline{(-x)}$. It is also trivial that multiplication satisfies axioms (v), (vi), (viii), and (ix); but axiom (vii) need not hold. However, for $n=p$, p prime, we proceed to show that axiom (vii) does hold.

Let $\bar{0} \neq \bar{x} \in \mathbf{Z}_p$. This implies that $p \nmid x$. So there exist integers u and v such that $pu+xv=1$ (Theorem 1.5). But then

$$\begin{aligned}1=pu+xv \Rightarrow \bar{1}&=\overline{pu+xv}=\overline{pu}+\overline{xv}=\bar{p}\bar{u}+\bar{x}\bar{v}=\bar{0}\bar{u}+\bar{x}\bar{v}\\ &=\bar{0}\bar{u}+\bar{x}\bar{v}=\bar{0}+\bar{x}\bar{v}=\bar{x}\bar{v}.\end{aligned}$$

Hence $\bar{x}^{-1}=\bar{v}$. Therefore, $(\mathbf{Z}_p,+,\cdot)$ is a field with p elements.

We shall return to the algebraic structure $(\mathbf{Z}_n,+,\cdot)$, where n is any positive integer, later in Chapter 9, for an example of a "finite ring."

CHAPTER 3

Matrices and determinants

1 Matrices

Matrices originated in the study of linear equations but have now acquired an independent status and constitute one of the most important systems in algebra. They are a rich source of examples of algebraic structures, which we shall study later.

As usual, for any positive integer n, the set $\{1,...,n\}$ will be denoted by $\mathbf{n}$. The cartesian product $\mathbf{m} \times \mathbf{n}$ is therefore the set

$$\mathbf{m} \times \mathbf{n} = \{(i,j) | i = 1,...,m; j = 1,...,n\}.$$

Definition. *Let F be a field and m and n positive integers. A mapping*

$$A: \mathbf{m} \times \mathbf{n} \to F$$

is called an m $\times$ n matrix *over F.*

Let A be an $m \times n$ matrix over F, and let $(i,j) \in \mathbf{m} \times \mathbf{n}$. Then $A(i,j)$, the image of the ordered pair (i,j) under the mapping A, is called the (i,j) entry (or element) of the matrix A and is denoted by A_{ij} or a_{ij}, the latter being more common. When the (i,j) entry of A is written a_{ij}, we say that $A = (a_{ij})$.

An $m \times n$ matrix A over F and a $p \times q$ matrix B over F are said to be *equal,* written $A = B$, if $m = p$, $n = q$, and $a_{ij} = b_{ij}$ for each $(i,j) \in \mathbf{m} \times \mathbf{n}$. If A is an $m \times n$ matrix, A is said to be of *size* $m \times n$. (Note that it is really the ordered pair (m,n) that is the size.)

An $m \times n$ matrix is usually exhibited by writing the mn entries a_{ij} in a rectangular array as shown below:

$$\begin{bmatrix} a_{11} & a_{12} & \cdots & a_{1n} \\ a_{21} & a_{22} & \cdots & a_{2n} \\ \cdot & \cdot & & \cdot \\ \cdot & \cdot & & \cdot \\ \cdot & \cdot & & \cdot \\ a_{m1} & a_{m2} & \cdots & a_{mn} \end{bmatrix}.$$

The list (n-tuple) of entries $a_{i1}, a_{i2}, \ldots, a_{in}$ is called the ith row (or row i) of A, denoted by A_i, and exhibited as $(a_{i1}\ a_{i2}\ \cdots\ a_{in})$. The list ($m$-tuple) $a_{1j}, a_{2j}, \ldots, a_{mj}$ is called the jth column (column j) of A, and denoted by A^j. In keeping with the topography of the rectangular array, A^j is shown as

$$\begin{bmatrix} a_{1j} \\ a_{2j} \\ \cdot \\ \cdot \\ \cdot \\ a_{mj} \end{bmatrix}.$$

Thus, A has m rows $A_1, \ldots, A_m$ and n columns $A^1, \ldots, A^n$. The (i,j) entry in A is common to the ith row and the jth column. An $m \times 1$ matrix is called a *column matrix* (or a *column vector*). Similarly, a $1 \times n$ matrix is called a *row matrix* (or a *row vector*).

An $n \times n$ matrix A is called a *square matrix of order n,* and the list of entries $a_{11}, a_{22}, \ldots, a_{nn}$ is called the *diagonal* of A. (A matrix that is not necessarily square is said to be *rectangular.*) If every entry in A that is not in its diagonal is zero, then A is called a *diagonal matrix* and denoted by $\operatorname{diag}(d_1, \ldots, d_n)$, where $d_i = a_{ii}$, $i = 1, \ldots, n$. A diagonal matrix $\operatorname{diag}(d_1, \ldots, d_n)$ is called a *scalar matrix* if $d_1 = d_2 = \cdots = d_n$.

A square matrix $A = (a_{ij})$ is called an *upper triangular* matrix if $a_{ij} = 0$ for all $i > j$, and *strictly upper triangular* if $a_{ij} = 0$ for all $i \geq j$. A (strictly) lower triangular matrix is defined similarly.

The set of all $m \times n$ matrices over a field F is denoted by $F^{m \times n}$. The set $F^{n \times n}$ of all square matrices of order n over F is, for the sake of convenience, written F_n. Again, for the sake of convenience, the set $F^{1 \times n}$ of all $1 \times n$ matrices (row vectors) and the set $F^{n \times 1}$ of all $n \times 1$ matrices (column vectors) over F are both written F^n when the context makes the meaning unambiguous.

2 Operations on matrices

In order to set up an algebraic system of matrices, we are going to introduce two binary operations, addition and multiplication, and a third operation called *scalar multiplication.* These three operations are induced, in a manner to be described hereafter, by the binary operations of addition and multiplication in the field F.

Definition. *Let* $A = (a_{ij})$ *and* $B = (b_{ij})$ *be two* $m \times n$ *matrices over* F. *The* sum *of* A *and* B, *written* $A + B$, *is the* $m \times n$ *matrix* $C = (c_{ij})$ *such that*

$$c_{ij} = a_{ij} + b_{ij}, \qquad (i,j) \in \mathbf{m} \times \mathbf{n}.$$

When the matrices are exhibited as rectangular arrays, the definition implies that

$$\begin{bmatrix} a_{11} & \cdots & a_{1n} \\ \cdot & & \cdot \\ \cdot & & \cdot \\ \cdot & & \cdot \\ a_{m1} & \cdots & a_{mn} \end{bmatrix} + \begin{bmatrix} b_{11} & \cdots & b_{1n} \\ \cdot & & \cdot \\ \cdot & & \cdot \\ \cdot & & \cdot \\ b_{m1} & \cdots & b_{mn} \end{bmatrix} = \begin{bmatrix} a_{11} + b_{11} & \cdots & a_{1n} + b_{1n} \\ \cdot & & \cdot \\ \cdot & & \cdot \\ \cdot & & \cdot \\ a_{m1} + b_{m1} & \cdots & a_{mn} + b_{mn} \end{bmatrix}.$$

Note that matrices A and B can be added if and only if they have the same size.

The $m \times n$ matrix in which every entry is zero is called the *$m \times n$ zero matrix* (or *null matrix*) and is denoted by $0_{m \times n}$. When the size of the $m \times n$ zero matrix is clear from the context, we denote it simply by 0.

Definition. *Let* $A = (a_{ij})$ *be an* $m \times n$ *matrix. The negative of the matrix* A *is the* $m \times n$ *matrix* $B = (b_{ij})$ *such that*

$$b_{ij} = -a_{ij}, \qquad (i,j) \in \mathbf{m} \times \mathbf{n}.$$

The negative of A *is written* $-A$. *That is,*

$$-\begin{bmatrix} a_{11} & \cdots & a_{1n} \\ \cdot & & \cdot \\ \cdot & & \cdot \\ \cdot & & \cdot \\ a_{m1} & \cdots & a_{mn} \end{bmatrix} = \begin{bmatrix} -a_{11} & \cdots & -a_{1n} \\ \cdot & & \cdot \\ \cdot & & \cdot \\ \cdot & & \cdot \\ -a_{m1} & \cdots & -a_{mn} \end{bmatrix}.$$

Next we define *multiplication* of matrices.

Definition. *Let $A = (a_{ij})$ and $B = (b_{ij})$ be, respectively, $m \times n$ and $n \times p$ matrices over F. The* product *AB is the $m \times p$ matrix $C = (c_{ij})$, where*

$$c_{ij} = \sum_{k=1}^{n} a_{ik} b_{kj}, \qquad (i,j) \in \mathbf{m} \times \mathbf{p}.$$

Note that matrices A and B can be multiplied (in that order) if and only if the number of columns in A is equal to the number of rows in B.

Let us define the *dot product* of two n-tuples of elements of F as follows. Given $a = (a_1,...,a_n)$ and $b = (b_1,...,b_n)$, we define

$$a \cdot b = a_1 b_1 + \cdots + a_n b_n.$$

Then the (i,j) entry in the product AB is the dot product of the n-tuples $(a_{i1},...,a_{in})$ and $(b_{1j},...,b_{nj})$ – that is, the dot product of the ith row in A and the jth column in B. Thus, we have the rule: Let $A = (a_{ij})$, $B = (b_{ij})$, and $C = (c_{ij})$ be their product AB. Then

$$c_{ij} = A_i \cdot B^j.$$

The reader may also find it helpful to keep in mind the following, which displays the fact that the (i,j) entry in AB is the dot product of the ith row in A and the jth column in B.

$$i \begin{bmatrix} \vdots & \vdots & & \vdots \\ a_{i1} & a_{i2} & \cdots & a_{in} \\ \vdots & \vdots & & \vdots \end{bmatrix} \begin{bmatrix} \cdots & \overset{j}{b_{1j}} & \cdots \\ \cdots & b_{2j} & \cdots \\ & \vdots & \\ \cdots & b_{nj} & \cdots \end{bmatrix} = i \begin{bmatrix} & \overset{j}{\vdots} & \\ \cdots & \sum_{k=1}^{n} a_{ik} b_{kj} & \cdots \\ & \vdots & \end{bmatrix}.$$

Let A and B be matrices. The products AB and BA both exist if and only if, for some m,n, A and B are $m \times n$ and $n \times m$ matrices, respectively. But then AB is an $m \times m$ matrix and BA is an $n \times n$ matrix; therefore, AB, BA have the same size if and only if $m = n$. But even when A and B are both $n \times n$ matrices, AB and BA are not necessarily equal. This is easily proved by the following example. Let

$$A = \begin{bmatrix} 1 & 1 \\ 0 & 0 \end{bmatrix} \quad \text{and} \quad B = \begin{bmatrix} 1 & 0 \\ 1 & 0 \end{bmatrix}.$$

Then

$$AB = \begin{bmatrix} 2 & 0 \\ 0 & 0 \end{bmatrix} \quad \text{and} \quad BA = \begin{bmatrix} 1 & 1 \\ 1 & 1 \end{bmatrix}.$$

Hence, $AB \neq BA$. This proves that multiplication of matrices is not commutative, except in the trivial case of 1×1 matrices.

2.1 Theorem. *Let A, B, and C be matrices.*

(i) If A and B can be multiplied, and B and C can be multiplied, then $A(BC) = (AB)C.$

(ii) If A and B can be multiplied and B and C can be added, then $A(B + C) = AB + AC.$

(iii) If A and B can be added, and B and C can be multiplied, then $(A + B)C = AC + BC.$

Proof. (i) Let $A = (a_{ij})$, $B = (b_{ij})$, $C = (c_{ij})$ be $m \times n$, $n \times p$, $p \times q$ matrices, respectively. Then $A(BC)$ and $(AB)C$ are both $m \times q$ matrices. Let $(i,j) \in \mathbf{m} \times \mathbf{q}$. Then the (i,j) entry of $A(BC)$ is

$$\sum_{k=1}^{n} a_{ik} \left(\sum_{l=1}^{p} b_{kl} c_{lj} \right) = \sum_{l=1}^{p} \left(\sum_{k=1}^{n} a_{ik} b_{kl} \right) c_{lj}$$
$$= \text{the } (i,j) \text{ entry of } (AB)C.$$

This proves $A(BC) = (AB)C$.

The proofs for (ii) and (iii) are similar to the one for (i) and are left as exercises. □

The *Kronecker delta* function is defined to be

$$\delta \colon \mathbf{N} \times \mathbf{N} \to \{0,1\}$$

such that

$$\delta_{ij} = \begin{cases} 1 & \text{if } i = j \\ 0 & \text{if } i \neq j \end{cases} \qquad \text{for } i,j \in \mathbf{N}.$$

It is worth noting that given an n-tuple $(a_1,\ldots,a_n)$,

$$\sum_{i=1}^{n} a_i \delta_{ij} = a_j.$$

Definition. *The* $n \times n$ *matrix in which, for each* $(i,j) \in \boldsymbol{n} \times \boldsymbol{n}$, *the* (i,j) *entry is* δ_{ij}, *is called the* identity matrix of order n *and denoted by* I_n.

In other words, each entry in the diagonal of I_n is 1, and every other entry is zero. Thus an identity matrix looks like

$$\begin{bmatrix} 1 & 0 & \cdots & 0 \\ 0 & 1 & \cdots & 0 \\ \cdot & \cdot & & \\ \cdot & \cdot & & \\ \cdot & \cdot & & \\ 0 & 0 & \cdots & 1 \end{bmatrix}.$$

When the order of the identity matrix I_n is clear from the context, we drop the subscript n and simply write I. The following theorem justifies the nomenclature of the identity matrix.

2.2 Theorem. *Let* A *be an* $m \times n$ *matrix. Then*

$$I_m A = A = A I_n.$$

Proof. Obvious. □

Definition. *Let* A *be an* $n \times n$ *matrix. If there is an* $n \times n$ *matrix* B *such that*

$$AB = I = BA,$$

then A *is said to be* invertible *and* B *is called an* inverse *of* A.

The inverse of a matrix A, if it exists, is denoted by A^{-1}.

Definition. *Let* $A = (a_{ij})$ *be an* $m \times n$ *matrix over* F, *and let* $\alpha \in F$. *The* product αA *(also written* $A\alpha$*) is the* $m \times n$ *matrix whose* (i,j) *element is* αa_{ij} *for all* $(i,j) \in \boldsymbol{m} \times \boldsymbol{n}$.

Remark. $\alpha A = \text{diag}(\alpha,\alpha,\ldots,\alpha)A$, where $\text{diag}(\alpha,\alpha,\ldots,\alpha)$ is an $m \times m$ scalar matrix.

In the context of matrices over F, the elements of F are called *scalars*. Therefore the multiplication of a matrix by an element of F is called *scalar multiplication*.

Like addition of matrices, scalar multiplication is also a *pointwise* (or *component-wise*) operation. By that term we mean that the (i,j) entry in αA is obtained by multiplying the (i,j) entry in A by α, and likewise, the

(i,j) entry in $A+B$ is obtained by adding the (i,j) entries in A and B. On the contrary, multiplication of matrices is not a pointwise operation.

The following theorem gives properties of scalar multiplication.

2.3 Theorem. *Let A and B be $n \times n$ matrices over F, and let $\alpha, \beta \in F$. Then*

(i) $\alpha(A+B) = \alpha A + \alpha B$.
(ii) $(\alpha + \beta)A = \alpha A + \beta A$.
(iii) $(\alpha\beta)A = \alpha(\beta A)$.
(iv) $\alpha(AB) = (\alpha A)B = A(\alpha B)$.
(v) $1A = A$.

Proof. Obvious. □

Let e_{ij} denote the $n \times n$ matrix in which the (i,j) entry is 1 and all other entries are 0. The n^2 matrices e_{ij}, $i,j = 1,\dots,n$, are called *matrix units* of order n. (The order n of the matrix e_{ij} is usually known from the context and therefore not included in the notation.) It is clear that any $n \times n$ matrix $A = (a_{ij})$ can be written as

$$A = \sum_{i=1}^{n} \sum_{j=1}^{n} a_{ij} e_{ij}.$$

The following result about the product of two matrix units will be used frequently.

For any two matrix units e_{ij}, e_{pq} of order n,

$$e_{ij} e_{pq} = \delta_{jp} e_{iq},$$

where δ_{jp} is the Kronecker delta.

Definition. *Let $A = (a_{ij})$ be an $m \times n$ matrix. The* transpose *of A is the $n \times m$ matrix $B = (b_{ij})$, where $b_{ij} = a_{ji}$, $(i,j) \in \mathbf{n} \times \mathbf{m}$.*

The transpose of A is generally denoted by ${}^tA$ and is obtained by interchanging rows and columns of A.

2.4 Theorem. *Let A and B be matrices over F and let $\alpha \in F$. Then*

(i) ${}^t({}^tA) = A$.
(ii) ${}^t(\alpha A) = \alpha({}^tA)$.
(iii) *If A and B can be added, then*

$${}^t(A+B) = {}^tA + {}^tB.$$

(iv) *If A and B can be multiplied, then*

$${}^t(AB) = {}^tB\,{}^tA.$$

(v) *If A is invertible, then ${}^tA$ is invertible and*

$$({}^tA)^{-1} = {}^t(A^{-1}).$$

Proof. (i), (ii), (iii) follow immediately from the definition. To prove (iv), let $A = (a_{ij})$ and $B = (b_{ij})$ be $m \times n$ and $n \times p$ matrices, respectively. Let ${}^tA = (a'_{ij})$ and ${}^tB = (b'_{ij})$. Then $a'_{ij} = a_{ji}$ and $b'_{ij} = b_{ji}$.

Suppose $AB = (c_{ij})$. Then the (i,j) entry of ${}^t(AB)$ is

$$c_{ji} = \sum_{k=1}^{n} a_{jk}b_{ki} = \sum_{k=1}^{n} a'_{kj}b'_{ik} = \sum_{k=1}^{n} b'_{ik}a'_{kj}$$
$$= (i,j) \text{ entry of } {}^tB\,{}^tA.$$

The proof of (v) is left as an exercise. □

Problems

1. Prove Theorem 2.1.
2. Prove Theorem 2.4.

3 Partitions of a matrix

Let A be an $m \times n$ matrix. If we delete any p rows and any q columns of A, the resulting $(m-p) \times (n-q)$ matrix is called a *submatrix* of A.

A matrix is said to be *partitioned* if it is divided into submatrices (called *blocks*) by horizontal and vertical lines between the rows and the columns. Sometimes the multiplication of two matrices A and B becomes simpler by partitioning them into suitable blocks. For example, let

$$A = \left[\begin{array}{ccc|c} a_{11} & a_{12} & a_{13} & a_{14} \\ a_{21} & a_{22} & a_{23} & a_{24} \\ \hline a_{31} & a_{32} & a_{33} & a_{34} \\ a_{41} & a_{42} & a_{43} & a_{44} \end{array}\right], \qquad B = \left[\begin{array}{cc} b_{11} & b_{12} \\ b_{21} & b_{22} \\ b_{31} & b_{32} \\ \hline b_{41} & b_{42} \end{array}\right]$$

be matrices that have been partitioned by the horizontal and vertical lines as shown. We can now consider A as the 2×2 matrix

$$\begin{bmatrix} A_{11} & A_{12} \\ A_{21} & A_{22} \end{bmatrix}$$

whose elements A_{ij} are the submatrices of A given by the partitioning. Similarly, we can consider B as the 2×1 matrix

$$\begin{bmatrix} B_{11} \\ B_{21} \end{bmatrix}$$

whose elements are the submatrices of B given by the partitioning. Then it is easily checked that AB is the same as the product

$$\begin{bmatrix} A_{11} & A_{12} \\ A_{21} & A_{22} \end{bmatrix}\begin{bmatrix} B_{11} \\ B_{21} \end{bmatrix} = \begin{bmatrix} A_{11}B_{11} + A_{12}B_{21} \\ A_{21}B_{11} + A_{22}B_{21} \end{bmatrix}.$$

Multiplying two matrices A and B in this manner is called *block multiplication* of A and B. The method of block multiplication is sometimes quite handy when multiplying large matrices. Of course, one must partition the two matrices in such a way that the product of the blocks occurring in the process of multiplication is meaningful.

4 The determinant function

Let **n** be the set $\{1,2,...,n\}$. The set of permutations of **n** is denoted by S_n.

For convenience a permutation $\sigma \in S_n$ is usually represented as

$$\begin{pmatrix} 1 & 2 & \cdots & n \\ \sigma(1) & \sigma(2) & \cdots & \sigma(n) \end{pmatrix}.$$

In this notation the first column expresses the fact that σ maps 1 to $\sigma(1)$; the second column, that σ maps 2 to $\sigma(2)$, and so on.

Note that the order of the columns in this representation of σ is immaterial. For example,

$$\begin{pmatrix} 1 & 2 & 3 & 4 \\ 2 & 4 & 1 & 3 \end{pmatrix} \quad \text{and} \quad \begin{pmatrix} 2 & 4 & 3 & 1 \\ 4 & 3 & 1 & 2 \end{pmatrix}$$

represent the same permutation. Furthermore, if in such a representation of σ, a certain number, say i, is absent, it is understood that σ keeps i fixed; that is, $\sigma(i) = i$. For example, $\left(\begin{smallmatrix} 1 & 2 & 5 \\ 2 & 5 & 1 \end{smallmatrix}\right)$ will be considered as a permutation σ of the set $\{1,2,3,4,5\}$ such that $\sigma(3) = 3$ and $\sigma(4) = 4$.

The product of two permutations of **n** is again a permutation of **n**. For example, if

$$\sigma = \begin{pmatrix} 1 & 2 & 3 & 4 \\ 2 & 3 & 4 & 1 \end{pmatrix}, \quad \tau = \begin{pmatrix} 1 & 2 & 3 & 4 \\ 4 & 3 & 2 & 1 \end{pmatrix}$$

are permutations of degree 4, then $\sigma\tau(1) = 1$ because τ maps 1 to 4 and σ maps 4 to 1; similarly, $\sigma\tau(2) = 4$, and so on; thus

$$\sigma\tau = \begin{pmatrix} 1 & 2 & 3 & 4 \\ 1 & 4 & 3 & 2 \end{pmatrix}.$$

In the same way

$$\tau\sigma = \begin{pmatrix} 1 & 2 & 3 & 4 \\ 3 & 2 & 1 & 4 \end{pmatrix}.$$

The inverse of a permutation of **n** is also a permutation. For example, if $\sigma = \binom{1\ 2\ 3\ 4}{2\ 3\ 4\ 1}$, then

$$\sigma^{-1} = \begin{pmatrix} 2 & 3 & 4 & 1 \\ 1 & 2 & 3 & 4 \end{pmatrix},$$

which can also be written as $\binom{1\ 2\ 3\ 4}{4\ 1\ 2\ 3}$. More generally, if

$$\sigma = \begin{pmatrix} 1 & 2 & \cdots & n \\ i_1 & i_2 & \cdots & i_n \end{pmatrix}$$

is a permutation of degree n, then σ^{-1}, the inverse of σ, is the permutation

$$\begin{pmatrix} i_1 & i_2 & \cdots & i_n \\ 1 & 2 & \cdots & n \end{pmatrix}.$$

Thus, if $\sigma, \tau \in S_n$, then $\sigma\tau, \sigma^{-1}, \tau^{-1} \in S_n$. The set S_n is called the *symmetric group of degree n.*

A permutation that interchanges two elements and keeps others fixed is called a *transposition.* For example,

$$\sigma = \begin{pmatrix} 1 & 2 & \cdots & i & \cdots & j & \cdots & n \\ 1 & 2 & \cdots & j & \cdots & i & \cdots & n \end{pmatrix},$$

which maps $i \to j$ and $j \to i$ and keeps other elements fixed, is a transposition. We write it as $\binom{i\ j}{j\ i}$ or, more usually, as $(i\ j)$. Note that the inverse of a transposition $(i\ j)$ is $(i\ j)$ itself.

A permutation σ of **n** is called a *cycle* of length k if σ maps a set of k elements, say $(a_1, a_2, \ldots, a_k)$, cyclically, keeping the remaining elements, if any, fixed; that is, $\sigma(a_1) = a_2$, $\sigma(a_2) = a_3, \ldots, \sigma(a_k) = a_1$, and $\sigma(a) = a$ if a is not one of the $a_1, \ldots, a_k$. We write such a cycle as $(a_1\ a_2\ \ldots\ a_k)$. A cycle of length 2 is a transposition. Observe that $(a_1\ a_2\ \ldots\ a_k)$, $(a_2\ a_3\ \ldots\ a_k\ a_1)$, $(a_3\ a_4\ \ldots\ a_1\ a_2)$, ... $(a_k\ a_1\ \ldots\ a_{k-1})$ are the same cycle.

It is easily seen that any permutation σ of **n** can be expressed as a product of disjoint cycles, that is, cycles for which no two have any common element. For example,

$$\sigma = \begin{pmatrix} 1 & 2 & 3 & 4 & 5 & 6 \\ 4 & 6 & 2 & 1 & 3 & 5 \end{pmatrix} = (1\ 4)(2\ 6\ 5\ 3).$$

Further, every cycle can be expressed as a product of transpositions, although not necessarily uniquely. For example, $(4\ 5\ 3\ 2\ 1) = (2\ 1)(3\ 1)(5\ 1)(4\ 1)$. Thus, every permutation in S_n can be expressed, not necessarily uniquely, as a product of transpositions. It can be shown that if a permutation is a product of r transpositions, and also of s transpositions, then r,s are both even or both odd (Theorem 2.1, Chapter 7). That is, the number of factors in any representation of σ as a product of transpositions is always even or else always odd. If a permutation is a product of an even number of transpositions, then it is called an *even permutation;* if it is a product of an odd number of permutations, it is called an *odd permutation.* The identity permutation I of $\mathbf{n}$ is even because $I = (i\ j)(j\ i)$, where $i,j \in \mathbf{n}$. Also, if a permutation σ is even (odd), then σ^{-1} is also even (odd). Let $\sigma = \eta_1\eta_2 \dots \eta_k$; then $\sigma^{-1} = \eta_k\eta_{k-1} \dots \eta_1$, because the inverse of any transposition η is η itself.

For any permutation σ we define sgn σ to be 1 or -1 according to whether σ is even or odd.

Definition. Let $A = (a_{ij})$ be an $n \times n$ matrix over a field F. Then the sum

$$\sum_{\sigma \in S_n} (\operatorname{sgn} \sigma) a_{1\sigma(1)} \cdots a_{n\sigma(n)}$$

is called the *determinant* of A and denoted by det A or $|A|$.

For example, consider a 3×3 matrix $A = (a_{ij})$. S_3 has $3! = 6$ elements given by

$$e = \begin{pmatrix} 1 & 2 & 3 \\ 1 & 2 & 3 \end{pmatrix}, \quad \sigma_1 = \begin{pmatrix} 1 & 2 & 3 \\ 2 & 3 & 1 \end{pmatrix}, \quad \sigma_2 = \begin{pmatrix} 1 & 2 & 3 \\ 3 & 1 & 2 \end{pmatrix},$$
$$\sigma_3 = \begin{pmatrix} 1 & 2 & 3 \\ 2 & 1 & 3 \end{pmatrix}, \quad \sigma_4 = \begin{pmatrix} 1 & 2 & 3 \\ 3 & 2 & 1 \end{pmatrix}, \quad \sigma_5 = \begin{pmatrix} 1 & 2 & 3 \\ 1 & 3 & 2 \end{pmatrix}.$$

Of these, the first three are even, and the others are odd. Hence,

$$\begin{aligned} \det A = {} & a_{11}a_{22}a_{33} + a_{12}a_{23}a_{31} + a_{13}a_{21}a_{32} \\ & - a_{12}a_{21}a_{33} - a_{13}a_{22}a_{31} - a_{11}a_{23}a_{32}. \end{aligned}$$

In the following sections we prove some results that enable us to evaluate the determinant of any square matrix in a less tedious manner.

For readers familiar with the concept of a ring (for the definition see Chapter 9), it is of interest to note that the determinant of a matrix over a commutative ring is defined exactly in the same manner. Indeed, most of the results proved for determinants over fields also hold for determinants over a commutative ring.

5 Properties of the determinant function*

5.1 Theorem. *Let $A = (a_{ij})$ be an $n \times n$ matrix and let $R_i = (a_{i1}\ a_{i2} \cdots a_{in})$ denote its ith row. Let $R_i' = (a_{i1}'\ a_{i2}' \cdots a_{in}')$ denote a fixed*

---

* The authors would like to thank Professor R. N. Gupta for his suggestions on this section.

$1 \times n$ matrix. Then

$$(i)\quad \det\begin{bmatrix} R_1 \\ R_2 \\ \vdots \\ R_i + R_i' \\ \vdots \\ R_n \end{bmatrix} = \det\begin{bmatrix} R_1 \\ R_2 \\ \vdots \\ R_i \\ \vdots \\ R_n \end{bmatrix} + \det\begin{bmatrix} R_1 \\ R_2 \\ \vdots \\ R_i' \\ \vdots \\ R_n \end{bmatrix}$$

$$(ii)\quad \det\begin{bmatrix} R_1 \\ R_2 \\ \vdots \\ \alpha R_i \\ \vdots \\ R_n \end{bmatrix} = \alpha \det\begin{bmatrix} R_1 \\ R_2 \\ \vdots \\ R_i \\ \vdots \\ R_n \end{bmatrix}$$

(*iii*) *If two rows of a matrix A are equal, then $\det A = 0$.*
(*iv*) *$\det {}^tA = \det A$*

Proof. (i) The left-hand side is

$$\sum_{\sigma\in S_n} (\operatorname{sgn}\sigma) a_{1\sigma(1)} \cdots (a_{i\sigma(i)} + a'_{i\sigma(i)}) \cdots a_{n\sigma(n)}$$

$$= \sum_{\sigma\in S_n} (\operatorname{sgn}\sigma) a_{1\sigma(1)} \cdots a_{i\sigma(i)} \cdots a_{n\sigma(n)} + \sum_{\sigma\in S_n} (\operatorname{sgn}\sigma) a_{1\sigma(1)} \cdots a'_{i\sigma(i)} \cdots a_{n\sigma(n)}$$

$$= \det\begin{bmatrix} R_1 \\ \vdots \\ R_i \\ \vdots \\ R_n \end{bmatrix} + \det\begin{bmatrix} R_1 \\ \vdots \\ R_i' \\ \vdots \\ R_n \end{bmatrix}, \text{ proving (i).}$$

(ii) Clear.

(iii) Suppose the ith and jth rows are equal. Suppose $i < j$. Let $S = \{\sigma\in S_n\colon \sigma(i) < \sigma(j)\}$, $T = \{\sigma\in S_n\colon \sigma(i) > \sigma(j)\}$. Let $\tau = (i, j)$. The mapping $\theta\colon S \to T$ such that $\theta(\sigma) = \sigma\tau$ is easily seen to be a $1-1$ onto mapping. Thus

$$|A| = \sum_{\sigma\in S} (\operatorname{sgn}\sigma)(a_{1\sigma(1)} \cdots a_{i\sigma(i)} \cdots a_{j\sigma(j)} \cdots a_{n\sigma(n)} - a_{1\sigma\tau(1)} \cdots a_{i\sigma\tau(i)} \cdots a_{j\sigma\tau(j)} \cdots a_{n\sigma\tau(n)})$$

$$= \sum_{\sigma \in S} (\operatorname{sgn} \sigma)(a_{1\sigma(1)} \cdots a_{i\sigma(i)} \cdots a_{j\sigma(j)} \cdots a_{n\sigma(n)}$$
$$- a_{1\sigma(1)} \cdots a_{i\sigma(j)} \cdots a_{j\sigma(i)} \cdots a_{n\sigma(n)})$$
$$= \sum_{\sigma \in S} (\operatorname{sgn} \sigma)(a_{1\sigma(1)} \cdots a_{i\sigma(i)} \cdots a_{i\sigma(j)} \cdots a_{n\sigma(n)}$$
$$- a_{1\sigma(1)} \cdots a_{i\sigma(j)} \cdots a_{i\sigma(i)} \cdots a_{n\sigma(n)}) \quad \text{(since } i\text{th row} = j\text{th row)}$$
$$= 0.$$

(iv) $$\det({}^tA) = \sum_{\sigma \in S_n} (\operatorname{sgn} \sigma) a_{\sigma(1)1} a_{\sigma(2)2} \cdots a_{\sigma(n)n}$$
$$= \sum_{\sigma \in S_n} (\operatorname{sgn} \sigma) a_{1\sigma^{-1}(1)} \cdots a_{n\sigma^{-1}(n)}$$
$$= \sum_{\sigma \in S_n} (\operatorname{sgn} \sigma^{-1}) a_{1\sigma^{-1}(1)} \cdots a_{n\sigma^{-1}(n)}$$
$$= \det A.$$

5.2 Corollary. *If a matrix B is obtained from a matrix A by interchanging the ith and jth rows, then* $\det B = -\det A$.

Proof.

$$0 = \det \begin{bmatrix} R_1 \\ \cdot \\ R_i + R_j \\ \cdot \\ R_j + R_i \\ \cdot \\ R_n \end{bmatrix}, \quad \text{by Theorem 5.1 (iii),}$$

$$= \det \begin{bmatrix} R_1 \\ \cdot \\ R_i \\ \cdot \\ R_j \\ \cdot \\ R_n \end{bmatrix} + \det \begin{bmatrix} R_1 \\ \cdot \\ R_i \\ \cdot \\ R_i \\ \cdot \\ R_n \end{bmatrix} + \det \begin{bmatrix} R_1 \\ \cdot \\ R_j \\ \cdot \\ R_j \\ \cdot \\ R_n \end{bmatrix} + \det \begin{bmatrix} R_1 \\ \cdot \\ R_j \\ \cdot \\ R_i \\ \cdot \\ R_n \end{bmatrix},$$

by Theorem 5.1 (i),

$$= \det B + 0 + 0 + \det A.$$

Thus $\det B = -\det A$

5.3 Corollary. *If σ is a permutation of $\{1,\dots,n\}$ and the rows of A are $R_1,\dots,R_n$, respectively, then*

$$\det \begin{bmatrix} R_{\sigma(1)} \\ \vdots \\ R_{\sigma(n)} \end{bmatrix} = (\operatorname{sgn}\sigma)\det A.$$

Proof. Suppose σ is a product of m transpositions. Then $A = \begin{bmatrix} R_1 \\ R_2 \\ \cdot \\ R_n \end{bmatrix}$ can be transformed to $\begin{bmatrix} R_{\sigma(1)} \\ \vdots \\ R_{\sigma(n)} \end{bmatrix}$ by m interchanges of rows. Hence

$$\det \begin{bmatrix} R_{\sigma(1)} \\ \vdots \\ R_{\sigma(n)} \end{bmatrix} = (-1)^m \det \begin{bmatrix} R_1 \\ R_2 \\ \cdot \\ R_n \end{bmatrix} = (\operatorname{sgn}\sigma)\det A, \text{ by Corollary 5.2}$$

5.4 Theorem. *Let X, Y be $n \times n$ matrices. Then*

$$\det(XY) = (\det X)(\det Y).$$

Proof. $\det(XY)$

$$= \sum_{\sigma\in S_n} (\operatorname{sgn}\sigma)\left\{\left(\sum_{i_1=1}^{n} x_{1i_1}y_{i_1\sigma(1)}\right)\left(\sum_{i_2=1}^{n} x_{2i_2}y_{i_2\sigma(2)}\right)\cdots\left(\sum_{i_n=1}^{n} x_{ni_n}y_{i_n\sigma(n)}\right)\right\}$$

$$= \sum_{1\leqslant i_1,i_2,\dots,i_n\leqslant n} (x_{1i_1}x_{2i_2}\cdots x_{ni_n})\left(\sum_{\sigma\in S_n}(\operatorname{sgn}\sigma)y_{i_1\sigma(1)}\cdots y_{i_n\sigma(n)}\right),$$

where $i_1, i_2, \dots, i_n$ may be assumed to be distinct because if $i_1,\dots,i_n$ are not all distinct, then

$$\det(Y) = \sum_{\sigma\in S_n}(\operatorname{sgn}\sigma)y_{i_1\sigma(1)}\cdots y_{i_n\sigma(n)} = 0, \text{ by Theorem 5.1 (iii).}$$

Thus $\det(XY) = \sum_{\tau\in S_n}(x_{1\tau(1)}\cdots x_{n\tau(n)})\left(\sum_{\sigma\in S_n}(\operatorname{sgn}\sigma)y_{\tau(1)\sigma(1)}\cdots y_{\tau(n)\sigma(n)}\right)$

$$= \left(\sum_{\tau\in S_n}(\operatorname{sgn}\tau)x_{1\tau(1)}\cdots x_{n\tau(n)}\right)(\det Y), \quad \text{by Corollary 5.3.}$$

Hence $\det(XY) = (\det X)(\det Y)$.

Problems

1. Show that if $A=(a_{ij})$ is an $n\times n$ upper (or lower) triangular matrix, then $\det A = a_{11}a_{22}\cdots a_{nn}$.
2. Evaluate the following determinants:

$$\text{(a)}\quad \begin{vmatrix} 1 & 2 & 3 \\ 2 & 3 & 4 \\ 3 & 4 & 5 \end{vmatrix}. \qquad \text{(b)}\quad \begin{vmatrix} 3 & 1 & 4 \\ 0 & 1 & 0 \\ 1 & 2 & -1 \end{vmatrix}.$$

3. Prove that $\begin{vmatrix} 1 & a & a^2 \\ 1 & b & b^2 \\ 1 & c & c^2 \end{vmatrix} = (a-b)(b-c)(c-a).$
4. If A is an invertible matrix, prove $\det A^{-1} = \dfrac{1}{\det A}$.
5. If A is a square matrix such that $A=A^2$, prove $\det A = 0$ or 1.
6. If A is a square matrix such that $A^k=0$ for some positive integer k, then $\det A = 0$.

6 Expansion of det A

Let A be an $n\times n$ matrix. Consider all the terms in the sum

$$\det A = \sum_{\sigma\in S_n} (-1)^k a_{1\sigma(1)}\cdots a_{n\sigma(n)}$$

that contain a given entry a_{ij} as a factor. They are given by those permutations σ that satisfy the condition $\sigma(i)=j$. Therefore, the sum of all such terms is

$$\sum_{\substack{\sigma\in S_n\\ \sigma(i)=j}} (-1)^k a_{1\sigma(1)}\cdots a_{i\sigma(i)}\cdots a_{n\sigma(n)} = a_{ij}A_{ij},$$

where

$$A_{ij} = \sum_{\substack{\sigma\in S_n\\ \sigma(i)=j}} (-1)^k a_{1\sigma(1)}\cdots a_{i-1,\sigma(i-1)}a_{i+1,\sigma(i+1)}\cdots a_{n\sigma(n)} \tag{1}$$

is called the *cofactor* of a_{ij} in det A. Now

$$\begin{aligned}\det A &= \sum_{\sigma\in S_n} (-1)^k a_{1\sigma(1)}\cdots a_{n\sigma(n)}\\ &= \sum_{j=1}^{n}\sum_{\substack{\sigma\in S_n\\ \sigma(i)=j}} (-1)^k a_{1\sigma(1)}\cdots a_{n\sigma(n)}.\end{aligned}$$

That is,

$$\det A = \sum_{j=1}^{n} a_{ij}A_{ij} = a_{i1}A_{i1} + \cdots + a_{in}A_{in}. \tag{2}$$

Similarly,

$$\det A = \sum_{i=1}^{n} \sum_{\substack{\sigma\in S_n \\ \sigma(i)=j}} (-1)^k a_{1\sigma(1)} \ldots a_{n\sigma(n)}$$
$$= \sum_{i=1}^{n} a_{ij}A_{ij}.$$

That is,

$$\det A = a_{1j}A_{1j} + \cdots + a_{nj}A_{nj}. \tag{3}$$

(2) is called the *expansion of det A according to the ith row*. Similarly, (3) is the *expansion of det A according to the jth column*.

Now consider the sum $\Sigma_{j=1}^{n} a_{pj}A_{ij}, p \neq i$. Substituting (1) for A_{ij}, we get

$$\sum_{j=1}^{n} a_{pj}A_{ij} = \sum_{j=1}^{n} a_{pj} \sum_{\substack{\sigma\in S_n \\ \sigma(i)=j}} (-1)^k a_{1\sigma(1)} \ldots a_{i-1,\sigma(i-1)} a_{i+1,\sigma(i+1)} \ldots a_{n\sigma(n)}$$
$$= \sum_{\sigma\in S_n} (-1)^k a_{1\sigma(1)} \ldots a_{i-1,\sigma(i-1)} a_{p\sigma(i)} a_{i+1,\sigma(i+1)} \ldots a_{n\sigma(n)}$$
$$= 0, \qquad \text{by Corollary 5.6.}$$

Similarly, $\Sigma_{i=1}^{n} a_{ip}A_{ij} = 0$ if $p \neq j$.

Combining the last two results and (2) and (3), we get

6.1 Theorem. *Let A be an n × n matrix. Then*

(i) $\sum_{j=1}^{n} a_{pj}A_{ij} = \delta_{pi} \det A.$

(ii) $\sum_{i=1}^{n} a_{ip}A_{ij} = \delta_{pj} \det A.$

In the following we develop the method to evaluate the cofactors A_{ij} for the matrix $A = (a_{ij})$. First we find A_{11}, the cofactor of a_{11}. By definition,

$$A_{11} = \sum_{\sigma} (-1)^k a_{2\sigma(2)} \ldots a_{n\sigma(n)}, \tag{1}$$

where σ is a permutation of $\{2,3,\ldots,n\}$ and $k = 0$ or 1 according to whether σ is even or odd.

The expression on the right side is the determinant of the $(n-1) \times$

$(n-1)$ submatrix of A, which is obtained by omitting the first row and the first column of A. To find the value of a general cofactor A_{ij} of a_{ij}, we first perform some row and column interchanges on A to bring a_{ij} to the (1,1) position, so that we can apply (1). Thus, to bring a_{ij} to the (1,1) position, we move the ith row up past row $i-1, \ldots$, row 1, so that after $i-1$ interchanges it becomes the first row. We next move the jth column to the left past the first $j-1$ columns in turn, so that it becomes the first column. We now have a matrix that is obtained from A by $i-1$ interchanges of rows and $j-1$ interchanges of columns. So by Theorem 5.1 the determinant of the new matrix is $(-1)^{i+j}$ det A. The cofactor of a_{ij} in this new determinant is the $(n-1)$th-order determinant obtained by omitting the new first row and first column, that is, the ith row and jth column of A. Thus,

$$A_{ij} = (-1)^{i+j} \cdot \det B,$$

where B is the submatrix of A obtained by omitting the ith row and jth column of A.

The foregoing discussion gives the following

6.2 Theorem. *Let A be an $n \times n$ matrix. The cofactor of a_{ij} in det A is equal to $(-1)^{i+j}$ times the determinant of the submatrix of A obtained by deleting the ith row and the jth column.*

6.3 Example

Let $A = (a_{ij})$ be a 3×3 matrix. If we expand det A by the first row, we get

$$\begin{aligned}\det A &= a_{11}(-1)^{1+1}\begin{vmatrix} a_{22} & a_{23} \\ a_{32} & a_{33}\end{vmatrix} + a_{12}(-1)^{1+2}\begin{vmatrix} a_{21} & a_{23} \\ a_{31} & a_{33}\end{vmatrix} \\ &\quad + a_{13}(-1)^{1+3}\begin{vmatrix} a_{21} & a_{22} \\ a_{31} & a_{32}\end{vmatrix} \\ &= a_{11}(a_{22}a_{33} - a_{32}a_{23}) - a_{12}(a_{21}a_{33} - a_{31}a_{23}) \\ &\quad + a_{13}(a_{21}a_{32} - a_{31}a_{22}).\end{aligned}$$

If we expand det A by the third row, we get

$$\begin{aligned}\det A &= a_{31}(-1)^{1+3}\begin{vmatrix} a_{12} & a_{13} \\ a_{22} & a_{23}\end{vmatrix} + a_{32}(-1)^{3+2}\begin{vmatrix} a_{11} & a_{13} \\ a_{21} & a_{23}\end{vmatrix} \\ &\quad + a_{33}(-1)^{3+3}\begin{vmatrix} a_{11} & a_{12} \\ a_{21} & a_{22}\end{vmatrix}.\end{aligned}$$

If we expand A by the second column, we get

$$\det A = a_{12}(-1)^{1+2}\begin{vmatrix} a_{21} & a_{23} \\ a_{31} & a_{33} \end{vmatrix} + a_{22}(-1)^{2+2}\begin{vmatrix} a_{11} & a_{13} \\ a_{31} & a_{33} \end{vmatrix}$$

$$+ a_{32}(-1)^{3+2}\begin{vmatrix} a_{11} & a_{13} \\ a_{21} & a_{23} \end{vmatrix}.$$

Problems

1. Prove the following result:

$$\begin{vmatrix} 1 & a & a^3 \\ 1 & b & b^3 \\ 1 & c & c^3 \end{vmatrix} = (b-c)(c-a)(a-b)(a+b+c).$$

2. Prove that

$$\begin{vmatrix} 1 & a_1 & a_1^2 & \cdots & a_1^{n-1} \\ 1 & a_2 & a_2^2 & \cdots & a_2^{n-1} \\ \cdot & \cdot & \cdot & & \cdot \\ \cdot & \cdot & \cdot & & \cdot \\ \cdot & \cdot & \cdot & & \cdot \\ 1 & a_n & a_n^2 & \cdots & a_n^{n-1} \end{vmatrix} = \prod_{1 \le q < p \le n} (a_p - a_q).$$

This determinant is known as the *Vandermonde determinant*.

3. Prove that

(a) $$\begin{vmatrix} (b+c)^2 & a^2 & a^2 \\ b^2 & (c+a)^2 & b^2 \\ c^2 & c^2 & (a+b)^2 \end{vmatrix} = 2abc(a+b+c)^3.$$

(b) $$\begin{vmatrix} a-b-c & 2a & 2a \\ 2b & b-c-a & 2b \\ 2c & 2c & c-a-b \end{vmatrix} = (a+b+c)^3.$$

4. Prove that

(a) $$\begin{vmatrix} 3 & a+b+c & a^2+b^2+c^2 \\ a+b+c & a^2+b^2+c^2 & a^3+b^3+c^3 \\ a^2+b^2+c^2 & a^3+b^3+c^3 & a^4+b^4+c^4 \end{vmatrix}$$

$$= \begin{vmatrix} 1 & a & a^2 \\ 1 & b & b^2 \\ 1 & c & c^2 \end{vmatrix}^2 = (a-b)^2(b-c)^2(c-a)^2.$$

(b)
$$\begin{vmatrix} a & b & c \\ c & a & b \\ b & c & a \end{vmatrix}^2 = \begin{vmatrix} 2bc-a^2 & c^2 & b^2 \\ c^2 & 2ca-b^2 & a^2 \\ b^2 & a^2 & 2ab-c^2 \end{vmatrix}$$
$$= \begin{vmatrix} a^2-bc & b^2-ca & c^2-ab \\ c^2-ab & a^2-bc & b^2-ca \\ b^2-ca & c^2-ab & a^2-bc \end{vmatrix}$$
$$= (a^3+b^3+c^3-3abc)^2.$$

PART II

Groups

CHAPTER 4

Groups

1 Semigroups and groups

An *algebraic structure* or *algebraic system* is a nonempty set together with one or more binary operations on that set. Algebraic structures whose binary operations satisfy particularly important properties are semigroups, groups, rings, fields, modules, and so on. The simplest algebraic structure to recognize is a *semigroup*, which is defined as a nonempty set S with an associative binary operation. Any algebraic structure S with a binary operation $+$ or $\cdot$ is normally written $(S,+)$ or $(S,\cdot)$. However, it is also customary to use an expression such as "the algebraic structure S under addition or multiplication." Examples of semigroups are

(a) The systems of integers, reals, or complex numbers under usual multiplication (or addition)
(b) The set of mappings from a nonempty set S into itself under composition of mappings
(c) The set of $n \times n$ matrices over complex numbers under multiplication (or addition) of matrices

Let $(S,\cdot)$ be a semigroup and let $a,b \in S$. We usually write ab instead of $a \cdot b$. An element e in S is called a *left identity* if $ea = a$ for all $a \in S$. A *right identity* is defined similarly. It is possible to have a semigroup with several left identities or several right identities. However, if a semigroup S has both a left identity e and a right identity f, then $e = ef = f$. Therefore, e is the unique two-sided identity of the semigroup.

Definition. *Let* $(S,\cdot)$ *be a semigroup. If there is an element e in S such that*

$$ex = x = xe \qquad \text{for all } x \in S,$$

then e is called the identity *of the semigroup* $(S,\cdot)$.

Again, let S be a semigroup and let e be a left (right, or two-sided) identity in S. Let $a \in S$. An element b in S is called a *left inverse* of a (with respect to e) if $ba = e$; b is a *right inverse* if $ab = e$. An element in S may have several left inverses or several right inverses. However, if e is a two-sided identity and a has a left inverse b and a right inverse c, then $b = be = bac = ec = c$.

Definition. *Let* $(S,\cdot)$ *be a semigroup with identity e. Let* $a \in S$. *If there exists an element b in S such that*

$$ab = e = ba,$$

then b is called the inverse *of a, and a is said to be* invertible.

Our objective in this chapter and the following four chapters is to give basic results and techniques in the theory of a somewhat richer algebraic structure called a *group*. A group may be defined as a semigroup with identity in which every element has an inverse (that must be unique). We now give an equivalent definition with weaker postulates.

Definition. *A nonempty set G with a binary operation* $\cdot$ *on G is called a* group *if the following axioms hold:*

(i) $a(bc) = (ab)c$ *for all* $a,b,c \in G$.
(ii) *There exists* $e \in G$ *such that* $ea = a$ *for all* $a \in G$.
(iii) *For every* $a \in G$ *there exists* $a' \in G$ *such that* $a'a = e$.

Suppose $(G,\cdot)$ is a group according to this definition. Let $a \in G$, let a' be a left inverse of a, and let a'' be a left inverse of a'. Then $a'a = e = a''a'$. Hence,

$$aa' = eaa' = a''a'aa' = a''ea' = a''a' = e.$$

Therefore a' is also a right inverse of a. Now $ae = aa'a = ea = a$ for every $a \in G$. Hence, e is also a right identity. This proves that G is a group according to the definition that it is a semigroup with identity in which every element has an inverse.

Henceforth, unless otherwise stated, the binary operation in a group will be written multiplicatively, and we shall denote the group $(G,\cdot)$ simply by G. The identity element is denoted by e or 1, and the inverse of an element a is written a^{-1}. If the binary operation is written additively, the identity element is denoted by 0, and the inverse of an element a is written $-a$, called the *negative* of a.

If the binary operation in a group $(G,\cdot)$ is commutative – that is, $ab = ba$ for all $a,b \in G$ – then the group is called *commutative* or *abelian*.

When a group A is abelian, its binary operation is generally written additively.

Let $(S,\cdot)$ be a semigroup. Since $\cdot$ is associative, Theorem 4.2 (Chapter 1) holds here, and therefore there is a unique product of any given elements $a_1,\ldots,a_n \in S$, in that order, that are written $a_1a_2\ldots a_n$. For any positive integer n we define, as before, $a^n = \prod_{i=1}^{n} a = aa\ldots a$ (n-fold product), which is called the *nth* power of a.

If S has an identity e, we also define

$$a^0 = e.$$

Further, if a is an invertible element in S, we define, for any positive integer n,

$$a^{-n} = (a^{-1})^n,$$

where a^{-1} is an inverse of a.

In the case of an additive semigroup $(S,+)$, we speak of *sums* and *multiples* instead of products and powers. For any positive integer n, and $a \in S$, we define

$$na = \sum_{i=1}^{n} a = a + a + \cdots + a \qquad (n\text{-fold sum}).$$

1.1 Theorem. *A semigroup G is a group if and only if for all a,b in G, each of the equations $ax = b$ and $ya = b$ has a solution.*

Proof. If G is a group, then clearly the equations $ax = b$ and $ya = b$ have solutions $x = a^{-1}b$ and $y = ba^{-1}$, respectively. Conversely, let G be a semigroup and suppose $ax = b$ and $ya = b$ have solutions for all $a,b \in G$. Let $a \in G$. Then, in particular, the equation $ya = a$ has a solution, say $y = e$, so that $ea = a$. Now for any b in G, the equation $ax = b$ has a solution, say $x = x_0$, so that $ax_0 = b$. Therefore, $eb = eax_0 = ax_0 = b$. Hence, e is a left identity.

Further, for every a in G, the equation $ya = e$ has a solution. Hence, a has a left inverse. Therefore, G is a group. □

The following theorem gives a simpler characterization for a finite group. A group (or a semigroup) $(G,\cdot)$ is said to be *finite* or *infinite* according to whether the set G is finite or infinite. The cardinality of the set G is called the *order* of the group G and denoted by $|G|$ or $o(G)$.

1.2 Theorem. *A finite semigroup G is a group if and only if the* cancellation laws *hold for all elements in G; that is,*

$$ab = ac \Rightarrow b = c \quad \textit{and} \quad ba = ca \Rightarrow b = c$$

for all $a,b,c \in G$.

Proof. If G is a group, the cancellation laws hold trivially.

Conversely, let G be a finite semigroup and suppose that the cancellation laws hold. We shall prove that for all a,b in G, the equations $ax = b$ and $ya = b$ are solvable, and therefore, by the last theorem, G is a group.

Let $a,b \in G$. Let $G = \{a_1, a_2, \ldots, a_n\}$, where $a_1, \ldots, a_n$ are distinct. Consider the set $H = \{aa_1, aa_2, \ldots, aa_n\}$. Since $aa_i \in G$ for each $i = 1, \ldots, n$, $H \subset G$. By the left cancellation law, $aa_i = aa_j \Rightarrow a_i = a_j$. Hence, H has n distinct elements, and therefore $H = G$. Hence, $b \in H$ and therefore $aa_i = b$ for some $a_i \in G$. This proves that the equation $ax = b$ is solvable. Similarly, it follows from the right cancellation law that the equation $ya = b$ is solvable. Hence, by Theorem 1.1, G is a group. □

1.3 Examples of groups

(a) Integers modulo n under addition

Let $x,y \in \mathbf{Z}$ and let n be a positive integer. Define $x \equiv y \pmod{n}$ if $x - y = qn$ for some $q \in \mathbf{Z}$. Clearly, $\equiv$ is an equivalence relation. Denote by $\bar{x}$ the equivalence class containing x. Let $\mathbf{Z}/(n)$ denote the set of equivalence classes. We define addition in $\mathbf{Z}/(n)$ as follows: $\bar{x} + \bar{y} = \overline{x + y}$. This is well defined, for if $\bar{x} = \bar{x}'$ and $\bar{y} = \bar{y}'$, then $n|x - x'$ and $n|y - y'$; hence, $n|(x + y) - (x' + y')$. Thus, $\overline{x + y} = \overline{x' + y'}$. Clearly, $\bar{0}$ is the identity, and $\overline{(-x)}$ is the inverse of $\bar{x}$. Thus, $\mathbf{Z}/(n)$ is an additive abelian group of order n. Indeed,

$$\mathbf{Z}/(n) = \{\bar{0}, \bar{1}, \ldots, \overline{(n-1)}\}.$$

(b) Integers modulo n under multiplication

Consider the set $\mathbf{Z}/(n)$ as in (a). We define multiplication in $\mathbf{Z}/(n)$ as $\bar{x}\bar{y} = \overline{xy}$. As proved for addition in example (a), we can prove that multiplication is well defined. Then $(\mathbf{Z}/(n), \cdot)$ is a semigroup with identity, and the set $(\mathbf{Z}/(n))^*$ consisting of the invertible elements in $\mathbf{Z}/(n)$ forms a multiplicative group of order $\phi(n)$, where ϕ is the Euler function.

(c) Permutations under usual composition

Let X be a nonempty set, and let G be the set of bijective mappings on X to X (i.e., permutations of X). Then G is a group under the usual composition of mappings. The unit element of G is the identity map of X, and the other group postulates are easily verified by direct applications of results on mappings (see Chapter 1).

This group is called the *group of permutations* of X (or the *symmetric group* on X) and is denoted as S_X. If $|X| = n, S_X$ is a group of order $n!$.

(d) Symmetries of a geometric figure

Consider permutations of the set X of all points of some geometric figures. Call a permutation σ: $X \rightarrow X$ a "symmetry" of S when it preserves distances, that is, when $d(a,b) = d(\sigma(a),\sigma(b))$, where $d(a,b)$ denotes the distance between the points $a,b \in X$. If σ,τ are two symmetries, then

$$d((\sigma\tau)(a),(\sigma\tau)(b)) = d(\sigma(\tau(a)),\sigma(\tau(b))) = d(\tau(a),\tau(b)) = d(a,b).$$

Thus, $\sigma\tau$ is also a symmetry. Further, if σ is a symmetry then

$$d(\sigma^{-1}(a),\sigma^{-1}(b)) = d(\sigma(\sigma^{-1}(a)),\sigma(\sigma^{-1}(b))) = d(a,b).$$

So σ^{-1} is also a symmetry. Clearly, the identity permutation is a symmetry. Hence, the set of symmetries of S forms a group under composition of mappings.

Let us consider a special case when X is the set of points on the perimeter of an equilateral triangle:

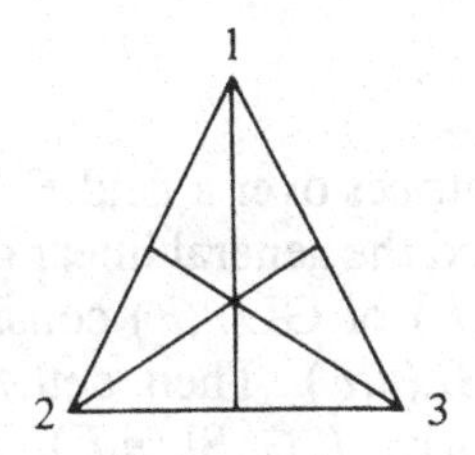

The counterclockwise rotations through 0, $2\pi/3$, and $4\pi/3$ are three of the symmetries that move the vertices in the following manner:

$$1 \to 1 \quad 1 \to 2 \quad 1 \to 3$$
$$2 \to 2, \quad 2 \to 3, \quad \text{and} \quad 2 \to 1,$$
$$3 \to 3 \quad 3 \to 1 \quad 3 \to 2$$

respectively. These are commonly written as

$$e = \begin{pmatrix} 1 & 2 & 3 \\ 1 & 2 & 3 \end{pmatrix}, \quad a = \begin{pmatrix} 1 & 2 & 3 \\ 2 & 3 & 1 \end{pmatrix}, \quad a^2 = \begin{pmatrix} 1 & 2 & 3 \\ 3 & 1 & 2 \end{pmatrix}.$$

(*Note.* Performing a rotation through $4\pi/3$ is equivalent to, or is a resultant of, performing a rotation through $2\pi/3$ and then again through $2\pi/3$. This explains our symbol a^2 for the rotation through $4\pi/3$.)

Three other symmetries are the reflections in the altitudes through the three vertices, namely,

$$1 \to 1 \quad 2 \to 2 \quad 3 \to 3$$
$$2 \to 3, \quad 3 \to 1, \quad \text{and} \quad 1 \to 2.$$
$$3 \to 2 \quad 1 \to 3 \quad 2 \to 1$$

These may be rewritten as

$$b = \begin{pmatrix} 1 & 2 & 3 \\ 1 & 3 & 2 \end{pmatrix}, \quad a^2b = \begin{pmatrix} 1 & 2 & 3 \\ 3 & 2 & 1 \end{pmatrix}, \quad \text{and} \quad ab = \begin{pmatrix} 1 & 2 & 3 \\ 2 & 1 & 3 \end{pmatrix},$$

respectively, where the "product" is composition of mappings.

Since any symmetry of the equilateral triangle is determined by its effect on three vertices, the set of six symmetries is a complete list of symmetries of an equilateral triangle. We denote this group by D_3, called the dihedral group of degree 3. Since D_3 is a subset of S_3 and each has six elements, $D_3 = S_3$.

Similar considerations apply to any regular polygon of n sides. This is discussed later in Section 5.

(e) Linear groups

Let $GL(n,F)$ be the set of $n \times n$ invertible matrices over a field F. Then $GL(n,F)$ is a group under multiplication, called the general linear group (in dimension n). Consider the subset $SL(n,F)$ of $GL(n,F)$ consisting of matrices of determinant 1. Let $A,B \in SL(n,F)$. Then $\det(AB) = (\det A)(\det B) = 1$, so $AB \in SL(n,F)$. Clearly, $I_n \in SL(n,F)$. Also, $\det(A^{-1})(\det A) = \det(I_n) = 1$ implies $\det(A^{-1}) = 1$, so $A^{-1} \in SL(n,F)$. Therefore, $SL(n,F)$ is also a group under multiplication.

(f) Direct product of groups

Let $G_1,...,G_n$ be a family of groups. Then the cartesian product $G_1 \times ... \times G_n$ is a group under the "pointwise" binary operation

$$(g_1,...,g_n)(g'_1,...,g'_n) = (g_1g'_1,...,g_ng'_n),$$

where $g_i, g'_i \in G_i$, $1 \le i \le n$. Clearly, if e_i is the identity of G_i, then $(e_1,...,e_n)$ is the identity of $G_1 \times ... \times G_n$ and $(g_1^{-1},...,g_n^{-1})$ is the inverse of $(g_1,...,g_n)$. Associativity of the binary operation is an immediate consequence of the associativity of the binary operations in each G_i.

This group is called the direct product of the groups G_i, $1 \le i \le n$, and is also written as $\Pi_{i=1}^n G_i$.

A finite group $(G,*)$ of order n can be exhibited by its *multiplication table*, which is an $n \times n$ matrix whose (i,j) entry is $a_i * a_j$, where $G = \{a_1,...,a_n\}$.

For example, the multiplication table for the group $(\mathbf{Z}/(4),+)$ is

+	$\bar{0}$	$\bar{1}$	$\bar{2}$	$\bar{3}$
$\bar{0}$	$\bar{0}$	$\bar{1}$	$\bar{2}$	$\bar{3}$
$\bar{1}$	$\bar{1}$	$\bar{2}$	$\bar{3}$	$\bar{0}$
$\bar{2}$	$\bar{2}$	$\bar{3}$	$\bar{0}$	$\bar{1}$
$\bar{3}$	$\bar{3}$	$\bar{0}$	$\bar{1}$	$\bar{2}$

We observe that the elements in each row (column) are distinct. This is a consequence of the cancellation law, as we saw in the proof of Theorem 1.2.

The multiplication table for the group D_3 $(= S_3)$, the symmetries of an equilateral triangle [Example 1.3(d)], is

$\cdot$	e	a	a^2	b	ab	a^2b
e	e	a	a^2	b	ab	a^2b
a	a	a^2	e	ab	a^2b	b
a^2	a^2	e	a	a^2b	b	ab
b	b	a^2b	ab	e	a^2	a
ab	ab	b	a^2b	a	e	a^2
a^2b	a^2b	ab	b	a^2	a	e

In the following examples we provide solutions to a couple of problems illustrating computational techniques.

1.4 Examples

(a) If G is a group such that $(ab)^n = a^n b^n$ for three consecutive integers, then $ab = ba$.

Solution. We have (1) $(ab)^n = a^n b^n$, (2) $(ab)^{n+1} = a^{n+1}b^{n+1}$, and (3) $(ab)^{n+2} = a^{n+2}b^{n+2}$. From (1) and (2) we get $(a^n b^n)(ab) = (ab)^{n+1} = a^{n+1}b^{n+1}$. This implies, by appropriate cancellations, $b^n a = ab^n$. Similarly, from (2) and (3), we get $b^{n+1}a = ab^{n+1}$. But then $b^{n+1}a = b(b^n a) = b(ab^n)$, so $ab^{n+1} = bab^n$. This gives $ab = ba$.

(b) Let G be a group such that $(ab)^2 = (ba)^2$ for all $a,b \in G$. Suppose G also has the property that $x^2 = e$, $x \in G$ implies $x = e$. Then G is abelian.

Solution. Let $a,b \in G$. Then, by hypothesis,

$$(ab^{-1}b)^2 = (bab^{-1})^2;$$

that is,

$$a^2 = ba^2b^{-1} \qquad \text{or} \qquad a^2b = ba^2;$$

so, in other words, the square of every element in G commutes with every element in G. Set $c = aba^{-1}b^{-1}$, and consider

$$\begin{aligned} c^2 &= ab(a^{-1}b^{-1}a)(ba^{-1}b^{-1}) = ab(aa^{-2}b^{-1}a)(ba^{-1}b^{-1}) \\ &= ab(ab^{-1}a^{-2}a)(ba^{-1}b^{-1}) = ab(ab^{-1}a^{-1})(ba^{-1}b^{-1}) \\ &= ab(abb^{-2}a^{-1})(ba^{-1}b^{-1}) = (ab)(aba^{-1}b^{-2})(ba^{-1}b^{-1}) \\ &= (ab)(aba^{-1})(b^{-1}a^{-1}b^{-1}) = (ab)^2(a^{-1}b^{-1})^2 = (ba)^2(a^{-1}b^{-1})^2 \\ &= e. \end{aligned}$$

Then, by hypothesis, $c = e$, and hence $ab = ba$.

Problems

1. Let A be the set of mappings from $\mathbf{R}$ to $\mathbf{R}$. Show that $(A,+)$ is a group under the usual addition of mappings.
2. Let H be the subset of $\mathbf{C}$ consisting of the nth roots of unity. Prove that H is a group under multiplication.
3. Let G be a group such that $a^2 = e$ for all $a \in G$. Show that G is abelian.
4. Let G be a group such that $(ab)^2 = a^2b^2$ for all $a,b \in G$. Show that G is abelian.

5. Prove that the matrices

$$\begin{bmatrix} 1 & 0 \\ 0 & 1 \end{bmatrix}, \begin{bmatrix} 1 & 0 \\ 0 & -1 \end{bmatrix}, \begin{bmatrix} -1 & 0 \\ 0 & 1 \end{bmatrix}, \text{ and } \begin{bmatrix} -1 & 0 \\ 0 & -1 \end{bmatrix}$$

form a multiplicative group. (This group is known as the *Klein four-group.*)

6. Prove that the following matrices form a nonabelian group of order 8 under multiplication of matrices:

$$\begin{bmatrix} 1 & 0 \\ 0 & 1 \end{bmatrix}, \begin{bmatrix} \sqrt{-1} & 0 \\ 0 & -\sqrt{-1} \end{bmatrix}, \begin{bmatrix} 0 & 1 \\ -1 & 0 \end{bmatrix}, \begin{bmatrix} 0 & \sqrt{-1} \\ \sqrt{-1} & 0 \end{bmatrix},$$

$$\begin{bmatrix} -1 & 0 \\ 0 & -1 \end{bmatrix}, \begin{bmatrix} -\sqrt{-1} & 0 \\ 0 & \sqrt{-1} \end{bmatrix}, \begin{bmatrix} 0 & -1 \\ 1 & 0 \end{bmatrix}, \begin{bmatrix} 0 & -\sqrt{-1} \\ -\sqrt{-1} & 0 \end{bmatrix}$$

Further, show that if

$$e = \begin{bmatrix} 1 & 0 \\ 0 & 1 \end{bmatrix}, \quad a = \begin{bmatrix} \sqrt{-1} & 0 \\ 0 & -\sqrt{-1} \end{bmatrix}, \quad b = \begin{bmatrix} 0 & \sqrt{-1} \\ \sqrt{-1} & 0 \end{bmatrix},$$

then $a^4 = e$, $b^2 = a^2$, and $b^{-1}ab = a^3$. (This group is known as the *quaternion group.*)

7. Let G be a finite semigroup. Prove that there exists $e \in G$ such that $e = e^2$.
8. Suppose G is a semigroup with the property that for each $a \in G$, there is a unique $a^* \in G$ such that $aa^*a = a$. Prove
   (i) If e is an idempotent in G, then $e^* = e$.
   (ii) If $a^*x = a^*$, x, $a \in G$, then $x = aa^*$.
   (iii) For all $a \in G$, $a^*aa^* = a^*$ and $a^{**} = a$.
   (iv) For all $a, b \in G$, $x = (ba^*)^*$ is a solution of $xb = a$.
   (v) G is a group.
9. Show that a group in which all the mth powers commute with each other and all the nth powers commute with each other, m and n relatively prime, is abelian.

2 Homomorphisms

Definition. *Let G,H be groups. A mapping*

$$\phi: G \to H$$

is called a homomorphism *if for all $x,y \in G$*

$$\phi(xy) = \phi(x)\phi(y).$$

Furthermore, if ϕ *is bijective, then* ϕ *is called an* isomorphism *of G* onto *H, and we write* $G \simeq H$. *If* ϕ *is just injective, that is,* 1-1, *then we say that* ϕ *is an isomorphism (or monomorphism) of G* into *H. If* ϕ *is surjective, that is, onto, then* ϕ *is called an* epimorphism. *A homomorphism of G into itself is called an* endomorphism *of G. An endomorphism of G that is both* 1-1 *and onto is called an* automorphism *of G.*

If $\phi: G \to H$ *is an onto homomorphism, then H is called a* homomorphic image *of G; also, G is said to be homomorphic to H. If* $\phi: G \to H$ *is a* 1-1 *homomorphism, then G is said to be* embeddable *in H, and we write* $G \subsetneq H$.

(Note that in the definition of homomorphism we have written both the groups multiplicatively, though the actual binary operations in G and H may be different.)

2.1 Examples

(a) Let G and H be groups, and let e' be the identity of H. The mapping

$$\theta: G \to H \qquad \text{given by} \qquad \theta(x) = e' \quad \text{for all } x \in G$$

is trivially a homomorphism.

(b) For any group G the identity mapping $i: G \to G$ is an automorphism of G.

(c) Let G be the group $(\mathbf{R}^+, \cdot)$ of positive real numbers under multiplication, and let H be the additive group $\mathbf{R}$. The mapping $\phi: \mathbf{R}^+ \to \mathbf{R}$ given by $\phi(x) = \log x$ is an isomorphism.

(d) Let G be a group. For a given $a \in G$ consider the mapping

$$I_a: G \to G \qquad \text{given by} \qquad I_a(x) = axa^{-1} \quad \text{for all } x \in G.$$

Since

$$I_a(xy) = axya^{-1} = (axa^{-1})(aya^{-1}) = I_a(x)I_a(y),$$

I_a is a homomorphism. By the cancellation laws,

$$axa^{-1} = aya^{-1} \Rightarrow x = y.$$

Hence, I_a is injective. For any x in G,

$$x = a(a^{-1}xa)a^{-1} = I_a(a^{-1}xa).$$

Hence, I_a is surjective. Consequently, I_a is an automorphism of G, called the *inner automorphism* of G determined by a.

2.2 Theorem. *Let G and H be groups with identities e and e', respec-*

tively, and let $\phi: G \to H$ *be a homomorphism. Then*

(i) $\phi(e) = e'$.
(ii) $\phi(x^{-1}) = (\phi(x))^{-1}$ *for each* $x \in G$.

Proof. (i) $\phi(e)\phi(e) = \phi(ee) = \phi(e) = e'\phi(e)$. Hence, by the cancellation law, $\phi(e) = e'$.

(ii) $\phi(x)\phi(x^{-1}) = \phi(xx^{-1}) = \phi(e) = e'$. Hence, $\phi(x^{-1}) = (\phi(x))^{-1}$. □

Thus, a homomorphism $\phi: G \to H$ not only "preserves" the group operation but also the group identity and the unary operation $x \mapsto x^{-1}$.

Definition. *Let G and H be groups, and let* $\phi: G \to H$ *be a homomorphism. The* kernel *of* ϕ *is defined to be the set*

$$\text{Ker } \phi = \{x \in G \mid \phi(x) = e'\},$$

where e' *is the identity in H.*

Because $\phi(e) = e'$, *it follows that* Ker ϕ *is not empty.*

2.3 Theorem. *A homomorphism* $\phi: G \to H$ *is injective if and only if* Ker $\phi = \{e\}$.

Proof. Suppose ϕ is injective, and let $x \in \text{Ker } \phi$. Then $\phi(x) = e' = \phi(e)$. Hence, $x = e$. Therefore, Ker $\phi = \{e\}$.

Conversely, suppose Ker $\phi = \{e\}$. Then

$$\begin{aligned} \phi(x) = \phi(y) &\Rightarrow \phi(xy^{-1}) = \phi(x)\phi(y)^{-1} = e' \\ &\Rightarrow xy^{-1} \in \text{Ker } \phi \Rightarrow xy^{-1} = e \Rightarrow x = y. \end{aligned}$$

Hence, ϕ is injective. □

It can be easily proved that if $f: G \to H$ and $g: H \to K$ are homomorphisms (isomorphisms) of groups, then the composite map $gf: G \to K$ is also a homomorphism (isomorphism). Further, if $f: G \to H$ is an isomorphism of G onto H, then the inverse mapping $f^{-1}: H \to G$ is also an isomorphism. We noted earlier that every group is isomorphic to itself. This proves that isomorphism of groups is an equivalence relation.

Problems

1. Let G be the multiplicative group of the nth roots of unity. Prove that $G \simeq (Z/(n), +)$.

2. Prove that a group G is abelian if and only if the mapping f: $G \to G$, given by $f(x) = x^{-1}$, is a homomorphism.
3. Show that a group G is abelian if and only if the mapping $f: G \to G$, given by $f(x) = x^2$, is a homomorphism.
4. Find the kernel of each of the following homomorphism:
   (a) $f: \mathbf{Z} \to \mathbf{Z}_n$, given by $f(x) = \bar{x}$.
   (b) $f: G \to \mathbf{Z}_2$, where G is the quaternion group (see Problem 6, Section 1) and $f(a) = \bar{0}$, $f(b) = \bar{1}$.
5. Show that there does not exist any nonzero homomorphism of the group D_3 (or S_3) (see Example 1.3(d)) to the group $\mathbf{Z}_3$.
6. Let F be a field with two elements. Show that $\mathrm{GL}(2, F) \simeq S_3$.

3 Subgroups and cosets

Definition. *Let* $(G, \cdot)$ *be a group, and let* H *be a subset of* G. H *is called a* subgroup *of* G, *written* $H < G$, *if* H *is a group relative to the binary operation in* G.

For any group G the singleton $\{e\}$ and G itself are subgroups of G, called *trivial* subgroups. A subgroup H of G is said to be a *proper* subgroup if $H \neq \{e\}$ and $H \neq G$.

It is easily seen that the identity element of a subgroup of a group must be same as that of the group.

3.1 Theorem. *Let* G *be a group. A nonempty subset* H *of* G *is a subgroup of* G *if and only if either of the following holds:*

(i) For all $a,b \in H$, $ab \in H$, *and* $a^{-1} \in H$.
(ii) For all $a,b \in H$, $ab^{-1} \in H$.

Proof. If H is a subgroup, (i) and (ii) are obviously true. Conversely, suppose H satisfies (i). Then for any $a \in H$, $a^{-1} \in H$. Hence, $e = aa^{-1} \in H$. Therefore, H is a subgroup. Next, suppose that H satisfies (ii). Let $a,b \in H$. Then $e = bb^{-1} \in H$. Hence, $b^{-1} = eb^{-1} \in H$. Therefore, $ab = a(b^{-1})^{-1} \in H$, which proves that H is a subgroup of G. □

There is a simpler criterion for a finite subgroup.

3.2 Theorem. *Let* $(G, \cdot)$ *be a group. A nonempty finite subset* H *of* G *is a subgroup if and only if* $ab \in H$ *for all* $a,b \in H$.

Proof. If H is closed under $\cdot$, then $(H,\cdot)$ is a finite semigroup. Since cancellation laws hold for all elements in G, they must hold for all elements in H. Therefore, by Theorem 1.2, H is a group and, hence, a subgroup of G. The converse is obvious. □

For an example of a subgroup, consider the set $n\mathbf{Z}$ of all multiples of n, where n is any given integer; that is, $n\mathbf{Z} = \{nk | k \in \mathbf{Z}\}$. Then for all $k,k' \in \mathbf{Z}$, $nk - nk' = n(k - k') \in n\mathbf{Z}$. Hence, $n\mathbf{Z}$ is a subgroup of the additive group $\mathbf{Z}$.

More generally, let $(G,\cdot)$ be any group, and let $a \in G$. Let H be the set of all powers of a: $H = \{a^k | k \in \mathbf{Z}\}$. Then $a^k(a^{k'})^{-1} = a^{k-k'} \in H$ for all $k,k' \in \mathbf{Z}$. Hence, H is a subgroup of G. H is called the *cyclic subgroup of G generated* by a, written $H = [a]$, and a is called a *generator* of H. If G is an additive group, then $[a] = \{ka | k \in \mathbf{Z}\}$ is the set of all multiples of a. The subgroup $n\mathbf{Z}$ in the previous example is thus the cyclic subgroup of $\mathbf{Z}$ generated by n. A group G is said to be *cyclic* if $G = [a]$ for some a in G.

The next two theorems give further examples of subgroups.

3.3 Theorem. *Let ϕ: $G \to H$ be a homomorphism of groups. Then* Ker ϕ *is a subgroup of G and* Im ϕ *is a subgroup of H.*

Proof. Ker ϕ and Im ϕ are both nonempty. Let $a,b \in$ Ker ϕ. Then $\phi(ab^{-1}) = \phi(a)\phi(b)^{-1} = e'e' = e'$ (where e' is the identity of H). Hence, $ab^{-1} \in$ Ker ϕ. This proves that Ker ϕ is a subgroup of G. Next, let $\alpha,\beta \in$ Im ϕ. Then $\alpha = \phi(x)$, $\beta = \phi(y)$ for some $x,y \in G$. Hence, $\alpha\beta^{-1} = \phi(x)\phi(y)^{-1} = \phi(xy^{-1}) \in$ Im ϕ, which proves that Im ϕ is a subgroup of H. □

Definition. *The* center *of a group G, written Z(G), is the set of those elements in G that commute with every element in G; that is,*

$$Z(G) = \{a \in G | ax = xa \text{ for all } x \in G\}.$$

3.4 Theorem. *The center of a group G is a subgroup of G.*

Proof. Since $ex = x = xe$ for all $x \in G$, $e \in Z(G)$. Let $a,b \in Z(G)$. Then for all $x \in G$,

$$\begin{aligned} ab^{-1}x &= ab^{-1}xe = ab^{-1}xbb^{-1} = ab^{-1}bxb^{-1} \\ &= aexb^{-1} = axb^{-1} = xab^{-1}. \end{aligned}$$

Hence, $ab^{-1} \in Z(G)$. Therefore, $Z(G)$ is a subgroup of G. □

Let us next see how new subgroups may be obtained from given subgroups. Let H and K be subgroups of a group G. Then $e \in H$, $e \in K$. Hence, $H \cap K$ is not empty. If $a,b \in H \cap K$, then $ab^{-1} \in H$, $ab^{-1} \in K$. Hence, $ab^{-1} \in H \cap K$. Therefore, $H \cap K$ is a subgroup of G. More generally, a similar argument shows that the intersection of any number of subgroups of G is again a subgroup of G. But the union of subgroups H and K is a subgroup of G if and only if $H \subset K$ or $K \subset H$. For suppose $H \cup K$ is a subgroup, but neither $H \subset K$ nor $K \subset H$. Then there exist elements $a \in H - K$ and $b \in K - H$. Now $a,b \in H \cup K$; hence, $ab \in H \cup K$. If $ab \in H$, then $b = a^{-1}ab \in H$, a contradiction. On the other hand, if $ab \in K$, then $a = abb^{-1} \in K$, again a contradiction.

A binary operation on a set S induces a binary operation on the power set $\mathcal{P}(S)$. Accordingly, for any subsets A,B of a group G, we define

$$AB = \{xy \in G | x \in A, y \in B\}.$$

(For an additive group G, we define $A + B = \{x + y | x \in A, y \in B\}$.)

3.5 Theorem. *Let H and K be subgroups of a group $(G, \cdot)$. Then HK is a subgroup of G if and only if $HK = KH$.*

Proof. Let $HK = KH$. Since $e = ee \in HK$, HK is not empty. Let $a,b \in HK$. Then $a = h_1k_1$, $b = h_2k_2$ for some $h_1,h_2 \in H$ and $k_1,k_2 \in K$. Hence,

$$ab^{-1} = h_1k_1k_2^{-1}h_2^{-1} = h_1k_3h_2^{-1},$$

where $k_3 = k_1k_2^{-1} \in K$. Now $k_3h_2^{-1} \in KH = HK$. Hence, $k_3h_2^{-1} = h_3k_4$ for some $h_3 \in H$, $k_4 \in K$. Therefore,

$$ab^{-1} = h_1h_3k_4 = h_4k_4,$$

where $h_4 = h_1h_3 \in H$. Hence, $ab^{-1} \in HK$. This proves that HK is a subgroup.

Conversely, suppose that HK is a subgroup. Let $a \in KH$, so that $a = kh$ for some $h \in H$, $k \in K$. Then $a^{-1} = h^{-1}k^{-1} \in HK$. Hence, $a \in HK$. Therefore, $KH \subset HK$. Next, let $b \in HK$. Then $b^{-1} \in HK$. Hence, $b^{-1} = h'k'$ for some $h' \in H$, $k' \in K$. Therefore, $b = k'^{-1}h'^{-1} \in KH$. Hence, $HK \subset KH$. This proves that $HK = KH$. □

It follows from the theorem that if G is an abelian group, then HK is a subgroup for any subgroups H,K of G.

The subgroups of a group G can be partially ordered by set inclusion. Let H and K be subgroups of G. Then $H \cap K$ is the largest subgroup of G contained in both H and K, in the sense that if L is any subgroup contained in both H and K, then $L \subset H \cap K$. If $HK = KH$, then HK is the smallest subgroup containing both H and K. For if M is any subgroup containing H and K, then $hk \in M$ for all $h \in H, k \in K$. Even if $HK \neq KH$, we can still find a smallest subgroup containing both H and K.

Let S be a subset of G. Consider the family $\mathscr{C}$ of all subgroups of G containing S. That is,

$$\mathscr{C} = \{A \mid A \text{ is a subgroup of } G, S \subset A\}.$$

Since G is in $\mathscr{C}$, $\mathscr{C}$ is not empty.

Let M be the intersection of all subgroups A in $\mathscr{C}$. Then M is a subgroup of G and $S \subset M$. If M' is a subgroup and $S \subset M'$, then $M' \in \mathscr{C}$. Hence, $M \subset M'$. Therefore M is the smallest subgroup containing S. It is called the subgroup *generated* by S and denoted by $[S]$. If $G = [S]$ for some subset S of G, then S is called a set of generators of G. (Trivially, the set G itself is a set of generators of the group G.) If S is empty, then $[S]$ is the trivial subgroup $\{e\}$. If S is a finite set and $G = [S]$, then G is said to be a finitely generated group.

To go back to the previous case of subgroups H and K, we see that the smallest subgroup containing H and K is the subgroup generated by $H \cup K$, which is denoted by $H \vee K$.

3.6 Theorem. *Let S be a nonempty subset of a group G. Then the subgroup generated by S is the set M of all finite products $x_1x_2...x_n$ such that, for each i, $x_i \in S$ or $x_i^{-1} \in S$.*

Proof. Clearly, $S \subset M$. For any two elements $a = x_1...x_m$ and $b = y_1...y_n$ in M,

$$ab^{-1} = x_1 \cdots x_m y_n^{-1} \cdots y_1^{-1} \in M.$$

Hence, M is a subgroup of G. Let M' be any subgroup of G containing S. Then for each $x \in S, x \in M'$. Hence, $x^{-1} \in M'$. Therefore M' contains all finite products $x_1 \cdots x_n$ such that $x_i \in S$ or $x_i^{-1} \in S$, $i = 1, \ldots, n$. Hence, $M \subset M'$. This proves that M is the smallest subgroup containing S and, therefore, the subgroup generated by S. □

If S is a singleton $\{a\}$, the subgroup generated by S is the cyclic subgroup $[a] = \{a^i \mid i \in \mathbf{Z}\}$, as shown earlier.

Definition. *Let G be a group, and $a \in G$. If there exists a least positive integer m such that $a^m = e$, then m is called the* order *of a, written $o(a)$. If no such positive integer exists, then a is said to be of infinite order.*

We now give some examples to illustrate this concept.

(a) In the additive group $\mathbf{Z}_4$, $o(\bar{2})=2$, since $\bar{2}+\bar{2}=\bar{0}$. Further, $o(\bar{3})=4$.

(b) In the multiplicative group $G=\{x\in\mathbf{C}\mid x^3=1\}$, the element $\omega=\dfrac{-1+\sqrt{-3}}{2}$ has order 3. The order of i in $G=\{x\in\mathbf{C}\mid x^4=1\}$ is 4.

(c) Each element different from e in the Klein four-group (see Problem 5, Section 1) is of order 2.

(d) In S_3, the order of $\begin{pmatrix}1&2&3\\2&1&3\end{pmatrix}$ is 2 and that of $\begin{pmatrix}1&2&3\\2&3&1\end{pmatrix}$ is 3.

(e) The order of any element of a finite group G is finite. For, if $a\in G$, then the elements $a, a^2, a^3,\ldots$ of G cannot all be distinct. So there exist distinct positive integers i, j such that $a^i=a^j$, which gives $a^{i-j}=e$. Hence $o(a)$ is finite.

(f) The order of any nonzero element in the group $(\mathbf{Z},+)$ is infinite. (There exist infinite groups in which each element is of finite order. Find one.)

3.7 Theorem. *Let G be a group and $a\in G$.*

(i) If $a^n=e$ for some integer $n\neq 0$, then $o(a)\mid n$.

(ii) If $o(a)=m$, then for all integers i, $a^i=a^{r(i)}$, where $r(i)$ is the remainder of i modulo m.

(iii) $[a]$ is of order m if and only if $o(a)=m$.

Proof. (i) If $a^n=e$, then $a^{-n}=e$. Hence $a^i=e$ for some $i>0$. Therefore by the well-ordering property of $\mathbf{N}$, there is a least positive integer $m=o(a)$ such that $a^m=e$. By the division algorithm, $n=mq+r, 0\leq r<m$. Hence, $e=a^n=(a^m)^q a^r=a^r$. Therefore $r=0$ and $m\mid n$.

(ii) Again by the division algorithm, for any $i\in\mathbf{Z}$, $i=mq+r$, $0\leq r<m$. Hence, $a^i=a^r$, where $r=r(i)$ is the remainder of i modulo m.

(iii) Let $o(a)=m$. Then $e,a,\ldots,a^{m-1}$ are distinct, for otherwise $a^i=a^j$ for some i,j, $0\leq i<j\leq m-1$. Hence, $a^{j-i}=e$, a contradiction. Let $H=[a]$ be the cyclic subgroup generated by a. For any $i\in\mathbf{Z}$, $a^i=a^{r(i)}$. This implies that H has exactly m elements $e,a,\ldots,a^{m-1}$.

Conversely, suppose that H is of finite order. Then a^i's are not distinct for all $i\in\mathbf{Z}$. Hence, $a^i=a^j$ for some $i,j\in\mathbf{Z}$, $i<j$. Then $a^{j-i}=e$. Hence, a is of finite order, say m. But then H has exactly m elements, as proved earlier. □

3.8 Corollary. *If G is a finite group, then there exists a positive integer k such that $x^k=e$ for all $x\in G$.*

Proof. Since G is finite, the cyclic subgroup $[a]$ is also finite. Hence, $o(a)$ is finite, say $n(a)$. Choose $k = \Pi_{a \in G} n(a)$. Then $x^k = e$ for all $x \in G$. □

Definition. *Let H be a subgroup of G. Given $a \in G$, the set*

$$aH = \{ah | h \in H\}$$

is called the left coset *of H determined by a. A subset C of G is called a left coset of H in G if $C = aH$ for some a in G. The set of all left cosets of H in G is written G/H.*

A right coset Ha is defined similarly. The set of all right cosets of H in G is written $H \backslash G$.

Let us confine our attention for the present to left cosets aH of H in G. For any $a \in G$ the mapping f: $H \to aH$, given by $f(h) = ah$, is clearly a bijection. Hence, every left coset of H has the same cardinality as H. Note that H itself is a left coset of H, since $eH = H$.

Consider the relation $\sim$ on G, where $a \sim b$ means $a^{-1}b \in H$. For all $a,b,c \in G$,

$$a^{-1}a = e \in H,$$
$$a^{-1}b \in H \Rightarrow b^{-1}a = (a^{-1}b)^{-1} \in H,$$

and

$$a^{-1}b,\ b^{-1}c \in H \Rightarrow a^{-1}c = (a^{-1}b)(b^{-1}c) \in H.$$

Hence, $\sim$ is an equivalence relation on G. It is easily verified that the equivalence classes of $\sim$ are precisely the left cosets of H in G. Therefore, the set G/H of left cosets of H in G is a partition of G. That is, the distinct left cosets of H are pairwise disjoint and their union is G.

Consider now the right cosets of H in G. By arguments similar to those given above, we can prove that any two right cosets of H have the same cardinality, and the set $H \backslash G$ of all right cosets of H is a partition of G. Two elements a,b in G belong to the same right coset if and only if $ab^{-1} \in H$.

Consider the mapping ψ: $G/H \to H \backslash G$ given by $aH \mapsto Ha^{-1}$. The mapping is well defined because

$$aH = bH \Rightarrow a^{-1}b \in H \Rightarrow a^{-1}(b^{-1})^{-1} \in H \Rightarrow Ha^{-1} = Hb^{-1}.$$

By a similar argument, $Ha^{-1} = Hb^{-1} \Rightarrow aH = bH$. Hence, ψ is injective. Moreover, ψ is obviously surjective; therefore ψ is bijective. Consequently, G/H and $H \backslash G$ have the same cardinality. Thus, we have the following definition.

Definition. *Let H be a subgroup of G. The cardinal number of the set of* *left (right) cosets of H in G is called the* index *of H in G and denoted by* $[G:H]$.

If H is the trivial subgroup $\{e\}$, each left (right) coset of H in G is a singleton. Hence, the index of H is equal to the order of G. Writing 1 for the trivial subgroup $\{e\}$, we get $|G| = [G:1]$.

For each nonzero subgroup K of the additive group $\mathbf{Z}$, $[\mathbf{Z}:K]$ is finite. Indeed it is easy to check that $K = n\mathbf{Z}$, $n \neq 0$, and so $\mathbf{Z}/n\mathbf{Z} = \mathbf{Z}_n$ (Example 1.3(a)).

Let G be a finite group, and let H be any subgroup of G. Let $|G| = n$, $|H| = m$. Then every left coset of H has m elements. Since the distinct left cosets of H are pairwise disjoint and their union is G, we must have $n = km$, where k is the number of left cosets of H in G. In other words,

$$[G:1] = [G:H][H:1].$$

This proves the following theorem, which is of fundamental importance in the theory of finite groups.

3.9 Theorem (Lagrange). *Let G be a finite group. Then the order of any subgroup of G divides the order of G.*

We shall now derive some important results from Lagrange's theorem.

3.10 Corollary. *Let G be a finite group of order n. Then for every $a \in G$, $o(a)|n$, and, hence, $a^n = e$.*

Consequently, every finite group of prime order is cyclic and, hence, abelian.

Proof. Let $a \in G$. By Lagrange's theorem the order of the cyclic subgroup $[a]$ divides n. So $o(a)|n$.

If n is prime and $a \neq e$, the order of $[a]$ must be n. Hence, $[a] = G$, and therefore G is cyclic. □

3.11 Examples

(a) (Groups of order <6)

It follows from Corollary 3.10 that groups of orders 2,3, or 5 are cyclic. Consider a group G of order 4. If G has an element a of order 4, then $[a] = G$. Hence, G is cyclic. Otherwise, every element $\neq e$ is of order 2;

that is, $G = \{e,a,b,c\}$ and $a^2 = b^2 = c^2 = e$. Consider the product ab. If $ab = e$, then $ab = aa$, which implies $b = a$, a contradiction. Hence, $ab \neq e$. Similarly, $ab \neq a$, $ab \neq b$. Hence, $ab = c$. By the same argument, $ba = c$, $bc = a = cb$, and $ca = b = ac$. Hence, G is abelian. (This group, or any group isomorphic to it, is called the *Klein four-group.*) Thus, all groups of order <6 are abelian.

(b) (Euler-Fermat theorem)

If a is an integer prime to the positive integer m, then $a^{\phi(m)} \equiv 1 \pmod{m}$, where ϕ is the Euler function.

Solution. Note that an element $\bar{x} \in \mathbf{Z}/(m)$ is invertible if and only if $(x,m) = 1$, and the order of the multiplicative group $(\mathbf{Z}/(m))^*$ of the invertible elements in $\mathbf{Z}/(m)$ is $\phi(m)$. Since $(a,m) = 1$, $\bar{a} \in (\mathbf{Z}/(m))^*$, so $\bar{a}^{\phi(m)} = \bar{1}$. This yields $a^{\phi(m)} \equiv 1 \pmod{m}$.

(c) (Poincaré's theorem)

The intersection of two subgroups of finite index is of finite index.

Solution. Let $H,K < G$, H,K of finite indices. Let $a \in G$. Then it is trivial that $(H \cap K)a = Ha \cap Ka$. Thus, any right coset of $H \cap K$ is the intersection of a right coset of H and a right coset of K; but the number of such intersections is finite. This proves that the number of right cosets of $H \cap K$ is finite.

(d)

Let G be a group and $a, b \in G$ such that $ab = ba$. If $o(a) = m$, $o(b) = n$, and $(m, n) = 1$, then $o(ab) = mn$.

Solution. Let $o(ab) = k$. Then $(ab)^k = e$. Consider

$$(ab)^{mn} = a^{mn}b^{mn} = e,$$

which gives $k|mn$ (Theorem 3.7). Also

$$(ab)^k = e \Rightarrow a^k b^k = e \Rightarrow a^k = b^{-k}.$$

Thus,

$$o(a^k) = o(b^{-k}) = o(b^k) \qquad \text{(Problem 10)}.$$

But

$$a^m = e \Rightarrow (a^k)^m = e \Rightarrow o(a^k)|m \qquad \text{(Theorem 3.7).}$$

Similarly, $o(b^k)|n$. Therefore, $o(a^k)$ divides $(m,n) = 1$. Hence, $o(a^k) = 1$, so $a^k = e$. But then $m|k$. Similarly, $n|k$. Consequently, $mn|k$. Since $k|mn$, it follows that $k = mn$.

The result contained in the following problem is sometimes quite handy when studying structure of finite groups.

(e)

Let G be a finite group, and $S, T < G$. Then

$$|ST| = \frac{|S||T|}{|S \cap T|}.$$

Solution. Consider $S \times T$ and define a relation $\sim$ in $S \times T$ by $(s,t) \sim (s',t')$ if and only if $s' = sa$ and $t' = a^{-1}t$ for some $a \in S \cap T$. Then $\sim$ is an equivalence relation. Let $\overline{(s,t)}$ denote the equivalence class of $(s,t) \in S \times T$, and let $S \times T/\sim$ be the set of equivalence classes. It follows from the definition of equivalence relation that

$$\overline{(s,t)} = \{(sa,a^{-1}t) | a \in S \cap T\}.$$

Thus,

$$|\overline{(s,t)}| = |S \cap T|.$$

Also, let $S \times T = \cup_{i=1}^{k} \overline{(s_i,t_i)}$ be the disjoint union of equivalence classes. Therefore,

$$|S \times T| = k|S \cap T|. \tag{1}$$

We now define a mapping $f\colon S \times T/\sim \to ST$ by $f(\overline{(s_i,t_i)}) = s_it_i$. f is well defined and injective, because

$$\overline{(s_i,t_i)} = \overline{(s_j,t_j)} \Leftrightarrow s_j = s_ia, \qquad t_j = a^{-1}t_i,$$

for some $a \in S \cap T \Leftrightarrow s_jt_j = s_it_i$.

Also, f is clearly surjective. Hence,

$$|S \times T/\sim| = |ST|.$$

Therefore, $k = |ST|$; so from (1) we get

$$|S \times T| = |ST||S \cap T|,$$

which gives

$$|S||T| = |ST||S \cap T| \quad \text{or} \quad |ST| = \frac{|S||T|}{|S \cap T|}.$$

Problems

1. Let H be a subgroup of a group G. Prove that $Ha = H$ if and only if $a \in H$. Also show that $HH = H$. More generally, if A is a nonempty subset of G then show that $AH = H$ if and only if $A \subset H$.
2. Let G be a group and $H < G$. Suppose $|G/H| = 2$. Prove that $aH = Ha$ for all $a \in G$.
3. Let G be a group and $H < G$. Show that for all $x \in G$, $x^{-1}Hx$ is a subgroup of G of the same cardinality as that of H.
4. Find the subgroups of the groups given in Problems 5 and 6 of Section 1.
5. Show that for any subgroup of a group the inverses of the elements of a left coset form a right coset.
6. Let V be the group of vectors in the plane, with vector addition as the binary operation. Show that the vectors that issue from the origin O and have endpoints on a fixed line through O form a subgroup. What are the cosets relative to this subgroup?
7. (a) If H and K are subgroups whose orders are relatively prime, prove that $H \cap K = \{e\}$.
   (b) If H and K are subgroups of order p and n, respectively, where p is prime, then either $H \cap K = \{e\}$ or $H < K$.
8. Show that the elements of finite order in any abelian group form a subgroup.
9. Show that a subset A of a group G cannot be a left coset of two distinct subgroups of G. If A is a left coset of some subgroup of G, then prove that A is also a right coset of some subgroup of G.
10. Show that $o(a) = o(a^{-1})$ for any element a in a group G.
11. If a,b are any two elements in a group G, show that ab and ba have the same order.
12. Let G be a group and $a,b \in G$. Show that $o(a) = o(b^{-1}ab)$.
13. If a group G has only one element a of order n, then $a \in Z(G)$ and $n = 2$.
14. Let G be a group and $a,b \in G$ such that $ab = ba$. Let $o(a) = m$ and $o(b) = n$. Show that there exists an element $c \in G$ such that $o(c)$ is the least common multiple of m and n.
15. Show that if G is a group of even order then there are exactly an odd number of elements of order 2.

16. Prove Wilson's theorem that, for any prime p, $(p-1)! \equiv -1 \pmod{p}$.
17. Let a be an element of a group G such that $o(a) = r$. Let m be a positive integer such that $(m,r) = 1$. Prove that $o(a^m) = r$.
18. Let a be an element of a group G such that $o(a) = r$. Let m be a positive integer. Prove that $o(a^m) = r/(m,r)$.
19. Let $g: G \to G'$ be a homomorphism of groups. Let $a \in G$. Prove that $o(g(a)) | o(a)$. Further prove that if g is injective, then $o(g(a)) = o(a)$.
20. Let G be a finite group, and let S, T be nonempty subsets of G such that $G \neq ST$. Prove that $|G| \geq |S| + |T|$.
21. Let S be a nonempty subset of a group G, and let

    $$C(S) = \{x \in G | xs = sx, \quad \text{for all } s \in S\}.$$

    Show that $C(S) < G$. What is $C(G)$? ($C(S)$ is called the *centralizer* of S in G.)
22. Find the centralizer of each of the following subsets of S_3:

    $$\{(1\ \ 2\ \ 3)\}, \quad \{(1\ \ 2)\}, \quad \text{and} \quad \{(1\ \ 2\ \ 3),\ (1\ \ 2)\}.$$

4 Cyclic groups

As already defined, a group G is said to be cyclic if $G = [a] = \{a^i | i \in \mathbf{Z}\}$ for some $a \in G$. The most important examples of cyclic groups are the additive group $\mathbf{Z}$ of integers and the additive groups $\mathbf{Z}/(n)$ of integers modulo n. In fact, these are the only cyclic groups up to isomorphism.

4.1 Theorem. *Every cyclic group is isomorphic to* $\mathbf{Z}$ *or to* $\mathbf{Z}/(n)$ *for some* $n \in \mathbf{N}$.

Proof. If $G = [a]$ is an infinite cyclic group, consider the mapping ψ: $\mathbf{Z} \to G$ given by $\psi(i) = a^i$. It is clear that ψ is a surjective homomorphism. Moreover, $i \neq j \Rightarrow a^i \neq a^j$, for otherwise a would be of finite order. Hence, ψ is injective. Therefore, ψ is an isomorphism.

Next, suppose $G = [a]$ is a cyclic group of finite order n. Then $G = \{e, a, \ldots, a^{n-1}\}$ and $o(a) = n$ (Theorem 3.7). Consider the mapping ψ: $\mathbf{Z}/(n) \to G$ given by $\psi(\bar{i}) = a^i$. ψ is well defined and also injective, for let $\bar{i}, \bar{i} \in \mathbf{Z}/(n)$. Then

$$\bar{i} = \bar{j} \Leftrightarrow n | i - j \Leftrightarrow a^{i-j} = e \Leftrightarrow a^i = a^j.$$

Clearly, ψ is surjective. Further,

$$\psi(\bar{i} + \bar{j}) = \psi(\overline{i + j}) = a^{i+j} = a^i a^j.$$

Hence, ψ is an isomorphism. $\square$

The following is immediate from the above theorem.

4.2 Theorem. *Any two cyclic groups of the same order (finite or infinite) are isomorphic.*

4.3 Theorem. *Every subgroup of a cyclic group is cyclic.*

Proof. Let $G = [a]$ be a cyclic group, and let H be a subgroup of G. If H is a trivial subgroup, the result is obvious. So let H be a proper subgroup of G. If $a^i \in H$, then $a^{-i} \in H$. Hence, there is a least positive integer m such that $b = a^m \in H$. We prove that $H = [b]$. Let $a^i \in H$. By the division algorithm $i = mq + r$, where $0 \leq r < m$. Then

$$a^r = a^{i-mq} = a^i b^{-q} \in H.$$

Hence, $r = 0$. Therefore, $a^i = b^q$, which proves that $H = [b]$. □

The converse of Lagrange's theorem is not true in general, but the following theorem shows that it does hold for cyclic groups.

4.4 Theorem. *Let G be a finite cyclic group of order n, and let d be a positive divisor of n. Then G has exactly one subgroup of order d.*

Proof. The result holds trivially if $d = 1$ or n. So let $1 < d < n$ and put $n/d = m$. Let $G = [a]$. Then $b = a^m$ is of order d. Hence, $[b]$ is a cyclic subgroup of order d. To prove the uniqueness, let H be any subgroup of G of order d. By Theorem 4.3, H is generated by an element $c = a^s$. Then $a^{sd} = c^d = e$. Hence, $n|sd$, that is, $md|sd$ and so $m|s$. Let $m\lambda = s$. This yields $[a^s] \subset [a^m] = [b]$. But since each of the subgroups $[a^s]$ and $[a^m]$ is of order d, $[a^s] = [a^m]$. This proves $H = [b]$.

4.5 Example

Let $H = [a]$ and $K = [b]$ be cyclic groups of orders m and n, respectively, such that $(m,n) = 1$. Then $H \times K$ is a cyclic group of order mn.

Solution. Let $o(a,b) = d$. Now $(a,b)^{mn} = (a^{mn},b^{mn}) = (e,e)$ implies $d|mn$. Also $(e,e) = (a,b)^d = (a^d,b^d)$ implies $a^d = e = b^d$, so $m|d$ and $n|d$. Therefore, $mn|d$. Consequently, $mn = d$. Since $|H \times K| = mn$, it follows that (a,b) generates the group $H \times K$.

Problems

1. Prove that the multiplicative group of the nth roots of unity is a cyclic group of order n.

2. Find all possible sets of generators of the subgroups of orders 3,4, and 12 of $\mathbf{Z}_{12}$.
3. Prove that there are exactly two elements of an infinite cyclic group G that can generate it.
4. Prove that there are exactly $\phi(n)$ elements of a cyclic group of order n that can generate it. [*Hint*: If $\bar{a}$ is a generator of $(\mathbf{Z}/(n), +)$, then $\bar{1} = m\bar{a}$ for some $m \in \mathbf{Z}$, so $ma \equiv 1 \pmod{n}$, which yields $(a, n) = 1$.]
5. Show that a group G has no proper subgroups if and only if it is a cyclic group of prime order.
6. Show that every finitely generated subgroup of $(\mathbf{Q}, +)$ is cyclic. Show also $(\mathbf{Q}, +) \not\simeq (\mathbf{Q}^+, \cdot)$, where $\mathbf{Q}^+$ represents positive rational numbers.
7. Find a homomorphism from S_3 onto a nontrivial cyclic group.
8. Show that $\mathbf{Z} \times \mathbf{Z}$ is not cyclic.
9. Is $\mathbf{Z}_2 \times \mathbf{Z}_2$ cyclic?
10. Give an example of a nonabelian group such that each of its proper subgroups is cyclic.

5 Permutation groups

Definition. *Let X be a nonempty set. The group of all permutations of X under composition of mappings is called the* symmetric group *on X and is denoted by S_X. A subgroup of S_X is called a* permutation group *on X.*

It is easily seen that a bijection $X \simeq Y$ induces in a natural way an isomorphism $S_X \simeq S_Y$. If $|X| = n$, S_X is denoted by S_n and called the *symmetric group of degree n*.

A permutation $\sigma \in S_n$ can be exhibited in the form

$$\begin{pmatrix} 1 & 2 & \cdots & n \\ \sigma(1) & \sigma(2) & \cdots & \sigma(n) \end{pmatrix},$$

consisting of two rows of integers; the top row has integers $1, \ldots, n$, usually (but not necessarily) in their natural order, and the bottom row has $\sigma(i)$ below i for each $i = 1, \ldots, n$. This is called a two-row notation for a permutation. There is a simpler, one-row notation for a special kind of permutation called a *cycle*.

Definition. *Let $\sigma \in S_n$. If there exists a list of distinct integers $x_1, \ldots, x_r \in$* $\mathbf{n}$, *such that*

$$\sigma(x_i) = x_{i+1}, \qquad i = 1,\dots,r-1,$$
$$\sigma(x_r) = x_1,$$
$$\sigma(x) = x \qquad \text{if } x \notin \{x_1,\dots,x_r\},$$

then σ is called a cycle of length r *and denoted by* $(x_1\dots x_r)$. *A cycle of length 2 is called a* transposition.

In other words, a cycle $(x_1\dots x_r)$ moves the integers $x_1\dots,x_r$ one step around a circle as shown in the diagram (for $r = 5$) and leaves every other integer in **n** unmoved. (If $\sigma(x) = x$, we say σ does not *move* x.)

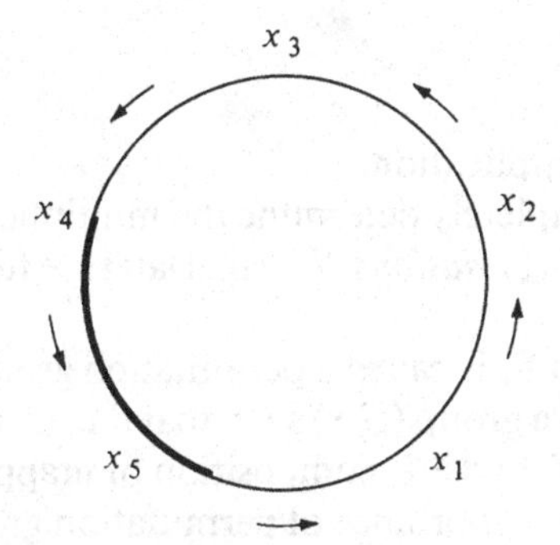

Trivially, any cycle of length 1 is the identity mapping e. Note that the one-row notation for a cycle does not indicate the degree n, which has to be understood from the context.

Two permutations $\sigma,\tau \in S_n$ are said to be *disjoint* if they do not both move the same integer; that is, for all $x \in \mathbf{n}$, $\sigma(x) = x$ or $\tau(x) = x$.

For the sake of illustration, let us consider the symmetric group S_3. There are six permutations of the set $\{1,2,3\}$ given by

$$e = \begin{pmatrix} 1 & 2 & 3 \\ 1 & 2 & 3 \end{pmatrix}, \quad \sigma_1 = \begin{pmatrix} 1 & 2 & 3 \\ 2 & 3 & 1 \end{pmatrix}, \quad \sigma_2 = \begin{pmatrix} 1 & 2 & 3 \\ 3 & 1 & 2 \end{pmatrix},$$
$$\tau_1 = \begin{pmatrix} 1 & 2 & 3 \\ 1 & 3 & 2 \end{pmatrix}, \quad \tau_2 = \begin{pmatrix} 1 & 2 & 3 \\ 3 & 2 & 1 \end{pmatrix}, \quad \tau_3 = \begin{pmatrix} 1 & 2 & 3 \\ 2 & 1 & 3 \end{pmatrix}.$$

On examination we find that they are all cycles; that is,

$$e = (1), \qquad \sigma_1 = (1 \ \ 2 \ \ 3), \quad \sigma_2 = (1 \ \ 3 \ \ 2),$$
$$\tau_1 = (2 \ \ 3), \quad \tau_2 = (3 \ \ 1), \qquad \tau_3 = (1 \ \ 2).$$

We can find the product of any two permutations by the usual rule for composition of mappings. Thus,

$$\tau_1\sigma_1 = \begin{pmatrix} 1 & 2 & 3 \\ 3 & 2 & 1 \end{pmatrix} = \tau_2.$$

(Recall that in the product $\tau\sigma$, σ acts first, followed by τ.) We can, however, give a simpler description of the multiplication structure in S_3. (This is not true for symmetric groups of degree >3.)

Let us write σ for $\sigma_1 = (1\ 2\ 3)$ and τ for $\tau_1 = (2\ 3)$. We easily verify that $\sigma^2 = \sigma_2$, $\sigma^3 = e$, $\tau^2 = e$, $\sigma\tau = \tau_3$, $\sigma^2\tau = \tau_2 = \tau\sigma$. Hence,

$$S_3 = \{e,\sigma,\sigma^2,\tau,\sigma\tau,\sigma^2\tau\},$$

and the product of any two elements in S_3 can be computed by using the relations

$$\sigma^3 = e = \tau^2, \qquad \tau\sigma = \sigma^2\tau \tag{*}$$

and the associative property of multiplication.

Relations of the type (*) that completely determine the multiplication table of the group are called its *defining relations*. We shall later see further examples of these relations.

A subgroup of a symmetric group S_X is called a permutation group. In other words, a permutation group is a group $(G,\cdot)$ such that the elements of G are permutations of some set X, and $\cdot$ is composition of mappings. The following theorem indicates the importance of permutation groups.

5.1 Theorem (Cayley). *Every group is isomorphic to a permutation group.*

Proof. Let G be a group. For any given $a \in G$, the mapping

$$f_a\colon G \to G, \qquad \text{given by} \qquad f_a(x) = ax \quad \text{for all } x \in G,$$

is a bijection, because $ax = ax' \Rightarrow x = x'$ and $y = f_a(a^{-1}y)$ for all $x,x',y \in G$. Consider the mapping

$$\phi\colon G \to S_G \qquad \text{given by} \qquad \phi(a) = f_a \quad \text{for all } a \in G,$$

where S_G is the symmetric group on the set G. For all $a,b,x \in G$,

$$f_{ab}(x) = abx = f_a(bx) = f_a(f_b(x)) = (f_af_b)(x).$$

Hence,

$$\phi(ab) = \phi(a)\phi(b).$$

Therefore, ϕ is a homomorphism, and $\operatorname{Im}\phi$ is a subgroup of S_G. Moreover,

$$\phi(a)=\phi(b)\Rightarrow ax=bx \quad \text{for all} \quad x\in G\Rightarrow a=b.$$

Hence, ϕ is an injective homomorphism. Therefore, G is isomorphic to a subgroup of S_G. □

The above isomorphism is called the *left regular representation* of G. Similarly, we have a right regular representation.

Groups of Symmetries

We now describe an important class of permutation groups known as *groups of symmetries*. Let X be a set of points in space, so that the distance $d(x,y)$ between points x,y is given for all $x,y\in X$. A permutation σ of X is called a *symmetry* of X if

$$d(\sigma(x),\sigma(y))=d(x,y) \qquad \text{for all } x,y\in X.$$

In other words, a symmetry is a permutation that preserves distance between every two points.

Let T_X denote the set of all symmetries of X. Then for all $\sigma,\tau\in T_X$ and $x,y\in X$,

$$\begin{aligned} d(\tau\sigma^{-1}(x),\tau\sigma^{-1}(y)) &= d(\sigma^{-1}(x),\sigma^{-1}(y))=d(\sigma\sigma^{-1}(x),\sigma\sigma^{-1}(y)) \\ &= d(x,y). \end{aligned}$$

Hence, $\tau\sigma^{-1}\in T_X$. This proves that T_X is a subgroup of S_X and, therefore, is itself a group under composition of mappings. It is called the *group of symmetries* of X.

Consider, in particular, the case when X is the set of points constituting a polygon of n sides in a plane. It is clear from geometrical considerations that any symmetry of X is determined uniquely by its effect on the vertices of the polygon. Therefore we need consider only the symmetries of the set of vertices, which can be labeled as $1,2,\ldots,n$. Thus, the group of symmetries of a polygon of n sides is a subgroup of S_n.

Definition. *The group of symmetries of a regular polygon P_n of n sides is called the* dihedral group *of degree n and denoted by D_n.*

The particular case $n=3$ was considered in Example 1.3(d).

Let us now consider the general case of a regular polygon P_n (see the figure). It is clear that a permutation $\sigma\in S_n$ is a symmetry of P_n if and only if σ takes any two adjacent vertices of P_n to adjacent vertices; that is, if and

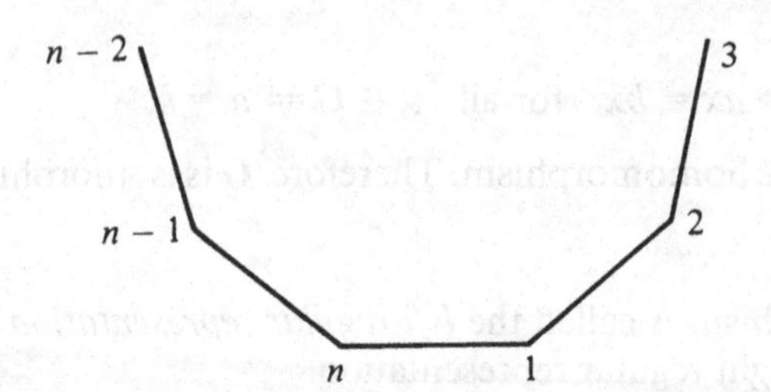

only if $\sigma(1),\sigma(2),\ldots,\sigma(n)$ are either in the cyclic order $1,2,\ldots,n$ or in the reverse cyclic order $n,n-1,\ldots,2,1$. Thus, the symmetries of P_n can be classified into two kinds: those that preserve the cyclic order $1,2,\ldots,n$, and those that reverse the order. Let σ be a symmetry that preserves the cyclic order. Now $\sigma(1)$ can have any one of the values $1,\ldots,n$, and once $\sigma(1)$ is fixed, $\sigma(2),\ldots,\sigma(n)$ are uniquely determined by the cyclic order. Therefore, there are exactly n symmetries that preserve the cyclic order of the vertices. Let us call them $\sigma_1,\sigma_2,\ldots,\sigma_n$, where σ_i is the symmetry taking 1 to i. By the same argument there are exactly n symmetries $\tau_1,\tau_2,\ldots,\tau_n$ that reverse the cyclic order, where τ_i takes 1 to i. This proves that the dihedral group D_n has $2n$ elements σ_i,τ_i $(i=1,\ldots,n)$.

We now give a simpler description of the elements of D_n that enables us to write the product of any two elements easily. Clearly, σ_1 is the identity permutation e, and σ_2 is the cycle $(1\quad 2\ldots n)$. Writing σ for σ_2, we easily verify that σ^i takes 1 to $i+1$ $(i=1,2,\ldots,n-1)$, and σ^n takes 1 to 1. Since σ^i $(i>1)$ preserves the cyclic order, it follows that

$$\sigma^i=\sigma_{i+1},\qquad i=1,\ldots,n-1,$$
$$\sigma^n=e.$$

Hence, the n symmetries that preserve the cyclic order of vertices are σ^i $(i=0,1,\ldots,n-1)$, where σ is the cycle $(1\quad 2\ldots n)$.

Writing τ for

$$\tau_1=\begin{pmatrix}1 & 2 & \cdots & n\\ 1 & n & \cdots & 2\end{pmatrix},$$

we clearly see that $\sigma^i\tau$ reverses the cyclic order and takes 1 to $i+1$. Hence, $\sigma^i\tau=\tau_{i+1}$ $(i=0,1,\ldots,n-1)$. Moreover, τ^2 takes 1 to 1 and preserves the order. Hence $\tau^2=e$.

Now consider the product $\tau\sigma$. Since $\tau\sigma$ reverses the cyclic order, and $\tau\sigma(1)=\tau(2)=n$, we conclude that $\tau\sigma=\sigma^{n-1}\tau$. We sum up the above discussion in

5.2 Theorem. *The dihedral group D_n is a group of order $2n$ generated*

by two elements σ,τ satisfying $\sigma^n = e = \tau^2$ and $\tau\sigma = \sigma^{n-1}\tau$, where

$$\sigma = (1\ 2\ \dots\ n), \qquad \tau = \begin{pmatrix} 1 & 2 & \cdots & n \\ 1 & n & \cdots & 2 \end{pmatrix}.$$

Geometrically, σ is a rotation of the regular polygon P_n through an angle $2\pi/n$ in its own plane, and τ is a reflection (or a turning over) in the diameter through the vertex 1.

Definition. *The dihedral group D_4 is called the* octic group.

As an example of the group of symmetries of a nonregular polygon, we shall consider the symmetries of a rectangle in Example 5.3.

5.3 Example

The symmetries of a rectangle

4 3

1 2

are easily seen to be

$$e = \begin{pmatrix} 1 & 2 & 3 & 4 \\ 1 & 2 & 3 & 4 \end{pmatrix}, \quad a = \begin{pmatrix} 1 & 2 & 3 & 4 \\ 3 & 4 & 1 & 2 \end{pmatrix},$$
$$b = \begin{pmatrix} 1 & 2 & 3 & 4 \\ 2 & 1 & 4 & 3 \end{pmatrix}, \quad c = \begin{pmatrix} 1 & 2 & 3 & 4 \\ 4 & 3 & 2 & 1 \end{pmatrix}.$$

Geometrically, a is a rotation through an angle π, and b and c are reflections in the axes of symmetry parallel to the sides. It is easily verified that $a^2 = b^2 = c^2 = e$, $ab = c$, $bc = a$, $ca = b$. Thus, the group of symmetries of a rectangle is the Klein four-group.

Problems

1. Find all subgroups of S_3 and S_4.
2. If $n \geq 3$, prove that $Z(S_n) = \{e\}$. [*Hint:* Let $e \neq \sigma \in S_n$. Suppose $\sigma(i) = j$, $i \neq j$. Let $k \neq i, j$. Choose $\tau \in S_n$ such that $\tau(i) = i$, $\tau(j) = k$, $\tau(k) = j$. Then $(\sigma\tau)(i) \neq (\tau\sigma)(i)$, so $\sigma\tau \neq \tau\sigma$.]
3. Show that $|Z(D_n)| = 1$ or 2 according to whether n is odd or even.
4. Let a and b be elements of order 2 in a group G. Suppose $o(ab) = 4$. Show that the subgroup generated by a and b is D_4.

6 Generators and relations

Sometimes it is convenient to define a group by a subset X of the group and a set of equations satisfied by elements of X. For example, the group $(\mathbf{Z}/(n),+)$ is generated by $\bar{1}$ such that $n\bar{1} = \bar{0}$. The group S_3 is generated by $\{a,b\}$ such that $a^3 = 1 = b^2$, $b^{-1}ab = a^2$. The dihedral group D_n is generated by $\{a,b\}$ such that $a^n = 1 = b^2$, $b^{-1}ab = a^{-1}$. A finite group, in general, may also be described by listing its elements and by writing down the multiplication table; but clearly this is an impossible act for groups of large orders (without computers!). Hence, we are naturally interested in a minimal subset X of the group that generates it, and also in the set of equations satisfied by elements in the generating set that would suffice to give us all products of elements in the group.

Definition. *Let G be a group generated by a subset X of G. A set of equations $(r_j = 1)_{j \in \Lambda}$ that suffice to construct the multiplication table of G is called a set of* defining relations *for the group (r_j are products of elements of X).*

The set X is called a set of generators. *The system $(X;(r_j = 1)_{j \in \Lambda})$ is called a* presentation *of the group.*

CHAPTER 5

Normal subgroups

1 Normal subgroups and quotient groups

We recall that multiplication in a group G induces a *product* of any two subsets A and B of G, given by $AB = \{xy | x \in A, y \in B\}$. If A or B is a singleton, we write aB for $\{a\}B$ and Ab for $A\{b\}$. Since multiplication in G is associative, the induced multiplication of subsets of G is also associative.

Definition. *Let G be a group. A subgroup N of G is called a* normal *subgroup of G, written $N \lhd G$, if $xNx^{-1} \subset N$ for every $x \in G$.*

Trivially, the subgroups $\{e\}$ and G itself are normal subgroups of G. If G is abelian, then every subgroup of G is a normal subgroup. But the converse is not true: A group in which every subgroup is normal is not necessarily abelian [Example 1.5(e)]. It is easily proved that the center of a group G is a normal subgroup of G, and if $\phi: G \to H$ is a homomorphism of groups, then Ker $\phi \lhd G$.

The following theorem gives several characterizations of a normal subgroup.

1.1 Theorem. *Let N be a subgroup of a group G. Then the following are equivalent.*

(i) $N \lhd G$.
(ii) $xNx^{-1} = N$ *for every* $x \in G$.
(iii) $xN = Nx$ *for every* $x \in G$.
(iv) $(xN)(yN) = xyN$ *for all* $x, y \in G$.

Proof. (i) $\Rightarrow$ (ii) Suppose $N \lhd G$. Let $x \in G$. Then, by definition, $xNx^{-1} \subset N$. Also, $x^{-1} \in G$. Hence, $x^{-1}Nx \subset N$. Therefore, $N = x(x^{-1}Nx)x^{-1} \subset xNx^{-1}$. Hence, $xNx^{-1} = N$.

(ii) $\Rightarrow$ (iii) $Nx = (xNx^{-1})x = xNx^{-1}x = xNe = xN$.

(iii) $\Rightarrow$ (iv) $(xN)(yN) = x(Ny)N = x(yN)N = xy(NN)$. Now $NN \subset N$ because N is closed under multiplication. On the other hand, $N = eN \subset NN$. Hence, $NN = N$. Therefore, $(xN)(yN) = (xy)N$.

(iv) $\Rightarrow$ (i) $xNx^{-1} = xNx^{-1}e \subset xNx^{-1}N = xx^{-1}N = eN = N$. Hence, $N \lhd G$. □

Let N be a normal subgroup of G. The above theorem has shown that any left coset of N in G is a right coset and vice versa. Hence, we need not distinguish between left cosets and right cosets of N. We shall write all cosets of N as left cosets and denote the set of all cosets of N in G by G/N.

1.2 Theorem. *Let N be a normal subgroup of the group G. Then G/N is a group under multiplication. The mapping ϕ: $G \to G/N$, given by $x \mapsto xN$, is a surjective homomorphism, and* Ker $\phi = N$.

Proof. By Theorem 1.1, $(xN)(yN) = (xy)N$ for all $x,y \in G$. Hence, G/N is closed under multiplication. Because multiplication is associative in G, multiplication is also associative in G/N. The coset $eN = N$ is the identity for multiplication in G/N, and for any $x \in G$, $(xN)(x^{-1}N) = (xx^{-1})N = eN$. This proves that G/N is a group.

The mapping ϕ is obviously surjective. For all $x,y \in G$,

$$\phi(xy) = (xy)N = (xN)(yN) = \phi(x)\phi(y).$$

Hence, ϕ is a homomorphism. Further, $xN = eN$ if and only if $x \in N$. Hence, $\text{Ker } \phi = \{x \in G | \phi(x) = eN\} = N$. □

Definition. *Let N be a normal subgroup of G. The group G/N is called the* quotient group *of G by N. The homomorphism $G \to G/N$, given by $x \mapsto xN$, is called the* natural *(or* canonical*) homomorphism of G onto G/N.*

Definition. *Let G be a group, and let S be a nonempty subset of G. The* normalizer *of S in G is the set*

$$N(S) = \{x \in G | xSx^{-1} = S\}.$$

The normalizer of a singleton $\{a\}$ is written $N(a)$.

1.3 Theorem. *Let G be a group. For any nonempty subset S of G, $N(S)$ is a subgroup of G. Further, for any subgroup H of G,*

(i) *$N(H)$ is the largest subgroup of G in which H is normal;*
(ii) *if K is a subgroup of $N(H)$, then H is a normal subgroup of KH.*

Proof. Clearly, $e \in N(S)$. If $x, y \in N(S)$, then

$$(x^{-1}y)S(x^{-1}y)^{-1} = x^{-1}(ySy^{-1})x = x^{-1}Sx = x^{-1}(xSx^{-1})x = S.$$

Hence $x^{-1}y \in N(S)$. Therefore, $N(S)$ is a subgroup of G. Let H be a subgroup of G. Then $hHh^{-1} = H$ for all $h \in H$. Therefore, H is a subset and, hence, a subgroup of $N(H)$. Further, by definition, $xHx^{-1} = H$ for all $x \in N(H)$. Hence, $H \lhd N(H)$. Let K be any subgroup of G such that $H \lhd K$. Then $kHk^{-1} = H$ for all $k \in K$. Hence, $K \subset N(H)$. This proves that $N(H)$ is the largest subgroup of G containing H as a normal subgroup.

Let K be a subgroup of $N(H)$. Then for all $k \in K$, $kHk^{-1} = H$. Hence, $kH = Hk$. Therefore $KH = HK$. Hence, by Theorem 3.5 of Chapter 4, KH is a subgroup of $N(H)$, and $H \subset KH$. Consequently, $H \lhd HK$. □

Definition. *Let G be a group. For any $a, b \in G$, $aba^{-1}b^{-1}$ is called a* commutator *in G. The subgroup of G generated by the set of all commutators in G is called the* commutator subgroup *of G (or the* derived group *of G) and denoted by G'.*

1.4 Theorem. *Let G be a group, and let G' be the derived group of G. Then*

(i) *$G' \lhd G$.*
(ii) *G/G' is abelian.*
(iii) *If $H \lhd G$, then G/H is abelian if and only if $G' \subset H$.*

Proof. (i) Let $x = aba^{-1}b^{-1}$ be any commutator in G. Then $x^{-1} = bab^{-1}a^{-1}$ is also a commutator. Moreover, for any g in G,

$$\begin{aligned} gxg^{-1} &= (gag^{-1})(gbg^{-1})(ga^{-1}g^{-1})(gb^{-1}g^{-1}) \\ &= (gag^{-1})(gbg^{-1})(gag^{-1})^{-1}(gbg^{-1})^{-1} \in G'. \end{aligned}$$

Now any element y in G' is a product of a finite number of commutators, say

$$y = x_1x_2...x_n,$$

where $x_1,...,x_n$ are commutators. Then for any $g \in G$,

$$gyg^{-1} = (gx_1g^{-1})(gx_2g^{-1}) \cdots (gx_ng^{-1}) \in G'.$$

Hence, G' is a normal subgroup of G.

(ii) For all $a, b \in G$,

$$(aG')(bG')(aG')^{-1}(bG')^{-1} = (aba^{-1}b^{-1})G' = G'.$$

Hence, $(aG')(bG') = (bG')(aG')$. Therefore, G/G' is abelian.

(iii) Suppose G/H is abelian. Then for all $a,b \in G$,

$$\begin{aligned}(aba^{-1}b^{-1})H &= (aH)(bH)(aH)^{-1}(bH)^{-1}\\ &= (aH)(aH)^{-1}(bH)(bH)^{-1} = H.\end{aligned}$$

Hence, $aba^{-1}b^{-1} \in H$. This proves that $G' \subset H$. The converse is proved similarly. □

1.5 Examples

(a) If $A < G$ and $B \triangleleft G$, then $A \cap B \triangleleft A$ and $AB < G$.

Solution. Obviously $A \cap B < A$. To prove its normality, let $a \in A$, $x \in A \cap B$. Then $axa^{-1} \in B$, because $B \triangleleft G$. Trivially, $axa^{-1} \in A$. Thus,

$$\forall a \in A, \quad \forall x \in A \cap B, \qquad axa^{-1} \in A \cap B \Rightarrow A \cap B \triangleleft A.$$

To prove $AB < G$, let a, $a_1 \in A$, b, $b_1 \in B$. Then $ab(a_1b_1)^{-1} = abb_1^{-1}a_1^{-1} \in AB$, because $(bb_1^{-1})a_1^{-1} = a_1^{-1}b_2$ for some $b_2 \in B$. Thus, $AB < G$.

(b) If $H_i \triangleleft G$, $i = 1,2,\dots,k$, then $H_1H_2\dots H_k < G$.

Solution. Follows by induction on k.

(c) If G is a group and H is a subgroup of index 2 in G, then H is a normal subgroup of G.

Solution. If $a \notin H$, then, by hypothesis, $G = H \cup aH$ and $aH \cap H = \varnothing$. Also, $G = H \cup Ha$, $Ha \cap H = \varnothing$. Thus, $aH = Ha$, $a \notin H$. But clearly $aH = Ha$, $\forall a \in H$. Hence, $\forall g \in G$, $gH = Hg$, proving that H is normal in G.

(d) If N and M are normal subgroups of G such that $N \cap M = \{e\}$, then $nm = mn$, $\forall n \in N$, $\forall m \in M$.

Solution. If $n \in N$, $m \in M$, then

$$n^{-1}m^{-1}\, nm = (n^{-1}m^{-1}n)m \in MM = M,$$

and also

$$n^{-1}m^{-1}nm = n^{-1}(m^{-1}nm) \in NN = N.$$

Thus,

$$n^{-1}m^{-1}nm \in N \cap M = \{e\}.$$

Hence, $nm = mn$.

(e) Give an example of a nonabelian group each of whose subgroups is normal.

Solution. Let G be the quaternion group (see Problem 6, Section 1, Chapter 4). The only subgroup of order 2, that is,

$$\left\{\begin{pmatrix}1 & 0\\0 & 1\end{pmatrix}, \begin{pmatrix}-1 & 0\\0 & -1\end{pmatrix}\right\},$$

is trivially normal. All the subgroups of order 4 have index 2 and, hence, are normal by Example (c). Hence, G is a desired group.

(f) Give an example of a group G having subgroups K and T such that $K \lhd T \lhd G$, but K is not a normal subgroup of G.

Solution. Let G be the octic group D_4 (see 5.2 in Chapter 4). Choose $T = \{e, \sigma^2, \tau, \sigma^2\tau\}$ and $K = \{e, \tau\}$. Clearly, from the defining relations of the octic group, T and K are subgroups. Since the index of T is 2, $T \lhd G$ by Example (c). Also K as a subgroup of T has index 2, and thus $K \lhd T$. However, K is not normal in G, because if we choose $\sigma \in G$ and $\tau \in K$, then $\sigma^{-1}\tau\sigma \notin K$.

(g) Let G be a finite group with a normal subgroup N such that $(|N|, |G/N|) = 1$. Then N is the unique subgroup of G of order $|N|$.

Solution. Let $K < G$ such that $|K| = |N|$. Then $KN/N < G/N$. Now

$$\left|\frac{KN}{N}\right| = \frac{|KN|}{|N|} = \frac{|K|}{|K \cap N|}.$$

So by Lagrange's theorem $|K|/|K \cap N|$ divides $|G/N|$. But since $|K| = |N|$ and $(|N|, |G/N|) = 1$, this implies $|K|/|K \cap N| = 1$. Hence, $K = K \cap N$, so $K = N$.

(h) If G is a group with center $Z(G)$, and if $G/Z(G)$ is cyclic, then G must be abelian.

Solution. Let $G/Z(G)$ be generated by $xZ(G)$, $x \in G$. Let $a,b \in G$. Then $aZ(G)$, as an element in $G/Z(G)$, must be of the form $x^mZ(G)$ for some integer m; that is,

$$aZ(G) = x^mZ(G).$$

Thus, $a = x^m y$ for some $y \in Z(G)$. Similarly, $b = x^n z$ for some $z \in Z(G)$ and some integer n, so that

$$ab = (x^m y)(x^n z) = x^m y x^n z = x^m x^n yz$$
$$= x^{m+n} yz, \qquad \text{since } y \in Z(G).$$

And

$$ba = (x^n z)(x^m y) = x^n z x^m y = x^{n+m} zy, \qquad \text{since } z \in Z(G).$$

Hence $ab = ba$, $\forall a,b \in G$.

Problems

1. Prove that the center $Z(G)$ of a group G is a normal subgroup.
2. If N is a normal subgroup of a group G, and H is any subgroup of G, prove that NH is a subgroup of G.
3. If N and M are normal subgroups of G, prove that NM is also a normal subgroup of G.
4. If there exists a unique subgroup H of order 10 (or 20) in a group G, show that $H \lhd G$. Generalize this result to subgroups of other orders. (*Hint*: See Problem 3, Section 3.)
5. Show that the cyclic subgroup of S_3 generated by $\begin{pmatrix} 1 & 2 & 3 \\ 2 & 1 & 3 \end{pmatrix}$ is not normal in S_3.
6. If N is a normal subgroup of a group G such that $N \cap G' = \{e\}$, prove that $N \subset Z(G)$.
7. Let H be a subgroup of a group G such that $x^2 \in H$ for every $x \in G$. Prove that $H \lhd G$ and G/H is abelian.
8. Find a subgroup N of order 12 in the symmetric group S_4 on four elements. Find the elements of S_4/N and show that they can be made into a group.
9. Prove that in S_3 the set $N = \{e, (1\ 2\ 3), (1\ 3\ 2)\}$ is a normal subgroup. Find the elements of S_3/N.
10. Let H be a subgroup of G. Prove that $H \lhd G$ if and only if the product of two left cosets of H in G is again a left coset.
11. Show that the set SL $(n, \mathbf{R})$ of invertible $n \times n$ matrices over $\mathbf{R}$ having determinant 1 forms a normal subgroup of the linear group GL $(n, \mathbf{R})$.
12. Let G/N be the quotient group of G. Suppose $o(gN)$ is finite. Show that $o(gN)$ divides $o(g)$. Also show that $g^m \in N$ if and only if $o(gN)$ divides m.
13. Show that the order of each element of the quotient group $\mathbf{Q}/\mathbf{Z}$ is finite.

14. Let G be a group such that for some integer $m > 1$,

$$(ab)^m = a^m \cdot b^m, \quad \text{for all} \quad a, b \in G,$$

and let

$$G^m = \{a^m \mid a \in G\}.$$

Show that

$$G^m \lhd G$$

and the order of each element of G/G^m is finite.

15. Show that there does not exist any group G such that $|G/Z(G)| = 37$.
16. For $G = S_3$, compute the first and second derived groups, $G^{(1)} = G'$ and $G^{(2)} = (G')'$.
17. If A, B are subgroups of a group G such that

$$B \lhd G, \quad A \cap B = \{e\}, \quad G = AB,$$

then G is called a *semidirect product* (or *split extension*) of A and B. Show that the dihedral group D_n can be expressed as a split extension of two suitable proper subgroups.

18. Show that the quaternion group (see Problem 4, Section 1, Chapter 4) cannot be expressed as a split extension of two nontrivial subgroups.
19. Let $L(G)$ be the set of normal subgroups of a group G. Show that $L(G)$ is a modular lattice. Show also that the lattice of all subgroups need not be modular. [A lattice L is called modular if for all $a, b, c \in L$, with $a \leq c$, $a \vee (b \wedge c) = (a \vee b) \wedge c$.]

2 Isomorphism theorems

In this section we prove some important theorems on group homomorphisms known as *isomorphism theorems.* We saw in Theorem 1.2 that every quotient group G/N of a group G is a homomorphic image of G. The first isomorphism theorem (which is also known as the *fundamental theorem of homomorphisms*) proves the converse; that is, every homomorphic image of G is isomorphic to a quotient group of G.

2.1 Theorem (first isomorphism theorem). *Let ϕ: $G \to G'$ be a homomorphism of groups. Then*

$$G/\text{Ker } \phi \simeq \text{Im } \phi.$$

Hence, in particular, if ϕ is surjective, then

$$G/\mathrm{Ker}\,\phi \simeq G'.$$

Proof. Consider the mapping

$$\psi: G/K \to \mathrm{Im}\,\phi \qquad \text{given by} \qquad xK \mapsto \phi(x),$$

where $K = \mathrm{Ker}\,\phi$. For any $x,y \in G$,

$$xK = yK \Leftrightarrow y^{-1}x \in K \Leftrightarrow \phi(y^{-1}x) = e' \Leftrightarrow \phi(x) = \phi(y).$$

Hence, ψ is well defined and injective. Further,

$$\psi((xK)(yK)) = \psi(xyK) = \phi(xy) = \phi(x)\phi(y).$$

Hence, ψ is a homomorphism. Since ψ is obviously surjective, we conclude that ψ is an isomorphism of groups. □

2.2 Corollary. *Any homomorphism $\phi: G \to G'$ of groups can be factored as*

$$\phi = j \cdot \psi \cdot \eta,$$

where $\eta: G \to G/\mathrm{Ker}\,\phi$ is the natural homomorphism, $\psi: G/\mathrm{Ker}\,\phi \to \mathrm{Im}\,\phi$ is the isomorphism obtained in the theorem, and $j: \mathrm{Im}\,\phi \to G'$ is the inclusion map.

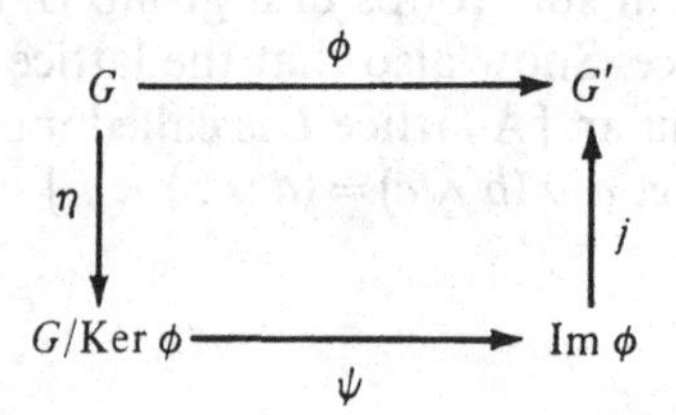

Proof. Clear. □

2.3 Theorem (second isomorphism theorem). *Let H and N be subgroups of G, and $N \lhd G$. Then*

$$H/H \cap N \simeq HN/N.$$

The inclusion diagram shown below is helpful in visualizing the theorem. Because of this, the theorem is also known as the "diamond isomorphism theorem."

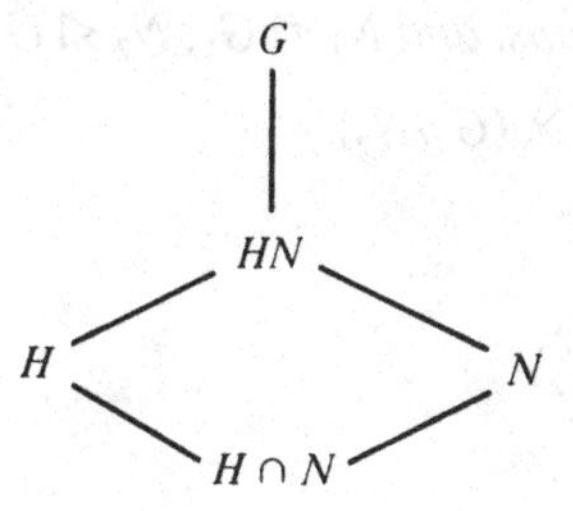

Proof. Since $N \lhd G$, $HN = NH$ is a subgroup of G and $N \lhd HN$. Consider the mapping

$$\phi: H \to HN/N \qquad \text{given by} \qquad h \mapsto hN \quad \text{for all } h \in H.$$

In fact, ϕ is the restriction of the natural homomorphism $p: G \to G/N$ to H. Hence, $\operatorname{Ker} \phi = H \cap N$. Moreover, ϕ is clearly surjective. Hence, by the first isomorphism theorem, $H/H \cap N \simeq HN/N$. □

2.4 Theorem (third isomorphism theorem). *Let H and K be normal subgroups of G and $K \subset H$. Then*

$$(G/K)/(H/K) \simeq G/H.$$

This theorem is also known as the "double quotient isomorphism theorem."

Proof. Consider the mapping

$$\phi: G/K \to G/H \qquad \text{given by} \qquad xK \mapsto xH.$$

The mapping is well defined, for

$$xK = yK \Rightarrow x^{-1}y \in K \Rightarrow x^{-1}y \in H \Rightarrow xH = yH.$$

Further, for all $x,y \in G$,

$$\phi((xK)(yK)) = \phi(xyK) = xyH = (xH)(yH).$$

Hence, ϕ is a homomorphism. Now ϕ is obviously surjective, and

$$\operatorname{Ker} \phi = \{xK \mid xH = H\} = \{xK \mid x \in H\} = H/K.$$

Hence, by the first isomorphism theorem, $(G/K)/(H/K) \simeq G/H$. □

The next theorem shows that the direct product of two quotient groups is isomorphic to a quotient group.

2.5 Theorem. *Let G_1 and G_2 be groups, and $N_1 \lhd G_1$, $N_2 \lhd G_2$. Then* $(G_1 \times G_2)/(N_1 \times N_2) \simeq (G_1/N_1) \times (G_2/N_2)$.

Proof. Consider

$$\phi: G_1 \times G_2 \to (G_1/N_1) \times (G_2/N_2)$$

given by

$$(x_1, x_2) \mapsto (x_1N_1, x_2N_2).$$

It is easily verified that ϕ is a surjective homomorphism and Ker $\phi = N_1 \times N_2$. Hence, the result follows by the first isomorphism theorem. □

We recall that if $\sigma: S \to T$ is a mapping from a set S into a set T, and if $X \subset T$, then $\sigma^{-1}(X)$ denotes the set $\{s \in S | \sigma(s) \in X\}$. $\sigma^{-1}(X)$ is known as the *inverse image* of X under σ. We may emphasize that σ^{-1} is a mapping not from T to S but rather from $\mathcal{P}(T)$ to $\mathcal{P}(S)$. However, if σ is bijection then σ^{-1} defined as above coincides with the usual inverse.

2.6 Theorem (correspondence theorem). *Let $\phi: G \to G'$ be a homomorphism of a group G onto a group G'. Then the following are true:*

(i) $H < G \Rightarrow \phi(H) < G'$.
(i)′ $H' < G' \Rightarrow \phi^{-1}(H') < G$.
(ii) $H \lhd G \Rightarrow \phi(H) \lhd G'$.
(ii)′ $H' \lhd G' \Rightarrow \phi^{-1}(H') \lhd G$.
(iii) $H < G$ *and* $H \supset \text{Ker}\, \phi \Rightarrow H = \phi^{-1}(\phi(H))$.
(iv) *The mapping $H \mapsto \phi(H)$ is a 1-1 correspondence between the family of subgroups of G containing* Ker ϕ *and the family of subgroups of G'; furthermore, normal subgroups of G correspond to normal subgroups of G'.*

Proof. (i) Let $a,b \in H$; so $\phi(a),\phi(b) \in \phi(H)$. Then

$$\phi(a)(\phi(b))^{-1} = \phi(a)\phi(b^{-1}) = \phi(ab^{-1}) \in \phi(H),$$

since $ab^{-1} \in H$. Thus, $\phi(H) < G'$.

(i)′ Let $a,b \in \phi^{-1}(H')$. Then $\phi(a),\phi(b) \in H'$. Now

$$\phi(ab^{-1}) = \phi(a)(\phi(b))^{-1} \in H';$$

so $ab^{-1} \in \phi^{-1}(H')$. Thus, $\phi^{-1}(H') < G$.

(ii) Let $\phi(h) \in \phi(H)$ and $g' \in G'$. Then $g' = \phi(g)$ for some $g \in G$. Now

$$g'^{-1}(\phi(h))g' = (\phi(g))^{-1} \cdot \phi(h) \cdot \phi(g) = \phi(g^{-1}hg) \in \phi(H),$$

since $H \lhd G$. Therefore, $\phi(H) \lhd G'$.

(ii)′ Let $h \in \phi^{-1}(H')$, $g \in G$. Then $\phi(h) \in H'$. Now

$$\phi(g^{-1}hg) = (\phi(g))^{-1} \cdot \phi(h) \cdot \phi(g) \in H',$$

since $H' \lhd G'$. Therefore, $g^{-1}hg \in \phi^{-1}(H')$. Hence, $\phi^{-1}(H') \lhd G$.

(iii) Trivially, $H \subset \phi^{-1}(\phi(H))$. Let $x \in \phi^{-1}(\phi(H))$. Then

$$\begin{aligned} \phi(x) \in \phi(H) &\Rightarrow \phi(x) = \phi(h) \quad \text{(for some } h \in H) \\ &\Rightarrow \phi(xh^{-1}) = \phi(e) \Rightarrow xh^{-1} \in \operatorname{Ker} \phi \\ &\Rightarrow xh^{-1} \in H \quad (\text{as } H \supset \operatorname{Ker} \phi) \\ &\Rightarrow x \in H. \end{aligned}$$

Thus, $H = \phi^{-1}(\phi(H))$.

(iv) Let $H' < G'$. Then by (i)′ $\phi^{-1}(H')$ is a subgroup of G containing $\operatorname{Ker} \phi$, so by (iii), $\phi(\phi^{-1}(H')) = H'$. So the mapping $H \mapsto \phi(H)$ is surjective. To show that the mapping is injective, let $\phi(H_1) = \phi(H_2)$, where H_1, H_2 are subgroups of G containing $\operatorname{Ker} \phi$. Then $\phi^{-1}(\phi(H_1)) = \phi^{-1}(\phi(H_2))$, so, by (iii), $H_1 = H_2$. The last part of (iv) follows from (ii). □

Remark. If $\phi: G \to G'$ is any homomorphism, then Theorem 2.6 remains true if G' is replaced by $\operatorname{Im} \phi$.

2.7 Corollary. *Let N be a normal subgroup of G. Given any subgroup H' of G/N, there is a unique subgroup H of G such that $H' = H/N$. Further, $H \lhd G$ if and only if $H/N \lhd G/N$.*

Proof. Consider the natural homomorphism

$$\phi: G \to G/N \qquad \text{given by} \qquad x \mapsto xN.$$

By Theorem 2.6 there is a unique subgroup H of G containing N such that $H' = \phi(H) = H/N$. □

Definition. *Let G be a group. A normal subgroup N of G is called a* maximal normal subgroup *if*

(i) $N \neq G$.
(ii) $H \lhd G$ *and* $H \supset N \Rightarrow H = N$ *or* $H = G$.

Definition. *A group G is said to be* simple *if G has no proper normal subgroups; that is, G has no normal subgroups except $\{e\}$ and G.*

The following result is an immediate consequence of Corollary 2.7.

2.8 Corollary. *Let N be a proper normal subgroup of G. Then N is a maximal normal subgroup of G if and only if G/N is simple.*

2.9 Corollary. *Let H and K be distinct maximal normal subgroups of G. Then $H \cap K$ is a maximal normal subgroup of H and also of K.*

Proof. By the diamond isomorphism theorem

$$H/H \cap K \simeq HK/K.$$

Now $K \lhd HK \lhd G$. Hence, $HK = K$ or $HK = G$, since K is maximal. But $HK = K \Rightarrow H \subset K$, a contradiction, since H and K are both maximal and distinct. Hence, $HK = G$. Therefore,

$$H/H \cap K \simeq G/K.$$

Hence, by Corollary 2.8, $H \cap K$ is a maximal normal subgroup of H and, similarly, of K. □

2.10 Examples

(a) Let G be a group such that for some fixed integer $n > 1$, $(ab)^n = a^n b^n$ for all $a,b \in G$. Let $G_n = \{a \in G | a^n = e\}$ and $G^n = \{a^n | a \in G\}$. Then

$$G_n \lhd G, \quad G^n \lhd G, \quad \text{and} \quad G/G_n \simeq G^n.$$

Solution. Let $a,b \in G_n$ and $x \in G$. Then $(ab^{-1})^n = a^n(b^n)^{-1} = e$, so $ab^{-1} \in G_n$. Also, $(xax^{-1})^n = xa^nx^{-1} = e$ implies $xax^{-1} \in G_n$. Hence, $G_n \lhd G$. Similarly, $G^n \lhd G$. Define a mapping $f\colon G \to G^n$ by $f(a) = a^n$. Then, for all $a,b \in G$, $f(ab) = (ab)^n = a^n b^n$. Thus, f is a homomorphism. Now Ker $f = \{a | a^n = e\} = G_n$. Therefore, by the first isomorphism theorem $G/G_n \simeq G^n$.

(b) Let G be a finite group, and let T be an automorphism of G with the property $T(x) = x$ if and only if $x = e$. Then every $g \in G$ can be expressed as $g = x^{-1}T(x)$ for some $x \in G$.

Solution. We assert that $x^{-1}T(x) = y^{-1}T(y)$ if and only if $x = y$.

$$\begin{aligned} x^{-1}T(x) = y^{-1}T(y) &\Leftrightarrow (yx^{-1}) = T(y)(T(x))^{-1} \\ &\Leftrightarrow yx^{-1} = T(yx^{-1}) \Leftrightarrow yx^{-1} = e \Leftrightarrow y = x, \end{aligned}$$

by hypothesis. Therefore

$$G = \{x^{-1}T(x) | x \in G\}.$$

(c) If in Example (b) we also have the condition that $T^2 = I$, then G is abelian.

Solution. Let $x \in G$. Then

$$x^{-1}(T(x)) = T^2(x^{-1}(T(x))) = T(T(x^{-1})T^2(x)) = T(T(x^{-1})x) = T((x^{-1}T(x))^{-1}).$$

Therefore, for all $g \in G$, $T(g^{-1}) = g$. Now if $a,b \in G$, then $T((ab)^{-1}) = ab$. On the other hand,

$$T((ab)^{-1}) = T(b^{-1}a^{-1}) = T(b^{-1})T(a^{-1}) = ba.$$

Therefore, $ab = ba$.

(d) A nonabelian group of order 6 is isomorphic to S_3.

Solution. If each element is of order 2, then G is abelian. So there must be an element a of order 3. Let $b \in G$ be such that $b \notin \{e,a,a^2\}$. Then it is easy to check that e,a,a^2,b,ab,a^2b are all distinct elements and, thus, constitute the whole group G. Now $b^2 \neq a$ or a^2. For let $b^2 = a$. Then $b^6 = e$. This implies that the order of $b = 2,3,6$. But then the order of $b = 2$ implies $a = e$, a contradiction. And the order of $b = 3$, along with $b^2 = a$, implies $ab = e$, a contradiction. Also the order of $b = 6$ implies that G is cyclic, a contradiction to the fact that G is not abelian. Hence, $b^2 \neq a$. Similarly, $b^2 \neq a^2$. Also $b^2 = b$, ab, or a^2b would imply $b = e$, a, or a^2, which is not the case. Therefore the only possibility for $b^2 \in G$ is that $b^2 = e$. Further, the subgroup $[a] = \{e,a,a^2\}$ generated by a is of index 2 and, thus, normal. Therefore, $bab^{-1} = e$, a, or a^2. But $bab^{-1} = e$ gives $a = e$, which is not so; and $bab^{-1} = a$ says that G is abelian, which is also not so. Thus, $bab^{-1} = a^2$. Hence, G is generated by a,b with the defining relations

$$a^3 = e = b^2, \qquad bab^{-1} = a^2.$$

On the other hand, S_3, is also generated by a' and b', where

$$a'^3 = e' = b'^2, \qquad b'a'b'^{-1} = a'^2.$$

Then the mapping

$$e \mapsto e', \quad a \mapsto a', \quad a^2 \mapsto a'^2, \quad b \mapsto b', \quad ab \mapsto a'b', \quad a^2b \mapsto a'^2b'$$

is an isomorphism of G onto S_3.

Problems

1. If σ is an isomorphism of a group G into a group H, prove that $(\sigma(a))^n = e'$ if and only if $a^n = e$. Give an example to show this is not true if σ is not an isomorphism.
2. If a group G is generated by a subset S, prove that $\sigma(S)$ also generates G, where σ is an automorphism of G.
3. Show that a cyclic group of order 8 is homomorphic to a cyclic group of order 4.
4. Write down all the homomorphic images of
   (i) the Klein four-group,
   (ii) the octic group.
5. Show that each dihedral group is homomorphic to the group of order 2.
6. Let $[a]$ be a cyclic group of order m and $[b]$ a cyclic group of order n. Show that there is a homomorphism σ of $[a]$ into $[b]$ such that $\sigma(a) = b^k$ if and only if mk is a multiple of n. Further, if $mk = qn$, show that σ is an isomorphism if and only if $(m,q) = 1$.

3 Automorphisms

Recall that an *automorphism* of a group G is an isomorphism of G onto G. The set of all automorphisms of G is denoted by $\mathrm{Aut}(G)$. We have seen (Example 2.1(d), Chapter 4) that every $g \in G$ determines an automorphism I_g of G (called an *inner* automorphism) given by $x \mapsto gxg^{-1}$. The set of all inner automorphisms of G is denoted by $\mathrm{In}(G)$.

3.1 Theorem. *The set* $\mathrm{Aut}(G)$ *of all automorphisms of a group G is a group under composition of mappings, and* $\mathrm{In}(G) \lhd \mathrm{Aut}(G)$. *Moreover,*

$$G/Z(G) \simeq \mathrm{In}(G).$$

Proof. Clearly, $\mathrm{Aut}(G)$ is nonempty. Let $\sigma,\tau \in \mathrm{Aut}(G)$. Then for all $x,y \in G$,

$$\sigma\tau(xy) = \sigma((\tau x)(\tau y)) = (\sigma\tau(x))(\sigma\tau(y)).$$

Hence, $\sigma\tau \in \mathrm{Aut}(G)$. Again,

$$\sigma(\sigma^{-1}(x)\sigma^{-1}(y)) = \sigma\sigma^{-1}(x)\sigma\sigma^{-1}(y) = xy.$$

Hence $\sigma^{-1}(x)\ \sigma^{-1}(y) = \sigma^{-1}(xy)$. Therefore, $\sigma^{-1} \in \mathrm{Aut}(G)$. This proves that $\mathrm{Aut}(G)$ is a subgroup of the symmetric group S_G and, hence, is itself a group.

Consider the mapping

$$\phi: G \to \text{Aut}(G) \quad \text{given by} \quad a \mapsto I_a.$$

For any $a,b \in G$,

$$I_{ab}(x) = abx(ab)^{-1} = a(bxb^{-1})a^{-1} = I_aI_b(x)$$

for all $x \in G$. Hence, ϕ is a homomorphism, and, therefore, $\text{In}(G) = \text{Im } \phi$ is a subgroup of $\text{Aut}(G)$. Further, I_a is the identity automorphism if and only if $axa^{-1} = x$ for all $x \in G$. Hence, $\text{Ker } \phi = Z(G)$, and by the fundamental theorem of homomorphisms

$$G/Z(G) \simeq \text{In}(G).$$

Finally, for any $\sigma \in \text{Aut}(G)$,

$$(\sigma I_a\sigma^{-1})(x) = \sigma(a\sigma^{-1}(x)a^{-1}) = \sigma(a)x\sigma(a)^{-1} = I_{\sigma(a)}(x);$$

hence, $\sigma I_a\sigma^{-1} = I_{\sigma(a)} \in \text{In}(G)$. Therefore, $\text{In}(G) \lhd \text{Aut}(G)$. □

It follows from the theorem that if the center of a group G is trivial, then $G \simeq \text{In}(G)$. A group G is said to be *complete* if $Z(G) = \{e\}$ and every automorphism of G is an inner automorphism; that is, $G \simeq \text{In}(G) = \text{Aut}(G)$.

When considering the possible automorphisms σ of a group G, it is useful to remember that, for any $x \in G$, x and $\sigma(x)$ must be of the same order (see Problem 19, Section 3, Chapter 4).

3.2 Examples

(a) The symmetric group S_3 has a trivial center $\{e\}$. Hence, $\text{In}(S_3) \simeq S_3$. Now consider all automorphisms of S_3.

We have seen that

$$S_3 = \{e,a,a^2,b,ab,a^2b\}$$

with the defining relations

$$a^3 = e = b^2, \qquad ba = a^2b.$$

The elements a and a^2 are of order 3, and b,ab, and a^2b are all of order 2. Hence, for any $\sigma \in \text{Aut}(S_3)$, $\sigma(a) = a$ or a^2, $\sigma(b) = b$, ab, or a^2b. Moreover, when $\sigma(a)$ and $\sigma(b)$ are fixed, $\sigma(x)$ is known for every $x \in S_3$. Hence, σ is completely determined. Thus, there cannot be more than six automorphisms of S_3. Hence, $\text{Aut}(S_3) = \text{In}(S_3) \simeq S_3$. Therefore, S_3 is a complete group.

(b) Let G be a finite abelian group of order n, and let m be a positive integer prime to n. Then the mapping $\sigma: x \mapsto x^m$ is an automorphism of G.

Solution. $(m,n) = 1 \Rightarrow \exists$ integers u and v such that $mu + nv = 1 \Rightarrow \forall x \in G$,

$$x = x^{mu+vn} = x^{mu}x^{vn} = x^{um},$$

since $o(G) = n$.

Now $\forall x \in G$, $x = (x^u)^m$ implies that σ is onto. Further, $x^m = e \Rightarrow x^{mu} = e \Rightarrow x = e$, showing that σ is 1-1. That σ is a homomorphism follows from the fact that G is abelian. Hence, σ is an automorphism of G.

(c) A finite group G having more than two elements and with the condition that $x^2 \neq e$ for some $x \in G$ must have a nontrivial automorphism. (Also see Problem 1.)

Solution. When G is abelian, then $\sigma: x \mapsto x^{-1}$ is an automorphism, and, clearly, σ is not an identity automorphism. When G is not abelian, there exists a nontrivial inner automorphism.

(d) Let $G = [a]$ be a finite cyclic group of order n. Then the mapping σ: $a \mapsto a^m$ is an automorphism of G iff $(m,n) = 1$.

Solution. If $(m,n) = 1$, then it has been shown in Example (b) that σ is an automorphism. So let us assume now that σ is an automorphism. Then the order of $\sigma(a) = a^m$ is the same as that of a, which is n. Further, if $(m,n) = d$, then $(a^m)^{n/d} = (a^n)^{m/d} = e$. Thus, the order of a^m divides n/d; that is, $n|n/d$. Hence, $d = 1$, and the solution is complete.

(e) If G is a finite cyclic group of order n, then the order of Aut(G), the group of automorphisms of G, is $\phi(n)$, where ϕ is Euler's function.

Solution. Let $G = [a]$ and $\sigma \in \text{Aut}(G)$. Since $\sigma(a^i) = (\sigma(a))^i$ for each integer i, σ is completely known if we know $\sigma(a)$. Now let $\sigma(a) = a^m$, $m \leq n$. Then as in Example (d), we have $(m,n) = 1$. Thus, each σ determines a unique integer m less than and prime to n, and conversely, which completes the solution.

Problems

1. Find the group of automorphisms of $(\mathbf{Z}, +)$.
2. Find the group of automorphism of $(\mathbf{Z}_n, +)$.

3. Find $\text{Aut}(K)$ where K is the Klein four-group. What about $\text{Aut}(\mathbf{Z}_2 \times \mathbf{Z}_2)$?
4. Show that the group of automorphism of D_4 is of order 8.
5. Show that $\text{Aut}(\mathbf{Z}_2 \times \mathbf{Z}_4)$ consists of eight elements sending $(1,0)$ to $(1,0)$ or $(1,2)$ and $(0,1)$ to $(0,1)$, $(0,3)$, $(1,1)$, or $(1,3)$.
6. Show that $\text{Aut}(\mathbf{Z}_2 \times \mathbf{Z}_3) \simeq \text{Aut}(\mathbf{Z}_2) \times \text{Aut}(\mathbf{Z}_3)$.

In the following problems 7 and 8, you may invoke Theorem 2.1, Chapter 8.

7. Let A be a noncyclic finite abelian group. Prove that $\text{Aut}(A)$ is not abelian.
8. Let G be a finite group such that $|\text{Aut}(G)| = p$. Prove $|G| \leqslant 3$. (*Hint*: Note G is abelian)

4 Conjugacy and G-sets

Definition. *Let G be a group and X a set. Then G is said to* act on X *if there is a mapping ϕ: $G \times X \to X$, with $\phi(a,x)$ written $a * x$, such that for all $a,b \in G$, $x \in X$,*

(i) $a * (b * x) = (ab) * x$,
(ii) $e * x = x$.

The mapping ϕ is called the action of G *on X, and X is said to be a* G-set.

One of the most important examples of G-sets is the action of the group G on itself by conjugation defined by

$$a * x = axa^{-1}$$

[see Example 4.1(b)]. The reader interested in conjugation and its applications only may go directly to Theorem 4.8 without breaking continuity.

To be more precise, we have just defined an action on the left. Similarly, an action of G on X on the right is defined to be a mapping $X \times G \to X$ with $(x,a) \mapsto x * a$ such that $x * (ab) = (x * a) * b$ and $x * e = x$ for all $a,b \in G$ and $x \in X$.

We shall confine ourselves to groups acting on the left, and for the sake of convenience we sometimes write ax instead of $a * x$.

4.1 Examples of G-sets

(a) Let G be the additive group $\mathbf{R}$, and X be the set of complex numbers z such that $|z| = 1$. Then X is a G-set under the action

$$\gamma * c = e^{i\gamma}c, \quad \text{where} \quad \gamma \in \mathbf{R} \quad \text{and} \quad c \in X.$$

Here the action of γ is the rotation through an angle $\theta = \gamma$ radians, anticlockwise.

(b) Let $G = S_5$ and $X = \{x_1, x_2, x_3, x_4, x_5\}$ be a set of beads forming a circular ring. Then X is a G-set under the action

$$g * x_i = x_{g(i)}, \quad g \in S_5.$$

(c) Let $G = D_4$ and X be the vertices $1, 2, 3, 4$ of a square. X is a G-set under the action

$$g * i = g(i), \quad g \in D_4, \quad i \in \{1,2,3,4\}.$$

(d) Let G be a group. Define

$$a * x = ax, \quad a \in G, x \in G.$$

Then, clearly, the set G is a G-set. This action of the group G on itself is called *translation*.

(e) Let G be a group. Define

$$a * x = axa^{-1}, \qquad a \in G, x \in G.$$

We show that G is a G-set. Let $a,b \in G$. Then

$$(ab) * x = (ab)x(ab)^{-1} = a(bxb^{-1})a^{-1}$$
$$= a(b * x)a^{-1} = a * (b * x).$$

Also, $e * x = x$. This proves G is a G-set.

This action of the group G on itself is called *conjugation*.

(f) Let G be a group acting on a set X. The action of G on X can be extended to the power set $\mathcal{P}(X)$ as follows. Define

$$a * S = \{a * x | x \in S\}$$

for all $a \in G$ and $S \subset X$. Then, clearly, $(ab) * S = a * (b * S)$ and $e * S = S$ for $a,b \in G$ and $S \subset X$. Therefore $\mathcal{P}(X)$ is a G-set.

(g) Let G be a group and $H < G$. Then the set G/H of left cosets can be made into a G-set by defining

$$a * xH = axH, \qquad a \in G, xH \in G/H.$$

(h) Let G be a group and $H \lhd G$. Then the set G/H of left cosets is a G-set if we define

$$a * xH = axa^{-1}H, \qquad a \in G, xH \in G/H.$$

To see this, let $a,b \in G$ and $xH \in G/H$. Then

$$(ab) * xH = abxb^{-1}a^{-1}H = a * bxb^{-1}H = a * (b * xH).$$

Also, $e * xH = xH$. Hence, G/H is a G-set.

If X is a G-set, then, as stated earlier, we generally write ax instead of $a * x$ for the sake of simplicity.

4.2 Theorem. *Let G be a group and let X be a set.*

(i) If X is a G-set, then the action of G on X induces a homomorphism $\phi: G \to S_X$.

(ii) Any homomorphism $\phi: G \to S_X$ *induces an action of G onto X.*

Proof. (i) We define $\phi: G \to S_X$ by $(\phi(a))(x) = ax$, $a \in G$, $x \in X$. Clearly $\phi(a) \in S_X$, $a \in G$. Let $a, b \in G$. Then

$$\begin{aligned}(\phi(ab))(x) &= (ab)x = a(bx) = a((\phi(b))(x)) = (\phi(a))((\phi(b))(x)) \\ &= (\phi(a)\phi(b))(x) \qquad \text{for all } x \in X.\end{aligned}$$

Hence, $\phi(ab) = \phi(a)\phi(b)$.

(ii) Define $a * x = (\phi(a))(x)$; that is, $ax = (\phi(a))(x)$. Then

$$\begin{aligned}(ab)x &= (\phi(ab))(x) = (\phi(a)\phi(b))(x) = \phi(a)(\phi(b)(x)) \\ &= \phi(a)(bx) = a(bx).\end{aligned}$$

Also, $ex = (\phi(e))(x) = x$. Hence, X is a G-set. □

Regarding G as a G-set as in Example 4.1(d), we obtain Cayley's theorem (see Theorem 5.1, Chapter 4).

4.3 Cayley's theorem. *Let G be a group. Then G is isomorphic into the symmetric group* S_G.

Proof. By Theorem 4.2 there exists a homomorphism $\phi: G \to S_G$, where $(\phi(a))(x) = ax$, $a \in G$, $x \in G$. Suppose $\phi(a) =$ the identity element in S_G. Then for all $x \in G$, $(\phi(a))(x) = x$. This implies $ax = x$ for all $x \in X$, and hence $a = e$, the identity in G. Therefore, ϕ is injective. □

Remark. An isomorphism of G into a group of permutations is called a *faithful representation of G by a group of permutations.*

Further, if $H < G$ the set G/H of left cosets of H in G is a G-set [Example 4.1(g)]. The action of G on G/H gives us another representation of G by a group of permutations, which is not necessarily faithful. This is contained in

4.4 Theorem. *Let G be a group and H < G of finite index n. Then there is a homomorphism* $\phi: G \to S_n$ *such that* Ker $\phi = \bigcap_{x \in G} xHx^{-1}$.

Proof. Since $|G/H| = n$, $S_{G/H} \simeq S_n$. By Theorem 4.2 there exists a homomorphism $\phi: G \to S_n$ such that $\phi(a)(xH) = axH$. Now

$$\begin{aligned}
\operatorname{Ker} \phi &= \{a \in G \mid \phi(a) = \text{identity permutation}\} \\
&= \{a \in G \mid axH = xH, \forall x \in G\} \\
&= \{a \in G \mid x^{-1}ax \in H, \forall x \in G\} \\
&= \{a \in G \mid a \in xHx^{-1}, \forall x \in G\} \\
&= \bigcap_{x \in G} xHx^{-1}. \quad \square
\end{aligned}$$

Remark. We obtain the Cayley representation of G by taking $H = \{e\}$.

4.5 Corollary. *Let G be a group with a normal subgroup H of index n. Then G/H is isomorphic into S_n.*

Proof. Follows from Theorem 4.4, since $H \lhd G$ implies $\operatorname{Ker} \phi = H$. □

4.6 Corollary. *Let G be a simple group with a subgroup $\neq G$ of finite index n. Then G is isomorphic into S_n.*

Proof. Follows from Theorem 4.4 by recalling that $\operatorname{Ker} \phi \lhd G$. □

Definition. *Let G be a group acting on a set X, and let $x \in X$. Then the set*

$$G_x = \{g \in G \mid gx = x\},$$

which can be easily shown to be a subgroup, is called the stabilizer (or isotropy) group *of x in G.*

For example, if G acts on itself by conjugation [Example 4.1(e)], then, for $x \in G$,

$$G_x = \{a \in G \mid axa^{-1} = x\} = N(x),$$

the normalizer of x in G. Thus, in this case the stabilizer of any element x in G is the normalizer of x in G.

Another example is the G-set G/H, where $H < G$ [Example 4.1(g)]. Here the stabilizer of a left coset xH is the subgroup

$$\begin{aligned}
\{g \in G \mid gxH = xH\} &= \{g \in G \mid x^{-1}gx \in H\} \\
&= \{g \in G \mid g \in xHx^{-1}\} = xHx^{-1}.
\end{aligned}$$

Definition. *Let G be a group acting on a set X, and let $x \in X$. Then the set*

$$Gx = \{ax \mid a \in G\}$$

is called the orbit *of x in G.*

In case G acts on itself as in Example 4.1(d), the orbit of $x \in G$ is $Gx = \{ax \mid a \in G\} = G$.

Further, considering G as a G-set as in Example 4.1(e), where the action of the group G is by conjugation, the orbit of $x \in G$ is $Gx = \{axa^{-1} \mid a \in G\}$, called the *conjugate class* of x and denoted by $C(x)$.

We now prove a very important result.

4.7 Theorem. *Let G be a group acting on a set X. Then the set of all orbits in X under G is a partition of X. For any $x \in X$ there is a bijection $Gx \to G/G_x$ and, hence,*

$$|Gx| = [G:G_x].$$

Therefore, if X is a finite set,

$$|X| = \sum_{x \in C} [G:G_x],$$

where C is a subset of X containing exactly one element from each orbit.

Proof. For any $x,y \in X$, let $x \sim y$ mean that $x = ay$ for some $a \in G$. Now $x = ex$ for all $x \in X$; and if $x = ay$, then $y = a^{-1}x$. If $x = ay$ and $y = bz$, then $x = (ab)z$. Hence, $\sim$ is an equivalence relation on X, and the equivalence class of $x \in X$ is the orbit Gx. Hence, the set of all orbits is a partition of X. Therefore,

$$X = \bigcup_{x \in C} Gx \qquad \text{(disjoint)},$$

where C is any subset of X containing exactly one element from each orbit.

Given $x \in X$, consider the mapping

$$\phi\colon Gx \to G/G_x \qquad \text{given by} \qquad ax \mapsto aG_x \quad \text{for all } a \in G.$$

For any $a,b \in G$,

$$ax = bx \Leftrightarrow a^{-1}bx = x \Leftrightarrow a^{-1}b \in G_x \Leftrightarrow aG_x = bG_x.$$

Hence, ϕ is well defined and injective. Moreover, ϕ is obviously surjective and, therefore, bijective. Hence, $|Gx| = |G/G_x| = [G:G_x]$.

Since X is the disjoint union of orbits Gx, it follows that if X is finite, then

$$|X| = \sum_{x \in C} |Gx| = \sum_{x \in C} [G:G_x]. \quad \square$$

The partition of X given by the orbits in X under G is called the *orbit decomposition of X under G.*

In Theorem 4.8 we give a version of Theorem 4.7 for the case when G acts on itself by conjugation. For the reader interested in conjugation only, we give an independent proof of Theorem 4.8.

Let $a \in G$. Then we define

$$C(a) = \{xax^{-1} | x \in G\},$$

which is called the conjugate class of a in G.

Recall that the normalizer of a in G is the set

$$N(a) = \{x \in G | xax^{-1} = a\},$$

which is a subgroup of G.

4.8 Theorem. *Let G be a group. Then the following are true:*

(i) The set of conjugate classes of G is a partition of G.
(ii) $|C(a)| = [G:N(a)]$.
(iii) If G is finite, $|G| = \Sigma[G:N(a)]$, *a running over exactly one element from each conjugate class.*

Proof. (i) Define a relation $\sim$ on G as follows:

$$a \sim b \quad \text{if} \quad a = xbx^{-1} \qquad \text{for some } x \in G.$$

Clearly, $\sim$ is an equivalence relation on G. The equivalence class of a in G is then the set $\{xax^{-1} | x \in G\}$, which is also the conjugate class $C(a)$ of a. This proves $G = \bigcup C(a)$, a disjoint union of conjugate classes.

(ii) Let $a \in G$. Then the mapping

$$\sigma: C(a) \rightarrow G/N(a) \qquad \text{given by} \qquad xax^{-1} \mapsto xN(a)$$

is trivially onto. Also, it is 1-1. To see this, let $x, y \in G$. Then

$$xN(a) = yN(a) \Rightarrow y^{-1}x \in N(a) \Rightarrow y^{-1}xa = ay^{-1}x \Rightarrow xax^{-1} = yay^{-1}.$$

Hence, σ is a bijection, proving $|C(a)| = |G/N(a)|$.

(iii) Follows from (i) and (ii). $\square$

The partition of G given by the conjugate classes in Theorem 4.8 (i) is called the *class decomposition* of G. The equation in part (iii) is called the *class equation* (or *class equation formula*) of G.

Let G be a group, and let S be a subset of G. If $x \in G$, then the set

$$x^{-1}Sx = \{x^{-1}sx \mid s \in S\}$$

is called a *conjugate* of S.

Definition. *Let S and T be two subsets of a group G. Then T is said to be conjugate to S if there exists $x \in G$ such that $T = xSx^{-1}$.*

Clearly, "being conjugate to" is an equivalence relation in the power set $\mathcal{P}(G)$ of the set G.

Following the method of proof of Theorem 4.8 or by specializing Theorem 4.7 to the G-set $\mathcal{P}(G)$, where G acts on $\mathcal{P}(G)$ by conjugation, we can easily prove

4.9 Theorem. *Let G be a group. Then for any subset S of G,*

$$|C(S)| = [G:N(S)] \qquad [N(S) = \{x \in G \mid x^{-1}Sx = S\}].$$

We now give some important applications of the class equation of a finite group G (Theorem 4.8). Let us first observe that $x \in Z(G)$ if and only if the conjugate class $C(x)$ has just one element, namely x itself. Therefore G is the disjoint union of $Z(G)$ and all conjugate classes having more than one element. Hence, the class equation can be expressed as

$$|G| = |Z(G)| + \sum_{x \in C} [G:N(x)],$$

where C contains exactly one element from each conjugate class with more than one element.

4.10 Theorem. *Let G be a finite group order of p^n, where p is prime and $n > 0$. Then*

(i) G has a nontrivial center Z.
(ii) $Z \cap N$ is nontrivial for any nontrivial normal subgroup N of G.
(iii) If H is a proper subgroup of G, then H is properly contained in $N(H)$; hence, if H is a subgroup of order p^{n-1}, then $H \lhd G$.

Proof. (i) Consider the class equation of G,

$$|G| = p^n = |Z| + \sum_{x \in C} [G:N(x)], \tag{1}$$

where $Z = Z(G)$, and C is a subset of G having exactly one element from each conjugate class not contained in Z. By Lagrange's theorem $|N(x)|$ divides p^n for every $x \in G$, and if $x \notin Z$, $|N(x)| < p^n$. Hence, p divides $[G:N(x)]$ for every $x \notin Z$. Therefore, p divides $\Sigma_{x \in C}[G:N(x)]$. It follows now from the class equation (1) that p divides $|Z|$. This proves that Z is not trivial.

(ii) Because $G = Z \cup (\bigcup_{x \in C} C(x))$ (disjoint),

$$N = G \cap N = (Z \cap N) \cup \left(\bigcup_{x \in C} C(x) \cap N \right) \qquad \text{(disjoint)}.$$

Hence,

$$|N| = |Z \cap N| + \sum_{x \in C} |C(x) \cap N|. \tag{2}$$

If $x \in N$, then $C(x) \subset N$. Hence, for every $x \in C$, $C(x) \cap N$ is either empty or equal to $C(x)$, and therefore $|C(x) \cap N|$ is either zero or equal to $[G:N(x)]$. Hence, p divides $\Sigma_{x \in C}|C(x) \cap N|$. Because p divides $|N|$, it follows from the equation (2) that p divides $|Z \cap N|$. This proves that $Z \cap N$ is not trivial.

(iii) Let K be a maximal normal subgroup of G contained in H. Then the quotient group G/K is of order p^r for some $r > 0$. Hence, by the first part of the theorem, G/K has a nontrivial center, say L/K. Since $L/K \lhd G/K$, it follows by Theorem 2.6 that $L \lhd G$. Now L is not contained in H, since K is a maximal normal subgroup of G contained in H, and $K \subsetneq L$.

Let $h \in H$, $l \in L$. Because L/K is the center of G/K, $(hK)(lK) = (lK)(hK)$. Hence, $l^{-1}hl \in hK \subset H$. Therefore, $L \subset N(H)$. This implies that $H \neq N(H)$. Because $H \subset N(H)$, it follows that H is properly contained in $N(H)$.

If H is a subgroup of order p^{n-1}, then $N(H)$ must be of order p^n. Hence, $N(H) = G$ and, therefore, $H \lhd G$. □

4.11 Corollary. *Every group of order p^2 (p prime) is abelian.*

Proof. Suppose G is a nonabelian group of order p^2. By Theorem 4.10, G has a nontrivial center Z that must be of order p. Let $a \in G$, $a \notin Z$. Then Z is a proper subset of $N(a)$. Hence, $N(a)$ must be of order p^2. This implies $N(a) = G$. Hence, $a \in Z$, a contradiction. □

We close this section by proving Burnside's theorem, which has applications in combinatorics. For a fixed $g \in G$, let $X_g = \{x \in X \mid gx = x\}$.

4.12 Theorem (Burnside). *Let G be a finite group acting on a finite set*

X. Then the number k of orbits in X under G is

$$k = \frac{1}{|G|} \sum_{g \in G} |X_g|.$$

Proof. Let $S = \{(g,x) \in G \times X \mid gx = x\}$. For any fixed $g \in G$ the number of ordered pairs (g,x) in S is exactly $|X_g|$. Similarly, for any fixed $x \in X$, $|G_x|$ is the number of ordered pairs (g,x) in S. Hence,

$$\sum_{g \in G} |X_g| = |S| = \sum_{x \in X} |G_x|.$$

By Theorem 4.7, $|Gx| = [G:G_x] = |G|/|G_x|$. Hence,

$$\sum_{x \in X} |G_x| = |G| \sum_{x \in X} \frac{1}{|Gx|} = |G| \sum_{a \in C} \sum_{x \in Ga} \frac{1}{|Gx|},$$

where C is a subset of X containing exactly one element from each orbit. Now $|Gx| = |Ga|$ for every $x \in Ga$. Hence,

$$\sum_{x \in Ga} \frac{1}{|Gx|} = 1.$$

Therefore, $\Sigma_{x \in X} |G_x| = |G| \cdot k$, which proves the theorem. □

4.13 Examples

(a) Let G be a group containing an element of finite order $n > 1$ and exactly two conjugacy classes. Prove that $|G| = 2$.

Solution. Let $e \neq a \in G$ such that $o(a) = n$. Because there are only two conjugate classes, these must be $\{e\}$ and $C(a)$. Therefore, if $e \neq b \in G$, then $b \in C(a)$. Thus $o(b) = o(a) = n$. Let $m|n$. Then there exists an element b such that $b \in [a]$, the cyclic group generated by a, and $o(b) = m$. It then follows that $m = n$. Hence, n is prime. We claim that $a^2 = e$. If $a^2 \neq e$, then $a^2 \in C(a)$. Therefore, $a^2 = xax^{-1}$ for some $x \in G$. This implies $a^{2^i} = x^i a x^{-i}$ for all $i > 0$. Choose $i = n$. Because $x^n = e$, we obtain $a^{2^n} = a$. This yields $2^n \equiv 1 \pmod{n}$, a contradiction. Hence, $a^2 = e$; so for all $g \in G$, $g^2 = e$. This implies G is abelian. But, because G has only two conjugate classes, $|G| = 2$.

(b) Let H be a subgroup of a finite group G. Let $A,B \in \mathcal{P}(G)$, the power set of G. Define A to be conjugate to B with respect to H if $B = hAh^{-1}$ for some $h \in H$. Then

(i) Conjugacy defined in $\mathcal{P}(G)$ is an equivalence relation.
(ii) If $C_H(A)$ is the equivalence class of $A \in \mathcal{P}(G)$ (called the conjugate class of A with respect to H), then

$$|C_H(A)| = [H:H \cap N(A)].$$

Solution. Follows from Theorem 4.7 by taking X to be the power set $\mathcal{P}(G)$ of G and H to be the group that acts on X by conjugation. Then the orbit of $A \in X$ is the conjugate class $C_H(A)$, and the stabilizer is $H \cap N(A)$.

The direct argument is as follows. First, (i) is obvious. For the proof of (ii), consider the mapping

$$\sigma: C_H(A) \to H/H \cap N(A),$$

where

$$\sigma(hAh^{-1}) = h(H \cap N(A)), \qquad h \in H.$$

σ is clearly onto. σ is also 1-1, for let $h_1, h_2 \in H$. Then

$$\begin{aligned} h_1(H \cap N(A)) = h_2(H \cap N(A)) &\Rightarrow h_1^{-1}h_2 \in H \cap N(A) \\ &\Rightarrow h_1^{-1}h_2A = Ah_1^{-1}h_2 \Rightarrow h_1Ah_1^{-1} = h_2Ah_2^{-1}. \end{aligned}$$

Hence, σ is bijective.

The example which follows gives a beautiful illustration of the Burnside Theorem.

(c) Find the number of different necklaces with p beads, p prime, where the beads can have any of n different colors.

Solution. Let X be the set of all possible necklaces. Clearly $|X| = n^p$. Let

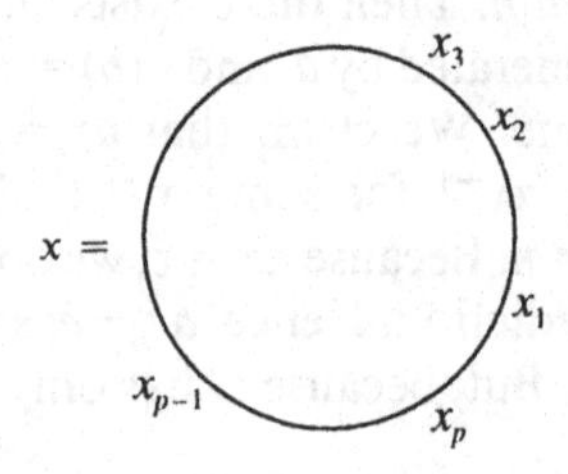

be a necklace. Let $G = [a]$ be the cyclic group of order p. Then X is a G-set under the action

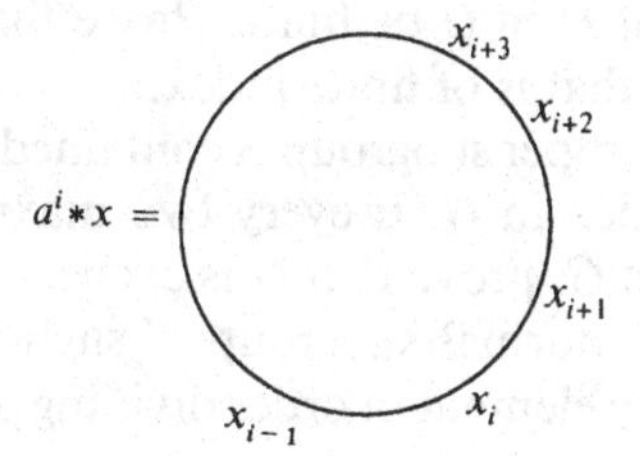

where subscripts are modulo p.

Clearly the action of any fixed element $a^i \in G$ on a given necklace yields the same necklace. (The beads are just being permuted cyclically.) So the number of orbits is the same as the number of different necklaces. We now compute X_g for each $g \in G$. First

$$X_e = \{x \in X \mid e * x = x\} = X.$$

So $|X_e| = n^p$. Let $g \neq e$. Since G is cyclic of prime order, g generates G. Then

$$\begin{aligned} X_g &= \{x \in X \mid g * x = x\} \\ &= \{x \in X \mid g^i * x = x, \forall i\} \\ &= \{x \in X \mid a * x = x, \forall a \in G\}. \end{aligned}$$

Therefore, X_g consists of all those necklaces which are unchanged by any permutation and these are precisely those which are of one color. Hence $|X_g| = n$, for all $g \in G$, $g \neq e$. The Burnside Theorem then gives the number of different necklaces as

$$\frac{1}{p}\Big[n^p + \overbrace{n + \cdots + n}^{(p-1 \text{ times})}\Big] = \frac{1}{p}\Big[n^p + n(p-1)\Big].$$

Problems

1. Let G be a group. Show that

   $$Z(G) = \bigcup_{|C(x)|=1} C(x), \ x \in G.$$

2. Find the number of conjugates of the element $(1 \quad 3)$ in D_4.
3. Determine the conjugate classes of the symmetric group of degree 3 and verify that the number of elements in each conjugate class is a divisor of the order of the group.
4. Let H be a proper subgroup of a finite group G. Show that G contains at least one element that is not in any conjugate of H. [*Hint*: $|C(H)| = [G:N(H)]$ and $|x^{-1}Hx| = |H|$.]

5. Let $H < G$, and let the index of H in G be finite. Prove that H contains a normal subgroup N that is of finite index.
6. Let G be a group in which each proper subgroup is contained in a maximal subgroup of finite index in G. If every two maximal subgroups of G are conjugate in G, prove that G is cyclic.
7. Let G be a finite group with a normal subgroup N such that $(|N|, |G/N|) = 1$. Show that every element of order dividing $|N|$ is contained in N.
8. Prove that in any group the subset of all elements that have only a finite number of conjugates is a subgroup.
9. If the commutator group G' of a group G is of order m, prove that each element in G has at most m conjugates.
10. Let G be the group of symmetries of a cube. Prove that each of the following sets is a G-set:

    (i) Set of vertices.
    (ii) Set of long diagonals.
    (iii) Set of faces.
    (iv) Set of edges.

11. Prove that if a finite group G has a normal subgroup N with $|N| = 3$ such that $N \nsubseteq Z(G)$, then G has a subgroup K such that $[G:K] = 2$. [*Hint*: G acts by conjugation on N.]
12. Let G be a group and X be a G-set. We say G acts transitively on X if for all $x_1, x_2 \in X$ there exists $g \in G$ such that $g * x_1 = x_2$. Show G acts transitively on X if and only if there is only one orbit. Also show if $H < G$ and $X = G/H$, the set of left cosets of H, then G acts on X transitively.
13. Let $G = S_3$ and X be the undirected graph with vertices 1,2,3 consisting of

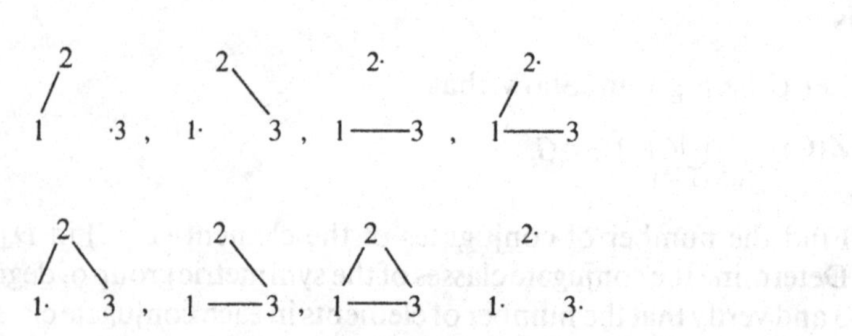

    Show G acts on X canonically. Use the Burnside Theorem to show that the number of orbits is four.
14. Let $X = \{\Delta_1, \Delta_2\}$ be the set consisting of diagonals of a square

with vertices 1,2,3, and 4. Regard X as a D_4-set under the action $g*\{a,a+2\}=\{g(a),g(a+2)\}$, where $\{a,a+2\}$ represents a diagonal with vertices $a,a+2$. Show that there exists a homomorphism $\varphi\colon G\to S_3$ with $\mathrm{Ker}\varphi=\{I,(2\ \ 4),(1\ \ 3),(1\ \ 3),(2\ \ 4)\}$.

15. Find the number of necklaces with six beads and two colors.
16. Show that there are six different necklaces with four beads and two colors and eight different ones with five beads and two colors.
17. Show that there are $\frac{1}{2}[k^n+k^{[(n+1)/2]}]$ neckties having n strips (of equal width) of k distinct colors.
18. Suppose X is a G-set. Show that $X\times X$ is also a G-set under the action $g*(x_1,x_2)=(g*x_1,g*x_2)$. Consider the Klein four-group K as the group of symmetries of a rectangle with vertices $1,2,3$, and 4, and let $X=\{1,2,3,4\}$. Regarding each of the sets X and $X\times X$ as a K-set in a canonical way, find the orbits of $1\in X$ and $(1,2)\in X\times X$.
19. Consider additive and multiplicative groups $(\mathbf{Z}_7,+)$ and $(\mathbf{Z}_7^*,\cdot)$. For convenience write $\mathbf{Z}_7=\{0,1,2,3,4,5,6\}$, and let $G=\left\{\begin{pmatrix} a & b \\ 0 & a^{-1}\end{pmatrix}\middle| b\in\mathbf{Z}_7, a=1,2,\text{or }4\right\}$. Show G is a group of order 21 and $N=\left\{\begin{pmatrix} 1 & b \\ 0 & 1\end{pmatrix}\middle| b\in\mathbf{Z}_7\right\}\lhd G$. Regard $X=\left\{\begin{pmatrix} x \\ y\end{pmatrix}\middle| x,y\in\mathbf{Z}_7\right\}$ and $X\times X$ as G-sets in a natural way (as ordinary matrix multiplication). Find the orbit of $\left(\begin{pmatrix}1\\0\end{pmatrix},\begin{pmatrix}3\\0\end{pmatrix}\right)\in X\times X$.

CHAPTER 6

Normal series

1 Normal series

Definition. *A sequence $(G_0, G_1, \ldots, G_r)$ of subgroups of a group G is called a* normal series *(or* subnormal series*) of G if*

$$\{e\} = G_0 \lhd G_1 \lhd G_2 \lhd \cdots \lhd G_{r-1} \lhd G_r = G.$$

The factors *of a normal series are the quotient groups G_i/G_{i-1}, $1 \le i \le r$.*

Definition. *A* composition series *of a group G is a normal series $(G_0, \ldots, G_r)$ without repetition whose factors G_i/G_{i-1} are all simple groups. The factors G_i/G_{i-1} are called* composition factors *of G.*

We often refer to a normal series $(G_0, G_1, \ldots, G_r)$ by saying that

$$\{e\} = G_0 \subset G_1 \subset \cdots \subset G_r = G$$

is a normal series of G.

For any group G, $\{e\} = G_0 \subset G_1 = G$ is trivially a normal series of G. If G is a simple group, then $\{e\} \subset G$ is the only composition series of G.

1.1 Lemma. *Every finite group has a composition series.*

Proof. If $|G| = 1$ or if G is simple, then the result is trivial (if $|G| = 1$, then (G) is a composition series of G without factors). Suppose G is not simple, and the result holds for all groups of order $< |G|$. Let H be a maximal normal subgroup of G. By the induction hypothesis H has a composition series $\{e\} \subset H_1 \subset \cdots \subset H$. Therefore, G has a composition series $\{e\} \subset H_1 \subset \cdots \subset H \subset G$. □

1.2 Examples

(a) $\{e\} \subset \{e,e'\} \subset \{e,e',i,i'\} \subset G$ is a composition series for the quaternion group G, where $e' = -e$ and

$$i = \begin{bmatrix} \sqrt{-1} & 0 \\ 0 & -\sqrt{-1} \end{bmatrix} = -i'.$$

(b) $\{e\} \subset \{e,(1\ 2\ 3),\ (1\ 3\ 2)\} \subset S_3$ is a composition series for S_3.

(c) $\{e\} \subset \{e,\sigma^2\} \subset \{e,\sigma,\sigma^2,\sigma^3\} \subset D_4$ is a composition series for the octic group D_4 (see Section 5, Chapter 4).

(d) $\{0\} \subset \{0,9\} \subset \{0,3,6,9,12,15\} \subset \{0,1,2,\ldots,17\} = \mathbf{Z}/(18)$ is a composition for $\mathbf{Z}/(18)$.

We remark that a composition series is not necessarily unique. For example, another composition series for the octic group D_4 is

$$\{e\} \subset \{e,\tau\} \subset \{e,\sigma^2,\tau,\sigma^2\tau\} \subset D_4.$$

However, it is shown in the next section that any two composition series of the same group are "equivalent," in the following sense.

Definition. *Two normal series $S = (G_0,G_1,\ldots,G_r)$ and $S' = (G'_0,G'_1,\ldots,G'_r)$ of G are said to be* equivalent, *written $S \sim S'$, if the factors of one series are isomorphic to the factors of the other after some permutation; that is,*

$$G'_i/G'_{i-1} \simeq G_{\sigma(i)}/G_{\sigma(i)-1}, \qquad i = 1,\ldots,r,$$

for some $\sigma \in S_r$.

Evidently, $\sim$ is an equivalence relation.

1.3 Theorem (Jordan–Hölder). *Any two composition series of a finite group are equivalent.*

Proof. Let G be a finite group and suppose the theorem holds for every group of order $< |G|$. Consider any two composition series of G, say

$$S_1: \ \{e\} = G_0 \subset G_1 \subset \cdots \subset G_r = G,$$
$$S_2: \ \{e\} = H_0 \subset H_1 \subset \cdots \subset H_s = G.$$

If $G_{r-1} = H_{s-1}$, then $S_1 \sim S_2$ by the induction hypothesis. If $G_{r-1} \neq H_{s-1}$, let $K = G_{r-1} \cap H_{s-1}$. By Corollary 2.9, Chapter 5, K is a maximal normal subgroup of G_{r-1} and also of H_{s-1}. Now K has a composition series, say $\{e\} = K_0 \subset K_1 \subset \cdots \subset K$. This gives two more composition series of G, namely,

$$S_3: \ \{e\} = K_0 \subset K_1 \subset \cdots \subset K \subset G_{r-1} \subset G,$$
$$S_4: \ \{e\} = K_0 \subset K_1 \subset \cdots \subset K \subset H_{s-1} \subset G.$$

Now $G_{r-1}H_{s-1} = G$. Hence, by the diamond isomorphism theorem,

$$G/G_{r-1} \simeq H_{s-1}/K \quad \text{and} \quad G/H_{s-1} \simeq G_{r-1}/K.$$

Therefore, $S_3 \sim S_4$. Further, by the induction hypothesis, $S_1 \sim S_3$ and $S_2 \sim S_4$. This proves that S_1 and S_2 are equivalent. □

Theorem 1.3 shows that the factors of any composition series of a finite group G – that is, the composition factors – are determined uniquely up to isomorphism and ordering.

1.4 Examples

(a) An abelian group G has a composition series if and only if G is finite.

Solution. Let G be an abelian group with a composition series $\{e\} = G_0 \subset G_1 \subset \cdots \subset G_r = G$. Then the quotient groups G_i/G_{i-1}, being abelian and simple, must be cyclic groups of prime order p_i (say), $i = 1,\ldots,r$. Hence,

$$|G| = \prod_{i=1}^{r} |G_i/G_{i-1}| = p_1 \cdots p_r.$$

Therefore, an abelian group G has a composition series if and only if G is finite. Moreover the composition factors of G are determined by the prime factors of $|G|$.

(b) If a cyclic group has exactly one composition series, then it is a p-group.

Solution. Suppose $G = [a]$ is a cyclic group of order $p_1 \cdots p_r$, where $p_1,\ldots,p_r$ are primes (not necessarily distinct). By Theorem 4.4 in Chapter 4, G has a unique subgroup G_i of order $p_1 \cdots p_i$, namely, $G_i = [a^{p_{i+1} \cdots p_r}]$, $i = 1,\ldots,r-1$. Hence, $\{e\} = G_0 \subset G_1 \subset \cdots \subset G_{r-1} \subset G_r = G$ is a unique composition series of G such that $|G_i/G_{i-1}| = p_i$, $i = 1,\ldots,r$. Thus, every permutation of the prime factors of $|G|$ determines a unique composition series of G. Hence, G has a unique composition series if and only if $p_1 = \cdots = p_r$; that is, $|G| = p^r$.

(c) The Jordan–Hölder theorem implies the fundamental theorem of arithmetic.

Solution. Let G be a cyclic group of order n. Suppose n has two factorizations into positive primes, say

$$n = p_1 \cdots p_r \quad \text{and} \quad n = q_1 \cdots q_s.$$

Then, as shown in Example (b), G has a composition series whose factors are cyclic groups of orders $p_1, \dots, p_r$, respectively, and also a composition series whose factors are cyclic groups of orders $q_1, \dots, q_s$, respectively. By the Jordan–Hölder theorem the two series must be equivalent. Thus, $r = s$ and $p_i = q_i$ (after reordering if necessary).

(d) Let G be a group of order p^n, p prime. Then G has a composition series such that all its composition factors are of order p.

Solution. Clearly, any composition factor is of order p^k for some $k > 0$. But since a group of prime power order has a nontrivial center, and each composition factor is a simple group, it follows that each composition factor is a simple abelian group and, therefore, of order p.

Problems

1. Write down all the composition series for the quaternion group.
2. Write down all the composition series for the octic group.
3. Write down a composition series for the Klein four-group.
4. Find the composition factors of the additive group of integers modulo 8.
5. Find all composition series for $\mathbf{Z}/(30)$. Verify that they are equivalent.
6. If G is a cyclic group such that $|G| = p_1 \cdots p_r$, p_i distinct primes, show that the number of distinct composition series of G is $r!$.
7. Let G be a finite group and $N \lhd G$. Show that G has a composition series in which N appears as a term.
8. Give an example of two nonisomorphic finite groups G, H which have isomorphic composition factors.
9. Let us define a composition series of any group G to be a descending chain (possibly infinite)

   $$G_0 = G \supset G_1 \supset G_2 \supset \cdots \supset \{e\}$$

   such that $G_i \lhd G_{i-1}$, $i = 1,2,3,\dots$, and G_{i-1}/G_i is simple. Show that $\mathbf{Z}$ has no finite composition series but does possess infinite composition series. Write two such series and show that the Jordan–Hölder Theorem does not generalize to infinite groups.

2 Solvable groups

Recall that the subgroup G' generated by the set of all commutators $aba^{-1}b^{-1}$ in a group G is called the derived group of G. For any positive integer n, we define the *nth derived group* of G, written $G^{(n)}$, as follows:

$$G^{(1)} = G', \qquad G^{(n)} = (G^{(n-1)})' \quad (n > 1).$$

Definition. *A group G is said to be* solvable *if $G^{(k)} = \{e\}$ for some positive integer k.*

It is obvious that if G is abelian, then $G' = \{e\}$. Thus, trivially, every abelian group is solvable.

2.1 Theorem. *Let G be a group. If G is solvable, then every subgroup of G and every homomorphic image of G are solvable. Conversely, if N is a normal subgroup of G such that N and G/N are solvable, then G is solvable.*

Proof. Let G be a solvable group, so that $G^{(k)} = \{e\}$. Then for any subgroup H of G, $H^{(n)} \subset G^{(n)}$ for every positive integer n. Hence, $H^{(k)} = \{e\}$. Therefore H is solvable.

Let $\phi: G \to H$ be a surjective homomorphism. For any $a,b \in G$,

$$\phi(aba^{-1}b^{-1}) = \phi(a)\phi(b)\phi(a)^{-1}\phi(b)^{-1}.$$

Hence, $H' = \phi(G')$, and, by induction, $H^{(n)} = \phi(G^{(n)})$ for any positive integer n. Therefore, $H^{(k)} = \phi(\{e\}) = \{e\}$. This proves that every homomorphic image of G is solvable.

Conversely, let $N \lhd G$ such that N and G/N are solvable. Then $N^{(k)} = \{e\}$ and $(G/N)^{(l)} = \{e\}$ for some positive integers k,l. Because G/N is a homomorphic image of G (under the natural homomorphism $G \to G/N$), $(G/N)^{(n)} = G^{(n)}N/N$ for every positive integer n. Hence, $G^{(l)} \subset N$. Therefore, $G^{(l+k)} \subset N^{(k)} = \{e\}$ and G is solvable. □

2.2 Theorem. *A group G is solvable if and only if G has a normal series with abelian factors. Further, a finite group is solvable if and only if its composition factors are cyclic groups of prime orders.*

Proof. If $G^{(k)} = \{e\}$, then

$$\{e\} = G^{(k)} \subset G^{(k-1)} \subset \cdots \subset G^{(1)} \subset G^{(0)} = G$$

is a normal series of G in which each factor $G^{(i-1)}/G^{(i)}$ is abelian. Conversely, suppose G has a normal series

$$\{e\} = H_0 \subset H_1 \subset \cdots \subset H_{r-1} \subset H_r = G$$

such that H_i/H_{i-1} is abelian, $i = 1,...,r$. Then by Theorem 1.4 of Chapter 5, $H_i' \subset H_{i-1}$. Hence, $G' = H_r' \subset H_{r-1}$, and, by induction, $G^{(i)} \subset H_{r-i}$, $i = 1,...,r$. Therefore, $G^{(r)} = \{e\}$, and G is solvable.

Now suppose G is a finite solvable group. Then G has a normal series $\{e\} = H_0 \subset \cdots \subset H_r = G$ in which each factor H_i/H_{i-1} is abelian. Now H_i/H_{i-1}, being a finite abelian group, has a composition series whose factors are cyclic of prime order. Inserting the corresponding subgroups of H_i between the terms H_{i-1} and H_i in the normal series $H_0 \subset \cdots \subset H_r$, we get a composition series of G in which every factor is a cyclic group of prime order. The converse follows from the first part of the theorem. □

2.3 Examples

(a) S_3 is solvable, for

$$\{e\} \subset N = \{e,(1\ 2\ 3),(1\ 3\ 2)\} \subset S_3$$

is a normal series such that S_3/N is abelian.

We shall see in the next chapter that S_4 is solvable, but S_n, $n \geq 5$, is not solvable.

(b) The dihedral group D_n is solvable, for

$$\{e\} \subset \{e,\sigma,...,\sigma^{n-1}\} \subset D_n$$

is a normal series with abelian factors.

(c) Let G be a finite solvable group and M be a minimal normal subgroup of G. Then M is cyclic of order p, for some prime p. We may first show that M is an abelian p-group.

Problems

1. Show that the group of all upper triangular matrices of the form

$$\begin{bmatrix} 1 & a & b \\ 0 & 1 & c \\ 0 & 0 & 1 \end{bmatrix}, \quad a, b, c \in \mathbf{R},$$

is solvable. Generalize this to $n \times n$ upper triangular matrices.

2. Show that a simple group is solvable if and only if it is cyclic.
3. Show that if A, B are groups, then $A \times B$ is solvable if and only if both $A.B$ are solvable.

4. Let A, B be normal solvable subgroups of a group G. Show that AB is also solvable.

Additional problems on solvable groups are given in the problem set at the end of Chapter 8.

3 Nilpotent groups

We define inductively the *nth center* of a group G as follows. For $n = 1$, $Z_1(G) = Z(G)$. Consider the center of the quotient group $G/Z_1(G)$. Because $Z(G/Z_1(G))$ is a normal subgroup of $G/Z(G)$, by Corollary 2.7 of Chapter 5 there is a unique normal subgroup $Z_2(G)$ of G such that $Z_2(G)/Z_1(G) = Z(G/Z_1(G))$. Thus, inductively we obtain a normal subgroup $Z_n(G)$ of G such that $Z_n(G)/Z_{n-1}(G) = Z(G/Z_{n-1}(G))$ for every positive integer $n > 1$. $Z_n(G)$ is called the *nth center* of G. Setting $Z_0(G) = \{e\}$, we have $Z_n(G)/Z_{n-1}(G) = Z(G/Z_{n-1}(G))$ for all positive integers n. It follows immediately from the definition that

$$Z_n(G) = \{x \in G | xyx^{-1}y^{-1} \in Z_{n-1}(G) \text{ for all } y \in G\}.$$

Hence, $(Z_n(G))' \subset Z_{n-1}(G)$.

The ascending series

$$\{e\} = Z_0(G) \subset Z_1(G) \subset \cdots \subset Z_n(G) \subset \cdots$$

of subgroups of a group G is called the *upper central* series of G.

Definition. *A group G is said to be* nilpotent *if $Z_m(G) = G$ for some m. The smallest m such that $Z_m(G) = G$ is called the* class of nilpotency *of G.*

If G is an abelian group, then $Z_1(G) = Z(G) = G$. Thus, trivially, every abelian group is nilpotent. For a nontrivial example we prove

3.1 Theorem. *A group of order p^n (p prime) is nilpotent.*

Proof. By Theorem 4.10 of Chapter 5, G has a nontrivial center $Z_1(G)$. Now $G/Z_1(G)$, being of order $p^r < p^n$, has a nontrivial center. Hence, $|Z_1(G)| < |Z_2(G)|$. By repeating the argument we get $|Z_m(G)| = p^n$ for some $m \leq n$. Hence, $Z_m(G) = G$. □

3.2 Theorem. *A group G is nilpotent if and only if G has a normal series*

$$\{e\} = G_0 \subset G_1 \subset \cdots \subset G_m = G$$

such that $G_i/G_{i-1} \subset Z(G/G_{i-1})$ for all $i = 1, \ldots, m$.

Proof. Let G be a nilpotent group of class m, so that $Z_m(G) = G$. Then

$$\{e\} = Z_0(G) \subset Z_1(G) \subset \cdots \subset Z_m(G) = G$$

is a normal series that satisfies the required condition, since $Z_i(G)/Z_{i-1}(G) = Z(G/Z_{i-1}(G))$. Conversely, suppose that G has a normal series $\{e\} = G_0 \subset G_1 \subset \cdots \subset G_m = G$ such that $G_i/G_{i-1} \subset Z(G/G_{i-1})$ for all i. Then $G_1 \subset Z(G)$. Now $G_2/G_1 \subset Z(G/G_1)$. Hence, for every $x \in G_2$, $y \in G$, $xyx^{-1}y^{-1} \in G_1 \subset Z_1(G)$. Therefore, $G_2 \subset Z_2(G)$. By repeating the argument we get $G_i \subset Z_i(G)$, $i = 1,\ldots,m$. Hence, $G = G_m \subset Z_m(G)$, and G is therefore nilpotent. □

3.3 Corollary. *Every nilpotent group is solvable.*

Proof. Let G be a nilpotent group. By Theorem 3.2, G has a normal series with abelian factors. Hence, by Theorem 2.2, G is solvable. □

The converse of this corollary is not true. For example, the symmetric group S_3 is not nilpotent because its center is trivial. But it is easily verified that $S_3^{(2)}$ is trivial. Hence, S_3 is solvable. Thus, we see that the class of nilpotent groups is contained strictly between the classes of abelian groups and solvable groups.

3.4 Theorem. *Let G be a nilpotent group. Then every subgroup of G and every homomorphic image of G are nilpotent.*

Proof. Let G be a nilpotent group of class m, so that $Z_m(G) = G$. Let H be a subgroup of G. Then, clearly, $H \cap Z(G) \subset Z(H)$. For all $x \in Z_2(G)$, and $y \in G$, $xyx^{-1}y^{-1} \in Z_1(G)$. Hence, for all $x \in H \cap Z_2(G)$, and $y \in H$, $xyx^{-1}y^{-1} \in H \cap Z_1(G)$. Therefore, $H \cap Z_2(G) \subset Z_2(H)$. By repeating the argument we see that $H \cap Z_i(G) \subset Z_i(H)$, $i = 1,\ldots,m$. Hence, $H = H \cap G = H \cap Z_m(G) \subset Z_m(H)$. Thus, H is nilpotent.

Let ϕ: $G \to H$ be a surjective homomorphism. Then

$$\phi(xyx^{-1}y^{-1}) = \phi(x)\phi(y)\phi(x)^{-1}\phi(y)^{-1} \qquad \text{for all } x,y \in G.$$

Hence, $\phi(Z(G)) \subset Z(H)$. Let $x \in Z_2(G)$. Then $xyx^{-1}y^{-1} \in Z(G)$ for all $y \in G$. Hence

$$\phi(x)\phi(y)\phi(x)^{-1}\phi(y)^{-1} \in \phi(Z(G)) \subset Z(H).$$

Because ϕ is surjective, it follows that $\phi(x) \in Z_2(H)$. Therefore, $\phi(Z_2(G)) \subset Z_2(H)$. By repeated use of the argument we see that $\phi(Z_i(G)) \subset Z_i(H)$, $i = 1,\ldots,m$. Hence, $H = \phi(G) = \phi(Z_m(G)) \subset Z_m(H)$. Therefore, H is nilpotent. □

3.5 Theorem. *Let $H_1, \ldots, H_n$ be a family of nilpotent groups. Then $H_1 \times \cdots \times H_n$ is also nilpotent.*

Proof. We show that if H and K are nilpotent groups, then $H \times K$ is also nilpotent. The proof of the theorem then follows by induction. It can be easily shown that $Z_m(H \times K) = Z_m(H) \times Z_m(K)$ for any m. Hence, $Z_m(H \times K) = H \times K$ for some m. Therefore, $H \times K$ is nilpotent. □

Problems

1. Show that S_n is not nilpotent for $n \geq 3$.
2. Is D_n nilpotent? solvable?
3. Give an example of a group G such that G has a normal subgroup H with both H and G/H nilpotent but G nonnilpotent.
4. Let a group G be a product of normal subgroups H and K of orders p^α and q^β, respectively, where p and q are distinct primes. Show that G is nilpotent.
5. Let G be a finite nilpotent group, and let p be prime. Show that if p divides $|G|$, then p also divides $|Z(G)|$.

CHAPTER 7

Permutation groups

1 Cyclic decomposition

Let us recall that a cycle $\sigma = (a_1 \cdots a_r)$ in S_n is a permutation such that $\sigma(a_i) = a_{i+1}$ for $i = 1, \ldots, r-1$, $\sigma(a_r) = a_1$, and $\sigma(x) = x$ for every other x in $\mathbf{n}$. Two cycles $(a_1 \cdots a_r)$, $(b_1 \cdots b_s)$ in S_n are *disjoint* permutations if and only if the sets $\{a_1, \ldots, a_r\}$ and $\{b_1, \ldots, b_s\}$ are disjoint. Note that a cycle of length r can be written in r ways, namely, as $(a_1 \cdots a_r)$ and $(a_i a_{i+1} \cdots a_r a_1 \cdots a_{i-1})$, $i = 2, \ldots, r$. A cycle of length r is also called an *r-cycle.*

1.1 Theorem. *Any permutation $\sigma \in S_n$ is a product of pairwise disjoint cycles. This cyclic factorization is unique except for the order in which the cycles are written and the inclusion or omission of cycles of length* 1.

Proof. We prove the theorem by induction on n. If $n = 1$ the theorem is obvious. Let $i_1 \in \{1,2,\ldots,n\}$. Then there exists a smallest positive integer r such that $\sigma^r(i_1) = i_1$. Let $i_2 = \sigma(i_1)$, $i_3 = \sigma(i_2) = \sigma^2(i_1),\ldots,i_r = \sigma(i_{r-1}) = \sigma^{r-1}(i_1)$. Next, let

$$X = \{1,2,\ldots,n\} - \{i_1, i_2,\ldots,i_r\}.$$

If $X = \varnothing$, then σ is a cycle and we are done. So let $X \neq \varnothing$, and let $\sigma^* = \sigma|X$. Then σ^* is a permutation of a set consisting of $n - r$ elements. By induction $\sigma^* = c_2 c_3 \cdots c_m$, where c_i are pairwise disjoint cycles. But since $\sigma = \sigma^* c_1$, where $c_1 = (i_1\, i_2 \cdots i_r)$, it follows that σ is a product of disjoint cycles.

To prove uniqueness, suppose σ has two decompositions into disjoint cycles of length > 1, say $\sigma = \gamma_1 \cdots \gamma_k = \beta_1 \cdots \beta_l$.

Let $x \in \mathbf{n}$. If x does not appear in any γ_i, then $\sigma(x) = x$. Hence, x does not appear in any β_i. If x is in some γ_i, then $\sigma(x) \neq x$. Hence, x must be in some β_j. But then $\sigma^r(x)$ occurs in both γ_i and β_j for every $r \in \mathbf{Z}$. Hence, γ_i and β_j are identical cycles. This proves that the two decompositions written earlier are identical, except for the ordering of factors. □

1.2 Corollary. *Every permutation can be expressed as a product of transpositions.*

Proof. We need only show that every cycle is a product of transpositions. It is easily verified that

$$(a_1\, a_2 \cdots a_m) = (a_1\, a_m)(a_1\, a_{m-1}) \cdots (a_1\, a_2). \qquad \square$$

Note that decomposition into transpositions is not unique, and the factors need not be pairwise disjoints. For example,

$$\begin{aligned}(1\ 2\ 3\ 4\ 5) &= (4\ 5)(3\ 5)(2\ 5)(1\ 5)\\ &= (5\ 2)(4\ 2)(3\ 2)(1\ 5)\\ &= (5\ 3)(2\ 1)(3\ 5)(4\ 5)(2\ 3)(3\ 5)\end{aligned}$$

Let $\sigma \in S_n$. Let us write the decomposition of σ into disjoint cycles, including cycles of length 1, and write the factors in order of increasing length. Suppose $\sigma = \gamma_1 \cdots \gamma_k$ and the cycles $\gamma_1,\ldots,\gamma_k$ are of lengths $n_1,\ldots,n_k$, respectively. Then $n_1 \le n_2 \le \cdots \le n_k$ and $n_1 + n_2 + \cdots + n_k = n$. Thus, σ determines a partition $(n_1,\ldots,n_k)$ of n, which is called the *cycle structure* of σ. Conversely, given a partition $(n_1,\ldots,n_k)$ of n, there always exists (not uniquely) a permutation $\sigma \in S_n$ such that the cycle structure of σ is $(n_1,\ldots,n_k)$, for the product of any disjoint cycles $\gamma_1,\ldots,\gamma_k$ of length $n_1,\ldots,n_k$, respectively, satisfies that condition.

1.3 Theorem. *If $\alpha,\sigma \in S_n$, then $\tau = \alpha\sigma\alpha^{-1}$ is the permutation obtained by applying α to the symbols in σ. Hence, any two conjugate permutations in S_n have the same cycle structure.*

Conversely, any two permutations in S_n with the same cycle structure are conjugate.

Proof. If $\sigma(i) = j$, then $\tau\alpha(i) = (\alpha\sigma\alpha^{-1})\alpha(i) = \alpha\sigma(i) = \alpha(j)$. Therefore, if $(a_1 \cdots a_m)$ is a cycle in the decomposition of σ, then $(\alpha(a_1) \cdots \alpha(a_m))$ is a cycle in the decomposition of τ. Hence, the cycle decomposition of τ is obtained by substituting $\alpha(x)$ for x everywhere in the decomposition of σ. Thus, σ and τ have the same cycle structure.

Conversely, suppose that σ and τ have the same cycle structure $(p,q,\ldots,r)$. Then σ and τ have cycle decompositions

$$\sigma = (a_1 \cdots a_p)(a_{p+1} \cdots a_{p+q}) \cdots (a_{n-r+1} \cdots a_n),$$
$$\tau = (b_1 \cdots b_p)(b_{p+1} \cdots b_{p+q}) \cdots (b_{n-r+1} \cdots b_n).$$

Define $\alpha \in S_n$ by $\alpha(a_i) = b_i$, $i = 1,\ldots,n$. Then it is easily verified that $\alpha\sigma\alpha^{-1}(b_i) = \tau(b_i)$, $i = 1,\ldots,n$. Hence, $\alpha\sigma\alpha^{-1} = \tau$, and σ and τ are conjugate. □

1.4 Corollary. *There is a one-to-one correspondence between the set of conjugate classes of S_n and the set of partitions of n.*

Proof. Obvious. □

1.5 Examples

(a) Express the following permutations as a product of disjoint cycles:

(i) $\begin{pmatrix} 1 & 2 & 3 & 4 & 5 & 6 \\ 2 & 5 & 1 & 6 & 4 & 3 \end{pmatrix}$

(ii) $(4\ \ 2\ \ 1\ \ 5)(3\ \ 4\ \ 2\ \ 6)(5\ \ 6\ \ 7\ \ 1)$

(iii) $\begin{pmatrix} 1 & 2 & 3 & 4 & 5 \\ 3 & 4 & 1 & 2 & 5 \end{pmatrix}$

Solution. (i) We start with 1 and obtain the cycle $(1\ 2\ 5\ 4\ 6\ 3)$.

(ii) The given permutation can be rewritten as

$$\sigma = \begin{pmatrix} 1 & 2 & 3 & 4 & 5 & 6 & 7 \\ 4 & 6 & 2 & 1 & 3 & 7 & 5 \end{pmatrix}$$

Starting with 1, we obtain the cycle $(1\ 4)$. Then start with any other element not in $(1\ 4)$, say 2. Starting with 2, we get a cycle $(2\ 6\ 7\ 5\ 3)$. These two cycles exhaust the list $\{1,2,3,4,5,6,7\}$. Thus, $\sigma = (2\ 6\ 7\ 5\ 3)(1\ 4)$.

(iii) Proceeding as before, we obtain that the given permutation is $(1\ 3)(2\ 4)(5)$ or, by the convention of omitting cycles of length 1, equal to $(1\ 3)(2\ 4)$.

(b) Let $\sigma = (1\ 5\ 3)(1\ 2)$, $\tau = (1\ 6\ 7\ 9)$. Compute $\sigma\tau\sigma^{-1}$.

Solution. By Theorem 1.3, $\sigma\tau\sigma^{-1} = (\sigma(1)\ \sigma(6)\ \sigma(7)\ \sigma(9)) = (2\ 6\ 7\ 9)$.

The interested reader may actually verify this by first computing σ^{-1}.

$$\sigma = \begin{pmatrix} 1 & 2 & 3 & 4 & 5 \\ 2 & 5 & 1 & 4 & 3 \end{pmatrix},$$

$$\sigma^{-1} = \begin{pmatrix} 2 & 5 & 1 & 4 & 3 \\ 1 & 2 & 3 & 4 & 5 \end{pmatrix} = \begin{pmatrix} 1 & 2 & 3 & 4 & 5 \\ 3 & 1 & 5 & 4 & 2 \end{pmatrix}.$$

Thus, by composing the mappings we obtain

$$\sigma\tau\sigma^{-1} = \begin{pmatrix} 1 & 2 & 3 & 4 & 5 & 6 & 7 & 8 & 9 \\ 1 & 6 & 3 & 4 & 5 & 7 & 9 & 8 & 2 \end{pmatrix} = (2 \ 6 \ 7 \ 9).$$

Problems

1. If α and β are disjoint cycles, then $\alpha\beta = \beta\alpha$.
2. If α is a cycle of length r, then α^r is the identity permutation.
3. Show that the order of any $\sigma \in S_n$ is the least common multiple of the lengths of its disjoint cycles.
4. Any two cycles in S_n are conjugate if and only if they are of the same length.

2 Alternating group A_n

We showed in Section 1 that every permutation in S_n can be expressed as a product of transpositions, and the number of transpositions in any given product is not necessarily unique. However, we have the following important result showing that the number of factors is always even or always odd.

2.1 Theorem. *If a permutation $\sigma \in S_n$ is a product of r transpositions and also a product of s transpositions, then r and s are either both even or both odd.*

Proof. Let $\sigma = \eta_1 \cdots \eta_r = \eta'_1 \cdots \eta'_s$, where η_i and η'_i are transpositions.

Let $P = \Pi_{i<j}(x_i - x_j)$ be a polynomial in the variables $x_1,\ldots,x_n$. Define

$$\eta(P) = \prod_{i<j}(x_{\eta(i)} - x_{\eta(j)}),$$

where $\eta \in S_n$. We show that if η is a transposition, then $\eta(P) = -P$. Let $\eta = (k\ l)$, $k < l$. Now one of the factors in the polynomial P is $x_k - x_l$, and in $\eta(P)$ the corresponding factor becomes $x_l - x_k$.

Any factor of P of the form $x_i - x_j$, where neither i nor j is equal to k or l, is unaltered under the mapping η. All other factors can be paired to form products of the form $\pm(x_i - x_k)(x_i - x_l)$, with the sign determined by the relative magnitudes of i, k, and l. But since the effect of η is just to interchange x_k and x_l, any such product of pairs is unaltered. Therefore, the above argument gives that the only effect of η is to change the sign of P. This proves our assertion.

Finally,

$$\sigma(P) = (\eta_1 \cdots \eta_r)P = (-1)^r P.$$

Also

$$\sigma(P) = (\eta'_1 \cdots \eta'_s)P = (-1)^s P.$$

Thus $(-1)^r = (-1)^s$, proving that r and s are both even or both odd. □

Definition. *A permutation in S_n is called an* even (odd) permutation *if it is a product of an even (odd) number of transpositions.*

The sign of a permutation σ, written $\text{sgn}(\sigma)$ or $\epsilon(\sigma)$, is defined to be $+1$ or -1, according to whether σ is even or odd. More generally, we define the sign of any mapping from **n** to **n** as follows.

Definition. *Let $\phi: \mathbf{n} \to \mathbf{n}$. Then*

$$\epsilon(\phi) = \begin{cases} +1 & \text{if } \phi \text{ is an even permutation,} \\ -1 & \text{if } \phi \text{ is an odd permutation,} \\ 0 & \text{if } \phi \text{ is not a permutation.} \end{cases}$$

2.2 **Lemma.** *Let ϕ, ψ be mappings from* **n** *to* **n**. *Then*

$$\epsilon(\phi\psi) = \epsilon(\phi)\epsilon(\psi).$$

Hence for any $\sigma \in S_n$, $\epsilon(\sigma^{-1}) = \epsilon(\sigma)$.

Proof. If ϕ and ψ are both permutations, the result follows from Theorem 2.1. If ϕ or ψ is not a permutation, then $\phi\psi$ is not a permutation. Hence, $\epsilon(\phi\psi) = 0 = \epsilon(\phi)\epsilon(\psi)$. If σ is a permutation, $\epsilon(\sigma)\epsilon(\sigma^{-1}) = \epsilon(\sigma\sigma^{-1}) = 1$. Hence, $\epsilon(\sigma^{-1}) = \epsilon(\sigma)$. □

Let A_n denote the set of all even permutations in S_n.

2.3 **Theorem.** *A_n is a normal subgroup of S_n. If $n > 1$, A_n is of index 2 in S_n, and, hence, $|A_n| = \frac{1}{2}n!$.*

Proof. If $n = 1$, $S_1 = \{e\} = A_1$. Hence, A_1 is trivially a normal subgroup. Let $n > 1$. Let $G = \{1, -1\}$ be the multiplicative group of integers $1, -1$.

Consider the mapping

$$\phi: S_n \to G \qquad \text{given by} \qquad \sigma \mapsto \epsilon(\sigma).$$

By Lemma 2.2, ϕ is a homomorphism. Because $n > 1$, the transposition $\tau = (1\ 2)$ is in S_n and $\phi(\tau) = -1$. Hence, ϕ is surjective. Therefore, by the fundamental theorem of homomorphisms, $S_n/\text{Ker}\ \phi \simeq G$. But Ker ϕ is precisely the set A_n of all even permutations. Hence, A_n is a normal subgroup of S_n, and $[S_n:A_n] = |G| = 2$. Consequently, $|A_n| = \frac{1}{2}|S_n| = \frac{1}{2}n!$. □

Definition. *The subgroup A_n of all even permutations in S_n is called the* **alternating group of degree** n.

2.4 Examples

(a) In S_3 the identity and the cycles $(1\ 2\ 3)$, $(1\ 3\ 2)$ are the only even permutations. Hence,

$$A_3 = \{(1),(1\ 2\ 3),(1\ 3\ 2)\}.$$

(b) A_n is nonabelian if $n > 3$. For, if a, b, c, d are distinct, then $(a\ b\ c)(a\ b\ d) = (a\ c)(b\ d)$, whereas $(a\ b\ d)(a\ b\ c) = (a\ d)(b\ c)$. Hence, $(a\ b\ c)(a\ b\ d) \neq (a\ b\ d)(a\ b\ c)$.

(c) The only possible even permutations in S_4 are 3-cycles and products of two transpositions, in addition to the identity. Hence, A_4 consists of the following elements:

$$\begin{aligned} &e = (1), \quad a = (1\ \ 2)(3\ \ 4), \quad b = (1\ \ 3)(2\ \ 4), \quad c = (1\ \ 4)(2\ \ 3), \\ &\alpha = (1\ \ 2\ \ 3), \quad \beta = (1\ \ 3\ \ 4), \quad \gamma = (1\ \ 4\ \ 2), \quad \delta = (2\ \ 4\ \ 3), \\ &\alpha^2 = (1\ \ 3\ \ 2), \quad \beta^2 = (1\ \ 4\ \ 3), \quad \gamma^2 = (1\ \ 2\ \ 4), \quad \delta^2 = (2\ \ 3\ \ 4). \end{aligned}$$

It is easily verified that a, b, and c are conjugate to one another. For example, $\alpha a \alpha^{-1} = c$. Similarly, α, β, γ, and δ are conjugate elements, and so are $\alpha^2, \beta^2, \gamma^2$, and δ^2. Thus, the conjugate classes in A_4 are $\{e\}$, $\{a,b,c\}$, $\{\alpha,\beta,\gamma,\delta\}$, and $\{\alpha^2,\beta^2,\gamma^2,\delta^2\}$.

Note that two elements of A_n may be conjugate in S_n but not in A_n. For example, α and α^2, being of the same length, are conjugate in S_4 but not in A_4.

(d) A_4 has only one proper normal subgroup, which is the Klein four-group. Consequently, A_4 has no subgroup of order 6. (This example shows that the converse of Lagrange's theorem does not hold.)

With the same notation as before, we easily verify that $a^2 = b^2 = c^2 = e$, $ab = c$, $bc = a$, $ca = b$. Hence, $H = \{e,a,b,c\}$ is a subgroup of A_4. Moreover, since a, b, and c constitute a conjugate class, H is normal. It is clear from the relations satisfied by a, b, and c that H is the Klein four-group.

Let K be any normal subgroup of A_4. If $x \in K$, then $x^2 \in K$, and because K is normal, every element conjugate to x or to x^2 is also in K. Hence, if K contains any of the elements $\alpha,\beta,\delta,\gamma,\alpha^2,\beta^2,\gamma^2,\delta^2$, then K must contain all of them. But then $|K| > 8$. Hence, by Lagrange's theorem, $K = A_4$. This proves that $H = \{e,a,b,c\}$ is the only normal subgroup of A_4.

Finally, any subgroup of order 6 of A_4, being of index 2, would be a normal subgroup. Hence, A_4 has no subgroup of order 6.

3 Simplicity of A_n

3.1 Lemma. *The alternating group A_n is generated by the set of all 3-cycles in S_n.*

Proof. The result holds trivially for $n = 1,2$, for then $A_n = \{e\}$ and S_n does not have any 3-cycle. So let $n > 2$. Now every 3-cycle is an even permutation and, hence, an element of A_n. Conversely, every element in A_n is a product of an even number of transpositions. Consider the product of any two transpositions σ and τ. If σ and τ are disjoint, say $\sigma = (a\ b)$, $\tau = (c\ d)$, then $\sigma\tau = (a\ b\ c)(b\ c\ d)$. Otherwise, $\sigma = (a\ b)$, $\tau = (b\ c)$ gives $\sigma\tau = (a\ b\ c)$. Thus, every even permutation is a product of 3-cycles. Hence, A_n is generated by the set of all 3-cycles in S_n. □

3.2 Lemma. *The derived group of S_n is A_n.*

Proof. For $n = 1,2$, $S'_n = \{e\} = A_n$. Consider $n > 2$ and let $\alpha = (1\ 2)$ and $\beta = (1\ 2\ 3)$. Then

$$\alpha\beta\alpha^{-1}\beta^{-1} = (1\quad 2)(1\quad 2\quad 3)(2\quad 1)(3\quad 2\quad 1) = (1\quad 2\quad 3).$$

Hence, $(1\ 2\ 3) \in S'_n$. Because $S'_n \lhd S_n$, S'_n must contain every conjugate of $(1\ 2\ 3)$. Hence, S'_n contains every 3-cycle; therefore, $A_n \subset S'_n$. On the other hand, every commutator $aba^{-1}b^{-1}$ in S'_n is an even permutation. Hence, $S'_n \subset A_n$. This proves that $S'_n = A_n$. □

3.3 Theorem. *The alternating group A_n is simple if $n > 4$. Consequently, S_n is not solvable if $n > 4$.*

Proof. Suppose H is a nontrivial normal subgroup of A_n. We first prove that H must contain a 3-cycle. Let $\sigma \neq e$ be a permutation in H that moves the least number of integers in **n**. Being an even permutation, σ cannot be a cycle of even length. Hence, σ must be a 3-cycle or have a decomposition of the form

$$\sigma = (a \quad b \quad c \quad \cdots)\cdots \tag{1}$$

or

$$\sigma = (a \quad b)(c \quad d)\cdots, \tag{2}$$

where a,b,c,d are distinct. Consider first case (1). Because σ cannot be a 4-cycle, it must move at least two more elements, say d and e. Let $\alpha = (c\ d\ e)$. Then

$$\begin{aligned}\alpha\sigma\alpha^{-1} &= (c \quad d \quad e)(a \quad b \quad c \quad \cdots)\cdots(e \quad d \quad c)\\ &= (a \quad b \quad d \quad \cdots)\cdots.\end{aligned}$$

Now let $\tau = \sigma^{-1}(\alpha\sigma\alpha^{-1})$. Then $\tau(a) = a$, and $\tau(x) = x$ whenever $\sigma(x) = x$. Thus, τ moves fewer elements than σ. But $\tau \in H$, a contradiction.

Consider now case (2). With $\alpha = (c\ d\ e)$ as before,

$$\begin{aligned}\alpha\sigma\alpha^{-1} &= (c \quad d \quad e)(a \quad b)(c \quad d)\cdots(e \quad d \quad c)\\ &= (a \quad b)(d \quad e)\cdots.\end{aligned}$$

Let $\beta = \sigma^{-1}(\alpha\sigma\alpha^{-1})$. Then $\beta(a) = a$, $\beta(b) = b$, and for every x other than e, $\beta(x) = x$ if $\sigma(x) = x$. Thus, $\beta \in H$ and moves fewer integers than σ, a contradiction. Hence, we conclude that σ must be a 3-cycle.

Let τ be any 3-cycle in S_n. Because any two cycles of the same length are conjugate in S_n, $\tau = \alpha\sigma\alpha^{-1}$ for some $\alpha \in S_n$. If α is odd, choose a transposition β in S_n such that σ and β are disjoint. (This is possible because $n > 4$.) Then $\alpha\beta \in A_n$, and

$$(\alpha\beta)(\sigma)(\alpha\beta)^{-1} = \alpha(\beta\sigma\beta^{-1})\alpha^{-1} = \alpha\sigma\alpha^{-1} = \tau.$$

Hence, σ and τ are conjugate in A_n. Therefore, $\tau \in H$. Thus, H contains every 3-cycle in S_n. Therefore, $H = A_n$. Hence, A_n has no proper normal subgroup, which proves that A_n is simple if $n > 4$.

Consider now the derived group of A_n. Because A_n' is a normal subgroup of A_n, and A_n is simple, either A_n' is trivial or $A_n' = A_n$. But A_n is nonabelian if $n > 3$. Hence, A_n' is not trivial. Therefore, $A_n' = A_n$ for every $n > 4$. Consequently, $S_n^{(2)} = A_n' = A_n$. Hence, $S_n^{(k)} = A_n$ for all $k \geq 1$. This proves that S_n is not solvable. □

3.4 Example

A_n, $n > 4$, is the only nontrivial normal subgroup of S_n.

Solution. Let $\{e\} \neq H \triangleleft S_n$. Suppose $H \cap A_n \neq \{e\}$. Then $H \cap A_n \triangleleft A_n$ gives $H \cap A_n = A_n$ because A_n is simple. This implies $A_n \subset H$. But then $|S_n/A_n| = 2$ implies $|H/A_n| \leq 2$, so $H = A_n$ or S_n.

If $H \cap A_n = \{e\}$, then $|H\,A_n| = \frac{1}{2}|H|n! \leq n!$. Thus, $|H| \leq 2$ or $|H| = 2$ because $H \neq \{e\}$. If $H = \{e,\sigma\}$, then $\eta\sigma\eta^{-1} = \sigma\ \forall \eta \in S_n$. This implies $\sigma \in Z(S_n) = \{e\}$, a contradiction. Hence, the only nontrivial normal subgroup of S_n is A_n.

Problems

1. Prove that S_n is generated by the cycles $(1\ 2\ \dots\ n-1)$ and $(n-1\ n)$.
2. Let $H < S_n$, $H \not\subset A_n$. Show that exactly half the permutations in H are even.
3. Show that any group of order $4n+2$ has a subgroup of index 2.
4. Let S_5 act on itself by conjugation. Let K and H be, respectively, the orbit and stabilizer of $(1\quad 2) \in S_5$. Show $|K| = 10$ and $|H| = 12$.
5. Show that there are 420 elements of S_7 having disjoint cyclic decomposition of the type $(a\quad b)(c\quad d\quad e)$. Find also the number of elements in the orbit of $(a\quad b)(c\quad d\quad e)$.
6. Show that there are 1120 elements of S_8 having disjoint cyclic decomposition of the type $(a\quad b\quad c)(d\quad e\quad f)(g\quad h)$.
7. Consider a cube with vertices numbered 1 through 8. Let H be the subgroup of the group G of rotations and reflections keeping the vertex 1 fixed. Show $H \simeq S_3$ and $|G| = 48$.
8. Consider a regular tetrahedron with vertices numbered 1 through 4. Let H be the subgroup of the group G of rotations and reflections keeping one vertex fixed. Show $H \simeq S_3$ and $|G| = 24$.
9. Prove that every finite group is isomorphic to a subgroup of the alternating group A_n for some $n > 1$.

CHAPTER 8

Structure theorems of groups

1 Direct products

If a group is isomorphic to the direct product of a family of its subgroups (known as summands) whose structures are known to us, then the structure of the group can generally be determined; these summands are like the "building blocks" of a structure. Theorem 1.1 answers the question, When is a group isomorphic to the direct product of a given finite family of its subgroups?

1.1 Theorem. *Let $H_1,\dots,H_n$ be a family of subgroups of a group G, and let $H = H_1 \cdots H_n$. Then the following are equivalent:*

(i) $H_1 \times \cdots \times H_n \simeq H$ under the canonical mapping that sends $(x_1,\dots,x_n)$ to $x_1 \cdots x_n$.

(ii) $H_i \triangleleft H$, and every element $x \in H$ can be uniquely expressed as $x = x_1 \cdots x_n$, $x_i \in H_i$.

(iii) $H_i \triangleleft H$, and if $x_1 \cdots x_n = e$, then each $x_i = e$.

(iv) $H_i \triangleleft H$, and $H_i \cap (H_1 \cdots H_{i-1}H_{i+1} \cdots H_n) = \{e\}$, $1 \le i \le n$.

Proof. (i) $\Rightarrow$ (ii) Let $H_i' = \{(e,\dots,h_i,\dots,e) \mid h_i \in H_i\}$. Trivially, $H_i' \triangleleft H_1 \times \cdots \times H_n$. Also $H_i' \simeq H_i$ under the given mapping. Thus, $H_i \triangleleft H_1 \cdots H_n$.

Further, let $x = x_1 \cdots x_n = x_1' \cdots x_n', x_i, x_i' \in H_i$. Then $(x_1,\dots,x_n)$ and $(x_1',\dots,x_n')$ have the same image under the given isomorphism. Hence, $x_i = x_i'$.

(ii) $\Rightarrow$ (iii) $x_1 \cdots x_n = e$. Also $e \cdots e = e$. By uniqueness of representation, $x_i = e$, $1 \le i \le n$.

(iii) $\Rightarrow$ (iv) The hypothesis implies $H_i \cap H_j = \{e\}$, $i \neq j$. For let $x_i = x_j, x_i \in H_i, x_j \in H_j$. Then $e = x_i x_j^{-1}$ implies $x_i = e = x_j^{-1}$. But then, since $H_i, H_j \lhd H$, it follows that $xy = yx$ for all $x \in H_i$, $y \in H_j$.

To prove (iv), let

$$x_i = x_1 \cdots x_{i-1} x_{i+1} \cdots x_n,$$

where $x_i \in H_i$, $1 \leq i \leq n$. Then since all x_i's commute, $e = x_1 \cdots x_{i-1} x_i^{-1} x_{i+1} \cdots x_n$, so by (iii) each $x_i = e$. This proves (iv).

(iv) $\Rightarrow$ (i) First observe that for all $x \in H_i, y \in H_j, i \neq j, xy = yx$. Then it follows immediately that the mapping $\sigma: H_1 \times \cdots \times H_n \to H$, given by $\sigma(x_1,\ldots,x_n) = x_1 \cdots x_n$, is a homomorphism. The mapping is injective; for let $x_1 \cdots x_n = e$. Then $x_1^{-1} = x_2 \cdots x_n$. This implies by hypothesis, that $x_1^{-1} = e$, so $x_1 = e$. The mapping is clearly surjective, proving (i). □

Under any one of the equivalent statements in Theorem 1.1, we say that H is an *internal direct product* of $H_1,\ldots,H_n$. The emphasis on the word internal is more psychological than mathematical because the "external direct product" $H_1 \times \cdots \times H_n$ is isomorphic to $H_1 \cdots H_n$ in our situation. Thus, we shall generally omit the word "internal" or "external" whenever the subgroups $H_1,\ldots,H_n$ satisfy any one of the conditions in Theorem 1.1. Note, however, that the external direct product $H_1 \times \cdots \times H_n$ always exists, whereas the internal direct product exists if and only if the canonical map $H_1 \times \cdots \times H_n \to H_1 \cdots H_n$ is an isomorphism.

If the group is additive, the (internal) direct product of subgroups $H_1,\ldots,H_n$ of G is also written as $H_1 \oplus \cdots \oplus H_n$ and called the *direct sum* of $H_1,\ldots,H_n$.

1.2 Examples

(a) If each element $\neq e$ of a finite group G is of order 2, then $|G| = 2^n$ and $G \simeq C_1 \times C_2 \times \cdots \times C_n$, where C_i are cyclic and $|C_i| = 2$.

Solution. Now $x^2 = e \Rightarrow x = x^{-1}$ for all $x \in G$. So if $a,b \in G$, then $ab = (ab)^{-1} = b^{-1}a^{-1} = ba$. Let $e \neq a_1 \in G$. Then either $G = [a_1]$, the cyclic group generated by a_1, or $G \supsetneq [a_1]$. In the second situation, $\exists\ a_2 \in G$ such that $a_2 \notin [a_1]$. Then the product $[a_1][a_2]$ is direct. If $G = [a_1][a_2]$, we are through; otherwise $G \supsetneq [a_1][a_2] \supsetneq [a_1]$. Continuing thus, we ultimately get $G = [a_1][a_2] \cdots [a_n]$, the direct product of cyclic groups each of order 2.

(b) A group G of order 4 is either cyclic of $G \simeq C_2 \times C_2'$, a direct product of two cyclic groups each of order 2.

Solution. If G is not cyclic, then by Lagrange's theorem G must contain an element a of order 2 and also another element b of order 2. Obviously, $G \simeq [a] \times [b]$.

(c) If G is a group of order pq, where p and q are distinct primes, and if G has a normal subgroup H of order p and a normal subgroup K of order q, then G is cyclic.

Solution. By Lagrange's theorem $H \cap K = \{e\}$. Since H and K are normal and $H \cap K = \{e\}$, $hk = kh$ for all $h \in H$ and $k \in K$. Also then $HK \simeq H \times K$. If $H = [h]$ and $K = [k]$, then $o(hk) = o(h) \cdot o(k) = pq$. Hence G is generated by hk.

(d) If G is a cyclic group of order mn, where $(m,n) = 1$, then $G \simeq H \times K$, where H is a subgroup of order m, and K is a subgroup of order n.

Solution. Because G is cyclic, there exist a unique subgroup H of order m and a unique subgroup K of order n. By Lagrange's theorem $H \cap K = \{e\}$. Then by Example 3.11(e) in Chapter 4, $|HK| = |H||K| = mn$. Hence, $G = HK \simeq H \times K$.

(e) If G is a finite cyclic group of order $n = p_1^{e_1} \cdots p_k^{e_k}$, p_i distinct primes, then G is a direct product of cyclic groups of orders $p_i^{e_i}$, $1 \leq i \leq k$.

Solution. Proceed as in (d).

Problems

1. Show that the group $\mathbf{Z}/(10)$ is a direct sum of $H = \{\bar{0}, \bar{5}\}$ and $K = \{\bar{0}, \bar{2}, \bar{4}, \bar{6}, \bar{8}\}$.
2. Show that the group $(\mathbf{Z}/(4), +)$ cannot be written as the direct sum of two subgroups of order 2.
3. Show that the group $(\mathbf{Z}/(8), +)$ cannot be written as the direct sum of two nontrivial subgroups.
4. Let $N \triangleleft G = H \times K$. Prove that either N is abelian or N intersects one of the subgroups $H \times \{e\}$, $\{e\} \times K$ nontrivially.

2 Finitely generated abelian groups

An important question in the study of an algebraic structure is to know if it can be decomposed into a direct sum of certain "well-behaved" direct summands. In this section we show that any finitely generated abelian group can be decomposed as a finite direct sum of cyclic groups. This decomposition, when applied to finite abelian groups, enables us to find the number of nonisomorphic abelian groups of a given order (Section 3). We may remark that not every finitely generated group is finite (consider, for example, $(\mathbf{Z},+)$).

2.1 Theorem (fundamental theorem of finitely generated abelian groups). *Let A be a finitely generated abelian group. Then A can be decomposed as a direct sum of a finite number of cyclic groups C_i. Precisely,*

$$A = C_1 \oplus \cdots \oplus C_k,$$

such that either $C_1,\dots,C_k$ are all infinite, or, for some $j \leq k$, $C_1,\dots,C_j$ are of order $m_1,\dots,m_j$, respectively, with $m_1 | m_2 | \cdots | m_j$, and $C_{j+1},\dots,C_k$ are infinite.

Proof. Let k be the smallest number such that A is generated by a set of k elements. We prove the theorem by induction on k. If $k = 1$, A is a cyclic group; hence, the theorem is trivially true. Let $k > 1$, and assume that the theorem holds for every group generated by a set of $k - 1$ elements.

First consider the case when A has a generating set $\{a_1,\dots,a_k\}$ with the property that, for all integers $x_1,\dots,x_k$, the equation

$$x_1a_1 + \cdots + x_ka_k = 0$$

implies

$$x_1 = \cdots = x_k = 0.$$

This implies that every $a \in A$ has a unique representation of the form

$$a = x_1a_1 + \cdots + x_ka_k, \qquad x_i \in \mathbf{Z}.$$

For if

$$a = x_1a_1 + \cdots + x_ka_k = x_1'a_1 + \cdots + x_k'a_k,$$

then

$$(x_1 - x_1')a_1 + \cdots + (x_k - x_k')a_k = 0.$$

Hence, $x_i = x_i'$, $i = 1,\dots,k$. Therefore, by Theorem 1.1, $A = C_1 \oplus \cdots \oplus C_k$, where $C_i = [a_i]$ is the cyclic subgroup generated by a_i, $i = 1,\dots,k$. Moreover, since $x_i a_i = 0$ implies $x_i = 0$, C_i is an infinite cyclic group. Hence, in this case, A is the direct sum of a finite number of infinite cyclic subgroups.

Now suppose that A has no generating set of k elements with the property stated in the last paragraph. Then, given any generating set $\{a_1,\dots,a_k\}$ of A, there exist integers $x_1,\dots,x_k$ (not all zero) such that $x_1a_1 + \cdots + x_ka_k = 0$. Since $\Sigma x_i a_i = 0 \Rightarrow \Sigma(-x_i)a_i = 0$, we can assume that $x_i > 0$ for some i. Consider now all possible generating sets of A with k elements, and let X denote the set of all k-tuples $(x_1,\dots,x_k)$ of integers such that

$$x_1a_1 + \cdots + x_ka_k = 0, \qquad x_i > 0 \text{ for some } i,$$

for some generating set $\{a_1,\dots,a_k\}$ of A. Let m_1 be the least positive integer that occurs as a component in any k-tuple in X. Without loss of generality, we may take m_1 to be the first component, so that, for some generating set $\{a_1,\dots,a_k\}$,

$$m_1a_1 + x_2a_2 + \cdots + x_ka_k = 0. \tag{1}$$

By the division algorithm we can write

$$x_i = q_i m_1 + r_i, \qquad \text{where } 0 \le r_i < m_1,$$

for each $i = 2,\dots,k$. Then (1) becomes

$$m_1b_1 + r_2a_2 + \cdots + r_ka_k = 0, \tag{2}$$

where $b_1 = a_1 + q_2a_2 + \cdots + q_ka_k$. Now $b_1 \neq 0$, for otherwise $a_1 = -q_2a_2 - \cdots - q_ka_k$, which implies that A is generated by a set of $k-1$ elements, a contradiction. Moreover, $a_1 = b_1 - q_2a_2 - \cdots - q_ka_k$. Hence, $\{b_1, a_2,\dots,a_k\}$ is a generating set of A. Therefore, by the minimal property of m_1, it follows from (2) that $r_2 = \cdots = r_k = 0$. Hence, $m_1b_1 = 0$. Let $C_1 = [b_1]$. Because m_1 is the least positive integer such that $m_1b_1 = m_1b_1 + 0a_2 + \cdots + 0a_k = 0$, C_1 is a cyclic subgroup of order m_1.

Let A_1 be the subgroup generated by $\{a_2,\dots,a_k\}$. We claim that $A = C_1 \oplus A_1$. So suppose $x_1b_1 \in A_1$ for some x_1, $0 \le x_1 < m_1$. Then $x_1b_1 = x_2a_2 + \cdots + x_ka_k$, $x_2,\dots,x_k \in \mathbf{Z}$. Hence, $x_1b_1 - x_2a_2 - \cdots - x_ka_k = 0$, which implies $x_1 = 0$ by the minimal property of m_1. Therefore, $C_1 \cap A_1 = \{0\}$, which proves our claim that $A = C_1 \oplus A_1$.

Now A_1 is generated by a set of $k-1$ elements, namely, $\{a_2,\dots,a_k\}$. Moreover, A_1 cannot be generated by a set with less than $k-1$ elements, for otherwise A would be generated by a set with less than k elements, a contradiction. Hence, by the induction hypothesis,

$$A_1 = C_2 \oplus \cdots \oplus C_k,$$

where $C_2,\ldots,C_k$ are cyclic subgroups that are either all infinite or, for some $j \le k$, $C_2,\ldots,C_j$ are finite cyclic groups of orders $m_2,\ldots,m_j$, respectively, with $m_2|m_3|\cdots|m_j$, and C_i are infinite for $i>j$.

Let $C_i = [b_i]$, $i = 2,\ldots,k$, and suppose that C_2 is of order m_2. Then $\{b_1,\ldots,b_k\}$ is a generating set of A, and

$$m_1b_1 + m_2b_2 + 0b_3 + \cdots + 0b_k = 0.$$

By repeating the argument given for equation (1), we conclude that $m_1|m_2$. This completes the proof of the theorem. □

Evidently, if A is a finite abelian group, then $C_1,\ldots,C_k$ are all finite. We prove in the next section that for a given A the positive integers $m_1,\ldots,m_k$ are determined uniquely.

3 Invariants of a finite abelian group

Throughout this section A denotes a finite abelian group written additively.

3.1 Theorem. *Let A be a finite abelian group. Then there exists a unique list of integers $m_1,\ldots,m_k$ (all >1) such that*

$$|A| = m_1 \cdots m_k, \qquad m_1|m_2| \cdots |m_k,$$

and $A = C_1 \oplus \cdots \oplus C_k$, where $C_1,\ldots,C_k$ are cyclic subgroups of A of order $m_1,\ldots,m_k$, respectively. Consequently,

$$A \simeq \mathbf{Z}_{m_1} \oplus \cdots \oplus \mathbf{Z}_{m_k}.$$

Proof. By Theorem 2.1, $A = C_1 \oplus \cdots \oplus C_k$, where $C_1,\ldots,C_k$ are cyclic subgroups of orders $m_1,\ldots,m_k$, respectively, such that $m_1|m_2| \cdots |m_k$. Further, since $|S \times T| = |S||T|$ for any finite sets S and T it follows that

$$|A| = |C_1| \cdots |C_k| = m_1 \cdots m_k.$$

Moreover, any cyclic group of order m is isomorphic to $\mathbf{Z}_m$. Hence,

$$A \simeq \mathbf{Z}_{m_1} \oplus \cdots \oplus \mathbf{Z}_{m_k}.$$

To prove the uniqueness of the list $m_1,\ldots,m_k$, suppose

$$A = C_1 \oplus \cdots \oplus C_k = D_1 \oplus \cdots \oplus D_l,$$

where C_i, D_j are all cyclic subgroups of A, with

$$|C_i| = m_i, \qquad m_1 | m_2 | \cdots | m_k,$$
$$|D_j| = n_j, \qquad n_1 | n_2 | \cdots | n_l.$$

Now D_l has an element of order n_l. But every element in A is of order $\leq m_k$. Hence, $n_l \leq m_k$. By the same argument $m_k \leq n_l$. Hence, $m_k = n_l$.

Now consider $m_{k-1}A = \{m_{k-1}a | a \in A\}$. From the two decompositions of A assumed above, we get

$$m_{k-1}A = (m_{k-1}C_1) \oplus \cdots \oplus (m_{k-1}C_{k-1}) \oplus (m_{k-1}C_k)$$
$$= (m_{k-1}D_1) \oplus \cdots \oplus (m_{k-1}D_{l-1}) \oplus (m_{k-1}D_l).$$

Because $m_i | m_{k-1}$ for $i = 1,...,k-1$, it follows that $m_{k-1}C_i$ is trivial for $i = 1,...,k-1$. Hence, $|m_{k-1}A| = |m_{k-1}C_k| = |m_{k-1}D_l|$. Therefore, $|m_{k-1}D_j| = 1$ for $j = 1,...,l-1$. Hence, $n_{l-1} | m_{k-1}$. By a symmetric argument, $m_{k-1} | n_{l-1}$. Therefore, $m_{k-1} = n_{l-1}$. Proceeding in this manner, we can show that $m_{k-r} = n_{l-r}$, $r = 0,1,2,...$. But $m_1 \cdots m_k = |A| = n_1 \cdots n_l$. Hence, $k = l$ and $m_i = n_i$, $i = 1,...,k$. □

Definition. *Let A be a finite abelian group. If*

$$A \simeq \mathbf{Z}_{m_1} \oplus \cdots \oplus \mathbf{Z}_{m_k}, \quad \textit{where } 1 < m_1 | m_2 | \cdots | m_k,$$

then A is said to be of type $(m_1,...,m_k)$, *and the integers $m_1,...,m_k$ are called the* invariants *of A.*

The fundamental theorem of finite abelian groups enables us to determine the number of isomorphism classes of abelian groups of order n. A partition of a positive integer k is an r-tuple $(k_1,...,k_r)$ of positive integers such that $k = \Sigma k_i$ and each $k_i \leq k_{i+1}$.

3.2 Lemma. *There is a* 1-1 *correspondence between the family F of nonisomorphic abelian groups of order p^e, p prime, and the set $P(e)$ of partitions of e.*

Proof. Let $A \in F$. Then A determines a unique type $(p^{e_1},...,p^{e_k})$, where $e_1 \leq e_2 \leq \cdots \leq e_k$ and $e_1 + e_2 + \cdots + e_k = e$. Define a mapping σ: $F \to P(e)$ by $\sigma(A) = (e_1,...,e_k)$. σ is clearly injective (Theorem 3.1). To show that σ is surjective, let $(e_1,...,e_s) \in P(e)$. Then the group $\mathbf{Z}_{p^{e_1}} \oplus \cdots \oplus \mathbf{Z}_{p^{e_s}}$ is obviously the preimage of $(e_1,...,e_s)$. This completes the proof. □

The following lemma is an immediate consequence of Sylow's theorems discussed in the next section. However, we give a proof based on the fundamental theorem of finitely generated abelian groups.

3.3 Lemma. *Let A be a finite abelian group of order $p_1^{e_1} \cdots p_k^{e_k}$, p_i distinct primes, $e_i > 0$. Then*

$$A = S(p_1) \oplus \cdots \oplus S(p_k), \quad \text{where } |S(p_i)| = p_i^{e_i}.$$

This decomposition of A is unique; that is, if

$$A = H_1 \oplus \cdots \oplus H_k, \quad \text{where } |H_i| = p_i^{e_i},$$

then $H_i = S(p_i)$.

Proof. We have $|A| = p_1^{e_1} \cdots p_k^{e_k}$, $e_i \geq 0$. By Theorem 3.1, $A = A_1 \oplus \cdots \oplus A_n$, where A_i are cyclic groups of orders, say, $p_1^{e_1} \cdots p_k^{e_{ki}}$, $e_{ji} \geq 0$. Because A_1 is cyclic, A_i contains unique subgroups $A_{1i}, \ldots, A_{ki}$ of orders $p_1^{e_{1i}}, \ldots, p_k^{e_{ki}}$, respectively. By Lagrange's theorem

$$A_{ji} \cap A_{1i} \cdots A_{j-1,i} A_{j+1,i} \cdots A_{ki} = \{e\}$$

for all $j = 1, \ldots, k$. Therefore, by Theorem 1.1,

$$A_i = A_{1i} \oplus \cdots \oplus A_{ki}.$$

So

$$A = [A_{11} \oplus \cdots \oplus A_{k1}] \oplus \cdots \oplus [A_{1n} \oplus \cdots \oplus A_{kn}].$$

Put

$$S(p_1) = [A_{11} \oplus \cdots \oplus A_{1n}], \ldots, S(p_k) = [A_{k1} \oplus \cdots \oplus A_{kn}].$$

Then $|S(p_1)| = p_1^{s_1}, \ldots, |S(p_k)| = p_k^{s_k}$, and $A = S(p_1) \oplus \cdots \oplus S(p_k)$. Then by comparing orders on both sides, it follows that $|S(p_1)| = p_1^{e_1}, \ldots, |S(p_k)| = p_k^{e_k}$. The last statement can be proved easily by observing that each of the subgroups $S(p_i)$ and H_i is the subgroup

$$\{x \in A \mid o(x) \text{ is a power of } p_i\}. \qquad \square$$

3.4 Theorem. *Let $n = \prod_{j=1}^{k} p_j^{f_j}$, p_j distinct primes. Then the number of nonisomorphic abelian groups of order n is $\prod_{j=1}^{k} |P(f_j)|$.*

Proof. Let Ab_n be the family of nonisomorphic abelian groups of order n. Let $A \in \text{Ab}_n$. By Lemma 3.3, $A = S(p_1) \oplus \cdots \oplus S(p_k)$. By Lemma 3.2

the number of nonisomorphic abelian groups $S(p_i)$ is $|P(f_i)|$. Hence,

$$|\mathrm{Ab}_n| = \prod_{i=1}^{k} |P(f_i)|. \qquad \square$$

3.5 Example

Find the nonisomorphic abelian groups of order $360 = 2^3 \cdot 3^2 \cdot 5^1$.

Solution. $n = 360 = 2^3 \cdot 3^2 \cdot 5^1$. $|P(3)| = 3, |P(2)| = 2, |P(1)| = 1$. Hence, there are six nonisomorphic abelian groups of order 360, and these are

$$\begin{aligned}
&\mathbf{Z}_8 \oplus \mathbf{Z}_9 \oplus \mathbf{Z}_5,\\
&\mathbf{Z}_8 \oplus \mathbf{Z}_3 \oplus \mathbf{Z}_3 \oplus \mathbf{Z}_5,\\
&\mathbf{Z}_2 \oplus \mathbf{Z}_4 \oplus \mathbf{Z}_9 \oplus \mathbf{Z}_5,\\
&\mathbf{Z}_2 \oplus \mathbf{Z}_4 \oplus \mathbf{Z}_3 \oplus \mathbf{Z}_3 \oplus \mathbf{Z}_5,\\
&\mathbf{Z}_2 \oplus \mathbf{Z}_2 \oplus \mathbf{Z}_2 \oplus \mathbf{Z}_9 \oplus \mathbf{Z}_5,\\
&\mathbf{Z}_2 \oplus \mathbf{Z}_2 \oplus \mathbf{Z}_2 \oplus \mathbf{Z}_3 \oplus \mathbf{Z}_3 \oplus \mathbf{Z}_5.
\end{aligned}$$

4 Sylow theorems

In the last section we obtained the decomposition of a finite abelian group A that provides a complete description of the structure of A. There is no such general result for finite nonabelian groups. The Sylow theorems, which occupy an important place in the theory of finite groups, are a powerful tool in studying the structure of finite nonabelian groups. In particular, the existence or nonexistence of a simple group of a given order can sometimes be determined by application of the Sylow theorems, among other machinery.

Definition. *Let G be a finite group, and let p be a prime. Let $p^m \,|\, |G|$, $p^{m+1} \nmid |G|$, $m > 0$. Then any subgroup of G of order p^m is called a* Sylow p-subgroup *of G.*

Definition. *Let p be prime. A group G is called a* p-group *if the order of every element in G is some power of p. Likewise, a subgroup H of any group G is called a* p-subgroup of G *if the order of every element in H is some power of p.*

4.1 Lemma (Cauchy's theorem for abelian groups). *Let A be a finite abelian group, and let p be a prime. If p divides $|A|$, then A has an element of order p.*

Proof. We prove the theorem by induction on n, the order of A, where $p|n$. Suppose that the result is true for all groups of orders $m, p|m$, and $m < n$. If $|A| = p$ or even if A is a cyclic group, then there exists a cyclic subgroup of A of order p, and therefore there exists an element of order p. Suppose now there exists $b\,(\neq e) \in A$ such that $A \neq [b]$, the cyclic subgroup generated by b. If $p\,|\,|[b]|$, we are done. But if $p \nmid |[b]|$, then p divides the order of $A/[b]$. Hence, by the induction hypothesis there exists $\bar{a} \in A/[b]$, $a \in A$, such that $o(\bar{a}) = p$. Let $k = o(a)$. Then $a^k = e$ implies $\bar{a}^k = \bar{e}$. Therefore, $p|k$, since $o(\bar{a}) = p$. This means p divides the order of $[a]$; hence, some power of the element a is of order p. □

4.2 Theorem (first Sylow theorem). *Let G be a finite group, and let p be a prime. If p^m divides $|G|$, then G has a subgroup of order p^m.*

Proof. We prove the theorem by induction on $n = |G|$. If $n = 1$, the result is obvious. So let us assume that the result holds for all groups of orders less than n.

If the order of the center of G is divisible by p, then it follows from Cauchy's theorem for abelian groups that the center of G contains an element a of order p. The cyclic group C generated by a is clearly normal in G, and the quotient group G/C has order n/p, which is divisible by p^{m-1}. Therefore, by the induction hypothesis, G/C contains a subgroup H/C of order p^{m-1}, proving the theorem in this case.

Now consider the case when the order of the center is not divisible by p. The class equation of G is

$$n = |G| = |Z(G)| + \sum_a [G:N(a)],$$

where summation runs over one element from each conjugate class having more than one element. Now $p|n$, but $p \nmid |Z(G)|$. Hence, $p \nmid [G:N(a)]$ for some $a \in G$, $a \notin Z(G)$. This implies that $p^m\,|\,|N(a)|$ and $|N(a)| < |G|$.

By the induction hypothesis $N(a)$ has a subgroup of order p^m, and this yields the required subgroup of G. □

A special case of the theorem is

4.3 Corollary (Cauchy's theorem). *If the order of a finite group G is divisible by a prime number p, then G has an element of order p.*

Proof. By Sylow's theorem there exists a subgroup of order p. Therefore, there are at least $p - 1$ elements of order p (namely, all but the identity element of the cyclic subgroup of order p). □

4.4 Corollary. *A finite group G is a p-group iff its order is a power of p.*

Proof. If G is of prime power order, then each element has, trivially, order a power of p. So let us assume that each element in G has order a power of p and prove that order of G is a power of p. If possible, let $q \neq p$ be a prime number dividing the order of G. Then by Cauchy's theorem there exists an element of order q, a contradiction. Hence, G is of order a power of p. □

4.5 Theorem (second and third Sylow theorems). *Let G be a finite group, and let p be a prime. Then all Sylow p-subgroups of G are conjugate, and their number* n_p *divides* $o(G)$ *and satisfies* $n_p \equiv 1 \pmod{p}$.

Proof. Let K be a Sylow p-subgroup in G. Let $C(K)$ denote the family of subgroups of G conjugate to K; that is, $C(K) = \{xKx^{-1} | x \in G\}$. Then

$$|C(K)| = |G/N(K)|. \tag{1}$$

Because K is a Sylow p-subgroup, $(|G/N(K)|, p) = 1$; therefore,

$$(|C(K)|, p) = 1.$$

Next let H be any Sylow p-subgroup. We want to show that H is conjugate to K. Regard the set $C(K)$ as an H-set by conjugation. Then for any $L \in C(K)$, let $C_H(L) = \{hLh^{-1} | h \in H\}$, the orbit of L. Obviously, $C_H(L) \subset C(K)$, and

$$C(K) = \bigcup_{L \in C(K)} C_H(L) \qquad \text{(disjoint)}, \tag{2}$$

where the union runs over one element L from each orbit [= conjugate class $C_H(L)$].

By Example 4.13 (b) in Chapter 5,

$$|C_H(L)| = \text{index of } H \cap N(L) \text{ in } H = p^e, \qquad e \geq 0, \tag{3}$$

because H is Sylow p-subgroup. We first show that $p^e = 1$ iff $H = L$. If $H = L$, then, trivially, $p^e = 1$. Conversely,

$$p^e = 1 \Rightarrow H = H \cap N(L) \Rightarrow H \subset N(L) \Rightarrow HL < G \text{ and } L \lhd HL.$$

But then

$$HL/L \simeq H/H \cap L \Rightarrow HL/L \text{ is a } p\text{-group} \Rightarrow HL = L,$$

since L is a Sylow p-subgroup. So our assertion that $p^e = 1$ iff $H = L$ is established.

By (2) and (3) we have

$$|C(K)| = \sum p^e. \tag{4}$$

But then (1) implies that in the summation Σp^e at least one term p^e must be equal to 1, and this in turn implies, by what we established in the last paragraph, that H must be equal to some conjugate L of K, proving the second Sylow theorem.

Because all Sylow p-subgroups of G are conjugate to K, it follows that their number n_p is equal to $|C(K)| = |G|/|N(K)|$. Hence, $n_p \mid |G|$.

It is also clear that there is one and only one term in the sum Σp^e that is equal to 1, so we have from (4) that the number of distinct conjugates of K is

$$n_p = 1 + \sum_{e>0} p^e = 1 + kp \equiv 1 \pmod p,$$

proving the third Sylow theorem. □

4.6 Corollary. *A Sylow p-subgroup of a finite group G is unique if and only if it is normal.*

4.7 Examples

(a) If a group of order p^n contains exactly one subgroup each of orders $p, p^2, \ldots, p^{n-1}$, then it is cyclic.

Solution. Let G be a group of order p^n, and let H be a subgroup of order p^{n-1}. By Sylow's first theorem H contains subgroups of orders $p, p^2, \ldots, p^{n-2}$. Because these subgroups are also subgroups of G, it follows that all the proper subgroups of G are contained in H. Now if $a \in G$, $a \notin H$, then the order of a must be p^n, otherwise a will generate a subgroup K of order $< p^n$. This would mean $a \in H$, because $a \in K \subset H$. Thus, the order of a is p^n, and G is therefore a cyclic group generated by a.

(b) If d is a divisor of the order n of a finite abelian group A, then A contains a subgroup of order d.

Solution. If $n = p_1^{e_1} \cdots p_r^{e_r}$, then $A = S(p_1) \oplus \cdots \oplus S(p_r)$, and

$$|(S(p_i))| = p_i^{e_i}, \quad 1 \leq i \leq r \qquad \text{(Lemma 3.3)}.$$

Let $d = p_1^{f_1} \cdots p_r^{f_r}$. By Sylow's first theorem each $S(p_i)$ contains a subgroup $S'(p_i)$ of order $p_i^{f_i}$. Then $B = S'(p_1) \oplus \cdots \oplus S'(p_r)$ is a subgroup of order d.

Note. Example (b) proves the converse of Lagrange's theorem for abelian groups.

(c) Prove that there are no simple groups of orders 63, 56, and 36.

Solution. The number of Sylow 7-subgroups is $n_7 = 1 + 7k$, and $1 + 7k$ divides $3^2 \cdot 7$. But then $1 + 7k$ divides 3^2. This yields $k = 0$. Therefore, there is a unique Sylow 7-subgroup that must be a normal subgroup.

For groups of order 56, the number of Sylow 7-subgroups $n_7 = 1 + 7k$ must divide $2^3 \cdot 7$. Thus, $(1 + 7k)|2^3$. This yields $n_7 = 1$ or 8. If $n_7 = 1$, we are done. If $n_7 = 8$, then there are $8 \cdot 6 = 48$ elements of order 7. But then the remaining eight elements must form a unique Sylow 2-subgroup. So again we have a normal subgroup, and therefore G cannot be a simple group.

The nonsimplicity of groups of order 36 may be proved similarly. Here is an alternative method, which is sometimes easier. Let H be a Sylow 3-subgroup of G. By Theorem 4.4, Chapter 5, there exists a homomorphism $\varphi: G \to S_4$, since the index of H in G is 4. Note $\text{Ker}\,\varphi = \{e\}$ implies $G \subsetneq S_4$, which is not possible since $|G| = 36$ and $|S_4| = 24$. Also $\text{Ker}\,\varphi = \cap_{x \in G} xHx^{-1} \neq G$. Thus $\text{Ker}\,\varphi$ is a nontrivial normal subgroup of G and so G is not simple.

(d) Let G be a group of order 108. Show that there exists a normal subgroup of order 27 or 9.

Solution. $|G| = 108 = 2^2 \cdot 3^3$. The number of Sylow 3-subgroups is $n_3 = 1 + 3k$ and $(1 + 3k)|2^2 \cdot 3^3$. Thus, $(1 + 3k)|2^2$. This implies $k = 0$ or 1. If $k = 0$, we have a unique Sylow 3-subgroup that is then a normal subgroup of order 27. If $k = 1$, we have four Sylow 3-subgroups. Let H and K be two distinct Sylow 3-subgroups. Then from the result $|HK| = |H||K|/|H \cap K|$ (Example 3.11(e), Chapter 4), $|HK| = (27 \cdot 27)/|H \cap K| \leq 108$. This gives $27/4 < |H \cap K|$, so $|H \cap K| = 9$.

Further, $H \cap K \lhd H$ and $H \cap K \lhd K$ because subgroups of order p^{n-1} are normal in a group of order p^n (or, alternatively, $H \cap K = H \cap N(K) \lhd H$). Now consider $N(H \cap K)$. Because $H \cap K$ is a normal subgroup in H as well as in K,

$$N(H \cap K) \supset H \quad \text{and} \quad N(H \cap K) \supset K.$$

Then $HK \subset N(H \cap K)$. Because

$$|HK| = \frac{|H| \cdot |K|}{|H \cap K|} = \frac{27 \cdot 27}{9} = 81,$$

and $N(H \cap K) < G$, $|N(H \cap K)| = 108$. Thus, $N(H \cap K) = G$. Hence, $H \cap K \lhd G$.

(e) If H is a normal subgroup of a finite group G, and if the index of H in G is prime to p, then H contains every Sylow p-subgroup of G.

Solution. Let $|G| = p^m q$, $(p,q) = 1$. Because the index of H – that is $|G|/|H|$ – is prime to p, $|H| = p^m q_1$, $(p,q_1) = 1$. By Sylow's first theorem, H (as a group by itself) contains a Sylow p-subgroup K, and, clearly, $|K| = p^m$. Because $|K| = p^m$, K is also a Sylow p-subgroup of G.

Let K_1 be another Sylow p-subgroup of G. Then by Sylow's second theorem $K_1 = xKx^{-1}$ for some $x \in G$. Then

$$K_1 = xKx^{-1} \subset xHx^{-1} \subset H, \qquad \text{since } H \lhd G.$$

Hence, all Sylow p-subgroups of G are contained in H.

(f) Let G be a group of order pq, where p and q are prime numbers such that $p > q$ and $q \nmid (p-1)$. Then G is cyclic.

Solution. By the third Sylow theorem the number n_p of Sylow p-subgroups is equal to $1 + kp$ for some positive integer k. Hence,

$$pq = \lambda(1 + kp),$$

for some positive integer λ. This implies $p|\lambda$, and thus

$$q = \lambda_1(1 + kp)$$

for some positive integer λ_1.

But since $p > q$, $q = \lambda_1(1 + kp)$ is not possible unless $k = 0$, and then we get by Corollary 4.6 that there is only one Sylow p-subgroup, say H, that must be normal.

Further, the number n_q of Sylow q-subgroups is equal to $1 + k'q$ for some integer k', and by the same argument

$$p = \mu_1(1 + k'q)$$

for some positive integer μ_1.

But since $q \nmid (p-1)$, $p = \mu_1(1 + k'q)$ is possible only if $k' = 0$, and this implies that there is a unique Sylow q-subgroup, say K. K is normal by Corollary 4.6.

It is easy to see that $K \cap H = \{e\}$, and, hence, $kh = hk$ $\forall h \in H$ and $\forall k \in K$. This implies that if x and y generate the cyclic groups H and K, respectively, then the order of the element xy is equal to $|H||K| = pq$. Hence, G is generated by the element xy, and, thus, G is cyclic.

(g) Any group of order 15 is cyclic.

Solution. Follows from (f).

(h) There are only two nonabelian groups of order 8.

Solution. Consider a nonabelian group G of order 8. Then G contains no element of order 8; otherwise G is abelian. If each element of G is of order 2, then G is abelian. Thus G contains an element a of order 4. Let $b \in G$ such that $b \notin [a]$. Then $G = [a] \cup [a]b$, and $b^2 \in [a]$. Now $b^2 = a$ or $a^3 \Rightarrow o(b)$ is 8, and G is abelian. Thus, we must have $b^2 = e$ or a^2. Because $[a]$ is of index 2 in G, $[a] \lhd G$. Hence, $b^{-1}ab \in [a]$. Because $o(b^{-1}ab) = o(a)$, we have $b^{-1}ab = a$ or a^3. Hence, $b^{-1}ab = a^3$.

Thus, we have two nonabelian groups of order 8:

(i) *The group G generated by a and b with defining relations*

$$a^4 = e, \quad b^2 = e, \quad b^{-1}ab = a^3.$$

(ii) *The group G generated by a and b with defining relations*

$$a^4 = e, \quad b^2 = a^2, \quad b^{-1}ab = a^3.$$

The first is the octic group, and the second is the quaternion group. That they are nonisomorphic is easily checked.

Incidently, there are three, and only three, abelian groups of order 8 that are of types (2,2,2), (2,4), (8).

5 Groups of orders p^2, pq

We recall that there are cyclic groups of every possible order, that two cyclic groups of the same order are isomorphic, and that a direct product of cyclic groups is abelian. We denote by $[x]$ the cyclic group generated by x, and, more generally, by $[x,y]$ the group generated by x,y. Throughout this section p and q are distinct primes.

We also recall the following properties of a finite p-group P:

(i) $Z(P)$ is nontrivial.
(ii) If H is a proper subgroup of P, then $N(H) \supsetneq H$.
(iii) If $|P| = p^n$, then every subgroup M of order p^{n-1} is normal.

In the following, two groups isomorphic to each other are identified.

5.1 Groups of order p^2

We have already shown (Corollary 4.11, Chapter 5) that a group of order p^2 is abelian. *Hence, there are two, and only two, groups of order p^2:*

(i) *abelian group of type* (p^2), *and*
(ii) *abelian group of type* (p,p).

5.2 Groups of order pq, $q > p$

Let G be a group of order pq. Then the number of Sylow q-subgroups of G is $1 + \lambda q$, which divides pq. This gives $1 + \lambda q = 1$, and thus there is exactly one Sylow q-subgroup, which, being of order q, is cyclic. Let $[b]$, where $b^q = 1$, be this unique Sylow q-subgroup.

Also the number of Sylow p-subgroups is $1 + \mu p$, which divides pq. This gives $1 + \mu p = 1$ or q.

Case (i). Suppose the number of Sylow p-subgroups in G is 1, and let this be $[a]$, where $a^p = 1$. Since $[a]$ and $[b]$ are normal in G, and $[a] \cap [b] = \{e\}$, we have $G = [a] \times [b]$. Because $o(ab)$ is pq, $G = [ab]$, and, thus, G is cyclic.

Case (ii). The number of Sylow p-subgroups is q. In this case $q \equiv 1 \pmod p$. Let $[a]$ be one of the Sylow p-subgroups of G, where $a^p = 1$. Then the subgroup $[a,b]$ contains a subgroup $[a]$ of order p and a subgroup $[b]$ of order q; thus, $|[a,b]| = pq$, whence $G = [a,b]$. Because $[b] \lhd G$, $a^{-1}ba = b^r$ for some integer r. Then $r \not\equiv 1 \pmod q$, for otherwise $ba = ab$, and G is abelian, which in turn implies that the number of Sylow p-subgroups is 1, a contradiction. Thus, G is generated by a and b with defining relations

$$a^p = 1 = b^q, \qquad a^{-1}ba = b^r. \tag{1}$$

The integer r in (1) is a solution of the congruence equation

$$Z^p \equiv 1 \pmod q \tag{2}$$

for

$$\begin{aligned} a^{-1}ba = b^r &\Rightarrow (a^{-1}ba)(a^{-1}ba) = b^{2r} \Rightarrow a^{-1}b^2a = b^{2r} \\ &\Rightarrow a^{-1}b^ra = b^{r^2} \qquad \text{by induction} \\ &\Rightarrow a^{-1}(a^{-1}ba)a = b^{r^2} \Rightarrow a^{-2}ba^2 = b^{r^2} \\ &\Rightarrow a^{-p}ba^p = b^{r^p} \qquad \text{by induction} \\ &\Rightarrow b = b^{r^p} \Rightarrow r^p \equiv 1 \pmod q. \end{aligned}$$

Conversely, if r is a solution of the congruence equation (2), then the defining relations (1) determine a group consisting of the pq distinct elements a^ib^j, $0 \le i < p-1$, $0 \le j < q-1$.

We note that the solutions of $Z^p \equiv 1 \pmod q$, $Z \not\equiv 1 \pmod q$ are $r, r^2, \ldots, r^{p-1}$, and all yield the same group, because replacing a by a^j as a generator of $[a]$ replaces r by r^j.

We conclude that there are at most two groups of order pq:

(i) *The cyclic group of order* pq.

(ii) *The nonabelian group generated by a and b with defining relations* $a^p = 1 = b^q$, $a^{-1}ba = b^r$, $r^p \equiv 1 \pmod q$, $r \not\equiv 1 \pmod q$, *provided p divides* $q - 1$.

Problems

1. Suppose a group $G = H \times K$ is a product of groups H and K. Prove that $G/H \times \{e\} \simeq K$.
2. Let G be a cyclic group of order m, and let H be a cyclic group of order n. Show that $G \times H$ is cyclic if and only if $(m,n) = 1$.
3. Prove that the group of all rational numbers under addition cannot be written as a direct sum of two nontrivial subgroups.
4. Show that the set of positive integers less than and prime to 21 form a group with respect to multiplication modulo 21. Prove that this group is isomorphic to $H \times K$, where H and K are cyclic groups of orders 6 and 2, respectively.
5. Prove that if the order of an abelian group is not divisible by a square, then the group must be cyclic.
6. Show that an abelian group of order p^k and of type $(p^{k_1}, p^{k_2}, \ldots, p^{k_t})$ contains $p^t - 1$ elements of order p.
7. Show that a finite abelian group that is not cyclic contains a subgroup of type (p,p).
8. Show that a group of order 1986 is not simple.
9. Prove that any group of order $2p$ must have a normal subgroup of order p, where p is prime.
10. Show that a group of order p^2q, p and q distinct primes, must contain a normal Sylow subgroup and be solvable.
11. Prove that a group of order 200 must contain a normal Sylow subgroup.
12. Let G be a simple group and let n_p be the number of Sylow p-subgroups, where p is prime. Show $|G|$ divides $(n_p)!$. Using this or otherwise, prove groups of order 48 cannot be simple.
13. If the order of a group is 42, prove that its Sylow 7-subgroup is normal.
14. If n is a divisor of the order of an abelian group, prove that the number of solutions of $x^n = e$ in G is a multiple of n.
15. If each Sylow subgroup of a finite group G is normal, prove that G is the direct product of its Sylow subgroups.
16. Prove that any group G of order 12 that is not isomorphic to A_4 contains an element of order 6. Deduce that G has a normal Sylow 3-subgroup.

17. Let G be a group. Prove that $|G/Z(G)| \neq 77$.
18. Let A be an abelian subgroup of G. Suppose $A \lhd G$ and $|G/A| = 91$. Prove that G is solvable.
19. Let G be a nonabelian group of order p^3, where p is prime. Show that $Z(G) = G'$. What is the order of $Z(G)$?
20. Suppose G is a finite group of order $p_1 p_2 \cdots p_k$, where p_i are distinct primes. Show that G contains a normal Sylow subgroup and G is solvable.
21. Let G be a finite group with a Sylow p-subgroup P. Let $N = N(P)$. Show that any subgroup of G that contains N (N included) is equal to its own normalizer.
22. Let G be a finite group, and let P be a Sylow p-subgroup of G. Let $N \lhd G$. Show that $N \cap P$ is a Sylow p-subgroup of N. Also, show that NP/N is a Sylow p-subgroup of G/N.
23. Prove that a finite group G is nilpotent if and only if it is the direct product of its Sylow subgroups. [*Hint*: First show that if $H < G$, then $N(H) \neq H$. Then use Problem 21 to obtain that if H is a Sylow p-subgroup, then $H \lhd G$.]
24. Prove that A_5 is the first nonabelian simple group.
25. Suppose P is a Sylow p-subgroup of a group G. Let $H < G$ such that $|H| = p^m$, $m > 0$. Show that $H \cap N(P) = H \cap P$.
26. Let P be a Sylow p-subgroup of a finite group G. If M is a normal p-subgroup, show that $M \subseteq P$. Show also that G contains a normal p-subgroup H that contains all normal p-subgroups of G.
27. Let G be a finite group such that $G = HK$, where $H < G, K < G$. For a prime p, if P is a Sylow p-subgroup of G, show that $x^{-1}Px \cap K$ is a Sylow p-subgroup of K for some $x \in H$.

PART III

Rings and modules

CHAPTER 9

Rings

1 Definition and examples

Definition. *By a* ring *we mean a nonempty set R with two binary operations + and ·, called addition and multiplication (also called product), respectively, such that*

(i) $(R,+)$ *is an additive abelian group.*
(ii) $(R,\cdot)$ *is a multiplicative semigroup.*
(iii) Multiplication is distributive (on both sides) over addition; that is, for all $a,b,c, \in R$,

$$a \cdot (b + c) = a \cdot b + a \cdot c, \qquad (a + b) \cdot c = a \cdot c + b \cdot c.$$

(The two distributive laws are respectively called the left distributive law and the right distributive law.)

We usually write ab instead of $a \cdot b$. The identity of the additive abelian group is called *zero element* of the ring R and is unique. We denote the zero element of a ring by 0. Further, the additive inverse of an element a of the additive abelian group $(R,+)$ shall, as usual, be denoted by $-a$. Thus, in a ring R,

$$a + (-a) = 0 \qquad \text{for all } a \in R.$$

Further, if $a,b \in R$, we denote $a + (-b)$ by $a - b$.

The most familiar examples of rings are

Z: the ring of all integers,
Q: the ring of all rational numbers,

R: the ring of all real numbers,
C: the ring of all complex numbers.

Other interesting examples are

$C[0,1]$: the ring of all continuous functions from the interval $[0,1]$ to **R**,
$\mathbf{R}^X$: the ring of all functions from a nonempty set X to the real numbers **R**.

The addition in each of the rings is the usual addition of mappings; that is,

$$(f+g)(x) = f(x) + g(x).$$

We define multiplication by

$$(fg)(x) = f(x)g(x).$$

Another example of a ring is an additive abelian group A in which multiplication is defined by

$$ab = 0 \qquad \text{for all } a,b \in A$$

The ring A so obtained is called a *trivial ring*. We define a trivial ring as one in which the product (or multiplication) of any two elements is 0.

An example of a finite ring – that is, a ring with finite number of elements – is $\mathbf{Z}_n$ (or $\mathbf{Z}/(n)$): the ring of integers modulo n. The elements of $\mathbf{Z}_n$ are the residue classes mod n; that is, $\mathbf{Z}_n = \{\bar{0},\bar{1},\ldots,\overline{(n-1)}\}$. Addition and multiplication in $\mathbf{Z}_n$ are defined as follows:

$$\bar{x} + \bar{y} = \overline{x+y}, \qquad \bar{x}\bar{y} = \overline{xy}.$$

(See Example 3.1 in Chapter 4.) To illustrate, we give the addition and multiplication tables of $\mathbf{Z}_4$.

$+$	$\bar{0}$	$\bar{1}$	$\bar{2}$	$\bar{3}$
$\bar{0}$	$\bar{0}$	$\bar{1}$	$\bar{2}$	$\bar{3}$
$\bar{1}$	$\bar{1}$	$\bar{2}$	$\bar{3}$	$\bar{0}$
$\bar{2}$	$\bar{2}$	$\bar{3}$	$\bar{0}$	$\bar{1}$
$\bar{3}$	$\bar{3}$	$\bar{0}$	$\bar{1}$	$\bar{2}$

$\cdot$	$\bar{0}$	$\bar{1}$	$\bar{2}$	$\bar{3}$
$\bar{0}$	$\bar{0}$	$\bar{0}$	$\bar{0}$	$\bar{0}$
$\bar{1}$	$\bar{0}$	$\bar{1}$	$\bar{2}$	$\bar{3}$
$\bar{2}$	$\bar{0}$	$\bar{2}$	$\bar{0}$	$\bar{2}$
$\bar{3}$	$\bar{0}$	$\bar{3}$	$\bar{2}$	$\bar{1}$

A *commutative* ring is a ring R in which multiplication is commutative; that is, $ab = ba$ for all $a,b \in R$. If a ring is not commutative it is called *noncommutative*.

A *ring with unity* is a ring R in which the multiplicative semigroup $(R,\cdot)$ has an identity element; that is, there exists $e \in R$ such that $ae = a = ea$ for all $a \in R$. The element e is called *unity* or the *identity* element of R. Generally, the unity or identity element is denoted by 1.

2 Elementary properties of rings

2.1 Theorem. *Let R be a ring. Then for all $a,b,c \in R$,*

(i) $a0 = 0 = 0a$.
(ii) $a(-b) = -(ab) = (-a)b$.
(iii) $a(b-c) = ab - ac$, $(a-b)c = ac - bc$.

Proof. (i)

$$a0 = a(0+0) = a0 + a0.$$

Thus,

$$a0 + (-(a0)) = a0.$$

So,

$$0 = a0.$$

Similarly,

$$0 = 0a.$$

(ii)

$$0 = a0 = a(b + (-b)) = ab + a(-b).$$

Thus,

$$-(ab) = a(-b).$$

Similarly,

$$-(ab) = (-a)b.$$

(iii)

$$a(b-c) = a(b + (-c)) = ab + a(-c) = ab - ac.$$

Similarly,

$$(a-b)c = ac - bc. \quad \square$$

Let $a_1, a_2, \ldots, a_n$ be a sequence of elements of a ring R. We define their product inductively:

$$\prod_{i=1}^{1} a_i = a_1,$$
$$\prod_{i=1}^{n} a_i = \left(\prod_{i=1}^{n-1} a_i\right) a_n, \qquad n > 1.$$

We then have the following result, which can be directly proved by induction on m and n. (See Theorem 4.2 of Chapter 1.)

2.2 Theorem

$$\left(\prod_{i=1}^{m} a_i\right)\left(\prod_{j=1}^{n} a_{m+j}\right) = \prod_{i=1}^{m+n} a_i,$$

for all $a_1, a_2, \ldots, a_{m+n}$ in a ring R.

This result, called the *generalized associative law* for products, essentially asserts that we can insert parentheses in any manner in our product without changing its value. The uniquely determined product of the sequence $a_1, a_2, \ldots, a_n$ is denoted $a_1 a_2 \cdots a_n$. In the same way we can define uniquely the sum of the sequence $a_1, a_2, \ldots, a_n$ and denote it $a_1 + a_2 + \cdots + a_n$. Also, by using induction on m and n, we can easily prove the generalized distributive law given in

2.3 Theorem

$$(a_1 + a_2 + \cdots + a_m)(b_1 + b_2 + \cdots + b_n) = a_1 b_1 + a_1 b_2 + \cdots + a_1 b_n + a_2 b_1 + \cdots + a_2 b_n + \cdots + a_m b_1 + a_m b_2 + \cdots + a_m b_n$$

for all $a_1, \ldots, a_m$ and $b_1, \ldots, b_n$ in a ring R.

If $a \in R$ and m is a positive integer, we write

$$a^m = \overbrace{a \cdot a \cdots a}^{m \text{ times}} \quad \text{and} \quad ma = \overbrace{a + a + \cdots + a}^{m \text{ times}}.$$

If m is a negative integer, then ma stands for

$$\overbrace{(-a) + (-a) + \cdots + (-a)}^{-m \text{ times}}.$$

If a has an inverse $a^{-1} \in R$, then a^{-m} denotes the element $(a^{-1})^m$. If $1 \in R$, then we define $a^0 = 1$. Also we define $0a$ to be the zero element of the ring.

2.4 Theorem. *(a) For all positive integers m and n and for all a in a ring R, the following hold:*

(i) $a^m a^n = a^{m+n}$.
(ii) $(a^m)^n = a^{mn}$.

(b) For all integers m and n and for all a,b in a ring R the following hold:

(iii) $ma + na = (m + n)a$.
(iv) $m(na) = (mn)a$.
(v) $(ma)(nb) = (mn)(ab) = (na)(mb)$.

The theorem may be proved by using induction on m and n. It is left as an exercise for the reader.

3 Types of rings

Definition. *A ring R whose nonzero elements form a group under multiplication is called a* division ring. *If, in addition, R is commutative, then R is called a* field.

The rational numbers **Q**, the real numbers **R**, and the complex numbers **C** are examples of fields.

An example of a division ring that is not a field is the *ring of real quaternions.*

Let H be the set of 2×2 matrices of the form

$$\begin{bmatrix} a & b \\ -\bar{b} & \bar{a} \end{bmatrix},$$

where a and b are complex numbers, and $\bar{a}$ and $\bar{b}$, respectively, denote their complex conjugates. Then H is a noncommutative division ring. For if

$$A = \begin{bmatrix} a & b \\ -\bar{b} & \bar{a} \end{bmatrix}$$

is a nonzero matrix, then its determinant $\delta = a\bar{a} + b\bar{b} \neq 0$, and, hence,

$$A^{-1} = \begin{bmatrix} \dfrac{\bar{a}}{\delta} & \dfrac{-b}{\delta} \\ \dfrac{\bar{b}}{\delta} & \dfrac{a}{\delta} \end{bmatrix}.$$

Definition. *A ring R is called an* integral domain *if* $xy = 0$, $x \in R$, $y \in R$, *implies* $x = 0$ *or* $y = 0$.

The ring of integers **Z** is an integral domain. Every field (therefore every division ring) is an integral domain. (Let F be a field, and let $xy = 0$, $x,y \in F$. If $x \neq 0$, then x^{-1} exists, so $x^{-1}(xy) = 0$. Thus, $y = 0$.)

However, an integral domain need not be a division ring or a field. (For example, the ring of integers **Z** is neither.)

Definition. *An element x in a ring R is called a* right zero divisor *if there exists a nonzero element $y \in R$ such that $yx = 0$.*

One can similarly define left zero divisor. An element is called a *zero divisor* if it is both a right and a left zero divisor.

We may note that a ring R is an integral domain if and only if it has no right (or left) zero divisors except 0.

3.1 Some important examples of rings

(a) The ring of matrices

Rudiments of matrices over any field have been given in Chapter 3. Matrices with entries in a ring R (i.e., matrices over R) and their addition and multiplication can be defined in exactly the same manner.

Let R be a ring, and let R_n be the set of all $n \times n$ matrices over R. Then R_n with the usual addition and multiplication of matrices is called the *ring of $n \times n$ matrices* over R. If $n > 1$ and R is not a trivial ring, then R_n is not commutative. For if

$$a = \begin{bmatrix} x & 0 \\ 0 & 0 \end{bmatrix}, \quad b = \begin{bmatrix} 0 & y \\ 0 & 0 \end{bmatrix}, \qquad xy \neq 0,$$

then $ab \neq ba$. It also follows that if $n > 1$, R_n contains nonzero zero divisors; that is, R_n is not an integral domain.

Suppose R has unity 1. We denote by e_{ij} the matrices in R_n whose (i,j) entry is 1 and whose other entries are zero. The e_{ij}'s, $1 \leq i,j \leq n$, are called *matrix units.* From the definition of multiplication of matrices, it follows that $e_{ij}e_{kl} = 0$ if $j \neq k$, and $e_{ij}e_{jk} = e_{ik}$; that is,

$$e_{ij}e_{kl} = \delta_{jk}e_{il},$$

where δ_{jk} is the Kronecker delta. Also, if $A = (a_{ij}) \in R_n$, then A can be uniquely expressed as a linear combination of e_{ij}'s over R; that is,

$$A = \sum_{1 \leq i,j \leq n} a_{ij}e_{ij}, \qquad a_{ij} \in R.$$

This representation of matrices in terms of matrix units is often a convenient tool in certain problems involving algebraic manipulations.

Let S be the set of $n \times n$ matrices, all of whose entries below the main diagonal are zero; that is, let S consist of matrices

$$\begin{bmatrix} a_{11} & a_{12} & \cdots & \cdots & a_{1n} \\ 0 & a_{22} & \cdots & \cdots & a_{2n} \\ \cdot & \cdot & & & \cdot \\ \cdot & \cdot & & & \cdot \\ \cdot & \cdot & & & \cdot \\ 0 & 0 & \cdots & \cdots & a_{nn} \end{bmatrix}, \qquad a_{ij} \in R.$$

Then S is a ring with the usual addition and multiplication of matrices and is called the ring of *upper triangular matrices*. Similarly, we have the ring of *lower triangular matrices*.

(b) The ring of polynomials

Let R be a ring. *A polynomial* with coefficients in R is a formal expression of the form

$$a_0 + a_1x + a_2x^2 + \cdots + a_mx^m, \qquad a_i \in R.$$

Two polynomials $a_0 + a_1x + \cdots + a_mx^m$ and $b_0 + b_1x + \cdots + b_nx^n$ are called equal if $m = n$ and $a_i = b_i$ for all i. Polynomials may be added and multiplied to form a ring in the following manner. Let

$$f(x) = a_0 + a_1x + a_2x^2 + \cdots + a_mx^m,$$
$$g(x) = b_0 + b_1x + b_2x^2 + \cdots + b_nx^n.$$

Then

$$f(x) + g(x) = (a_0 + b_0) + (a_1 + b_1)x + (a_2 + b_2)x^2 + \cdots,$$
$$f(x)g(x) = c_0 + c_1x + c_2x^2 + \cdots + c_{m+n}x^{m+n},$$

where

$$c_i = \sum_{j+k=i} a_jb_k, \qquad 0 \leq i \leq m + n.$$

Alternatively, one may look upon

$$a_0 + a_1x + \cdots + a_mx^m, \qquad a_i \in R,$$

as a sequence

$$(a_0, a_1, \ldots, a_m, 0, 0, \ldots), \qquad a_i \in R;$$

that is, a function $f: \mathbf{N} \to R$, where $f(i) = a_{i-1}$, $1 \leq i \leq m + 1$, and $f(i) = 0$, $i > m + 1$.

Let $b_0 + b_1x + \cdots + b_nx^n$, $b_i \in R$, be another polynomial, and let

$g: \mathbf{N} \to R$ be the corresponding function, so that $g(i) = b_{i-1}$, $1 \le i \le n+1$ and $g(i) = 0$, $i > n+1$. The sum $f + g$ of the functions f and g is defined to be the usual sum of functions, namely,

$$(f+g)(i) = f(i) + g(i),$$

or, equivalently,

$$(a_0 + a_1x + \cdots + a_mx^m) + (b_0 + b_1x + \cdots + b_nx^n) \\ = (a_0 + b_0) + (a_1 + b_1)x + (a_2 + b_2)x^2 + \cdots.$$

The product of two functions f and g is defined to be the function h, where

$$h(i) = \sum_{j+k=i+1} f(j)g(k), \qquad i \in \mathbf{N},$$

or, equivalently,

$$(a_0 + a_1x + \cdots + a_mx^m)(b_0 + b_1x + \cdots + b_nx^n) \\ = c_0 + c_1x + \cdots + c_{m+n}x^{m+n},$$

where

$$c_i = \sum_{j+k=i} a_jb_k.$$

It is now straightforward to verify that the set of polynomials (i.e., sequences with finitely many nonzero entries) forms a ring called the *polynomial ring over R*, denoted by $R[x]$.

Further, the mapping $a \to (a,0,0,\ldots)$ of the ring R into the ring $R[x]$ preserves binary operations; that is, if $b \to (b,0,0,\ldots)$, then

$$a + b \to (a + b,0,0,\ldots) = (a,0,0,\ldots) + (b,0,0,\ldots), \\ ab \to (ab,0,0,\ldots) = (a,0,0,\ldots)(b,0,0,\ldots),$$

and the mapping is also 1-1. Thus, $R[x]$ contains a carbon copy of R. By identifying the element $a \in R$ with the sequence $(a,0,0,\ldots) \in R[x]$, we say that R is a subring of $R[x]$.

In what follows we assume $1 \in R$, and we denote by x the sequence $(0,1,0,\ldots)$. Then $x^2 = (0,0,1,0,\ldots)$, $x^3 = (0,0,0,1,0,\ldots)$, and so on. Consider a sequence $(a_0,a_1,\ldots,a_n,0,\ldots)$ with finitely many nonzero entries. We find

$$(a_0,a_1,\ldots,a_n,0,\ldots) = (a_0,0,\ldots) + (0,a_1,0,\ldots) + \cdots + (0,\ldots,0,a_n,0,\ldots) \\ = a_0 + a_1x + \cdots + a_nx^n,$$

which yields the familiar expression of a polynomial in x. We call x an *indeterminate* and refer to the polynomials in $R[x]$ as polynomials in one indeterminate over R (sometimes we also call them polynomials in a variable x).

(c) Ring of endomorphisms of an abelian group

Let M be an additive abelian group. Let $\text{End}(M)$ [or $\text{Hom}(M, M)$] denote, as usual, the set of endomorphisms of the group M into itself. We define addition and multiplication in $\text{End}(M)$ as follows:

(i) $(f+g)(x) = f(x) + g(x)$,
(ii) $(fg)(x) = f(g(x))$,

for all $f, g \in \text{End}(M)$ and $x \in M$.

We show that the set $\text{End}(M)$ together with the above binary operations of addition and multiplication forms a ring. First we show that if $f, g \in \text{End}(M)$, then $f+g \in \text{End}(M)$ and $fg \in \text{End}(M)$. Let $x, y \in M$. Then

$$\begin{aligned}(f+g)(x+y) &= f(x+y) + g(x+y) \qquad \text{[by (i)]}\\ &= [f(x)+f(y)] + [g(x)+g(y)]\\ &\qquad\qquad (\text{because } f \text{ and } g \text{ are homomorphisms})\\ &= [f(x)+g(x)] + [f(y)+g(y)]\\ &\qquad\qquad (\text{because } M \text{ is abelian})\\ &= (f+g)(x) + (f+g)(y) \qquad \text{[by (i)]}.\end{aligned}$$

This shows $(f+g) \in \text{End}(M)$.
Further

$$\begin{aligned}(fg)(x+y) &= f(g(x+y)) \qquad \text{[by (ii)]}\\ &= f[g(x)+g(y)] \qquad (\text{because } g \text{ is a homomorphism})\\ &= f(g(x)) + f(g(y)) \qquad (\text{because } f \text{ is a homomorphism})\\ &= (fg)(x) + (fg)(y) \qquad \text{[by (ii)]},\end{aligned}$$

which shows that $fg \in \text{End}(M)$. Also it is easily shown that the two mappings

$$\bar{0}: x \mapsto 0 \qquad (\text{zero endomorphism of } M)$$

and

$$1: x \mapsto x \qquad (\text{identity endomorphism of } M)$$

of M into M are also endomorphisms of M.

If $f \in \text{End}(M)$, define $-f$ to be the mapping given by $(-f)(x) = -f(x)$ for all $x \in M$. Then $-f \in \text{End}(M)$ because, for any $x, y \in M$,

$$\begin{aligned}(-f)(x+y) &= -[f(x+y)] = -[f(x)+f(y)]\\ &= -f(x) - f(y) = (-f)(x) + (-f)(y).\end{aligned}$$

Obviously, $f + \bar{0} = f$ and $f + (-f) = \bar{0}$ for all $f \in \text{End}(M)$.

Furthermore, along the same lines, it can be easily shown that for all $f,g,h \in \text{End}(M)$,

$$\begin{aligned}(f+g)+h &= f+(g+h),\\ f+g &= g+f,\\ (fg)h &= f(gh),\\ (f+g)h &= fh+gh,\end{aligned}$$

and

$$f(g+h) = fg+fh.$$

Hence $(\text{End}(M),+,\cdot)$ is a ring.

(d) Boolean ring

Let $\mathscr{B}$ be the set of all subsets of a nonempty set A [called the power set of A and denoted by $\mathscr{P}(A)$].

If $a,b \in \mathscr{B}$ we define

$$\begin{aligned}a+b &= (a \cup b) - (a \cap b),\\ ab &= a \cap b.\end{aligned}$$

Then $(\mathscr{B},+,\cdot)$ is a commutative ring with identity (which is the whole set A). The zero element of this ring is the empty set. Some of the important properties of the ring R are

(i) $a^2 = a$, and
(ii) $2a = 0$,

for all $a \in \mathscr{B}$.

A ring R is called a *Boolean ring* if $x^2 = x$ for all $x \in R$. It can be easily shown that $x^2 = x$ for all $x \in R$ implies $2x = 0$. It is interesting to point out that any Boolean ring is known to be a subring of the ring of all subsets of a set.

4 Subrings and characteristic of a ring

Definition. *Let $(R,+,\cdot)$ be a ring, and let S be a nonempty subset of R. Then S is called a* subring *if $(S,+,\cdot)$ is itself a ring.*

We note that addition and multiplication of elements of S are to coincide with addition and multiplication of these elements considered as elements of the larger ring R. Similarly, we can define a *subdivision ring* of a division ring and a *subfield* of a field.

Every ring R has two trivial subrings, (0) or simply 0 and R. Let R be a ring (with or without unity). A subring of R may have unity different from the unity of R [see Problem 10(b)]. The following result is frequently useful:

4.1 Theorem. *A nonempty subset S of a ring R is a subring if and only if for all $a,b \in S$ we have $a - b \in S$ and $ab \in S$.*

Proof (If part). The condition $a - b \in S$ for all $a,b \in S$ implies $(S,+)$ is an additive subgroup of the group $(R,+)$. The condition $ab \in S$ for all $a,b \in S$ implies $(S,\cdot)$ is a semigroup. Because the two distributive laws hold in R, they also hold in S. Hence, $(S,+,\cdot)$ is a subring of $(R,+,\cdot)$. The converse follows from the definition of a subring. □

Definition. *Let R be a ring. Then the set $Z(R) = \{a \in R | xa = ax$ for all $x \in R\}$ is called the* center of the ring R.

4.2 Theorem. *The center of a ring is a subring.*

Proof. Let R be a ring. Write $Z(R) = S$. Because $0 \in S$, then $S \neq \emptyset$. Let $a,b \in S$, $x \in R$. Then

$$(a - b)x = ax - bx = xa - xb = x(a - b).$$

Thus, $a - b \in S$. Further,

$$(ab)x = a(bx) = a(xb) = (ax)b = (xa)b = x(ab).$$

Hence, $ab \in S$. Therefore, by Theorem 4.1, S is a subring. □

Definition. *Let S be a subset of a ring R. Then the smallest subring of R containing S is called the* subring generated by S.

Because the intersection of a family of subrings is again a subring, it follows that the subring generated by a subset S of R is the intersection of all subrings of R containing S.

The subring generated by the empty set is clearly (0), and the subring generated by a single element a in R consists of all elements of the form $n_1a + n_2a^2 + \cdots + n_ka^k$, $n_i \in \mathbf{Z}$, and k a positive integer.

We now define the characteristic of a ring.

Definition. *If there exists a positive integer n such that $na = 0$ for each element a of a ring R, the smallest such positive integer is called the* characteristic *of R. If no such positive integer exists, R is said to have characteristic zero. The characteristic of R is denoted char R.*

For example, the ring of integers has characteristic zero. The ring $\mathbf{Z}/(n)$ of integers modulo n has characteristic n.

4.3 Theorem. *Let F be a field. Then the characteristic of F is either 0 or a prime number p.*

Proof. Let $n \neq 0$ be the characteristic of F. If e is the identity of F, then $ne = 0$. If possible, let $n = n_1 n_2$, $n_1 < n$, $n_2 < n$. Then $(n_1 n_2)e = 0$ yields $(n_1 e)(n_2 e) = 0$. But since F is a field, either $n_1 e = 0$ or $n_2 e = 0$. But then either $n_1 ea = 0$ (i.e., $n_1 a = 0$) or $n_2 ea = 0$ (i.e., $n_2 a = 0$) for all $a \in R$. Thus, the characteristic of F is $\leq \max(n_1, n_2) < n$, a contradiction. This proves n is prime. □

Definition. *An element a in a ring R is called* nilpotent *if there exists a positive integer n such that $a^n = 0$.*

0 is always nilpotent. The element $\left(\begin{smallmatrix} 0 & 1 \\ 0 & 0 \end{smallmatrix}\right)$ in a 2×2 matrix ring is also nilpotent because

$$\begin{pmatrix} 0 & 1 \\ 0 & 0 \end{pmatrix}^2 = \begin{pmatrix} 0 & 0 \\ 0 & 0 \end{pmatrix}.$$

If R is an integral domain, then R does not possess any nonzero nilpotent elements, because for $0 \neq a \in R$, we have $a^n = 0 \Rightarrow a \cdot a^{n-1} = 0$, so $a^{n-1} = 0$. Continuing like this we obtain $a = 0$.

Definition. *An element e in a ring R is called* idempotent *if $e = e^2$.*

Clearly, 0 and 1 (if R has unity) are idempotent elements.

Definition. *Let A be an arbitrary ring, and let F be a field. Then we say that A is an* algebra *over F if there exists a mapping $(\alpha, x) \mapsto \alpha x$ of $F \times A$ into A such that*

(i) $(\alpha + \beta)x = \alpha x + \beta x, \quad \alpha, \beta \in F, x \in A;$
(ii) $\alpha(x + y) = \alpha x + \alpha y, \quad \alpha \in F, x, y \in A;$
(iii) $(\alpha\beta)x = \alpha(\beta x), \quad \alpha, \beta \in F, x \in A;$
(iv) $1x = x, \quad 1 \in F, x \in A;$
(v) $\alpha(xy) = (\alpha x)y = x(\alpha y), \quad \alpha \in F, x, y \in A.$

The element αx is called the scalar multiplication *of α and x.*

Any field F can be regarded as an algebra over itself by defining scalar multiplication as multiplication of elements in F. More generally, if R is a ring containing a subring $F \subset Z(R)$ such that F is a field, then R can be regarded as an algebra over F. Thus, the field of real numbers $\mathbf{R}$ can be regarded as an algebra over the field of rational numbers $\mathbf{Q}$. Also, the polynomial ring $\mathbf{Q}[x]$ is an algebra over $\mathbf{Q}$.

We now introduce the concepts of direct product and direct sum of a family of rings. These will be discussed in somewhat greater detail in a later chapter. Because of their many-sided importance and also for our present need to construct some interesting examples, we define them here.

Let R_i, $i = 1,2,...$, be a family of rings. We construct from this family a new ring as follows. Consider the set

$$R = \{(a_1, a_2, ...) | a_i \in R_i\}$$

of sequences $(a_1, a_2, ...)$, which we write as (a_i). We define binary operations on R in the following natural way:

$$(a_1, a_2, ...) + (b_1, b_2, ...) = (a_1 + b_1, a_2 + b_2, ...)$$

and

$$(a_1, a_2, ...)\,(b_1, b_2 ...) = (a_1 b_1, a_2 b_2, ...).$$

It is then easy to verify that R is a ring. This ring R is called the *direct product* of the rings R_i, $i = 1,2,...$, and often designated by $\Pi_i R_i$. The rings R_i may be called the *component rings* of this direct product. Moreover, if $(a_1, a_2, ..., a_i, ...) \in R$, it is convenient to call a_i the ith component of this element and to refer to addition and multiplication in R as componentwise addition and multiplication.

The zero element of R is the element whose ith component is the zero of R_i for all $i = 1,2,....$ Also, R will have unity if every R_i has unity. The set R_i^* of elements of the form $(0,0,...,0,a_i,0...)$ of R forms a subring of R.

Consider a subset S of R such if $(a_i) \in S$, then all but a finite number of the components of (a_i) are zero. It is easy to verify that S is a subring of R. This ring S is called the *direct sum* of the family R_i, $i = 1,2,...$, and is denoted by $\oplus\, \Sigma_i R_i$. If the family of rings R_i is infinite, then $\oplus\, \Sigma_i R_i$ will not have unity even though each R_i may have unity.

Note that if the family of rings R_i is finite, then direct product and direct sum coincide.

4.4 Examples

(a) Let R be a ring and let $x \in R$. If there exists a unique $a \in R$ such that $xa = x$, then $ax = x$.

In particular, if R has a unique right unity e, then e is the unity of R.

Solution. Clearly, $x(a + ax - x) = x$. So $a + ax - x = a$ and, hence, $ax = x$.

(b) Let x be a nonzero element of a ring R with 1. If there exists a unique $y \in R$ such that $xyx = x$, then $xy = 1 = yx$; that is, x is invertible in R.

Solution. Let $xr = 0$, $r \in R$. Then $x(y + r)x = x$ and, hence, $y + r = y$. Thus, $xr = 0$ implies $r = 0$. Now $xyx = x$ gives $x(yx - 1) = 0$. Thus, $yx - 1 = 0$. Similarly, $xy - 1 = 0$, which completes the solution.

(c) Let R be a ring with 1. If an element a of R has more than one right inverse, then it has infinitely many right inverses.

Solution. Let $A = \{a' \in R \mid aa' = 1\}$. By hypothesis A has more than one element. It follows then that a does not have a two-sided inverse. Fix $a_0 \in A$ and consider the mapping $a' \mapsto a'a - 1 + a_0$ of A to A. It can be easily checked that this mapping is 1-1 but not onto (a_0 does not have a preimage). Hence, A must be an infinite set because every 1-1 mapping of a finite set into itself is an onto mapping (Theorem 3.3, Chapter 1).

(d) Let R be a ring with more than one element. If $ax = b$ has a solution for all nonzero $a \in R$ and for all $b \in R$, then R is a division ring.

Solution. First we show R is an integral domain. If possible, let $a \neq 0$, $b \neq 0$, and $ab = 0$. Then $abx = 0$ for all $x \in R$. Now, by hypothesis, for all $c \in R$, $bx = c$ has a solution. Thus, $ac = 0$ for every $c \in R$. Because, for some $c \in R$, $ac = a$, we get $a = 0$, a contradiction. Hence, R must be an integral domain.

Next let $x = e \in R$ be a solution of $ax = a$, $a \neq 0$. Then $e \neq 0$ and $a(e - e^2) = 0$. But this implies $e - e^2 = 0$. Thus, e is a nonzero idempotent.

We assert that e is an identity element in R. Now for all $x \in R$, we have $(xe - x)e = xe^2 - xe = 0$, so $xe - x = 0$. Similarly, $e(ex - x) = 0$ yields $ex = x$. This proves our assertion. Finally, if $0 \neq a \in R$, then there exists $b \in R$ such that $ab = e$. Also, $(ba - e)b = bab - b = 0$; thus, $ba = e$. This proves that each nonzero element $a \in R$ has an inverse.

(e) Let R be a ring such that $x^3 = x$ for all $x \in R$. Then R is commutative.

Solution. $x^3 = x$ for all $x \in R$ implies $(x + x)^3 = (x + x)$ for all $x \in R$. Thus, $6x = 0$ for all $x \in R$. Also, $(x^2 - x)^3 = x^2 - x$ implies $3x^2 = 3x$ after simplifications.

Consider $S = \{3x | x \in R\}$. It can be easily checked that S is a subring of R and, for $y \in S$,

$$y^2 = (3x)^2 = 9x^2 = 6x^2 + 3x^2 = 3x^2 = 3x = y.$$

Thus, $y^2 = y$ for all $y \in S$ implies S is commutative, so $(3x)(3y) = (3y)(3x)$, that is, $9xy = 9yx$, which implies $3xy = 3yx$. Now $(x + y)^3 = x + y$ implies

$$xy^2 + x^2y + xyx + yx^2 + yxy + y^2x = 0, \tag{1}$$

and $(x - y)^3 = x - y$ implies

$$xy^2 - x^2y - xyx - yx^2 + yxy + y^2x = 0. \tag{2}$$

By adding (1) and (2) we get

$$2xy^2 + 2yxy + 2y^2x = 0.$$

Multiply the last equation by y on right and by y on left:

$$2xy + 2yxy^2 + 2y^2xy = 0. \tag{3}$$
$$2yxy^2 + 2y^2xy + 2yx = 0. \tag{4}$$

Subtract (3) from (4) to get

$$2xy = 2yx.$$

Because $3xy = 3yx$, we have $xy = yx$ for all $x,y \in R$. Hence, R is commutative.

Problems

1. Let $S = C[0,1]$ be the set of real-valued continuous functions defined on the closed interval $[0,1]$, where we define $f + g$ and fg, as usual, by $(f + g)(x) = f(x) + g(x)$ and $(fg)(x) = f(x)g(x)$. Let 0 and 1 be the constant functions 0 and 1, respectively. Show that
   (a) $(S, +, \cdot)$ is a commutative ring with unity.
   (b) S has nonzero zero divisors.
   (c) S has no idempotents $\neq 0,1$.
   (d) Let $a \in [0,1]$. Then the set $T = \{f \in S | f(a) = 0\}$ is a subring such that $fg, gf \in T$ for all $f \in T$ and $g \in S$.

2. Let R be an integral domain and $a,b \in R$. If $a^m = b^m$, $a^n = b^n$, and $(m,n) = 1$, show that $a = b$.
3. (a) Show that the following are subrings of **C**.
   (i) $A = \{a + b\sqrt{-1} \mid a,b \in \mathbf{Z}\}$.
   (ii) $B = \{a + b\sqrt{-3} \mid$ either $a,b \in \mathbf{Z}$ or both a,b are halves of odd integers$\}$.

   The set A is called the ring of *Gaussian integers.*
   (b) Let e be an idempotent in a ring R. Show that the set $eRe = \{eae \mid a \in R\}$ is a subring of R with unity e.
4. Show that an integral domain contains no idempotents except 0 and 1 (if 1 exists).
5. (a) Determine the idempotents, nilpotent elements, and invertible elements of the following rings:
   (i) $\mathbf{Z}/(4)$ (ii) $\mathbf{Z}/(20)$
   (b) Show that the set $U(R)$ of units of a ring R with unity forms a multiplicative group (cf. Problem 18(c)).
   (c) Prove that an element $\bar{x} \in \mathbf{Z}/(n)$ is invertible if and only if $(x,n) = 1$. Show also that if $(x,n) = 1$, then $x^{\phi(n)} \equiv 1 \pmod{n}$, where $\phi(n)$ is Euler's function (this is called the *Euler–Fermat theorem*).
6. Let S be the set of 2×2 matrices over **Z** of the form $\left(\begin{smallmatrix} a & b \\ 0 & c \end{smallmatrix}\right)$. Show that
   (a) S is a ring.
   (b) $\begin{pmatrix} a & b \\ 0 & c \end{pmatrix}^k = \begin{pmatrix} a^k & * \\ 0 & c^k \end{pmatrix}$, where $*$ denotes some integer.

   Also find the idempotents and nilpotent elements of S. Show that nilpotent elements form a subring.
7. If a and b are nilpotent elements of a commutative ring, show that $a + b$ is also nilpotent. Give an example to show that this may fail if the ring is not commutative.
8. Prove that the following statements for a ring R are equivalent:
   (a) R has no nonzero nilpotent elements.
   (b) If $a \in R$ such that $a^2 = 0$, then $a = 0$.
   Furthermore, show that under any of the conditions (a) or (b) all idempotents are central.
9. Show that the characteristic of an integral domain is either 0 or a prime number.
10. (a) If a ring R with 1 has characteristic 0, show that R contains a subring that is in 1-1 correspondence with **Z** (this subring is called the *prime subring* of R).
    (b) Let $S = \{\bar{0},\bar{2},\bar{4},\bar{6},\bar{8}\}$. Show that S is a subring of $\mathbf{Z}_{10}$ with unity different from the unity of $\mathbf{Z}_{10}$.

11. Show that a finite integral domain is a division ring.
12. If $x^4 = x$ for all x in a ring R, show that R is commutative.
13. Let a be an element in a ring R with unity. If there exists an invertible element $u \in R$ such that $aua = a$, show that $ab = 1$ for some $b \in R$ implies $ba = 1$.
14. Let R be a ring with unity such that for each $a \in R$ there exists $x \in R$ such that $a^2x = a$. Prove the following:
    (a) R has no nonzero nilpotent elements.
    (b) $axa - a$ is nilpotent and, hence, $axa = a$.
    (c) $ax = xa$.
    (d) ax and xa are idempotents in the center of R.
    (e) There exists $y \in R$ such that $a^2y = a, y^2a = y$, and $ay = ya$.
    (f) $aua = a$, where $u = 1 + y - ay$ is invertible, and $y \in R$ is chosen as in (e).
15. Let R be a ring with left identity e. Suppose, for each nonzero element a in R, there exists b in R such that $ab = e$. Prove that R is a division ring.

In Problems 16–23, by a right (left) *quasi-regular element* in a ring R, we mean an element $a \in R$ such that the equation

$$a + x - ax = 0 \qquad (a + x - xa = 0)$$

has a solution $x = b \in R$.

An element $a \in R$ is called *quasi-regular* if it is both right and left quasi-regular in R.

16. Show that if $a \in R$ is nilpotent, then a is quasi-regular. Further, if R has a unity, then $1 - a$ is invertible.
17. Let $a,b \in R$ such that ab is nilpotent. Show that ba is nilpotent. Show also that if ab is right quasi-regular, then ba is also right quasi-regular.
18. Let R be a ring. Define a circle composition $\circ$ in R by $a \circ b = a + b - ab$, $a,b \in R$.
    (a) Show that $(R,\circ)$ is a semigroup with identity.
    (b) If S is the set of all quasi-regular elements in R, then $(S,\circ)$ is a group.
    (c) Suppose R has unity. Let $(U(R),\cdot)$ denote the group of units (i.e., invertible elements) of R. Show that $(U(R),\cdot) \simeq (S,\circ)$.
19. Let $a,b \in R$ such that $ba = b$ and a is right quasi-regular. Show that $b = 0$.
20. Let $a \in R$ and let a^n be right quasi-regular, where n is some positive integer. Show that a is right quasi-regular.
21. Let e be an idempotent in a ring R. If e is right quasi-regular, show that $e = 0$.

22. Call an element a in a ring R (von Neumann) *regular* if there exists an element $b \in R$ such that $aba = a$ (for example, all invertible elements and idempotent elements are von Neumann regular). Let R be a ring with 1 such that each regular element different from 1 is right quasi-regular. Show the following:
    (a) $0 \neq a \in R$ is invertible if and only if a is regular.
    (b) $0, 1 \neq a \in R$ is invertible if and only if a is right quasi-regular.
    (c) R has no nonzero nilpotent elements.
    (d) R has no idempotents different from 0 and 1.
    (e) Either $2 = 0$ or 2 is invertible in R.
    (f) The set of invertible elements in R along with 0 forms a division ring.
23. Let R be a ring with more than one element. Show that R is a division ring if and only if all but one element are right quasi-regular.
24. Let R be the direct sum of a family of rings R_i, $i = 1,2,...,n$. Show that the invertible elements of R are the elements $(u_1,u_2,...,u_n)$, u_i an invertible element of R_i. In other words, show that

    $$U(R) \simeq U(R_1) \times U(R_2) \times \cdots \times U(R_n),$$

    where $U(R)$, $U(R_i)$ denote, as usual, the group of units of the rings R and R_i, respectively.
25. Let R be the direct sum of a family of rings $R_1,R_2,...,R_n$, $n > 1$. Show that R cannot be an integral domain. Also show that $a = (a_1,a_2,...,a_n) \in R$ is nilpotent if and only if each a_i, $i = 1,2,...,n$, is nilpotent.

5 Additional examples of rings

(a) Ring of formal power series

Let R be a ring, and let

$$R[[x]] = \{(a_i) = (a_0,a_1,a_2,...) | a_i \in R\}.$$

Define addition and multiplication in $R[[x]]$ as in $R[x]$; that is,

$$(a_i) + (b_i) = (a_i + b_i),$$
$$(a_i)(b_i) = (c_i),$$

where $c_i = \Sigma_{j+k=i}\, a_j b_k$. Then $(R[[x]], +, \cdot)$ becomes a ring, called the *ring of formal power series over R*. As in a polynomial ring, the elements of R

themselves may be considered to be elements of $R[[x]]$. It is an interesting exercise to show that if R is an integral domain, then $R[[x]]$ is also an integral domain. Further, the elements of $R[[x]]$ are generally written as $\Sigma_{i=0}^{\infty} a_i x^i$,

(b) Ring of formal Laurent series

Let R be a ring, and let

$$R\langle x\rangle = \{(\ldots,0,0,a_{-k},a_{-k+1},\ldots,a_0,a_1,a_2,\ldots) \mid a_i \in R\}.$$

In other words, the elements of $R\langle x\rangle$ are infinite sequences, with the understanding that at most a finite number of the components a_i, with i a negative integer, are different from zero. If addition and multiplication of elements of $R\langle x\rangle$ are defined precisely as in $R[x]$ or in $R[[x]]$, then $R\langle x\rangle$ becomes a ring containing $R[[x]]$ as a subring. $R\langle x\rangle$ is called the ring of formal Laurent series over R. Straightforward computations yield that $R\langle x\rangle$ is a field if R is a field.

(c) Group algebra

Let G be a group, R a ring with unity, and $R[G]$ the set of mappings $a \mapsto f(a)$ of G into R such that $f(a)=0$ for all but a finite number of elements of G. Define addition, multiplication, 0, and 1 in $R[G]$ by

$$\begin{aligned}
(f+g)(a) &= f(a) + g(a),\\
(fg)(a) &= \sum_{bc=a} f(b)g(c),\\
0(a) &= 0,\\
1(e) &= 1, \quad 1(a)=0 \qquad \text{if } a \neq e.
\end{aligned}$$

Show that $R[G]$ is a ring called the group algebra of G over R. Show that if R is commutative, then R is contained in the center of $R[G]$. As an exercise, write the multiplication table and list some properties for the group algebra $F[G]$, where $F=\mathbf{Z}/(2)$ and G is a group of order 3.

(d) Quaternions over any field

Readers not familiar with the notion of a basis of a vector space may skip this. Let Q be a vector space of dimension 4 over a field K. Let $\{1,i,j,k\}$ be a basis of Q over K. We shall make Q into an algebra over K by suitably defining multiplication of these basis elements in Q as follows:

$\cdot$	1	i	j	k
1	1	i	j	k
i	i	-1	k	$-j$
j	j	$-k$	-1	i
k	k	j	$-i$	-1

(Recall that the product of any element in the left column by any element in the top row (in this order) is to be found at the intersection of the respective row and column.) We extend multiplication to the whole set Q in the obvious manner. This yields that Q is an algebra over K, called the algebra of quaternions over K.

If $K=\mathbf{R}$, then by taking

$$e=\begin{pmatrix}1 & 0\\ 0 & 1\end{pmatrix}, i=\begin{pmatrix}\sqrt{-1} & 0\\ 0 & -\sqrt{-1}\end{pmatrix}, j=\begin{pmatrix}0 & 1\\ -1 & 0\end{pmatrix}, k=\begin{pmatrix}0 & \sqrt{-1}\\ \sqrt{-1} & 0\end{pmatrix},$$

we see that Q is the same as the ring of real quaternions H, discussed earlier in this chapter.

As an exercise show that if $K=\mathbf{C}$, then Q is not a division ring.

CHAPTER 10

Ideals and homomorphisms

The concept of an "ideal" in a ring is analogous to the concept of a "normal subgroup" in a group. Rings modulo an ideal are constructed in the same canonical fashion as groups modulo a normal subgroup. The role of an ideal in a "homomorphism between rings" is similar to the role of a normal subgroup in a "homomorphism between groups." Theorems proved in this chapter on the direct sum of ideals in a ring and on homomorphisms between rings are parallel to the corresponding theorems for groups proved in Chapters 5 and 8.

1 Ideals

Definition. *A nonempty subset S of a ring R is called an* ideal *of R if*

(i) *$a,b \in S$ implies $a - b \in S$.*
(ii) *$a \in S$ and $r \in R$ imply $ar \in S$ and $ra \in S$.*

Definition. *A nonempty subset S of a ring R is called a* right (left) ideal *if*

(i) *$a,b \in S$ implies $a - b \in S$.*
(ii) *$ar \in S$ ($ra \in S$) for all $a \in S$ and $r \in R$.*

Clearly, a right ideal or a left ideal is a subring of R, and every ideal is both right and left, so an ideal is sometimes called a *two-sided ideal.* Trivially, in a commutative ring every right ideal or left ideal is two-sided.

In every ring R, (0) and R are ideals, called *trivial ideals.*

1.1 Examples of ideals

(a) In the ring of integers **Z** every subring is an ideal. To see this, let I be a subring of **Z** and $a \in I$, $r \in \mathbf{Z}$. Then

$$
\begin{aligned}
ar &= \overbrace{a + a + \cdots + a}^{r \text{ times}}, && \text{if } r > 0,\\
&= 0 && \text{if } r = 0,\\
&= \overbrace{-a - a - \cdots - a}^{-r \text{ times}}, && \text{if } r < 0.
\end{aligned}
$$

So in every case $ar \in I$. Hence, I is an ideal of **Z**.

(b) The right as well as left ideals of a division ring are trivial ideals only. Indeed, if a right or a left ideal I of a ring R contains a unit of R, then I is the whole ring, because a unit $u \in I$ implies $uu^{-1} \in I$; that is, $1 \in I$, so for all $r \in R$, we have $r = r1 \in I$. Hence, $I = R$.

(c) Let R be a ring and $a \in R$. Then

$$aR = \{ax \mid x \in R\}$$

is a right ideal of R, and

$$Ra = \{xa \mid x \in R\}$$

is a left ideal of R. If R is commutative, then aR is an ideal of R.

We note that a need not belong to aR. A sufficient condition for a to be in aR is that $1 \in R$; in this case aR is the smallest right ideal containing a.

(d) Let R be the $n \times n$ matrix ring over a field F. For any $1 \leq i \leq n$ let A_i (or B_i) be the set of matrices in R having all rows (columns), except possibly the ith, zero. Then A_i is a right ideal, and B_i is a left ideal in R. We later show that R has no nontrivial ideals.

(e) Let R be the ring of 2×2 upper triangular matrices over a field F. Then the subset

$$I = \left\{ \begin{bmatrix} 0 & a \\ 0 & 0 \end{bmatrix} \middle| \, a \in F \right\}$$

is an ideal in R.

(f) Let R be the ring of all functions from the closed interval $[0,1]$ to the field of real numbers. Let $c \in [0,1]$ and $I = \{f \in R \mid f(c) = 0\}$. Then I is an ideal of R.

(g) Let $R = F_2$, the 2×2 matrix ring over a field F. Let $S = \left[\begin{smallmatrix} F & F \\ 0 & F \end{smallmatrix}\right]$ be the set of upper triangular matrices over F. Then S is a subring of R. If $I = \left[\begin{smallmatrix} 0 & F \\ 0 & 0 \end{smallmatrix}\right]$, then I is an ideal of S, but I is neither a right ideal nor a left ideal of R.

Next we prove an important result about ideals in the matrix ring R_n. If A is an ideal in the ring R, then it can be easily shown that the ring A_n of all $n \times n$ matrices with entries from A is an ideal in R_n. The following theorem states that the converse is true for rings with unity.

1.2 Theorem. *If a ring R has unity, then every ideal I in the matrix ring R_n is of the form A_n, where A is some ideal in R.*

Proof. Let (e_{ij}), $i,j = 1,2,\ldots,n$, denote the matrix units in R_n. Set $A = \{a_{11} \in R \mid \Sigma a_{ij}e_{ij} \in I\}$. Then we claim A is an ideal in R. For let $a_{11}, b_{11} \in A$. Then there exist matrices $\alpha = \Sigma a_{ij}e_{ij}$ and $\beta = \Sigma b_{ij}e_{ij}$ in I. Because I is an ideal, $\alpha - \beta \in I$; so $a_{11} - b_{11} \in A$. Next, let $r \in R$ and $a_{11} \in A$ with $\Sigma a_{ij}e_{ij} \in I$. Then the matrix $(\Sigma a_{ij}e_{ij})(re_{11}) = \Sigma a_{i1}re_{i1} \in I$, and, hence, $a_{11}r \in A$. Similarly, $ra_{11} \in A$. So A is an ideal in R. We now proceed to show that $I = A_n$. Let $x = \Sigma a_{ij}e_{ij} \in I$ and let r and s be some fixed integers between 1 and n. Our purpose is to show that we can multiply x suitably on the right and left to produce an element in I whose entry at the (1,1) position is a_{rs}. This will mean that all $a_{ij} \in A$ and, hence, $x \in A_n$. Clearly,

$$e_{1r}\left(\sum a_{ij}e_{ij}\right)e_{s1} = a_{rs}e_{11},$$

which shows $a_{rs} \in A$, as desired. Thus, $I \subseteq A_n$. On the other hand, let $x = \Sigma a_{ij}e_{ij} \in A_n$. Here we will show that each term $a_{ij}e_{ij}$ of x is in I. Because $a_{ij} \in A$, there exists a matrix $\Sigma b_{rs}e_{rs} \in I$ such that $b_{11} = a_{ij}$. Then $e_{i1}(\Sigma b_{rs}e_{rs})e_{1j} = b_{11}e_{ij} \in I$; that is, $a_{ij}e_{ij} \in I$ for each $i,j = 1,2,\ldots,n$. Hence, $x \in I$. This completes the proof. □

As an immediate consequence, we get an important result contained in

1.3 Corollary. *If D is a division ring, then $R = D_n$ has no nontrivial ideals.*

Proof. Let I be a nonzero ideal in D_n. Then $I = A_n$, where A is some nonzero ideal in D. But the only ideals in any division ring D are (0) and D. Thus, $A = D$; so $I = D_n$, proving that (0) and D_n are the only ideals in the ring D_n. □

Remark. If $n > 1$, D_n has nontrivial right, as well as left, ideals [see Example 1.1(d)]. Because (0) and D are also the only right (or left) ideals of D, it follows that Theorem 1.2 does not hold, in general, if we replace the word "ideal" by "right ideal" (or "left ideal") in the theorem.

Further, the following example shows that Theorem 1.2 is not necessarily true if R is a ring without unity.

Example. Let $(R,+)$ be an additive group of order p, where p is prime. Define multiplication by $ab = 0$ for all $a,b \in R$. Then R is a ring without unity. If X is an additive subgroup of $(R,+)$, then X is also an ideal, because for all $x \in X$ and $r \in R$, we have $xr = 0 \in X$ and $rx = 0 \in X$. It then follows that a nonempty subset X of R is an ideal of R iff $(X,+)$ is a subgroup of $(R,+)$. But since $(R,+)$ has no proper subgroups, (0) and R are the only ideals of R. Now, it is straightforward to verify that

$$I = \left\{ \begin{pmatrix} a & b \\ 0 & 0 \end{pmatrix} \middle| \, a,b \in R \right\}$$

is an ideal in R_2. Suppose $I = A_2$ for some ideal A in R. This implies $I = (0)$ or R_2, since (0) or R are the only ideals in R. But clearly $I \neq 0$, and $I \subsetneq R_2$. Hence, Theorem 1.2 does not hold, in general, for matrix rings over rings without unity.

1.4 Theorem. *Let $(A_i)_{i \in \Lambda}$ be a family of right (left) ideals in a ring R. Then $\bigcap_{i \in \Lambda} A_i$ is also a right (left) ideal.*

Proof. Let $a,b \in \bigcap_{i \in \Lambda} A_i$, $r \in R$. Then for all $i \in \Lambda$, $a - b \in A_i$, and ar $(ra) \in A_i$, because A_i are right (left) ideals. Thus $a - b$ and ar $(ra) \in \bigcap_{i \in \Lambda} A_i$, proving that $\bigcap_{i \in \Lambda} A_i$ is a right (left) ideal. □

Next, let S be a subset of a ring R. Let $\mathscr{A} = \{A \mid A$ is a right ideal of R containing $S\}$. Then $\mathscr{A} \neq \emptyset$ because $R \in \mathscr{A}$. Let $I = \bigcap_{A \in \mathscr{A}} A$. Then I is the smallest right ideal of R containing S and is denoted by $(S)_r$. The smallest right ideal of R containing a subset S is called a *right ideal generated* by S. If $S = \{a_1, ..., a_m\}$ is a finite set, then $(S)_r$ is also written $(a_1, ..., a_m)_r$. Similarly, we define the *left ideal* and the *ideal generated by a subset S*, denoted, respectively, by $(S)_l$ and (S).

Definition. *A right ideal I of a ring R is called* finitely generated *if $I = (a_1, ..., a_m)_r$ for some $a_i \in R$, $1 \leq i \leq m$.*

Definition. *A right ideal I of a ring R is called* principal *if $I = (a)_r$ for some $a \in R$.*

In a similar manner we define a finitely generated left ideal, a finitely generated ideal, a principal left ideal, and a principal ideal.

We leave it as an exercise to verify that

$$(a) = \left\{ \sum_{\substack{i \\ \text{finite sum}}} r_i a s_i + ra + as + na \mid r,s,r_i,s_i \in R,\ n \in \mathbf{Z} \right\},$$

$$(a)_r = \{ar + na \mid r \in R,\ n \in \mathbf{Z}\},$$
$$(a)_l = \{ra + na \mid r \in R,\ n \in \mathbf{Z}\}.$$

If $1 \in R$, these can be simplified to

$$(a) = \left\{ \sum_{\text{finite sum}} r_i a s_i \mid r_i,s_i \in R \right\},$$

$$(a)_r = \{ar \mid r \in R\},$$
$$(a)_l = \{ra \mid r \in R\}.$$

In this case the symbols RaR, aR, and Ra are also used for (a), $(a)_r$, and $(a)_l$, respectively.

Definition. *A ring in which each ideal is principal is called a* principal ideal ring *(PIR).*

Similarly, we define *principal right (left) ideal rings.* A commutative integral domain with unity that is a principal ideal ring is called a *principal ideal domain* (PID).

We note that all ideals in the ring of integers $\mathbf{Z}$ and the polynomial ring $F[x]$ over a field F are principal. Let I be a nonzero ideal in $\mathbf{Z}$, and let n be the smallest positive integer in I (note that I does contain positive integers). Then for any element $m \in I$, we write $m = qn + r$, where $0 \leq r < n$. Then $r = m - qn \in I$. Because n is the smallest positive integer in I, r must be zero. Hence, $m = qn$, so $I = (n)$. Similarly, by using the division algorithm for $F[x]$ (Theorem 4.1, Chapter 11), we can show that every ideal in $F[x]$ is principal. Thus, $\mathbf{Z}$ and $F[x]$ are principal ideal rings.

We now proceed to define the quotient ring of a ring R. Let I be an ideal in a ring R. For $a,b \in R$ define $a \equiv b \pmod{I}$ if $a - b \in I$. Then $\equiv$ is an equivalence relation in R. Let R/I denote the set of equivalence classes, and let $\bar{a} \in R/I$ be the equivalence class containing a. It can be easily checked that the class $\bar{a}$ consists of elements of the form $a + x$, $x \in I$, and thus, it is also denoted by $a + I$.

We can make R/I into a ring by defining addition and multiplication as follows:

$$\bar{a} + \bar{b} = \overline{a + b}, \qquad \bar{a}\bar{b} = \overline{ab}.$$

We can easily show that these binary operations are well defined. For example, we proceed to show that multiplication is well defined. Therefore, let $\bar{a} = \bar{c}$ and $\bar{b} = \bar{d}$. Then $a - c \in I$ and $b - d \in I$. Write

$$ab - cd = a(b - d) + (a - c)d.$$

Because $a - c, b - d \in I$, which is an ideal, we have $ab - cd \in I$. Hence, $\overline{ab} = \overline{cd}$. This proves that multiplication is well defined.

We show $(R/I, +, \cdot)$ is a ring. First, $(R/I, +)$ is an additive abelian group:

$$\bar{a} + (\bar{b} + \bar{c}) = \bar{a} + (\overline{b + c}) = \overline{(a + (b + c))} = \overline{(a + b) + c}$$
$$= \overline{a + b} + \bar{c} = (\bar{a} + \bar{b}) + \bar{c}. \tag{1}$$
$$\bar{a} + \bar{0} = \overline{a + 0} = \bar{a}. \tag{2}$$
$$\bar{a} + (\overline{-a}) = \overline{a + (-a)} = \bar{0}. \tag{3}$$
$$\bar{a} + \bar{b} = \overline{a + b} = \overline{b + a} = \bar{b} + \bar{a}. \tag{4}$$

Second, $(R/I \cdot)$ is a semigroup:

$$\bar{a}(\bar{b}\,\bar{c}) = \bar{a}(\overline{bc}) = \overline{a(bc)} = \overline{(ab)c} = \overline{(ab)}\bar{c} = (\bar{a}\bar{b})\bar{c}. \tag{5}$$

Third, the distributive laws hold:

$$\bar{a}(\bar{b} + \bar{c}) = \bar{a}(\overline{b + c}) = \overline{a(b + c)} = \overline{ab + ac}$$
$$= \overline{ab} + \overline{ac} = \bar{a}\bar{b} + \bar{a}\bar{c}. \tag{6}$$
$$(\bar{b} + \bar{c})\bar{a} = \bar{b}\bar{a} + \bar{c}\bar{a}, \quad \text{similarly.} \tag{7}$$

Further, if $1 \in R$, then $\bar{1}$ is the unity of R/I; and if R is commutative, then R/I is also commutative.

Definition. *Let I be an ideal in a ring R. Then the ring $(R/I, +, \cdot)$ is called the* quotient ring *of R modulo I.*

The ring R/I is also sometimes denoted by $\bar{R}$ if there is no ambiguity about I.

If $I = R$, then R/I is the zero ring. If $I = (0)$, then R/I is, abstractly speaking, the same as the ring R by identifying $a + (0)$, $a \in R$, with a.

1.5 Examples

(a) Let $(n) = \{na \mid a \in \mathbf{Z}\}$ be an ideal in $\mathbf{Z}$. If $n \neq 0$, then the quotient ring $\mathbf{Z}/(n)$ is $\mathbf{Z}_n$, the usual ring of integers modulo n. If $n = 0$, then $\mathbf{Z}/(n)$ is the same as $\mathbf{Z}$.

(b) Let R be a ring with unity, and let $R[x]$ be the polynomial ring over R. Let $I = (x)$ be the ideal in $R[x]$ consisting of the multiples of x. Then the quotient ring

$$R[x]/I = \{\bar{a} \mid a \in R\}.$$

Solution. Let $x \in I$. Then $\bar{x} = \bar{0}$. Therefore, if

$$\overline{a + bx + cx^2 \cdots} \in R[x]/I,$$

then

$$\overline{a + bx + cx^2 + \cdots} = \bar{a} + \bar{b}\bar{x} + \bar{c}\bar{x}^2 + \cdots = \bar{a},$$

showing that $R[x]/I = \{\bar{a} \mid a \in R\}$.

(c) Consider the quotient ring $\mathbf{R}[x]/(x^2+1)$. Then

$$x^2 + 1 \in (x^2+1) \Rightarrow \overline{x^2+1} = \bar{0} \Rightarrow \bar{x}^2 + \bar{1} = \bar{0} \Rightarrow \bar{x}^2 = -\bar{1}.$$

This yields $\bar{x}^3 = -\bar{x}, \bar{x}^4 = \bar{1}$, and so on; so in general, $\bar{x}^n = \pm\bar{1}$ if n is even and $\bar{x}^n = \pm\bar{x}$ if n is odd. Let

$$\overline{a + bx + cx^2 + \cdots} \in \mathbf{R}[x]/(x^2+1).$$

Then

$$\begin{aligned}\overline{a + bx + cx^2 + \cdots} &= \bar{a} + \bar{b}\bar{x} + \bar{c}\bar{x}^2 + \cdots \\ &= \bar{\alpha} + \bar{\beta}\bar{x} \qquad \text{for some } \alpha, \beta \in \mathbf{R}.\end{aligned}$$

Thus,

$$\mathbf{R}[x]/(x^2+1) = \{\bar{\alpha} + \bar{\beta}\bar{x} \mid \alpha, \beta \in \mathbf{R}\} \qquad \text{where } \bar{x}^2 = -\bar{1}.$$

Indeed, $\mathbf{R}[x]/(x^2+1)$ is the field of complex numbers, where $\bar{\alpha}$, $\alpha \in \mathbf{R}$, is identified with α, and $\bar{x}$ is identified with $\sqrt{-1}$.

(d) Let $R = \left(\begin{smallmatrix}\mathbf{Z} & \mathbf{Q} \\ 0 & 0\end{smallmatrix}\right)$, and let $A = \left(\begin{smallmatrix}0 & \mathbf{Q} \\ 0 & 0\end{smallmatrix}\right)$ be an ideal in R. Then

$$R/A = \left\{ \overline{\begin{pmatrix} n & 0 \\ 0 & 0 \end{pmatrix}} \,\middle|\, n \in \mathbf{Z} \right\}.$$

Solution. Let

$$\begin{pmatrix} 0 & x \\ 0 & 0 \end{pmatrix} \in \begin{pmatrix} 0 & \mathbf{Q} \\ 0 & 0 \end{pmatrix}.$$

Then

$$\overline{\begin{pmatrix} 0 & x \\ 0 & 0 \end{pmatrix}} = \bar{0}.$$

Therefore,

$$\overline{\begin{pmatrix} n & x \\ 0 & 0 \end{pmatrix}} \in R/A \Rightarrow \overline{\begin{pmatrix} n & 0 \\ 0 & 0 \end{pmatrix}} + \overline{\begin{pmatrix} 0 & x \\ 0 & 0 \end{pmatrix}} \in R/A \Rightarrow \overline{\begin{pmatrix} n & 0 \\ 0 & 0 \end{pmatrix}} \in R/A.$$

Hence,

$$R/A = \left\{ \overline{\begin{pmatrix} n & 0 \\ 0 & 0 \end{pmatrix}} \,\middle|\, n \in \mathbf{Z} \right\}.$$

R/A is, indeed, the ring of integers, where $\overline{\begin{pmatrix} n & 0 \\ 0 & 0 \end{pmatrix}}$ is identified with n.

(e) Find the nontrivial (i) right ideals and (ii) ideals of the ring $R = \begin{pmatrix} \mathbf{Z} & \mathbf{Q} \\ 0 & 0 \end{pmatrix}$.

Solution. (i) Let A be a nonzero right ideal of R. Let

$$X = \left\{ n \in \mathbf{Z} \,\middle|\, \begin{pmatrix} n & a \\ 0 & 0 \end{pmatrix} \in A \text{ for some } a \in \mathbf{Q} \right\}.$$

Clearly, X is an additive subgroup of $\mathbf{Z}$; hence, $X = n_0\mathbf{Z}$ for some $n_0 \in \mathbf{Z}$.

Case 1. $X \neq (0)$; that is, $n_0 \neq 0$.

We show

$$A = \begin{pmatrix} n_0\mathbf{Z} & \mathbf{Q} \\ 0 & 0 \end{pmatrix}.$$

Let $a \in \mathbf{Q}$ be such that

$$\begin{pmatrix} n_0 & a \\ 0 & 0 \end{pmatrix} \in A.$$

Let $z \in \mathbf{Z}$ and $q \in \mathbf{Q}$. Then

$$\begin{pmatrix} n_0 z & q \\ 0 & 0 \end{pmatrix} = \begin{pmatrix} n_0 & a \\ 0 & 0 \end{pmatrix} \begin{pmatrix} z & q/n_0 \\ 0 & 0 \end{pmatrix} \in A.$$

Hence,

$$\begin{pmatrix} n_0\mathbf{Z} & \mathbf{Q} \\ 0 & 0 \end{pmatrix} \subset A.$$

Trivially,

$$A \subset \begin{pmatrix} n_0\mathbf{Z} & \mathbf{Q} \\ 0 & 0 \end{pmatrix}.$$

Therefore,

$$A = \begin{pmatrix} n_0\mathbf{Z} & \mathbf{Q} \\ 0 & 0 \end{pmatrix},$$

as claimed. The proof also shows that A is a principal right ideal of R.

Case 2. $X = (0)$.

Let

$$K = \left\{ q \in \mathbf{Q} \,\middle|\, \begin{pmatrix} 0 & q \\ 0 & 0 \end{pmatrix} \in A \right\}.$$

Clearly, K is an additive subgroup of $\mathbf{Q}$, and $A = \begin{pmatrix} 0 & K \\ 0 & 0 \end{pmatrix}$. Conversely, if K is an additive subgroup of $\mathbf{Q}$, it is easy to verify that $\begin{pmatrix} 0 & K \\ 0 & 0 \end{pmatrix}$ is indeed a right ideal of R. Therefore, in this case any right ideal of R is of the form $\begin{pmatrix} 0 & K \\ 0 & 0 \end{pmatrix}$, where K is an additive subgroup of $\mathbf{Q}$.

(ii) It is straightforward to verify that the nontrivial right ideals

$$A = \begin{pmatrix} n_0\mathbf{Z} & \mathbf{Q} \\ 0 & 0 \end{pmatrix}, \qquad 0 \neq n_0 \in \mathbf{Z}, \tag{1}$$

and

$$A = \begin{pmatrix} 0 & K \\ 0 & 0 \end{pmatrix}, \qquad K \text{ an additive subgroup of } \mathbf{Q}, \tag{2}$$

of R are also left ideals. Therefore, we have an example of a ring that is not commutative in which each right ideal is an ideal. It is interesting to note that each left ideal of R is not an ideal. (Consider

$$A = \left\{ \begin{pmatrix} n_0 m & ma \\ 0 & 0 \end{pmatrix} \middle| \, m \in \mathbf{Z} \right\},$$

where n_0 and a are fixed elements in $\mathbf{Z}$ and $\mathbf{Q}$, respectively.)

Problems

1. Let R be a commutative ring with unity. Suppose R has no nontrivial ideals. Prove that R is a field.
2. Prove the converse of Problem 1.
3. Generalize Problem 1 to noncommutative rings with unity having no nontrivial right ideals.
4. Find all ideals in $\mathbf{Z}$ and also in $\mathbf{Z}/(10)$.
5. Find all ideals in a polynomial ring $F[x]$ over a field F.
6. Find right ideals, left ideals, and ideals of a ring $R = [\begin{smallmatrix} \mathbf{Q} & \mathbf{Q} \\ 0 & 0 \end{smallmatrix}]$.
7. Show that every nonzero ideal in the ring of formal power series $F[[x]]$ in an indeterminate x over a field F is of the form (x^m) for some nonnegative integer m.
8. Let $L(R)$ be the set of all right (left) ideals in a ring R. Show that $L(R)$ is a modular lattice (see Problem 17, Section 1, Chapter 5). Give an example to show that the lattice $L(R)$ need not be distributive.

[A lattice is called distributive if $a \vee (b \wedge c) = (a \vee b) \wedge (a \vee c)$ for all $a, b, c \in L$.]

2 Homomorphisms

Let R be a ring, and let I be an ideal in R. Let R/I be the quotient ring of R modulo I. Then there exists a natural mapping

$$\eta \colon R \to R/I$$

that sends $a \in R$ into $\bar{a} \in R/I$. This mapping preserves binary operations in the sense that

$$\begin{aligned} \eta(a+b) &= \eta(a)+\eta(b), \\ \eta(ab) &= \eta(a)\eta(b), \end{aligned} \qquad a,b \in R.$$

A mapping with these two properties is called a *homomorphism* of R into R/I. The mapping η is called a *natural* or *canonical homomorphism.*

Definition. *Let f be a mapping from a ring R into a ring S such that*

(i) $f(a+b) = f(a)+f(b), \quad a,b \in R,$
(ii) $f(ab) = f(a)f(b), \quad a,b \in R.$

Then f is called a homomorphism *of R into S.*

If f is 1-1, then f is called an *isomorphism* (or a *monomorphism*) of R *into* S. In this case f is also called an *embedding* of the ring R into the ring S (or R is *embeddable* in S); we also say that S contains a copy of R, and R may be identified with a subring of S. The symbol $R \subseteq S$ means that R is embeddable in S.

If a homomorphism f from a ring R into a ring S is both 1-1 and onto, then there exists a homomorphism g from S into R that is also 1-1 and onto. In this case we say that the two rings R and S are *isomorphic,* and, abstractly speaking, these rings can be regarded as the same. We write $R \simeq S$ whenever there is a 1-1 homomorphism of R onto S. As stated above $R \simeq S$ implies $S \simeq R$. Also, the identity mapping gives $R \simeq R$ for any ring R. It is easy to verify that if $f: R \to S$ and $g: S \to T$ are isomorphisms of R onto S and S onto T, respectively, then gf is also an isomorphism of R onto T. Hence, $R \simeq S$ and $S \simeq T$ imply $R \simeq T$. Thus, we have shown that

Isomorphism is an equivalence relation in the class of rings.

Let us now list a few elementary but fundamental properties of homomorphisms.

2.1 Theorem. *Let $f: R \to S$ be a homomorphism of a ring R into a ring S. Then we have the following:*

(i) *If 0 is the zero of R, then $f(0)$ is the zero of S.*
(ii) *If $a \in R$, then $f(-a) = -f(a)$.*
(iii) *The set $\{f(a) | a \in R\}$ is a subring of S called the* homomorphic image *of R by the mapping f and is denoted by* Im f *or* $f(R)$.
(iv) *The set $\{a \in R | f(a) = 0\}$ is an ideal in R called the* kernel *of f and is denoted by* Ker f *or* $f^{-1}(0)$.

(v) *If* $1 \in R$, *then* $f(1)$ *is the unity of the subring* $f(R)$.
(vi) *If* R *is commutative, then* $f(R)$ *is commutative.*

Proof. (i) Let $a \in R$. Then $f(a) = f(a+0) = f(a) + f(0)$, and thus $f(0)$ is the zero of S. For convenience, $f(0)$ will also be written as 0.

(ii) $f(0) = f(a + (-a)) = f(a) + f(-a)$; therefore, $f(-a) = -f(a)$.

(iii) Let $f(a), f(b) \in f(R)$. Then

$$f(a) - f(b) = f(a) + f(-b) = f(a-b) \in f(R).$$

Also,

$$f(a)f(b) = f(ab) \in f(R).$$

Hence, $f(R)$ is a subring of S.

(iv) Let $a,b \in f^{-1}(0) = \text{Ker} f$, $r \in R$. Then

$$f(a-b) = f(a) - f(b) = 0 - 0 = 0.$$

Therefore, $a - b \in f^{-1}(0)$. Also

$$f(ar) = f(a)f(r) = 0f(r) = 0.$$

Thus, $ar \in f^{-1}(0)$. Similarly, $ra \in f^{-1}(0)$. Hence, $f^{-1}(0)$ is an ideal in R.

(v) If $a \in R$, then $f(a)f(1) = f(a1) = f(a)$. Similarly, $f(1)f(a) = f(a)$. Hence, $f(1)$ is the unity of $f(R)$.

(vi) If $a,b \in R$, then

$$f(a)f(b) = f(ab) = f(ba) = f(b)f(a).$$

Thus, $f(R)$ is commutative. □

It is clear that if f is 1-1, then $\text{Ker} f = (0)$.

Conversely, if the kernel of a homomorphism $f: R \to S$ is (0), then f is necessarily 1-1. For if $f(a) = f(b)$, it follows that $f(a-b) = 0$; thus, $a - b \in \text{Ker} f = (0)$. This shows $a = b$, and therefore f is 1-1. We record this fact separately in

2.2 Theorem. *Let* $f: R \to S$ *be a homomorphism of a ring* R *into a ring* S. *Then* $\text{Ker} f = (0)$ *if and only if* f *is* 1-1.

Let N be an ideal in a ring R. Form the quotient ring R/N of R modulo N. We can define a mapping

$$\eta: R \to R/N$$

in an obvious way (called the *canonical* or *natural homomorphism*) by the rule

$$\eta(a) = a + N = \bar{a}.$$

Then η is a homomorphism of R onto R/N because

$$\eta(a+b) = \overline{a+b} = \bar{a} + \bar{b} = \eta(a) + (b),$$
$$\eta(ab) = \overline{ab} = \bar{a}\bar{b} = \eta(a)\eta(b).$$

Also, if $\bar{x} \in R/N$, there exists $x \in R$ such that $\eta(x) = \bar{x}$. Thus, η is a homomorphism of R onto R/N; therefore R/N is a homomorphic image of R. It is an important fact that, in a certain sense, every homomorphic image of a ring R is of this type – a quotient ring of R modulo some ideal of R.

This is stated more precisely in the following

2.3 Theorem (fundamental theorem of homomorphisms). *Let f be a homomorphism of a ring R into a ring S with kernel N. Then*

$$R/N \simeq \operatorname{Im} f.$$

Proof. Let us define $g(a+N) = f(a)$. Then g is a mapping of R/N into $\operatorname{Im} f$, for $a + N = b + N$ implies $a - b \in N$, or $f(a-b) = 0$. This gives $f(a) - f(b) = 0$, so g is well defined. Next, g is a homomorphism. Writing $\bar{a}$ for $a + N$, we have

$$g(\bar{a} + \bar{b}) = g(\overline{a+b}) = f(a+b) = f(a) + f(b) = g(\bar{a}) + g(\bar{b}).$$

Similarly, $g(\bar{a}\bar{b}) = g(\bar{a})g(\bar{b})$. Clearly, g is an onto mapping. We show g is 1-1. Let $f(a) = f(b)$. Then $f(a-b) = 0$, so $a - b \in N$. But then $\bar{a} = \bar{b}$. This shows g is 1-1. Hence, $R/N \simeq \operatorname{Im} f$. □

As remarked earlier, we regard isomorphic rings (abstractly) as the same algebraic systems. In this sense this theorem states that the only homomorphic images of a ring R are the quotient rings of R. Accordingly, if we know all the ideals N in R, we know all the homomorphic images of R.

The fundamental theorem of homomorphisms between rings is also stated as follows.

2.4 Theorem (fundamental theorem of homomorphisms). *Given a homomorphism of rings $f\colon R \to S$, there exists a unique injective homomorphism $g\colon R/\operatorname{Ker} f \to S$ such that the diagram*

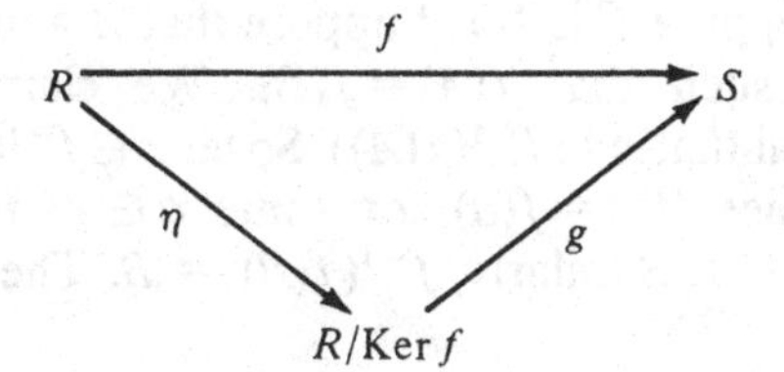

commutes; that is, $f = g\eta$, *where* η *is the canonical homomorphism.*

Clearly, g defined as $g(a+N) = f(a)$ is injective. Also, $f = g\eta$, as proved in Theorem 2.3. To observe that g is unique, let $f = h\eta$, where $h{:}R/\operatorname{Ker} f \to S$ is a homomorphism. Then $g\eta(a) = h\eta(a)$ for all $a \in R$; so $g(a+N) = h(a+N)$. This proves $g = h$. □

Next, let f be a mapping of a ring R into a ring S, and let A be a subset of S. We denote by $f^{-1}(A)$ the set of all elements of R whose images under the mapping f are in A; that is,

$$f^{-1}(A) = \{r \in R \mid f(r) \in A\}.$$

We remark that f^{-1} is a mapping of subsets of S into subsets of R.

If, however, f is a 1-1 mapping of R onto S, then the symbol f^{-1} is used to denote the mapping of S onto R defined by $f^{-1}(s) = r$, where r is the unique element of R such that $f(r) = s$. The context will make it clear in which sense the symbol f^{-1} is used.

2.5 Theorem (correspondence theorem). *Let* $f\colon R \to S$ *be a homomorphism of a ring* R *onto a ring* S, *and let* $N = \operatorname{Ker} f$. *Then the mapping* $F\colon A \mapsto f(A)$ *defines a* 1-1 *correspondence from the set of all ideals (right ideals, left ideals) in* R *that contain* N *onto the set of all ideals (right ideals, left ideals) in* S. *It preserves ordering in the sense that* $A \subsetneq B$ *iff* $f(A) \subsetneq f(B)$.

Proof. Let X be an arbitrary ideal in S, and set $A = f^{-1}(X)$. Then A is an ideal in R. For if $a,b \in A$ and $r \in R$, then $f(a-b) = f(a) - f(b) \in X$, and $f(ar) = f(a)f(r) \in X$, since $f(a) \in X$ and X is an ideal. Thus, $a - b \in f^{-1}(X)$ and $ar \in f^{-1}(X)$. Similarly, $ra \in f^{-1}(X)$. Hence, $A = f^{-1}(X)$ is an ideal in R. For any ideal A in R it is easy to check that $f(A)$ is an ideal in S.

To show that the mapping F is onto, we show that any ideal X in S is of the form $f(A)$ for some ideal A in R. In fact, we show $X = f(f^{-1}(X))$. Trivially, $X \supset f(f^{-1}(X))$. So let $x \in X$. Because f is onto, there exists an $a \in R$ such that $f(a) = x$. Then $a \in f^{-1}(X)$; so $x = f(a) \in f(f^{-1}(X))$ as desired.

We now show that the mapping F is 1-1. Suppose that A and B are ideals in R containing N such that $f(A)=f(B)$. We claim that $f^{-1}(f(A))=A$. Here it is trivial that $A \subset f^{-1}(f(A))$. So let $x \in f^{-1}(f(A))$. Then $f(x) \in f(A)$. This implies $f(x)=f(a)$ for some $a \in A$; that is, $x-a \in \operatorname{Ker} f \subset A$. Hence, $x \in A$. Similarly, $f^{-1}(f(B))=B$. Therefore, $f(A)=f(B)$ implies $A=B$.

Finally, let A and B be ideals in R such that $A \subsetneq B$. Then $f(A) \subset f(B)$. If $f(A)=f(B)$, we get, as before, $A=B$. Thus, $f(A) \subseteq f(B)$. Conversely, let $f(A) \subsetneq f(B)$. Then $A=f^{-1}(f(A)) \subseteq f^{-1}(f(B))=B$. But since $f(A) \subsetneq f(B)$, we cannot have $A=B$. Hence, $A \subsetneq B$. The 1-1 correspondence between the set of right or left ideals can be established exactly in the same way. □

An important consequence of Theorem 2.5 is the following immediate result, which we state as a theorem in view of its importance throughout our study.

2.6 Theorem. *If K is an ideal in a ring R, then each ideal (right or left ideal) in R/K is of the form A/K, where A is an ideal (right or left ideal) in R containing K.*

Proof. Let $\eta: R \rightarrow R/K$ be the natural homomorphism of R onto R/K. Then any ideal (right or left ideal) in R/K is of the form $\eta(A)$, where A is an ideal (right or left ideal) in R containing $\operatorname{Ker} \eta = K$. Then K is also an ideal in A (regarded as a ring by itself), and $\eta(A)=\{\eta(a)=a+K \mid a \in A\}$ is precisely A/K. This proves the theorem. □

Theorem 2.6, in particular, shows that if an ideal M in a ring R is such that $M \neq R$, and no ideal of R (other than R itself) contains M properly, then R/M has no nontrivial ideals. Such ideals (called "maximal") exist, in general, in abundance and play a very important role in the theory of rings and fields. The examples are ideals (p) in $\mathbf{Z}$, where p is prime, and ideal (x^2+1) in $\mathbf{R}[x]$. We return to this discussion in Section 4.

A concept opposite to that of homomorphism is that of antihomomorphism, defined next.

Definition. *Let R and S be rings. A mapping $f: R \rightarrow S$ is an* antihomomorphism *if, for all $x,y \in R$,*

$$f(x+y)=f(x)+f(y) \quad \text{and} \quad f(xy)=f(y) \cdot f(x).$$

Thus, an antihomomorphism preserves addition but reverses multiplication. An antihomomorphism that is both 1-1 and onto is called an *anti-isomorphism.*

Let $R = (R,+,\cdot)$ be a ring. Define binary operation $\circ$ in R as follows:

$$x \circ y = y \cdot x \qquad \text{for all } x,y \in R.$$

It is easy to verify that the system $(R,+,\circ)$ is a ring.

Definition. *Let $(R,+,\cdot)$ be a ring. Then the* opposite ring *of R, written R^{op}, is defined to be the ring $(R,+,\circ)$, where $x \circ y = y \cdot x$, for all $x,y \in R$.*

2.7 Examples

(a) The only homomorphisms from the ring of integers **Z** to **Z** are the identity and zero mappings.

Solution. Let f be a nonzero homomorphism. Then $(f(1))^2 = f(1)1f(1) = f(1 \cdot 1) = f(1) \neq 0$. So $f(1)$ is a nonzero idempotent in $f(\mathbf{Z}) \subset \mathbf{Z}$. But the only nonzero idempotent in **Z** is 1. Therefore $f(1) = 1$; hence,

$$\begin{aligned} f(n) &= f(1+1+\cdots+1) = \overbrace{f(1)+\cdots+f(1)}^{n \text{ times}} && (\text{if } n>0) \\ &= f(-1-1-\cdots-1) = \overbrace{f(-1)+\cdots+f(-1)}^{-n \text{ times}} && (\text{if } n<0) \\ &= 0 && (\text{if } n=0). \end{aligned}$$

In each case $f(n) = n$, which completes the solution.

(b) Let A and B be ideals in R such that $B \subseteq A$. Then

$$R/A \simeq (R/B)/(A/B).$$

Solution. Define a mapping f: $R/B \to R/A$ by setting $f(x+B) = x+A$. f is well defined, for $x_1 + B = x_2 + B$ implies $x_1 - x_2 \in B \subseteq A$. Thus, $x_1 + A = x_2 + A$.

Now $\operatorname{Ker} f = \{x+B \mid x+A = \bar{0}\} = \{x+B \mid x \in A\} = A/B$. Hence, by the fundamental theorem of homomorphisms

$$(R/B)/(A/B) \simeq R/A.$$

(c) Any ring R can be embedded in a ring S with unity.

Solution. Let S be the cartesian product of R and the set of integers **Z**; that is $S = R \times \mathbf{Z}$. S is made into a ring by defining

$$(a,m) + (b,n) = (a+b, m+n),$$
$$(a,m)(b,n) = (ab + na + mb, mn),$$

where $a,b \in R$ and $m,n \in \mathbf{Z}$. It can be easily checked that S is a ring with binary operations defined above, and $(0,1)$ is the unity of S. Also the mapping $f\colon R \to S$ defined by $f(a) = (a,0)$ is a 1-1 homomorphism. Hence, f is an embedding of R into S.

(d) Let R be a ring. Then $(R_n)^{\text{op}} \simeq (R^{\text{op}})_n$.

Solution. Define a mapping $f\colon (R_n)^{\text{op}} \to (R^{\text{op}})_n$ by $f(A) = {}^tA$, the transpose of A. Recall that, as sets (and indeed as additive groups), $R = R^{\text{op}}$ and $R_n = (R_n)^{\text{op}}$. Then by the definition of the transpose of a matrix, ${}^t(A+B) = {}^tA + {}^tB$; so $f(A+B) = f(A) + f(B)$. We now show $f(A \circ B) = f(A)f(B)$, where the multiplication of matrices $f(A)$ and $f(B)$ is in the ring $(R^{\text{op}})_n$. Assume $A = (a_{ij})$, $B = (b_{ij})$, $f(A) = {}^tA = (a'_{ij})$, and $f(B) = {}^tB = (b'_{ij})$. Then $a'_{ij} = a_{ji}$ and $b'_{ij} = b_{ji}$ for all $1 \le i,j \le n$. Now $f(A \circ B) = f(BA) = {}^t(BA)$. The (i,j) entry of ${}^t(BA)$ is the (j,i) entry of BA, which is given by

$$\sum_{k=1}^{n} b_{jk}a_{ki} = \sum_{k=1}^{n} a_{ki} \circ b_{jk} = \sum_{k=1}^{n} a'_{ik} \circ b'_{kj}$$
$$= (i,j) \text{ entry of } {}^tA\,{}^tB \in (R^{\text{op}})_n.$$

Hence, ${}^t(BA) = {}^tA\,{}^tB$. This proves that $f(A \circ B) = f(A)f(B)$. f is clearly 1-1 and onto. Hence, $(R_n)^{\text{op}} \simeq (R^{\text{op}})_n$.

Problems

1. Find the ideals of the ring $\mathbf{Z}/(n)$.
2. Prove that $\mathbf{Z}[x]/(x^2+1) \simeq \mathbf{Z}[i]$, where $\mathbf{Z}[i] = \{a + b\sqrt{-1} \mid a,b \in \mathbf{Z}\}$.
3. Show that there exists a ring homomorphism $f\colon \mathbf{Z}/(m) \to \mathbf{Z}/(n)$ sending $\bar{1}$ to $\bar{1}$ if and only if $n \mid m$.
4. Show that the set N of all nilpotent elements in a commutative ring R forms an ideal. Also show that R/N has no nonzero nilpotent elements. Give an example to show that N need not be an ideal if R is not commutative.
5. Let S be a nonempty subset of R. Let

$$r(S) = \{x \in R \mid Sx = 0\} \quad \text{and} \quad l(S) = \{x \in R \mid xS = 0\}.$$

Then show that $r(S)$ and $l(S)$ are right and left ideals, respectively [$r(S)$ and $l(S)$ are called right and left *annihilators* of S, respectively].

6. In Problem 5 show that $r(S)$ and $l(S)$ are ideals in R if S is an ideal in R.
7. Show that any nonzero homomorphism of a field F into a ring R is 1-1.
8. Let $f: F \to F$ be a nonzero homomorphism of a field F into itself. Show that f need not be onto.
9. In any ring R show that the set $A = \{x \in R | r(x) \cap B \neq (0)$ for all nonzero right ideals B of $R\}$ is an ideal in R.
10. Show that the intersection of two ideals is an ideal, but that the union need not be ideal.
11. If A, B, and C are right (left) ideals in a ring R such that $A \subseteq B \cup C$, show that $A \subseteq B$ or $A \subseteq C$.
12. Show that a ring R with unity is a division ring if and only if R has no nontrivial right ideals.
13. Let R_1 and R_2 be rings and $R = R_1 \times R_2$, the direct product of R_1 and R_2. Show that

    $R_1^* = \{(x,0) | x \in R_1\}, \qquad R_2^* = \{(0,y) | y \in R_2\}$

    are ideals in R. Also show that $R_1^* \simeq R_1$, $R_2^* \simeq R_2$ as rings.
14. Let A be a right ideal of R. Define

    $I(A) = \{x \in R | xa \in A \text{ for all } a \in A\}.$

    Prove that $I(A)$ is the largest subring of R in which A is contained as an ideal. ($I(A)$ is called the *idealizer* of A in R.)
15. Show that $(R/A)_n \simeq R_n/A_n$, where A is an ideal in a ring R.
16. Let R be a ring. Show that R is anti-isomorphic to R^{op}.
17. Let $f,g: R \to R$ be antihomomorphisms (anti-isomorphisms) of a ring R into itself. Show that fg is a homomorphism (isomorphism).
18. Let R be a ring with unity. If the set of all noninvertible elements forms an ideal M, show that R/M is a division ring. Also show that this is so if, for each $r \in R$, either r or $1 - r$ is invertible.
19. Show that the following rings R satisfy the property of Problem 18: that is, the set of all noninvertible elements forms a unique maximal ideal M. Find the ideal M in each case.

    (a) $R = \mathbf{Z}/(p^2)$, p prime.

    (b) $R = F[[x]]$.

    (Such rings are called *local*, as defined in Chapter 12.)

3 Sum and direct sum of ideals

In this section we discuss the concept of the sum of ideals (right ideals, left ideals) in a ring R.

Definition. *Let $A_1, A_2, \ldots, A_n$ be a family of right ideals in a ring R. Then the smallest right ideal of R containing each A_i, $1 \le i \le n$ (that is, the intersection of all the right ideals in R containing each A_i), is called the* sum *of $A_1, A_2, \ldots, A_n$.*

3.1 Theorem. *If $A_1, A_2, \ldots, A_n$ are right ideals in a ring R, then $S = \{a_1 + a_2 + \cdots + a_n | a_i \in A_i,\ i = 1,2,\ldots,n\}$ is the sum of right ideals $A_1, A_2, \ldots, A_n$.*

Proof. It is clear that

$$S = \{a_1 + a_2 + \cdots + a_n | a_i \in A_i, i = 1,\ldots,n\}$$

is a right ideal in R. Also, if $a_1 \in A_1$, then $a_1 = a_1 + 0 + \cdots + 0$ is in S, and, hence, $A_1 \subset S$. Similarly, each A_i, $i = 2,\ldots,n$, is contained in S. Further, if T is any right ideal in R containing each A_i, then obviously $T \supseteq S$. Thus, S is the intersection of all the right ideals in R containing each A_i. □

Notation. The sum of right (or left) ideals $A_1, A_2, \ldots, A_n$ in a ring R is denoted by $A_1 + A_2 + \cdots + A_n$. From the definition of the sum it is clear that the order of A_i's in $A_1 + A_2 + \cdots + A_n$ is immaterial. We write $A_1 + A_2 + \cdots + A_n$ as $\Sigma_{i=1}^n A_i$.

Definition. *A sum $A = \Sigma_{i=1}^n A_i$ of right (or left) ideals in a ring R is called a* direct sum *if each element $a \in A$ is uniquely expressible in the form $\Sigma_{i=1}^n a_i$, $a_i \in A_i$, $1 \le i \le n$. If the sum $A = \Sigma A_i$ is a direct sum, we write it as*

$$A = A_1 \oplus A_2 \oplus \cdots \oplus A_n = \oplus \sum_{i=1}^{n} A_i.$$

Note. One can similarly define the sum and direct sum of an infinite family of right (left) ideals in a ring R. Although no extra effort is needed to talk about this, we prefer to postpone it to Chapter 14, where we discuss sum and direct sum of a more general family.

3.2 Theorem. *Let $A_1, A_2, \ldots, A_n$ be right (or left) ideals in a ring R. Then the following are equivalent:*

(i) $A = \Sigma_{i=1}^{n} A_i$ *is a direct sum.*
(ii) *If* $0 = \Sigma_{i=1}^{n} a_i$, $a_i \in A_i$, *then* $a_i = 0$, $i = 1,2,\ldots,n$.
(iii) $A_i \cap \Sigma_{j=1, j\neq i}^{n} A_j = (0)$, $i = 1,2,\ldots,n$.

Proof. (i) $\Rightarrow$ (ii) Follows from definition of direct sum.

(ii) $\Rightarrow$ (iii) Let $x \in A_i \cap \Sigma_{j=1, j\neq i}^{n} A_j$. Then

$$x = a_1 + \cdots + a_{i-1} + a_{i+1} + \cdots + a_n \in A_i.$$

Thus,

$$0 = a_1 + \cdots + a_{i-1} + (-x) + a_{i+1} + \cdots + a_n.$$

Then by (ii) we get $x = 0$.

(iii) $\Rightarrow$ (i) Let $a = a_1 + a_2 + \cdots + a_n$ and $a = b_1 + b_2 + \cdots + b_n$, where $a_i, b_i \in A_i$, $i = 1,2,\ldots,n$. Then $0 = (a_1 - b_1) + (a_2 - b_2) + \cdots + (a_n - b_n)$. This gives

$$a_1 - b_1 \in A_1 \cap \sum_{i=2}^{n} A_i = (0).$$

Hence, $a_1 = b_1$. Similarly, $a_2 = b_2, \ldots, a_n = b_n$. Hence, $A = \Sigma_{i=1}^{n} A_i$ is a direct sum. □

3.3 Theorem. *Let* $R_1, R_2, \ldots, R_n$ *be a family of rings, and let* $R = R_1 \times R_2 \times \cdots \times R_n$ *be their direct product. Let* $R_i^* = \{(0,\ldots,0,a_i,0,\ldots,0) | a_i \in R_i\}$. *Then* $R = \oplus \Sigma_{i=1}^{n} R_i^*$ *is a direct sum of ideals* R_i^*, *and* $R_i^* \simeq R_i$ *as rings. On the other hand, if* $R = \oplus \Sigma_{i=1}^{n} A_i$, *a direct sum of ideals of* R, *then* $R \simeq A_1 \times A_2 \times \cdots \times A_n$, *the direct product of the* A_i*'s considered as rings on their own right.*

Proof. Clearly, R_i^* are ideals in R, and $R = R_1^* + \cdots + R_n^*$. We prove that R is a direct sum of ideals R_i^*. Let

$$x \in R_i^* \cap \sum_{\substack{j=1 \\ j\neq i}}^{n} R_j^*.$$

Then

$$x = (0,0,\ldots,a_i,0,\ldots,0) = (a_1, a_2, \ldots, a_{i-1}, 0, a_{i+1}, \ldots, a_n).$$

This gives $a_i = 0$ and, hence, $x = 0$. Therefore,

$$R = \oplus \sum_{i=1}^{n} R_i^*.$$

For the second part we note that if $x \in R$, then x can be uniquely expressed as $a_1 + a_2 + \cdots + a_n$, $a_i \in A_i$, $1 \le i \le n$. Define a mapping

$$f: \oplus \sum_{i=1}^{n} A_i \to A_1 \times A_2 \times \cdots \times A_n$$

by

$$f(a_1 + a_2 + \cdots + a_n) = (a_1, a_2, \ldots, a_n).$$

Because $\Sigma_{i=1}^{n} A_i$ is direct, f is well defined. It is also clear that f is both 1-1 and onto. Also, if $x,y \in \oplus \Sigma_{i=1}^{n} A_i$, then $f(x+y) = f(x) + f(y)$. To show $f(xy) = f(x)f(y)$, we need to note that if $x = a_1 + a_2 + \cdots + a_n$ and $y = b_1 + b_2 + \cdots + b_n$, with $a_i, b_i \in A_i$, $1 \le i \le n$, then, for $i \ne j$, $a_i b_j = 0$, since $a_i b_j \in A_i \cap A_j = (0)$. This remark then immediately yields that $f(xy) = f(x)f(y)$. Hence, f is an isomorphism. □

The direct sum $R = \oplus \Sigma_{i=1}^{n} A_i$ is also called the *(internal) direct sum* of ideals A_i in R, and the direct product $A_1 \times A_2 \times \cdots \times A_n$ is called the *(external) direct sum* of the family of rings A_i, $i = 1,2,\ldots,n$. In the latter case the notation $A_1 \oplus A_2 \oplus \cdots \oplus A_n$ is also frequently used. The context will make clear the sense in which the term "direct sum" is used.

Definition. *A right (left) ideal I in a ring R is called* minimal *if (i) $I \ne (0)$, and (ii) if J is a nonzero right (left) ideal of R contained in I, then $J = I$.*

It is clear that if I is a minimal right ideal, then I is generated by any nonzero element of I. Indeed, if I is a right ideal of R with the property that each nonzero element generates I, then I is minimal. To see this, let J be a right ideal of R such that $0 \ne J \subset I$. Suppose $0 \ne a \in J$. By assumption, $I = (a)_r \subset J$, so $I = J$.

If R is a division ring, then R itself is a minimal right ideal as well as a minimal left ideal. A nontrivial example of a minimal right ideal is

$$e_{11}R = \left\{ \begin{bmatrix} a_{11} & \cdots & a_{1n} \\ 0 & \cdots & 0 \\ \cdot & & \cdot \\ \cdot & & \cdot \\ \cdot & & \cdot \\ 0 & \cdots & 0 \end{bmatrix} \middle| \; a_{ij} \in D \right\},$$

where $R = D_n$ is the $n \times n$ matrix ring over a division ring D. Let J be a nonzero right ideal of R contained in $e_{11}R$. Let $0 \ne \Sigma_{j=1}^{n} a_{1j} e_{1j} \in J$. Then for some l, $1 \leq l \leq n$, $a_{1l} \neq 0$; so a_{1l}^{-1} exists. This gives

$$e_{11} = \left(\sum_{j=1}^{n} a_{1j}e_{1j}\right)(a_{1l}^{-1}e_{l1}) \in J.$$

Therefore, $e_{11}R \subset J \subset e_{11}R$. Hence, $J = e_{11}R$, showing that $e_{11}R$ is minimal.

3.4 Examples

(a) For any two ideals A and B in a ring R,

(i) $$\frac{A+B}{B} \simeq \frac{A}{A \cap B}.$$

(ii) $$\frac{A+B}{A \cap B} \simeq \frac{A+B}{A} \times \frac{A+B}{B} \simeq \frac{B}{A \cap B} \times \frac{A}{A \cap B}.$$

In particular, if $R = A + B$, then

$$\frac{R}{A \cap B} \simeq \frac{R}{A} \times \frac{R}{B}.$$

More generally, if $A_1, A_2, ..., A_n$ is a family of ideals in a ring R with unity such that $A_i + A_j = R$ for all $1 \leq i, j \leq n$, $i \neq j$, then

$$\frac{R}{\bigcap_{i=1}^{n} A_i} \simeq \frac{R}{A_1} \times \frac{R}{A_2} \times \cdots \times \frac{R}{A_n}.$$

Solution. (i) It is clear that B is an ideal in $A + B$ and $A \cap B$ is an ideal in A. Then the mapping $f: A \to (A + B)/B$, given by $f(a) = a + B$, is clearly a homomorphism. Also, if $x + B \in (A + B)/B$ and $x = a + b$, $a \in A$, $b \in B$, then $f(a) = a + B = x + B$. Thus, f is onto. It is easy to check that Ker $f = A \cap B$. Hence, by the fundamental theorem of homomorphisms, $A/(A \cap B) \simeq (A + B)/B$.

(ii) Note that A and B are also ideals in $A + B$ considered as a ring by itself. The mapping

$$g: A + B \to (A + B)/A \times (A + B)/B,$$

given by $g(x) = (x + A, x + B)$, $x \in A + B$, is clearly a homomorphism whose kernel is $A \cap B$. We show that g is onto.

Let $(x + A, 0 + B) \in (A + B)/A \times (A + B)/B$. Write $x = a + b$, $a \in A$, $b \in B$. Then $g(b) = (b + A, b + B) = (x + A, 0 + B)$. Thus, each element of the type $(x + A, 0 + B)$ has a preimage in $A + B$ by the mapping g. Similarly, elements of the type $(0 + A, y + B)$ have preimages. Hence, g is an onto mapping. Then by the fundamental theorem of homomorphisms we get

$$\frac{A+B}{A \cap B} \simeq \frac{A+B}{A} \times \frac{A+B}{B}.$$

By (i)

$$\frac{A+B}{A} \simeq \frac{B}{A \cap B} \quad \text{and} \quad \frac{A+B}{B} \simeq \frac{A}{A \cap B},$$

proving (ii).

To prove the last part we first prove by induction that

$$R = A_i + \bigcap_{\substack{j=1 \\ j \neq i}}^{n} A_j, \qquad 1 \leq i \leq n,\ n \geq 2.$$

Assume that this is true for $n - 1$ ideals. Then

$$R = R^2 = (A_i + A_n)(A_i + \bigcap_{\substack{j=1 \\ j \neq i}}^{n-1} A_j) \subset A_i + \bigcap_{\substack{j=1 \\ j \neq i}}^{n} A_j,$$

so

$$R = A_i + \bigcap_{\substack{j=1 \\ j \neq i}}^{n} A_j.$$

To complete the proof, we assume the result is true for $n - 1$ ideals. Then by (ii)

$$\frac{R}{\bigcap_{i=1}^{n} A_i} \simeq \frac{R}{\bigcap_{i=1}^{n-1} A_i} \times \frac{R}{A_n} \simeq \frac{R}{A_1} \times \cdots \times \frac{R}{A_n}.$$

(b) For any positive integer n,

$$\mathbf{Z}/(n) \simeq \mathbf{Z}/(p_1^{e_1}) \times \mathbf{Z}/(p_2^{e_2}) \times \cdots \times \mathbf{Z}/(p_k^{e_k}),$$

where $n = p_1^{e_1} p_2^{e_2} \cdots p_k^{e_k}$, p_i are distinct primes.

Solution. Let $A_i = (p_i^{e_i})$, $1 \leq i \leq k$. We show $A_i + A_j = \mathbf{Z}$, $i \neq j$. Because p_i and p_j are relatively prime, we can find integers a and b such that

$ap_i^{e_i} + bp_j^{e_j} = 1$. Then for all $x \in \mathbf{Z}$,

$$x = xap_i^{e_i} + xbp_j^{e_j} \in (p_i^{e_i}) + (p_j^{e_j}).$$

Hence, $A_i + A_j = \mathbf{Z}$, $i \neq j$.

We further show

$$(n) = \bigcap_{i=1}^{k} (p_i^{e_i}) = \bigcap_{i=1}^{k} A_i.$$

Recall that $n = p_1^{e_1} p_2^{e_2} \cdots p_k^{e_k}$. Let $x \in (n)$. Then $x = \lambda p_1^{e_1} p_2^{e_2} \cdots p_k^{e_k}$, $\lambda \in \mathbf{Z}$. This implies $x \in (p_i^{e_i})$ for all $i = 1,2,...,k$, and thus $x \in \bigcap_{i=1}^{k} A_i$. On the other hand, let $x \in \bigcap_{i=1}^{k} A_i$. Then

$$x = p_1^{e_1} a_1 = p_2^{e_2} a_2 = \cdots = p_k^{e_k} a_k, \qquad a_i \in \mathbf{Z}.$$

This shows $p_i^{e_i} | x$, $1 \leq i \leq k$. Because the p_i's are all distinct, we get $n = p_1^{e_1} p_2^{e_2} \cdots p_k^{e_k}$ divides x. Hence, $x \in (n)$. This proves that $(n) = \bigcap_{i=1}^{k} (p_i^{e_i})$. By example (a), the solution is then complete.

(c) Let (e_i), $1 \leq i \leq n$, be a family of idempotents in a ring R such that $e_i e_j = 0$, $i \neq j$. Then $A = e_1 R \oplus \cdots \oplus e_n R$ is a direct sum of right ideals $e_i R$, $1 \leq i \leq n$. (A family of idemopotents (e_i), $1 \leq i \leq n$, is called an *orthogonal family of idempotents* if $e_i e_j = 0$, $i \neq j$.)

Solution. Let $0 = e_1 a_1 + \cdots + e_n a_n$, $e_i a_i \in e_i R$. Multiplying both sides on the left by e_1, we get $0 = e_1 a_1$. Similarly, $e_i a_i = 0$, $2 \leq i \leq n$. Thus, by Theorem 3.2, A is a direct sum of right ideals $e_i R$, $1 \leq i \leq n$.

(d) Let R be a ring with unity, and let (A_i), $1 \leq i \leq n$, be a family of right ideals of R. Suppose $R = A_1 \oplus \cdots \oplus A_n$. Then there exists an orthogonal family of idempotents (e_i), $1 \leq i \leq n$, such that $A_i = e_i R$.

Solution. Because $1 \in R$, we can write

$$1 = e_1 + \cdots + e_n, \qquad e_i \in A_i.$$

Then

$$e_i = e_1 e_i + \cdots + e_i^2 + e_{i+1} e_i + \cdots + e_n e_i.$$

This yields

$$0 = e_1 e_i + \cdots + (e_i^2 - e_i) + e_{i+1} e_i + \cdots + e_n e_i.$$

But then by Theorem 3.2, each term is zero; that is, $e_j e_i = 0, j \neq i, e_i^2 = e_i$. Further, $e_i \in A_i$ implies $e_i R \subset A_i$. Next, let $x \in A_i$. Then $1 = e_1 + \cdots + e_n$

gives $x = e_1x + \dots + e_nx$, which gives, as before, $x = e_ix$. Hence, $x \in e_iR$, proving that $A_i = e_iR$.

(e) Let R be a ring with 1 such that R is a finite direct sum of minimal right (left) ideals. Then every right (left) ideal A in R is of the form eR (Re) where e is an idempotent.

Solution. By hypothesis $R = M_1 \oplus \cdots \oplus M_k$, where M_i are minimal right ideals in R. Assume that $A \neq (0)$ and $A \neq R$. Because $A \cap M_i \subseteq M_i$ and M_i are minimal right ideals, either $A \cap M_i = (0)$ or $A \cap M_i = M_i$. Because $A \neq R$, there exists, say, M_1 (after renumbering if necessary) such that $M_1 \not\subset A$. But then $M_1 \cap A = (0)$. So $A_1 = A + M_1$ is a direct sum. If $A_1 \neq R$, then there exists some M_i, say M_2, such that $A_1 \cap M_2 = (0)$ and $A \subseteq A_2 = A_1 \oplus M_2 = A \oplus M_1 \oplus M_2$. By continuing this process we must come to a right ideal, say A_s, that contains all M_i and therefore coincides with R. Thus, there exists a right ideal B such that $R = A \oplus B$. By Example (d), $A = eR$.

Problems

1. Show that $A + A = A$ for any ideal A in a ring R.
2. Show directly that $\mathbf{Z}/(12) \simeq \mathbf{Z}/(3) \times \mathbf{Z}/(4)$.
3. Let $A_1, A_2, \dots, A_n$ be ideals in a ring R with unity such that $A_i + A_j = R$ for all $i \neq j$. Show that if $x_1, x_2, \dots, x_n \in R$, then there exists an element $x \in R$ such that $x \equiv x_i \pmod{A_i}$ for all i. (This is called the *Chinese remainder theorem.*)
4. If e is any idempotent in R show that $R = eR \oplus (1 - e)R$ is a direct sum of right ideals. (In case $1 \notin R$, then $(1 - e)R$ denotes the set $\{a - ea \mid a \in R\}$.)
5. Let $R = R_1 \oplus R_2$ be the direct sum of the rings R_1 and R_2 each with unity. Denote by f_i the *projections* of R onto R_i, $i = 1,2$, given by $f_1(a_1, a_2) = a_1$ and $f_2(a_1, a_2) = a_2$. Prove the following:
   (a) Each f_i is an onto homomorphism.
   (b) If A is an ideal (right ideal) in R, then

   $A = f_1(A) \oplus f_2(A)$.

   (c) The ideals (right ideals) in R are precisely those ideals (right ideals) that are of the form $A_1 \oplus A_2$, where A_1 and A_2 are ideals (right ideals) in R_1 and R_2, respectively.
6. Let $R = D_n$ be the ring of $n \times n$ matrices over a division ring D. Let (e_{ij}), $1 \leq i, j \leq n$, be the set of matrix units. Show that
   (a) $e_{ii}R$ (Re_{ii}) is a right ideal (left ideal) in R, $i = 1,2,\dots,n$.

(b) Let A be a nonzero right ideal of R contained in $e_{ii}R$. Then $A = e_{ii}R$. Similarly, for left ideals, Re_{ii}.

(c) $R = \oplus\Sigma_{i=1}^{n} e_{ii}R$ is a direct sum of right ideals $e_{ii}R$. Also show this for left ideals.

(d) If A is right (left) ideal of R, then $A = eR$ (Re) for some idempotent $e \in R$ [see Example 3.4(e)].

7. Let R and S be rings, and let A and B be ideals in R and S, respectively. Show that

$$\frac{R \times S}{A \times B} \simeq \frac{R}{A} \times \frac{S}{B}.$$

4 Maximal and prime ideals

If $A = (p_1^{e_1})$ and $B = (p_2^{e_2})$ are ideals in **Z** generated by $p_1^{e_1}$ and $p_2^{e_2}$, respectively, where p_1, p_2 are distinct primes and e_1, e_2 are positive integers, then $A + B = \mathbf{Z}$. Such ideals are called *comaximal ideals* in **Z**.

Definition. *Two ideals A, B in any ring R are called* comaximal *if $A + B = R$.*

Definition. *An ideal A in a ring R is called* maximal *if*

(i) $A \neq R$; and

(ii) for any ideal $B \supseteq A$, either $B = A$ or $B = R$.

Clearly, an ideal A is maximal if and only if the pair X, A, for all ideals $X \not\subset A$, is comaximal.

4.1 Theorem. *For any ring R and any ideal $A \neq R$, the following are equivalent:*

(i) A is maximal.

(ii) The quotient ring R/A has no nontrivial ideals.

(iii) For any element $x \in R$, $x \notin A$, $A + (x) = R$.

Proof. (i) $\Rightarrow$ (ii) By Theorem 2.6 the ideals of R/A are of the form B/A, where B is an ideal in R containing A. If B/A is not zero, then $B \neq A$ and, therefore, $B = R$. Therefore, the only ideals of R/A are R/A and the zero ideal.

(ii) $\Rightarrow$ (iii) Let $x \in R$ and $x \notin A$. Then $A + (x)$ is an ideal of R properly containing A. This implies that $(A + (x))/A$ is a nonzero ideal in R/A and, hence, $A + (x) = R$.

(iii) $\Rightarrow$ (i) Let $B \supseteq A$. If $B \neq A$, choose an element $x \in B$, $x \notin A$. Now by (iii) $A + (x) = R$. But $B \supseteq A + (x) = R$. Hence, $B = R$. □

Definition. *A ring R is called a* simple *ring if the only ideals of R are the zero ideal and R itself.*

A field is clearly a simple ring. Indeed, a commutative simple ring with unity must be a field (Problem 1, Section 1). An example of a noncommutative simple ring is F_n, the $n \times n$ matrix ring over a field F, $n > 1$.

An immediate but important consequence of Theorem 4.1 is

4.2 Theorem. *In a nonzero commutative ring with unity, an ideal M is maximal if and only if R/M is a field.*

Proof. Suppose R is a nonzero commutative ring with unity. Then, for any ideal M, R/M is also a commutative ring with unity. If M is a maximal ideal, then, by Theorem 4.1, R/M is also a simple ring. Let $\bar{0} \neq \bar{a} \in \bar{R} = R/M$. Then $\bar{a}\bar{R}$ is a nonzero ideal in $\bar{R}$. Because $\bar{R}$ is simple, $\bar{a}\bar{R} = \bar{R}$. This implies that there exists $\bar{b} \in \bar{R}$ such that $\bar{a}\bar{b} = \bar{1}$. This shows that $\bar{R}$ is a field. Conversely, let $\bar{R} = R/M$ be a field. Then $\bar{R}$ has no nontrivial ideals. So if $K \supsetneq M$ is an ideal in R, the nonzero ideal K/M in R/M must be R/M. Hence, $K = R$, so M is a maximal ideal. □

4.3 Examples of maximal ideals

(a) An ideal M in the ring of integers $\mathbf{Z}$ is a maximal ideal if and only if $M = (p)$, where p is some prime.

Recall that each ideal in $\mathbf{Z}$ is of the form (n), where $n \in \mathbf{Z}$. Now, clearly, $(n) = (-n)$. So we can always assume that any ideal in $\mathbf{Z}$ is of the form (n), where n is a nonnegative integer.

Suppose (n) is a maximal ideal. Then $\mathbf{Z}/(n)$ is a field. But if n is composite, say $n = n_1 n_2$, where $n_1 > 1, n_2 > 1$, then $\bar{n}_1\bar{n}_2 = \bar{0}$ in $\mathbf{Z}/(n)$ and $\bar{0} \neq \bar{n}_1$, $\bar{0} \neq \bar{n}_2$, which is a contradiction because $\mathbf{Z}/(n)$ is a field. Hence, n is prime. To prove the converse, let $a \in \mathbf{Z}$ such that $p \nmid a$. Then there exist integers $x, y, \in \mathbf{Z}$ such that $ax + py = 1$. This implies $\bar{a}\bar{x} + \bar{p}\bar{y} = \bar{1}$, so $\bar{a}\bar{x} = \bar{1}$, which proves that $\mathbf{Z}/(p)$ is a field.

(b) If R is the ring of 2×2 matrices over a field F of the form $\begin{bmatrix} a & b \\ 0 & 0 \end{bmatrix}$, $a, b \in F$, then the set

$$M = \left\{ \begin{bmatrix} 0 & b \\ 0 & 0 \end{bmatrix} \middle| \, b \in F \right\}$$

is a maximal ideal in R. Clearly, M is an ideal in R. Let

$$S=\left\{\begin{bmatrix} a & 0 \\ 0 & 0 \end{bmatrix} \middle| a \in F\right\}.$$

Then S is a subring of R that is isomorphic to F under the mapping $\left[\begin{smallmatrix} a & 0 \\ 0 & 0 \end{smallmatrix}\right] \mapsto a$. Further, the mapping $f\colon R \to S$, where

$$f\begin{bmatrix} a & b \\ 0 & 0 \end{bmatrix} = \begin{bmatrix} a & 0 \\ 0 & 0 \end{bmatrix}$$

is an onto homomorphism whose kernel is M. Thus, by the fundamental theorem of homomorphisms, $R/M \simeq S$. Because S is a field, it follows that M is a maximal ideal in R.

Definition. *Let A and B be right (left) ideals in a ring R. Then the set*

$$\left\{\sum_{\text{finite sum}} a_i b_i \mid a_i \in A, b_i \in B\right\},$$

which is easily seen to be a right (left) ideal in R, is called the product *of A and B (in this order) and written AB.*

Note that if A and B are ideals in R, then their product AB is an ideal in R.

4.4 Theorem. *Let A, B, and C be right (left) ideals in a ring R. Then*

(i) $(AB)C = A(BC)$,
(ii) $A(B+C) = AB + AC$ *and* $(B+C)A = BA + CA$.

Proof. (i) Follows from the associativity of multiplication in R.

(ii) Clearly AB, $AC \subset A(B+C)$. Also, if $a \in A$, $b \in B$, $c \in C$, then $a(b+c) = ab + ac \in AB + AC$. Hence, $AB + AC = A(B+C)$. □

If $A_1, \ldots, A_n$ are right (left) ideals in a ring R, then their product is defined inductively. Because associativity [Theorem 4.4(i)] holds, the product of $A_1, \ldots, A_n$ may be written as $A_1 A_2 \cdots A_n$ (without any parentheses). It may be verified that

$$A_1 \cdots A_n = \left\{\sum_{\text{finite sum}} a_1 \cdots a_n \mid a_i \in A_i\right\}.$$

A prime integer p has the fundamental property that if $p|ab$, where a and b are integers, then $p|a$ or $p|b$. In terms of the ideals in $\mathbf{Z}$, this property may be restated as follows: If $ab \in (p)$, then $a \in (p)$ or $b \in (p)$. Equivalently, if $(a)(b) \subseteq (p)$, then $(a) \subseteq (p)$ or $(b) \subseteq (p)$. This suggests the following

Definition. *An ideal P in a ring R is called a* prime ideal *if it has the following property: If A and B are ideals in R such that $AB \subseteq P$, then $A \subseteq P$ or $B \subseteq P$.*

Clearly, in any integral domain (0) is a prime ideal. In fact, a commutative ring R is an integral domain if and only if (0) is a prime ideal. For each prime integer p, the ideal (p) in $\mathbf{Z}$ is a prime ideal.

4.5 Theorem. *If R is a ring with unity, then each maximal ideal is prime. But the converse, in general, is not true.*

Proof. Let M be a maximal ideal, and let A and B be ideals in R such that $AB \subseteq M$. Suppose $A \not\subset M$. Then $A + M = R$. Write $1 = a + m$, $a \in A$, $m \in M$. Let $b \in B$. Then $b = ab + mb \in AB + M \subseteq M$. Hence, $B \subseteq M$. This proves M is prime.

The ideal (0) in the ring of integers is prime but not maximal. □

For commutative rings we have

4.6 Theorem. *If R is a commutative ring, then an ideal P in R is prime if and only if $ab \in P$, $a \in R$, $b \in R$ implies $a \in P$ or $b \in P$.*

Proof. Let $ab \in P$ and P be a prime ideal. We know $(a)(b)$ consists of finite sums of products of elements of the type $na + ar$ and $mb + bs$, where $n,m \in \mathbf{Z}$ and $r,s \in R$. Now

$$(na + ar)(mb + bs) = nmab + nabs + mabr + abrs.$$

Because $ab \in P$ and P is an ideal in R, we get $(na + ar)(mb + bs)$ or finite sums of such like products are in P. Hence, each element of $(a)(b)$ is in P. Thus, $(a)(b) \subseteq P$. But since P is prime, $(a) \subseteq P$ or $(b) \subseteq P$. This implies $a \in P$ or $b \in P$. Conversely, if $ab \in P$, $a \in R$, $b \in R$ implies $a \in P$ or $b \in P$, then we show P is prime. Let A and B be ideals in R such that $AB \subseteq P$. Suppose $A \not\subset P$. Choose an element $a \in A$ such that $a \notin P$. Then $AB \subseteq P$ implies $aB \subseteq P$, so $ab \in P$ for all $b \in B$. But then by our hypothesis $b \in P$. Hence, $B \subset P$. □

4.7 Theorem. *Let R be a commutative principal ideal domain with identity. Then any nonzero ideal $P \neq R$ is prime if and only if it is maximal.*

Proof. Let $P = aR$ be a nonzero prime ideal in R. If P is not maximal, then there exists an ideal M such that $P \subsetneq M$ and $M \neq R$. Now, $M = bR$ for some $b \in R$. Thus, $aR \subsetneq bR$ implies $a = bx, x \in R$, and $b \notin aR$. Because $P = aR$ is a prime ideal and $b \notin aR$, we have $x \in aR$. Then $x = ay$ for some $y \in R$. Then

$$a = bx = bay$$

implies $1 = by$, because R is a domain. Hence, $M = R$, a contradiction. Therefore P is maximal. The converse follows from Theorem 4.5. □

4.8 Examples

(a) Let R be a commutative ring with unity in which each ideal is prime. Then R is a field.

Solution. If $ab = 0$, then $(a)(b) = (0)$. But since (0) is a prime ideal, $(a) = (0)$ or $(b) = 0$. Hence, $ab = 0$ implies $a = 0$ or $b = 0$; that is, R is an integral domain. Next, if $a \in R$, then $(a)(a) = (a^2)$ because R is commutative. But (a^2) is prime. Thus, $(a) \subseteq (a^2)$. Because (a^2) is trivially contained in (a), we get $(a) = (a^2)$. This implies $a \in (a^2)$, and, hence, there exists an element $x \in R$ such that $a = a^2x$ or $a(1 - ax) = 0$. If $a \neq 0$, we get $1 = ax$; that is, a is invertible. Hence, R is a field.

(b) Let R be a Boolean ring. Then each prime ideal $P \neq R$ is maximal.

Solution. Consider R/P. Because P is prime, $ab \in P$ (equivalently, $\bar{a}\bar{b} = \bar{0}$) implies $a \in P$ or $b \in P$ (i.e., $\bar{a} = \bar{0}$ or $\bar{b} = \bar{0}$). So R/P is an integral domain. Now for all $x \in R$, $(x + P)(x + P) = x^2 + P = x + P$. So R/P is also a Boolean ring in addition to being an integral domain. But we know that an integral domain has no idempotents except 0 and possibly unity. Thus, $R/P = (\bar{0})$ or $\{\bar{0}, \bar{1}\}$. However, $R/P = (\bar{0})$ implies $R = P$, which is not true. So $R/P = \{\bar{0}, \bar{1}\}$, which obviously forms a field. Hence, P is a maximal ideal.

(c) Let a be a nonnilpotent element in a ring R, and let $S = \{a, a^2, a^3, \ldots\}$. Suppose P is maximal in the family F of all ideals in R that are disjoint from S. Then P is a prime ideal. (Note that the statement does not say that P is maximal in R. Precisely, it means that there does not exist any ideal $X \in F$ such that $X \supsetneq P$.)

Solution. Let $AB \subseteq P$, where A and B are ideals in R. If possible, let $A \not\subseteq P$ and $B \not\subseteq P$. Then $A + P \supsetneq P$ and $B + P \supsetneq P$. By maximality of P, $(A + P) \cap S \neq \emptyset$ and $(B + P) \cap S \neq \emptyset$. Thus, there exist positive integers i and j such that $a^i \in A + P$ and $a^j \in B + P$. Then

$$a^i a^j \in (A + P)(B + P) = AB + AP + PB + PP \subseteq P,$$

because $AB \subseteq P$ and because P is an ideal in R. Thus, $P \cap S \neq \emptyset$, a contradiction. Hence, $AB \subseteq P$ implies either $A \subseteq P$ or $B \subseteq P$, which proves that P is a prime ideal.

(d) Let $R = C[0,1]$ be the ring of all real-valued continuous functions on the closed unit interval. If M is a maximal ideal of R, then there exists a real number r, $0 \leq r \leq 1$, such that $M = M_r = \{f \in R \mid f(r) = 0\}$, and conversely.

Solution. Let M be a maximal ideal of $C[0,1]$. We claim that there exists $r \in [0,1]$ such that, for all $f \in M$, $f(r) = 0$. Otherwise for each $x \in [0,1]$ there exists $f_x \in M$ such that $f_x(x) \neq 0$. Because f_x is continuous, there exists an open interval, say I_x, such that $f_x(y) \neq 0$ for all $y \in I_x$.

Clearly, $[0,1] = \bigcup_{x \in [0,1]} I_x$. By the Heine–Borel theorem in analysis there exists a finite subfamily, say, $I_{x_1}, I_{x_2}, \ldots, I_{x_n}$, of this family of open intervals I_x, $x \in [0,1]$, such that $[0,1] = I_{x_1} \cup I_{x_2} \cup \cdots \cup I_{x_n}$.

Consider $f = \Sigma_{i=1}^n f_{x_i}^2$. Suppose that $f(z) = 0$ for some z in $[0,1]$. Now $[0,1] = \bigcup_{i=1}^n I_{x_i}$ implies that there exists I_{x_k} such that $z \in I_{x_k}$ $(1 \leq k \leq n)$. Then $f_{x_k}(z) \neq 0$. But

$$f(z) = 0 \Rightarrow \sum (f_{x_i}(z))^2 = 0 \Rightarrow f_{x_k}(z) = 0,$$

a contradiction. Thus, $f(z) \neq 0$ for any $z \in [0,1]$, which in turn yields that f is invertible and a fortiori $M = C[0,1]$, which is not true.

Conversely, we show that M_r is a maximal ideal of $C[0,1]$ for any $r \in [0,1]$. It is easy to check that M_r is an ideal. To see that it is maximal, we note that $C[0,1]/M_r$ is a field isomorphic to $\mathbf{R}$. Alternatively, we may proceed as follows.

Let J be an ideal of $C[0,1]$ properly containing M_r. Let $g \in J$, $g \notin M_r$. Then $g(r) \neq 0$. Let $g(r) = \alpha$. Then $h = g - \alpha$ is such that $h(r) = 0$; that is, $h \in M_r$, so $\alpha = g - h \in J$. But $\alpha \neq 0$ implies that α is invertible. Consequently, $J = \mathbf{R}$, which proves the converse.

Problems

1. Let F be a field. Prove that (0) is a prime ideal in F_m.
2. Prove that the ideal $(x^3 + x + 1)$ in the polynomial ring $\mathbf{Z}/(2)[x]$ over $\mathbf{Z}/(2)$ is a prime ideal.
3. Prove that the ideal $(x^4 + 4)$ is not a prime ideal in the polynomial ring $\mathbf{Q}[x]$ over the field of rational numbers.
4. Show that the ideal $(x^3 - x - 1)$ in the polynomial ring $\mathbf{Z}/(3)[x]$ over the field $\mathbf{Z}/(3)$ is a maximal ideal and therefore, prime.
5. Show that the following are equivalent for any ring R.
   (a) P is a prime ideal in R.
   (b) For all $a,b \in R$, $aRb \subseteq P$ implies $a \in P$ or $b \in P$.
   (c) For all right ideals A, B in R, $AB \subseteq P$ implies $A \subseteq P$ or $B \subseteq P$.
   (d) For all left ideals A, B in R, $AB \subseteq P$ implies $A \subseteq P$ or $B \subseteq P$.
   Let P be a prime ideal in R. Show that (0) is a prime ideal in R/P. (A ring S is called a *prime ring* if (0) is a prime ideal in S.)
6. Let P be a prime ideal in a ring R such that the quotient ring R/P has no nonzero nilpotent elements. Show that R/P is an integral domain.
7. Let $(P_i)_{i \in \Lambda}$ be the family of all prime ideals in R and let $A = \bigcap_{i \in \Lambda} P_i$. Show that A is a nil ideal. [*Hint:* Use Example (c) and assume that a maximal member in the family F exists. The existence of a maximal member in F follows by Zorn's lemma, given in Section 6.]

5 Nilpotent and nil ideals

Let $A_1, A_2, \ldots, A_n$ be right (or left) ideals in a ring R. If $A_1 = A_2 = \cdots = A_n = A$ (say), then their product is written as A^n. It is possible that a right (or left) ideal A is not zero, but $A^n = (0)$ for some positive integer $n > 1$. For example, in the ring $R = \mathbf{Z}/(4)$ the ideal $A = \{0,2\}$ is not zero, but $A^2 = (0)$. As another example, consider the ring $R = \left(\begin{smallmatrix} \mathbf{Z} & \mathbf{Z} \\ 0 & \mathbf{Z} \end{smallmatrix}\right)$ of 2×2 (upper) triangular matrices. It has a nonzero ideal $A = \left(\begin{smallmatrix} 0 & \mathbf{Z} \\ 0 & 0 \end{smallmatrix}\right)$ such that $A^2 = (0)$. Such right (or left) ideals are called *nilpotent.*

Definition. *A right (or left) ideal A in a ring R is called* nilpotent *if $A^n = (0)$ for some positive integer n.*

Clearly, every zero ideal is a nilpotent ideal, and every element in a nilpotent ideal is a nilpotent element. However, the set of nilpotent elements in a ring R does not necessarily form a nilpotent ideal. (Indeed this set may not even be an ideal.)

A ring R may have nonzero nilpotent elements, but it may not possess a nonzero nilpotent ideal, as the following example shows.

5.1 Example. Let $R = F_n$ be the ring of $n \times n$ matrices over a field F. Then R has nonzero nilpotent elements, such as e_{ij}, $i \neq j$, $1 \leq i,j \leq n$. Let I be a nilpotent right ideal in R with $I^k = 0$, where k is some positive integer. Then consider the ideal

$$\overbrace{(RI)(RI)\cdots(RI)}^{k \text{ times}} = R\overbrace{(IR)\cdots(IR)}^{k-1 \text{ times}}I \subseteq R\overbrace{II\cdots I}^{k \text{ times}} = RI^k = (0).$$

Hence, RI is a nilpotent ideal in R. But we know that the ring $R = F_n$ has no nontrivial ideals (Corollary 1.3). Hence, $RI = (0)$ or $RI = R$. But RI cannot be equal to R, since R has unity $\neq 0$. Therefore, $RI = (0)$. But then for any $a \in I$, $a = 1a \in RI = (0)$. Hence, $I = (0)$.

Definition. *A right (or left) ideal A in a ring R is called a* nil ideal *if each element of A is nilpotent.*

Clearly, every nilpotent right (or left) ideal is nil. However, the converse is not true, as the following example shows.

5.2 Example. Let $R = \oplus\Sigma\mathbf{Z}/(p^i)$, $i = 1,2,\ldots$, be the direct sum of the rings $\mathbf{Z}/(p^i)$, p is prime. Then R contains nonzero nilpotent elements, such as $(0 + (p), p + (p^2), 0 + (p^3), \ldots)$. Let I be the set of all nilpotent elements. Then I is an ideal in R because R is commutative. So I is a nil ideal. But I is not nilpotent, for if $I^k = 0$ for some positive integer $k > 1$, then the element

$$x = (0 + (p), 0 + (p^2), \ldots, 0 + (p^k), p + (p^{k+1}), 0 + (p^{k+2}), \ldots)$$

is nilpotent, so it belongs to I. But $x^k \neq 0$, a contradiction. So I is not nilpotent.

Problems

1. Show that if A and B are nilpotent ideals, their sum $A + B$ is nilpotent.
2. If R is a commutative ring, show that the sum of nil ideals $A_1, A_2, \ldots, A_n$ is a nil ideal.
3. Show that the $n \times n$ matrix ring F_n over a field F has no nonzero nil right or left ideals.

6 Zorn's lemma

A partially ordered set (also called a *poset*) is a system consisting of a nonempty set S and a relation, usually denoted by $\leq$, such that the following conditions are satisfied: For all $a,b,c \in S$,

(i) $a \leq b$ and $b \leq a \Rightarrow a = b$ (antisymmetric).
(ii) $a \leq a$ (reflexive).
(iii) $a \leq b$ and $b \leq c \Rightarrow a \leq c$ (transitive).

A *chain* C in a poset $(S,\leq)$ is a subset of S such that, for every $a,b \in C$, either $a \leq b$ or $b \leq a$. An element $u \in S$ is an *upper bound* of C if $a \leq u$ for every $a \in C$; an element $m \in S$ is a *maximal element* of a poset $(S,\leq)$ if $m \leq a$, $a \in S$, implies $m = a$.

The following axiom from set theory about posets is widely used in mathematics.

Zorn's lemma. *If every chain C in a poset $(S,\leq)$ has an upper bound in S, then $(S,\leq)$ has a maximal element.*

As a simple example of the use of Zorn's lemma, we prove the following result.

6.1 Theorem. *If R is a nonzero ring with unity 1, and I is an ideal in R such that $I \neq R$, then there exists a maximal ideal M of R such that $I \subseteq M$.*

Proof. Let S be the set of all ideals in R such that if $X \in S$ then $X \neq R$ and $X \supseteq I$. $(S, \subseteq)$ is a partially ordered set under inclusion. We assert that the union U of an arbitrary chain C in S is an element of S. For if $a,b \in U$, then there exist elements A and B of S such that $a \in A$ and $b \in B$. But since C is a chain, we have $A \subseteq B$ or $B \subseteq A$, and, accordingly, $a - b \in B$ or $a - b \in A$. Also, $ra \in B$ or $ra \in A$ for all $r \in R$. In any case, $a - b \in U$ and $ra \in U$. Similarly, $ar \in U$. Hence, U is an ideal in R. If $U = R$, then $1 \in U$, so $1 \in X$ for some $X \in C$, which is not true. Hence, $U \neq R$ and, thus, $U \in S$. Then by Zorn's lemma S contains a maximal member M. We claim that M is a maximal ideal in R. So let $M \subsetneq N$, where N is an ideal in R. If $N \neq R$, then clearly $N \in S$, a contradiction to the maximality of M in S. Hence, $N = R$, and thus M is a maximal ideal in R, as desired. □

Problem

Use Zorn's lemma to show that every proper right ideal A in a ring R with unity is contained in a maximal right ideal M. (M is called a maximal right ideal if $M \neq R$, and if $M \subseteq N$, for some right ideal N, implies $N = M$ or $N = R$.)

CHAPTER 11

Unique factorization domains and euclidean domains

1 Unique factorization domains

Throughout this chapter R is a commutative integral domain with unity. Such a ring is also called a domain.

If a and b are nonzero elements in R, we say that b *divides* a (or b is a *divisor* of a) and that a is *divisible* by b (or a is a *multiple* of b) if there exists in R an element c such that $a = bc$, and we write $b|a$ or $a \equiv 0 \pmod{b}$. Clearly, an element $u \in R$ is a unit if and only if u is a divisor of 1.

Two elements a,b in R are called *associates* if there exists a unit $u \in R$ such that $a = bu$. Of course, then $b = av$, where $v = u^{-1}$. This means that if a and b are associates, then $a|b$ and $b|a$. In fact, if R is a commutative integral domain, then the converse is also true; that is, if $a|b$ and $b|a$, then a and b are associates. For let $b = ax$ and $a = by$. Then $b = byx$. Because R is an integral domain with 1, this gives $yx = 1$. Therefore x and y are units, and, hence, a and b are associates.

We call an element b in R an *improper divisor* of an element $a \in R$ *if* b is either a unit or an associate of a.

Definition. *A nonzero element a of an integral domain R with unity is called an* irreducible element *if (i) it is not a unit, and (ii) every divisor of a is improper; that is, $a = bc$, $b,c \in R$, implies either b or c is a unit.*

Definition. *A nonzero element p of an integral domain R with unity is called a prime element if (i) it is not a unit, and (ii) if $p|ab$, then $p|a$ or $p|b$, $a,b \in R$.*

It is quite easy to verify that a prime element is an irreducible element. But an irreducible element need not be prime (see Problem 5). However, we have the following

1.1 Theorem. *An irreducible element in a commutative principal ideal domain (PID) is always prime.*

Proof. Let R be a PID, and let $p \in R$ be an irreducible element. Suppose $p|ab$, $a,b \in R$. Assume $p \nmid a$. Because R is a PID, there exists $c \in R$ such that

$$pR + aR = cR.$$

Then $p \in cR$, which implies $p = cd$, $d \in R$. But since p is irreducible, c or d must be a unit. Suppose d is a unit. Then $pR = cR$; so $pR + aR = pR$. This gives $a \in pR$, which contradicts the assumption that $p \nmid a$. Hence, c is a unit; therefore, $cR = R$, so $pR + aR = R$. Then there exists $x,y \in R$ such that $px + ay = 1$. This implies $pbx + aby = b$. Hence, $p|b$ because $p|ab$. □

We now define an important class of integral domains that contains the class of PIDs, as shown later.

Definition. *A commutative integral domain R with unity is called a* unique factorization domain *(or briefly, a UFD) if it satisfies the following conditions:*

(i) Every nonunit of R is a finite product of irreducible factors.
(ii) Every irreducible element is prime.

The factorization into irreducible elements is unique, as will be shown in Theorem 1.3.

1.2 Examples of UFDs

(a) The ring of integers **Z** is a UFD.

(b) The polynomial ring $F[x]$ over a field F is a UFD.

(c) Each commutative principal ideal ring with unity that is also an integral domain [called a principal ideal domain (PID)] is a UFD (see Theorem 2.1).

(d) The commutative integral domain $R = \{a + b\sqrt{-5} | a,b \in \mathbf{Z}\}$ is not a UFD. for let $r = a + b\sqrt{-5} \in R$. Define a function N: $R \to \mathbf{Z}$ such that

$N(r) = r\bar{r} = a^2 + 5b^2$; $N(r)$ is called the *norm* of r and has the properties that (i) $N(r) \geq 0$, (ii) $N(r) = 0$ iff $r = 0$, (iii) $N(rs) = N(r)N(s)$.

It follows from property (iii) that if $rs = 1$, then $N(r)N(s) = N(rs) = N(1) = 1$. Thus, $N(r) = 1$, so $a^2 + 5b^2 = 1$. This gives $a = \pm 1$, $b = 0$. Hence, the only units are ± 1. In particular, the only associates of an element are the element itself and its negative.

Now consider the factorizations $9 = 3 \cdot 3 = (2 + \sqrt{-5})(2 - \sqrt{-5})$. We claim that 3, $2 + \sqrt{-5}$, and $2 - \sqrt{-5}$ are all irreducible. Let $2 + \sqrt{-5} = rs$. Then $N(2 + \sqrt{-5}) = N(r)N(s)$. This gives $9 = N(r)N(s)$. Hence, $N(r) = 1, 3$, or 9. But if $N(r) = 3$ and $r = a + b\sqrt{-5}$, then $a^2 + 5b^2 = 3$, which is impossible for integers a and b. Therefore, either $N(r) = 1$ or $N(r) = 9$ and $N(s) = 1$. In the former case r is a unit, and in the latter case s is a unit. So $2 + \sqrt{-5}$ is irreducible. Similarly, $2 - \sqrt{-5}$ and 3 are irreducible. Thus, R is not a UFD in view of the next theorem.

1.3 Theorem. *If R is a UFD, then the factorization of any element in R as a finite product of irreducible factors is unique to within order and unit factors.*

Proof. More explicitly, the theorem states that if $a = p_1 p_2 \cdots p_m = q_1 q_2 \cdots q_n$, where p_i and q_i are irreducible, then $m = n$, and on renumbering the q_i, we have that p_i and q_i are associates, $i = 1,2,\ldots,m$.

Because the theorem is obvious for factorizations of irreducible elements, we assume it is true for any element of R that can be factored into s irreducible factors. We then prove it is true for any element that can be factored into $s + 1$ irreducible factors. Let

$$a = \prod_{i=1}^{s+1} p_i = \prod_{j=1}^{m} p'_j \tag{1}$$

be two factorizations of a into irreducible factors, one of which involves exactly $s + 1$ factors. We have that p_i divides the product of the p'_j's, and, hence, by axiom (ii) for a UFD, p_i must divide one of the elements $p'_1, p'_2, \ldots, p'_m$. Let p_1 divide p'_k. Because p'_k is irreducible, p_1 and p'_k are associates. Then $p'_k = up_1$, where u is a unit, and after cancellation of the common factor p_1, (1) gives

$$\prod_{i=2}^{s+1} p_i = u \prod_{\substack{j=1 \\ j \neq k}}^{m} p'_j. \tag{2}$$

Hence, by the induction hypothesis, the two factorizations in (2) can differ only in the order of the factors and by unit factors. Because we already know that p_1 and p'_k differ by a unit factor, the proof is complete. □

In the next theorem we show that in a UFD any pair of elements a and b has a *greatest common divisor* (g.c.d) defined as follows.

Definition. *An element d in an integral domain R is called a* greatest common divisor *of elements a and b in R if*

(i) $d|a$ and $d|b$, and
(ii) if for $c \in R$, $c|a$, and $c|b$, then $c|d$.

1.4 Theorem. *Let R be UFD, and $a,b \in R$. Then there exists a greatest common divisor of a and b that is uniquely determined to within an arbitrary unit factor.*

Proof. Let us write

$$a = p_1^{e_1} p_2^{e_2} \cdots p_m^{e_m}, \qquad b = p_1^{f_1} p_2^{f_2} \cdots p_m^{f_m},$$

where p_i are irreducible, e_i, f_i are nonnegative integers, and by p_i^0 we mean a unit. Set $g_i = \min(e_i, f_i)$, $i = 1,2,\ldots,m$, and $d = p_1^{g_1} p_2^{g_2} \cdots p_m^{g_m}$. Then clearly, $d|a$ and $d|b$. Let $c \in R$ be such that $c|a$ and $c|b$. Then $c|a$ implies

$$c = p_1^{\lambda_1} p_2^{\lambda_2} \cdots p_m^{\lambda_m},$$

where λ_i, $i = 1,2,\ldots,m$, are nonnegative integers. But then $c|a$ and $c|b$ imply, by Theorem 1.3, that $\lambda_i \leq e_i$ and $\lambda_i \leq f_i$, $i = 1,2,\ldots,m$. Hence, $\lambda_i \leq \min(e_i, f_i) = g_i$. This proves $c|d$, as desired.

Now suppose d and d' are the two greatest common divisors of a and b. Then $d|d'$ and $d'|d$, and, therefore, they are associates because R is a commutative integral domain. □

The uniquely determined (within an arbitrary unit factor) greatest common divisor of a and b is denoted by (a,b).

Note that (a,b) is a set in which every two elements are associates. We write $(a,b) = c$ to mean that (a,b) consists of all unit multiples of c.

Definition. *In a UFD two elements a and b are called relatively prime if* $(a,b) = 1$.

Problems

In the following problems R is a commutative integral domain with unity in which for each pair a, $b \in R$, g.c.d. (a, b) exists. Let $a, b, c \in R$.

1. Show that $c(a,b)$ and (ca,cb) are associates.
2. Show that if $(a,b) = 1$ and if $a|c$ and $b|c$, then $ab|c$.

3. Show that if $(a,b) = 1$ and $b|ac$, then $b|c$.
4. Show that if $(a,b) = 1$ and $(a,c) = 1$, then $(a,bc) = 1$.
5. Prove that $2 + \sqrt{-5}$ is irreducible but not prime in $\mathbf{Z}[\sqrt{-5}]$.
6. Show that 3 is irreducible but not prime in the ring $\mathbf{Z}[\sqrt{-5}]$.
7. Show that in the ring $\mathbf{Z}[\sqrt{-3}]$, $(1+\sqrt{-3}, 1-\sqrt{-3}) = 1$.
8. Show that in the ring $\mathbf{Z}[\sqrt{-3}]$ the g.c.d. of 4 and $2+2\sqrt{-3}$ does not exist.
9. Find the g.c.d., if it exists, of $10+11i$ and $8+i$ in $\mathbf{Z}[i]$.
10. Show that there exist irreducible elements which are not prime in each of the rings $\mathbf{Z}[\sqrt{-6}]$ and $\mathbf{Z}[\sqrt{10}]$.
11. Let $\mathbf{Q}[\sqrt{-3}] = \left\{ \dfrac{a+b\sqrt{-3}}{2} \,\middle|\, a,b \in \mathbf{Z}, \text{ and } a,b \text{ are both even or both odd} \right\}$. Show that the units of $\mathbf{Q}[\sqrt{-3}]$ are ± 1, $\dfrac{\pm 1 \pm \sqrt{-3}}{2}$.

2 Principal ideal domains

Recall that a commutative integral domain R with unity is a principal ideal domain (PID) if each ideal in R is of the form $(a) = aR$, $a \in R$.

2.1 Theorem. *Every PID is a UFD, but a UFD is not necessarily a PID.*

Proof. First we show that if R is a principal ideal ring, then R cannot have any infinite properly ascending chain of ideals. Therefore, let $a_1R \subset a_2R \subset a_3R \subset \cdots$ be a chain of ideals in R. Let $A = \bigcup a_iR$ and $a,b \in A$, $r \in R$. Then $a \in a_iR$, $b \in a_jR$ for some i,j. Because either $a_iR \subset a_jR$ or $a_jR \subset a_iR$, both a and b lie in one of the two ideals a_iR, a_jR, say, in a_iR. Then $a - b \in a_iR \subset A$. Also, $ar \in a_iR \subset A$. Hence, A is an ideal in R. Because R is a principal ideal ring, $A = aR$ for some $a \in R$. Now $a \in A$ implies $a \in a_kR$ for some k. Further, $A = aR \subset a_kR \subset A$ gives that $A = aR = a_kR$; hence, $a_kR = a_{k+1}R = \cdots$, proving our assertion.

Next, we show that each element $a \in R$ is a finite product of irreducible elements. If a is irreducible, we are done. So let $a = bc$, where neither b nor c is a unit. If both b and c are products of irreducible elements, we are done. So let b not be a product of irreducible elements, and write $b = xy$, where x, say, is not a product of irreducible elements. This process leads to a properly ascending chain of ideals $(a) \subset (b) \subset (x) \subset \cdots$ that will continue indefinitely if a is not a finite product of irreducible elements. But since R cannot possess any infinite properly ascending chain of ideals, we conclude that a must be a finite product of irreducible elements.

To complete the proof that R is a UFD, we need to show that if $p|ab$, where p is an irreducible element in R, and $a,b \in R$, then $p|a$ or $p|b$. This has been shown in Theorem 1.1.

Finally, we give an example to show that a UFD need not be a PID. Consider the polynomial ring $R = F[x,y]$ over a field F in variables x and y. The ideal $I = (x) + (y)$ in $F[x,y]$ cannot be of the form $(f(x,y))$ for any polynomial $f(x,y) \in F[x,y]$, because

$$(x) + (y) = (f(x,y)) \Rightarrow x = cf(x,y),\ y = df(x,y)$$

for some $c,d\,(\neq 0)$ in F. This gives $x/c = y/d$; that is, $dx - cy = 0$, which is absurd because x and y are independent variables over F. Thus, $F[x,y]$ is not a PID. But $F[x,y]$ is a UFD (Theorem 4.3, proved in Section 4). □

3 Euclidean domains

An important class of unique factorization domains is given by *euclidean domains* or rings admitting a division algorithm. These rings are defined as follows:

Definition. *A commutative integral domain E with unity is called a* euclidean domain *if there exists a function $\phi: E \to \mathbf{Z}$ satisfying the following axioms:*

(i) *If $a,b \in E^* = E - \{0\}$ and $b|a$, then $\phi(b) \leq \phi(a)$.*
(ii) *For each pair of elements $a,b \in E$, $b \neq 0$, there exist elements q and r in E such that $a = bq + r$, with $\phi(r) < \phi(b)$.*

3.1 Examples of euclidean domains

(a) The ring of integers $\mathbf{Z}$ is a euclidean domain if we set $\phi(n) = |n|$, $n \in \mathbf{Z}$. Then for any two integers a and $b\,(\neq 0)$, the ordinary division algorithm yields integers q (= quotient) and r (= remainder) satisfying axiom (ii). Axiom (i) is clear.

(b) The polynomial ring $F[x]$ over a field F in a variable x is a euclidean domain if, for any nonzero polynomial $f = f(x) \in F[x]$, we set $\phi(f) =$ degree of f and $\phi(0) = -1$.

(c) The *ring of Gaussian integers* $R = \{m + n\sqrt{-1} | m,n \in \mathbf{Z}\}$ is a euclidean domain if we set $\phi(m+n\sqrt{-1}) = m^2+n^2$. Note that for $x,y \in R$, we have $\phi(xy) = \phi(x)\,\phi(y)$, and thus $b|a$ implies $\phi(b) \le \phi(a)$, which proves (i).

To prove (ii), let $0 \ne b \in R$ and write $ab^{-1} = \alpha + \beta\sqrt{-1}$, where α and β are rational numbers. Choose integers α_0 and β_0 such that $|\alpha-\alpha_0| \le \frac{1}{2}$, $|\beta-\beta_0| \le \frac{1}{2}$. Then

$$\begin{aligned} a &= b(\alpha+\beta\sqrt{-1}) \\ &= b((\alpha-\alpha_0)+\alpha_0+(\beta-\beta_0)\sqrt{-1}+\beta_0\sqrt{-1}) \\ &= b(\alpha_0+\beta_0\sqrt{-1})+b(\alpha-\alpha_0)+b(\beta-\beta_0)\sqrt{-1}. \end{aligned}$$

Because $\alpha_0,\beta_0 \in Z$, $q = \alpha_0+\beta_0\sqrt{-1} \in R$. Thus, $a - b(\alpha_0+\beta_0\sqrt{-1}) \in R$. Hence, $r = b(\alpha-\alpha_0)+b(\beta-\beta_0)\sqrt{-1} \in R$. We can then write $a = bq+r$, where $q,r \in R$. Further

$$\phi(r) = \phi(b)[(\alpha-\alpha_0)^2+(\beta-\beta_0)^2] \le \tfrac{1}{2}\,\phi(b) < \phi(b).$$

Hence, R is a euclidean domain.

3.2 Theorem. *Every euclidean domain is a PID.*

Proof. Let A be a nonzero ideal in a euclidean domain R. Because for all $a \in A$, $1|a$, then $\phi(a) \ge \phi(1)$. Then the set $\{\phi(a) | 0 \ne a \in A\}$ is a nonempty set of integers with $\phi(1)$ as a lower bound. So by the principle of well-ordering of integers there exists an element $d \in A$ such that $\phi(d)$ is the smallest in this set. We claim $A = (d)$. If $a \in A$, write $a = qd+r$, $q,r \in R$ and $\phi(r) < \phi(d)$. But since $r = a - qd \in A$ and $\phi(r) \ge \phi(d)$ by the choice of d, r must be zero. Hence, $a \in (d)$. □

3.3 Theorem. *Every euclidean domain is a UFD.*

Proof. Follows from Theorems 2.1 and 3.2. □

Remark. There exists an interesting class of principal ideal domains that are not euclidean domains. Consider, for example,

$$\mathbf{Z}[\sqrt{-19}] = \{a+b\sqrt{-19} \,|\, a,b \in \mathbf{Z} \text{ and } a,b \text{ have the same parity}\}.$$

(a and b have the same parity if both are odd or both are even.) It requires quite tedious computations to show that $\mathbf{Z}[\sqrt{-19}]$ is a PID but not a euclidean domain. The interested reader is referred to J. C. Wilson, A principal ideal ring that is not a Euclidean ring, *Math. Mag.* **46**, 34–8 (1973).

Problems

1. Let R be a euclidean domain. Prove the following:
   (a) If $b \neq 0$, then $\phi(0) < \phi(b)$.
   (b) If a and b are associates, then $\phi(a) = \phi(b)$.
   (c) If $a|b$ and $\phi(a) = \phi(b)$, then a and b are associates.
2. Show that each of the rings $\mathbf{Z}[\sqrt{2}]$ and $\mathbf{Z}[\sqrt{-2}]$ is a (i) euclidean domain and (ii) UFD. Explain why in the UFD $\mathbf{Z}[\sqrt{2}]$, $(5+\sqrt{2})(2-\sqrt{2}) = (11-7\sqrt{2})(2+\sqrt{2})$ even though each of the factors is irreducible.
3. Let R be a commutative integral domain with unity that is not a field; show that the polynomial ring $R[x]$ in a variable x is not a PID.
4. Let $a = 3 + 2i$ and $b = 2 - 3i$ be two elements in $\mathbf{Z}[i]$. Find q and r in $\mathbf{Z}[i]$ such that $a = bq + r$ and $\varphi(r) < \varphi(b)$, where $\varphi(x + iy) = x^2 + y^2$.
5. Prove that every nonzero prime ideal in a euclidean domain is maximal.
6. Prove that if D is a domain that is not a field, then $D[x]$ is not a euclidean domain.
7. Prove that in any PID every ideal is a unique product of prime ideals
8. Show that $\mathbf{Z}[\sqrt{-6}]$ is not a euclidean domain.
9. Let R be a euclidean domain. Show that $a \in R$ is a unit if and only if $\varphi(a) = \varphi(1)$. Conclude that the only units of $\mathbf{Z}[i]$ are ± 1, $\pm i$.
10. An ideal P in a commutative ring R is called primary if for all $x, y \in R$, $xy \in P$, $x \notin P$ implies $y^n \in P$ for some positive integer n. Show that if P is a primary ideal then

$$\sqrt{P} = \{x \in R \mid x^n \in P \text{ for some } n > 0\}$$

is a prime ideal. ($\sqrt{P}$ is called the radical of P.)

11. Let R be a PID. Show each primary ideal P is of the form (p^e), $e > 0$, for some prime element p. Give an example to show that this may not be true if R is not a PID. [Consider the ideal (x^2, y) in $F[x, y]$.]

4 Polynomial rings over UFD

Recall that the ring of polynomials $R[x]$ over a ring R in x (called a variable or an indeterminate) is the set of formal expressions $a_0 +$

$a_1x + \cdots + a_nx^n$, $a_i \in R$, or, equivalently, the set of sequences $(a_0,a_1,a_2,\ldots)$, $a_i \in R$, where all but a finite number of a_i's are zero.

If $f = (a_i)$ is a nonzero polynomial and n is the greatest integer such that $a_n \neq 0$, then n is called the *degree* of f, and a_n is called the *leading coefficient* of f. Also, if $1 \in R$ and $a_n = 1$, then f is called *monic;* in other words, a polynomial f is monic if its leading coefficient is 1. The degree of a zero polynomial is generally defined to be $-\infty$.

The natural mapping $a \mapsto (a,0,0,\ldots)$ is an isomorphism of R into $R[x]$. Hence, R can be embedded in $R[x]$. We shall regard R as a subring of $R[x]$ by identifying a with $(a,0,0,\ldots)$.

Let R be a ring, and let $S = R[x]$ be a polynomial ring over R. If we start with S and construct the polynomial ring $S[y]$ over S in the indeterminate, y, for example, then $S[y]$ is called a *polynomial ring in two indeterminates or variables* x,y over R. We write this ring as $R[x,y]$. It follows from the definition that $R[x,y] = R[y,x]$. Also, a typical element of $R[x,y]$ is of the form

$$\sum_{i=0}^{m}\sum_{j=0}^{n} a_{ij}x^iy^j, \qquad a_{ij} \in R,$$

where $a_{ij}x^iy^j$ denotes $a_{i0}x^i$ if $j = 0$, $a_{0j}y^j$ if $i = 0$, and a_{00} if i and j are both zero. Inductively, we can define a polynomial ring in a finite number of variables.

The polynomial rings over a commutative integral domain have a *division algorithm,* which is given in the following

4.1 Theorem. *Let $R = F[x]$ be a polynomial ring over a commutative integral domain F. Let $f(x)$ and $g(x) \neq 0$ be polynomials in $F[x]$ of degrees m and n, respectively. Let $k = \max(m - n + 1, 0)$, and let a be the leading coefficient of $g(x)$. Then there exist unique polynomials $q(x)$ and $r(x)$ in $F[x]$ such that*

$$a^kf(x) = q(x)g(x) + r(x),$$

where $r(x) = 0$ or $r(x)$ has degree less than the degree of $g(x)$.

Proof. If $m < n$, we take $q(x) = 0$ and $r(x) = f(x)$. So let $m > n$ and $k = m - n + 1$. We prove the theorem by induction on m. We assume it is true for all polynomials of degree $< m$, and we prove it for polynomials of degree m.

Now the polynomial $af(x) - bx^{m-n}g(x)$ has degree at most $m - 1$, where b is the leading coefficient of f. By the induction hypothesis there exist polynomials $q_1(x)$ and $r_1(x)$ such that

$$a^{(m-1)-n+1}(af(x) - bx^{m-n}g(x)) = q_1(x)g(x) + r_1(x).$$

Then

$$a^k f(x) = (ba^{m-n}x^{m-n} + q_1(x))g(x) + r_1(x),$$

as desired. Uniqueness follows immediately. □

Now our aim is to show that polynomial rings over a UFD are also UFD. Because a polynomial ring $F[x]$ over a field F, which is trivially a UFD, is also a UFD [Example 3.1(b) and Theorem 3.3], our next theorem may be regarded as a generalization of this fact.

We first need some preliminaries.

Definition. *Let R be a UFD. Then $f(x) \in R[x]$ is called* primitive *if the g.c.d. of its coefficients is a unit.*

We observe that it is possible to write any nonzero polynomial $f = f(x) \in R[x]$ in the form $f = cf_1$, where $c \in R$ and $f_1(x)$ is primitive, by choosing c to be equal to the g.c.d. of the coefficients of $f(x)$. Any element c satisfying the condition $f = cf_1$, where f_1 is primitive, is necessarily a g.c.d. of the coefficients of $f(x)$; therefore, c is determined to within a unit factor. The factor c is called the *content* of $f(x)$ and is denoted by $c(f)$. Note that $f(x)$ is primitive if and only if $c(f)$ is a unit.

In what follows we assume that R is a UFD.

4.2 Lemma (Gauss). *If $f(x), g(x) \in R[x]$, then $c(fg) = c(f)c(g)$. In particular, the product of two primitive polynomials is primitive.*

Proof. Writing $c = c(f)$ and $d = c(g)$, we have $f(x) = cf_1(x)$ and $g(x) = dg_1(x)$, where f_1 and g_1 are primitive. Because $fg = (cd)(f_1g_1)$, we need only prove that f_1g_1 is primitive. Suppose f_1g_1 is not primitive and let p be an irreducible element of R that divides all the coefficients of f_1g_1. If $f_1(x) = \Sigma\, a_ix^i$ and $g_1(x) = \Sigma\, b_jx^j$, $a_i, b_j \in R$, let a_s and b_t be the first coefficients of f_1 and g_1, respectively, that are not divisible by p (the existence of a_s and b_t follows from the primitivity of f_1 and g_1). Now the coefficient of x^{s+t} in $f_1(x)g_1(x)$ is

$$\cdots + a_{s-1}b_{t+1} + a_sb_t + a_{s+1}b_{t-1} + \cdots.$$

Because R is a UFD, $p \nmid a_sb_t$. Because p divides all terms of the above sum that precede and follow $a_s b_t$, it does not divide the sum itself, a contradiction. Hence, $f_1(x)g_1(x)$ is primitive. □

We now prove

4.3 Theorem. *Let R be a unique factorization domain. Then the polynomial ring R[x] over R is also a unique factorization domain.*

Proof. We first show that each nonunit element of $R[x]$ factors into irreducible ones. Without any loss of generality, it suffices to prove the theorem for polynomials $f(x) \in R[x]$ that are primitive. We prove the result by induction on the degree of a polynomial. If the degree of $f(x) \leq 1$, then the result clearly holds because R is a UFD. Now assume that the degree of $f(x) = n > 1$, and the result is true for all polynomials of degree $< n$. If $f(x)$ is irreducible over R, we are done. Otherwise, write $f(x) = f_1(x)f_2(x)$, where $f_1(x)$ and $f_2(x)$ are nonconstant polynomials over R of degrees n_1 and n_2, respectively. Thus, $n_1 < n$ and $n_2 < n$; therefore, the result follows by induction.

We now show that if $p(x), f(x), g(x) \in R[x]$ and $p(x)$ is an irreducible element in $R[x]$ such that $p(x)|f(x)g(x)$, then $p(x)|f(x)$ or $p(x)|g(x)$. If the degree of $p(x)$ is zero, then $p(x) = a \in R$. Then

$$p(x)|f(x)g(x) \Rightarrow ah(x) = f(x)g(x), \qquad h(x) \in R[x].$$

This implies $ac(h) = c(f)c(g)$. But since a is irreducible, this yields $a|c(f)$ or $a|c(g)$. Hence, $a|f(x)$ or $a|g(x)$. Assume that the degree of $p(x)$ is positive, and let $p(x)$ not divide $f(x)$. Consider the ideal S generated by $f(x)$ and $p(x)$ in $R[x]$. The elements of S are of the form $A(x)p(x) + B(x)f(x)$, where $A(x), B(x) \in R[x]$. Let $0 \neq \phi(x) \in S$ be of the smallest degree, and let a be its leading coefficient. By Theorem 4.1 there exists a nonnegative integer k and polynomials $h(x)$ and $r(x)$ such that $a^k f(x) = \phi(x)h(x) + r(x)$, where either $r(x) = 0$ or the degree of $r(x)$ is less than that of $\phi(x)$. This implies $r(x) \in S$, a contradiction unless $r(x) = 0$; so

$$a^k f(x) = \phi(x)h(x) = c(\phi)\phi_1(x)h(x),$$

where $\phi_1(x)$ is primitive. Now $\phi_1(x)$ divides $a^k f(x)$ implies $\phi_1(x)t(x) = a^k f(x)$, which yields $c(t) = a^k c(f)$. Therefore, $a^k|c(t)$ and, hence, $a^k|t(x)$. This, along with $\phi_1(x)t(x) = a^k f(x)$ and the fact that $R[x]$ is an integral domain, yields $\phi_1(x)$ divides $f(x)$. Similarly, $\phi_1(x)$ divides $p(x)$. Because $p(x)$ is irreducible and does not divide $f(x)$, it follows that $\phi_1(x)$ is a unit in $R[x]$. Hence, $\phi_1(x) \in R$. Thus, $\phi(x) = c(\phi)\phi_1(x) \in R$; so $\phi(x) = a \in R$, and we can write

$$a = A(x)p(x) + B(x)f(x)$$

for some $A(x)$, $B(x)$ in $R[x]$. Then by multiplying by $g(x)$ on both sides of the above equation, we get

$$ag(x) = A(x)p(x)g(x) + B(x)f(x)g(x).$$

This yields $p(x)|ag(x)$, since, by hypothesis, $p(x)|f(x)g(x)$. Thus, $ag(x) = p(x)t(x)$, $t(x) \in R[x]$. By equating the contents of the polynomials on both sides and using Lemma 4.2, we obtain

$$ac(g) = c(p)c(t) = c(t),$$

because $p(x)$ is irreducible. Thus, $a|c(t)$ and therefore $a|t(x)$. Hence,

$$g(x) = p(x)t'(x), \qquad t'(x) \in R[x].$$

This proves that $p(x)|g(x)$, as desired. □

Remark. It follows immediately from Theorem 4.3 that if R is a UFD, then the polynomial ring $R[x_1, x_2, \ldots, x_n]$ over R in a finite number of variables x_i, $i = 1, 2, \ldots, n$, is also a UFD.

Problems

1. Show that the polynomial ring $F[x,y]$ in two variables over a field F is a UFD but not a PID.
2. Let $F[x]$ be polynomial ring over a field F. Show that a nonzero polynomial $f(x) \in F[x]$ is a unit if and only if $f(x) \in F$.
3. Let R be a commutative ring with unity. Show that an element $f(x) \in R[x]$ is a zero divisor if and only if there exists an element $0 \neq b \in R$ such that $bf(x) = 0$.
4. Show that the $n \times n$ matrix ring $(R[x])_n$ over a polynomial ring $R[x]$ is isomorphic to the polynomial ring $R_n[x]$ over the $n \times n$ matrix ring R_n.

CHAPTER 12

Rings of fractions

Let R be a commutative ring containing *regular* elements; that is, elements $a \in R$ such that $a \neq 0$ and a is not a zero divisor. In this chapter we show that any commutative ring R with regular elements can be embedded in a ring Q with unity such that every regular element of R is invertible in Q. In particular, any integral domain can be embedded in a field. Indeed, by defining the general notion of ring of fractions with respect to a multiplicative subset S, we obtain a ring R_S such that there is a canonical homomorphism from R to R_S. The conditions under which a noncommutative integral domain can be embedded in a division ring are also discussed.

1 Rings of fractions

Definition. *A nonempty subset S of a ring R is called a* multiplicative set *if for all $s_1, s_2 \in S$, we have $s_1 s_2 \in S$. If, in addition, each element of S is regular, then S is called a* regular multiplicative set.

Clearly, the set of all regular elements is a regular multiplicative set. In particular, if R is an integral domain, then $S = R - \{0\}$ is a regular multiplicative set.

Let R be a commutative ring and S a multiplicative set. Define a relation $\sim$ on $R \times S$ by $(r,s) \sim (r',s')$ if there exists $s'' \in S$ such that $s''(rs' - r's) = 0$.

1.1 Lemma. $\sim$ *is an equivalence relation.*

Proof. The proof follows by routine verification. □

Let a/s denote the equivalence class determined by (a,s), and let R_S denote the set of all equivalence classes. Our purpose is to make R_S into a ring by defining addition and multiplication in a natural manner, reminiscent of addition and multiplication of "fractions" in arithmetic. We define

$$a/s_1 + b/s_2 = (as_2 + bs_1)/s_1 s_2,$$
$$a/s_1 \cdot b/s_2 = ab/s_1 s_2.$$

1.2 Theorem. $(R_S,+,\cdot)$ *is a ring with unity.*

Proof. It is routine to verify that addition and multiplication as defined preceding the theorem are well defined and that $(R_S,+,\cdot)$ is a ring. The reader is encouraged to verify these facts for the sake of getting familiar with the notion. It may be noted that the zero element of R_S is $0/s$ for any $s \in S$, and s/s is the unity for any $s \in S$. □

The ring R_S is called the *ring of fractions* of R with respect to S or the *localization* of R at S; R_S is also called the *quotient ring* of R with respect to S. (*Warning:* The quotient ring of a ring R modulo an ideal I, that is, R/I, is altogether a different notion, as defined in Chapter 10.) If S consists of all the regular elements of R, then R_S is usually known as the *total quotient ring* of R. Finally, if R_S is a ring of fractions where $0 \in S$, then R_S is the zero ring. Therefore, we assume throughout this section that $0 \notin S$.

1.3 Theorem. *Let R be a commutative ring, and let S be a multiplicative subset of R. Then there exists a natural homomorphism $f: R \to R_S$ defined by $f(a) = as/s$, where $a \in R$, and $s \in S$ is some fixed element. Further, f is a monomorphism iff $xa = 0$, $x \in S$, $a \in R$ implies $a = 0$.*

Proof. Let $a,b \in R$. Then

$$f(a+b) = (a+b)s/s = as/s + bs/s = f(a) + f(b),$$
$$f(ab) = abs/s = as/s\ bs/s = f(a)f(b).$$

Hence, f is a homomorphism. Next,

$$a \in \operatorname{Ker} f \Leftrightarrow as/s = 0/s \Leftrightarrow t(as^2 - 0s) = 0 \quad \text{for some } t \in S$$
$$\Leftrightarrow ats^2 = 0 \Leftrightarrow ax = 0 \quad \text{for some } x \in S.$$

Therefore, $\operatorname{Ker} f = (0)$ iff $ax = 0$ implies $a = 0$. □

As a consequence of Theorem 1.3, we have

1.4 Theorem. *Let R be a commutative ring with regular elements. Let S be the set of all regular elements in R. Then the ring R_S has the following properties.*

(i) *R is embeddable in R_S.*

Regarding R as a subring of R_S, we have

(ii) *each regular element of R is invertible in R_S, and*
(iii) *each element of R_S is of the form as^{-1}, where $a \in R$, $s \in S$.*

Proof. (i) follows from Theorem 1.3. Further, if a is a regular element in R, then identifying a with $ab/b \in R_S$, where b is some element in S, it is clear that $b/ab \in R_S$ is the inverse of ab/b, because $ab \in S$. Finally, to prove (iii), let $a/s \in R_S$. Then $a/s = (ab/b)(b/bs) = as^{-1}$. □

Definition. *Let R be any ring (not necessarily commutative), and let Q be a ring with unity containing R as a subring such that each regular element of R is invertible in Q and each element of Q is of the form ab^{-1} ($a^{-1}b$), where $a \in R$ and b is a regular element of R. Then Q is called a* right (left) quotient ring of R. *If R is commutative, then Q is called a quotient ring of R.*

1.5 Theorem. *Any commutative integral domain R can be embedded in a field R_S (called the* field of fractions *of R).*

Proof. Let $S = R - \{0\}$. Then S is a regular multiplicative set. By Theorem 1.4, R_S is a field containing R as a subring. □

Definition. *A ring R with unity is called a* local *ring if it has a unique maximal right ideal.*

1.6 Theorem. *Let R be a commutative ring and P a prime ideal. Then $S = R - P$ is a multiplicative set, and R_S is a local ring with unique maximal ideal $P_S = \{a/s \mid a \in P, s \notin P\}$.*

Proof. That S is a multiplicative set follows from the definition of a prime ideal. We first show that P_S is an ideal in R_S. Let a_1/s_1, $a_2/s_2 \in P_S$

and $r/s \in R_S$. Then $a_1/s_1 - a_2/s_2 = (a_1s_2 - a_2s_1)/s_1s_2 \in P_S$. Further, $(a_1/s_1)(r/s) = a_1r/s_1s \in P_S$. Hence, P_S is an ideal. To show that P_S is maximal, let $r/s \notin P_S$. Then $r \notin P$. Thus, $s/r \in R_S$, so r/s is invertible. This shows that all the elements of R_S that lie outside P_S are invertible. Hence, P_S is the unique maximal ideal in R_S. This proves that R_S is a local ring. □

1.7 **Example.** Let R be a UFD, and let S be a multiplicative subset of R containing the unity of R. Then R_S is also a UFD.

Solution. First we show that if $a \in R$ is irreducible in R, then $a/1$ is irreducible in R_S. For, let $a/1 = (b/s_1)(c/s_2)$, where b/s_1 and c/s_2 are nonunits in R_S. Then $as_1s_2 = bc$. Because a is irreducible and, hence, prime, it follows that $a|b$ or $a|c$. For definiteness let $a|b$; so $ab_1 = b$ for some $b_1 \in R$. Then $a/1 = (b/s_1)(c/s_2) \Rightarrow 1 = (b_1/s_1)(c/s_2)$, a contradiction, because c/s_2 is not a unit. Now let $a/s \in R_S$. Write $a = a_1a_2 \cdots a_r$ as a product of irreducible elements in R. Then $a/s = (1/s)(a_1/1)(a_2/1)\cdots(a_r/1)$ is a product of irreducible elements in R_S. This proves one of the conditions for R_S to be a UFD.

We proceed to prove the second condition, that if $a/s \in R_S$ is irreducible then it is prime. Now a/s irreducible implies a is irreducible in R, so a is prime in R. We prove $a/1$ is prime in R_S. Let $a/1$ divide $(b/s_1)(c/s_2)$. Then $(a/1)(d/s_3) = (b/s_1)(c/s_2)$ for some $d/s_3 \in R_S$. This implies $ads_1s_2 = bcs_3$, so $a|bcs_3$. But then a divides b,c, or s_3. If $a|s_3$, it follows that $a/s|1$, so a/s is a unit in R_S, a contradiction. Thus, $a|b$ or $a|c$. This implies a/s divides b/s_1 or c/s_2 as desired. □

Problems

In Problems 1–3, S is a multiplicative subset of a commutative ring R with unity and $1 \in S$.

1. Let $f: R \to R_S$ be the canonical homomorphism given by $f(a) = as/s$ for some fixed $s \in S$. If $g: R \to R'$ is any homomorphism of a ring R into a ring R' such that every element $g(s)$, $s \in S$, is a unit, then there exists a homomorphism $h: R_S \to R'$ such that the following diagram is commutative:

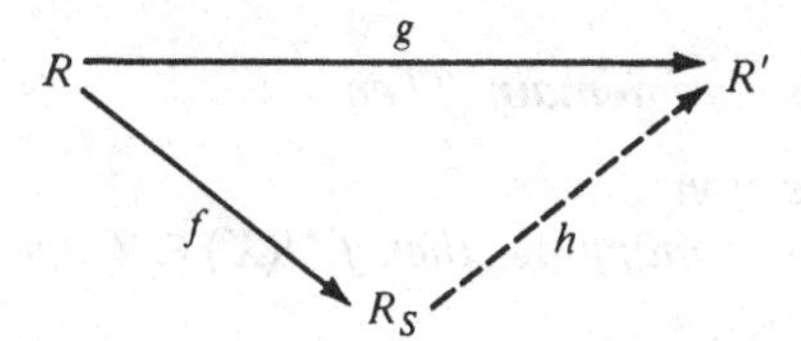

This is also expressed by saying that g factors through f.

2. Let R be a PID. Show that R_S is also a PID.
3. Let R be a commutative ring with 1 having no infinite properly ascending chain of ideals in R. Show that the same holds for R_S.
4. Let p be prime. Show that the ring

$$\mathbf{Z}_{(p)} = \{a/b \mid a,b \in \mathbf{Z}, p \nmid b\}$$

is a local ring. What is the unique maximal ideal?

5. Let $\mathbf{R}(x)$ and $\mathbf{R}(x, y)$ be fields of rational functions over $\mathbf{R}$. Is it true that $\mathbf{R}(x)$ and $\mathbf{R}(x,y)$ are isomorphic fields?

2 Rings with Ore condition

We now consider the question of embedding any integral domain in a division ring.

Definition. *An integral domain R is called a* right (left) Ore domain *if for any pair of nonzero elements $a,b \in R$ there exist nonzero elements $x, y \in R$ such that $ax = by$ ($xa = yb$).*

We note that an integral domain R is a right Ore domain if and only if every pair of nonzero right ideals has a nonzero intersection. Suppose, first, that R is a right Ore domain. Let I and J be nonzero right ideals in R. Choose $0 \neq a \in I$ and $0 \neq b \in J$. Then there exist $0 \neq x,y \in R$ such that $ax = by$. Since $ax = by \in I \cap J$, it follows that $I \cap J \neq (0)$. The converse is similar.

We also remark that it follows from the definition that any commutative integral domain is a right Ore domain as well as a left Ore domain.

Let R be an integral domain, and let I be a nonzero right ideal of R. Denote by $\mathrm{Hom}_R(I,R)$ the set of mappings $f\colon I \to R$ such that (i) $f(x+y) = f(x) + f(y)$, and (ii) $f(xr) = f(x)r$ for all $x,y \in I, r \in R$. A mapping with these two properties is known as an *R-linear mapping.* We note that if f is an R-linear mapping, then f is an additive group homomorphism. Thus, $f(0) = 0$ and $f(-x) = -f(x)$.

Throughout this section $\mathscr{C}$ will denote the set of nonzero right ideals of a ring R.

2.1 Lemma. *Let R be a right Ore domain. Then*

(i) $\mathscr{C}$ is closed under intersection.

(ii) If $f\colon I \to R$ is an R-linear mapping, then $f^{-1}(X) \in \mathscr{C}$ for any $X \in \mathscr{C}$.

(Recall that $f^{-1}(X) = \{a \in I \mid f(a) \in X\}$.)

Proof. (i) Let $I, J \in \mathscr{C}$. Then because R is a right Ore domain, $I \cap J \neq (0)$; so $I \cap J \in \mathscr{C}$.

(ii) Let $a,b \in f^{-1}(X)$ and $r \in R$. Then $f(a), f(b) \in X$, so $f(a) - f(b) \in X$. Thus $a - b \in f^{-1}(X)$. Also, because X is a right ideal, $f(a)r \in X$. So $f(ar) \in X$. This implies $ar \in f^{-1}(X)$. This proves that $f^{-1}(X)$ is a right ideal.

By using the hypothesis that R is a right Ore domain, we show that $f^{-1}(X) \neq (0)$. For, if $f(I) = 0$, then $f(I) \subset X$. So $I \subset f^{-1}(X)$. On the other hand, if $f(I) \neq 0$, then because R is a right Ore domain, $f(I) \cap X \neq (0)$. This implies that there exists $0 \neq a \in I$ such that $f(a) \in X$. Therefore, $0 \neq a \in f^{-1}(X)$. This proves that $f^{-1}(X) \in \mathscr{C}$. □

Let R be a right Ore domain and $H = \bigcup_{I \in \mathscr{C}} \mathrm{Hom}(I,R)$. Define a relation $\sim$ on H by $f \sim g$ if $f = g$ on some $A \in \mathscr{C}$.

2.2 Lemma. *$\sim$ is an equivalence relation.*

Proof. Clearly, $\sim$ is reflexive and symmetric. We now show $\sim$ is transitive. Let $f \sim g$ and $g \sim h$. Then $f = g$ on some $A \in \mathscr{C}$ and $g = h$ on some $B \in \mathscr{C}$. But then $f = h$ on $A \cap B \in \mathscr{C}$. Hence $f \sim h$. This proves that $\sim$ is an equivalence relation. □

Denote by $[f]$ the equivalence class determined by $f \in H$, and by Q the set $H/\sim$ of equivalence classes. Our purpose now is to make $Q = H/\sim$ into a ring by suitably defining addition and multiplication. In what follows let Dom f denote the domain of a mapping f. We define addition in H as follows:

$$[f] + [g] = [f + g],$$

where

$$f + g\colon \mathrm{Dom}\, f \cap \mathrm{Dom}\, g \to R.$$

First, note that $\mathrm{Dom}\, f \cap \mathrm{Dom}\, g \in \mathscr{C}$, so $f + g \in H$. To show that addition is well defined, let $[f] = [f']$ and $[g] = [g']$. Then $f = f'$ on some $A \in \mathscr{C}$ and $g = g'$ on some $B \in \mathscr{C}$. This yields $f + g = f' + g'$ on $A \cap B \in \mathscr{C}$. Hence, $[f + g] = [f' + g']$. Also, the addition is commutative and associative, $[0]$ is the identity element in $(Q,+)$ and $[-f]$ is the additive inverse of $[f]$. Thus, $(Q,+)$ is an additive abelian group.

Next, we define multiplication in H as follows:

$$[f][g] = [fg],$$

where

$$fg\colon g^{-1}(\operatorname{Dom} f) \to R.$$

By Lemma 2.1, $g^{-1}(\operatorname{Dom} f) \in \mathscr{C}$, so $fg \in H$. To show that multiplication is well defined, let $[f] = [f']$ and $[g] = [g']$. Then $f = f'$ on, say, $A \in \mathscr{C}$, and $g = g'$ on, say, $B \in \mathscr{C}$. We claim that $fg = f'g'$ on $B \cap g^{-1}(A) \in \mathscr{C}$. Let $x \in B \cap g^{-1}(A)$. Then $x \in B$ implies $g(x) = g'(x)$ and $g(x) \in A$ implies $f(g(x)) = f'(g(x))$. Hence, $(fg)(x) = (f'g')(x)$. This proves $[fg] = [f'g']$. It is straightforward to check that $(Q, \cdot)$ is a semigroup and that the distributive laws

$$[f]([g] + [h]) = [f][g] + [f][h],$$
$$([f] + [g])[h] = [f][h] + [g][h]$$

hold. Also, if $I\colon R \to R$ is the identity mapping, then $[I]$ is the unity of Q. Thus, Q is a ring with unity. We now show that Q is a division ring.

2.3 Theorem. *Let R be a right Ore domain. Then $Q = \bigcup_{I \in \mathscr{C}} Hom(I,R)/\sim$ is a division ring.*

Proof. Let $[f]$ be a nonzero element in Q. If $\operatorname{Ker} f \neq (0)$, then $\operatorname{Ker} f \in \mathscr{C}$. Now $f =$ the zero mapping on $\operatorname{Ker} f$. Therefore, $[f] = [0]$, a contradiction. This gives $\operatorname{Ker} f = (0)$. Define $g\colon \operatorname{Im} f \to R$ by $g(f(x)) = x$, $x \in \operatorname{Dom} f$. The mapping g is well defined because $\operatorname{Ker} f = (0)$. But then $gf =$ identity mapping I on $\operatorname{Dom} f$; therefore $[gf] = [I]$. This gives $[g][f] = [I]$, so $[f]$ has a left inverse. This proves that Q is a division ring. □

2.4 Theorem. *Let R be an integral domain. Then R is a right Ore domain if and only if there exists a division ring Q such that*

(i) *R is a subring of Q.*
(ii) *Every element of Q is of the form ab^{-1}, for some $a,b, \in R$.*

(Recall that the ring Q with properties (i) and (ii) is called a right quotient ring of R.)

Proof. Let R be a right Ore domain. Then by Theorem 2.3, $Q = \bigcup_{I \in \mathscr{C}} \operatorname{Hom}(I,R)/\sim$ is a division ring. We assert that the mapping $a \mapsto [a*]$ of R into Q, where $a* \in \operatorname{Hom}_R(R,R)$, given by $a*(x) = ax$, is a 1-1 homomorphism. For,

$$(a+b)*(x) = (a+b)x = ax + bx = a*(x) + b*(x) = (a* + b*)(x),$$

so $(a+b)* = a* + b*$. This implies $[(a+b)*] = [a* + b*] = [a*] + [b*]$. Further,

$$(ab)*(x) = (ab)x = a(bx) = a(b*x) = a*(b*x) = (a*b*)(x).$$

Therefore, $(ab)* = a*b*$, so $[(ab)*] = [a*][b*]$. To show that the mapping is 1-1, let $[a*] = [0]$. Then $a* = 0$ on some $I \in \mathscr{C}$. This implies $a*I = 0$; that is, $aI = 0$. But because R is an integral domain, this implies $a = 0$. This proves that the mapping is 1-1. Hence, R can be embedded in a division ring Q. We may then identify R with its copy in Q and regard R as a subring of Q. This proves (i). We may remark that under this identification if $q \in Q$ and $a \in R$, then $q(a)$ is identified with qa, the composition of mappings q and a. For, $q(a)(b) = q(ab)$, by definition of an R-linear mapping q. Also, considering q and a $(=a*)$ as mappings, $(qa)(b) = (qa*)(b) = q(a*b) = q(ab)$. Therefore, under the stated identification if $q \in Q$, $a \in R$, then $q(a) = qa$.

To prove (ii), let $q \in Q$. By construction of Q there exists $I \in \mathscr{C}$ such that $q(I) \subset R$. Let $0 \neq b \in I$. Then there exists $a \in R$ such that $q(b) = a$; that is, $qb = a$. This implies $q = ab^{-1}$ because every nonzero element in Q is invertible.

Conversely, let a and b be nonzero elements in R. Then $a^{-1}b \in Q$, so $a^{-1}b = xy^{-1}$ for some nonzero elements $x,y \in R$. This implies $ax = by$. Hence, R is a right Ore domain. □

Problems

1. If R is an integral domain with unity in which each right ideal is of the form aR, $a \in R$ (i.e., R is a right principal ideal domain), show that R is a right Ore domain.
2. Let R be an integral domain satisfying the standard identity of degree 3; that is, for every ordered set (a_1, a_2, a_3) of elements in R,

   $$\sum \pm a_{i_1} a_{i_2} a_{i_3} = 0,$$

   where the summation runs over every permutation (i_1, i_2, i_3) of $(1,2,3)$, and the sign of the corresponding term $a_{i_1} a_{i_2} a_{i_3}$ is positive or negative according to whether (i_1, i_2, i_3) is an even permutation or an odd permutation. Show that R is a right Ore domain and a left Ore domain.

3. Show that an integral domain satisfying a standard identity of any degree n is both a right and a left Ore domain.
4. Let F be a field, and let $R = \{\left(\begin{smallmatrix} a & b \\ 0 & a \end{smallmatrix}\right) | a,b \in F\}$. Show that $\left(\begin{smallmatrix} a & b \\ 0 & a \end{smallmatrix}\right)$ is regular if and only if $a \neq 0$, and every regular element in R is invertible in R. Conclude that R is its own right and left ring of quotients. (We may remark that a right, as well as a left, ring of quotients of $R = \{\left(\begin{smallmatrix} a & b \\ 0 & c \end{smallmatrix}\right) | a,b,c \in F\}$ is the 2×2 matrix ring over F.)

CHAPTER 13

Integers

In this chapter our purpose is to establish all the familiar properties of the natural numbers (= positive integers) and to obtain the ring of integers by starting from the five axioms of Peano. We also demonstrate that the five axioms are equivalent to the axioms of an ordered integral domain whose positive elements are well-ordered. Either of these sets of axioms determines a unique ring (up to isomorphism) called the ring of integers.

1 Peano's axioms

The traditional method of describing the set **N** of natural numbers axiomatically is by means of the following axioms of Peano:

(i) $1 \in \mathbf{N}$.
(ii) For each $a \in \mathbf{N}$ there exists a unique $a' \in \mathbf{N}$ called the *successor* of a. (In other words there exists a map $a \mapsto a'$ of **N** into itself, called the *successor map*.)
(iii) $a' \neq 1$ for any $a \in \mathbf{N}$.
(iv) For all $a,b \in \mathbf{N}$, $a' = b' \Rightarrow a = b$. (In other words, the successor map $a \mapsto a'$ of **N** into itself is 1-1.)
(v) Let S be a subset of **N** such that (a) $1 \in S$; (b) if $a \in S$, then $a' \in S$. Then $S = \mathbf{N}$.

The fifth axiom is called the axiom of induction or the *first principle of induction* and is the basis of the proofs of many theorems in mathematics.

We write $1' = 2$, $2' = 3$, $3' = 4$, and so on.

We may restate the first principle of induction (also called the *principle of mathematical induction*) as follows:

(v′) Suppose that for each natural number $n \in \mathbf{N}$ we have associated a statement $S(n)$ such that

(a) $S(1)$ is true.
(b) $S(n)$ is true implies $S(n')$ is also true.

Then $S(n)$ is true for all $n \in \mathbf{N}$.

The following are some of the immediate consequences of Peano's axioms.

1.1 Corollary. *For any $a \in \mathbf{N}$, $a' \neq a$.*

Proof. Let $S = \{n \in \mathbf{N} \mid n' \neq n\}$. By axioms (i) and (iii), $1 \in S$. Let $n \in S$. Then $n' \neq n$; so $(n')' \neq n'$, for by axiom (iv), $(n')' = n'$ implies $n' = n$. Consequently, $n' \in S$. Axiom (v) then gives $S = N$. □

1.2 Corollary. *For any $a \in \mathbf{N}$, $a \neq 1$, there exists a unique $b \in \mathbf{N}$ such that $a = b'$.*

Proof. Let $S = \{n \in \mathbf{N} \mid n = 1 \text{ or } n = m' \text{ for some } m \in \mathbf{N}\}$. Then by definition of S, $1 \in S$. Let $n \in S$. Then $n = m'$ for some $m \in \mathbf{N}$, so $n' = (m')'$. Thus $n' \in S$. Hence, by axiom (v), $S = \mathbf{N}$. Further, if $a = b'$ and $a = c'$, then by axiom (iv), $b = c$. □

Addition and multiplication of natural numbers can be defined in a "recursive manner." The next theorem shows the existence and uniqueness of a binary operation in $\mathbf{N}$ (called plus) satisfying certain conditions.

1.3 Theorem. *There exists one and only one binary operation "+" in $\mathbf{N}$ satisfying*

(i) $a + 1 = a'$,
(ii) $a + b' = (a + b)'$

for all $a, b \in \mathbf{N}$.

Proof. We first show that there exists a binary operation "+" in $\mathbf{N}$ satisfying (i) and (ii).

Consider the set

$$S = \{a \in \mathbf{N} \mid a + b \text{ can be defined for all } b \text{ in } \mathbf{N} \text{ satisfying conditions (i) and (ii)}\}.$$

To show that $1 \in S$, we define

$$1 + b = b' \qquad \text{for each } b \in \mathbf{N}.$$

Then $1 + 1 = 1'$ by our definition, and $1 + b' = (b')' = (1 + b)'$ by repeatedly using our definition. Thus, (i) and (ii) are satisfied for the case $a = 1$; so $1 \in S$.

Let $a \in S$. Hence, $a + b$ is defined for all $b \in \mathbf{N}$. We now define

$$a' + b = (a + b)'.$$

Then

$$\begin{aligned} a' + 1 &= (a + 1)' && \text{(by definition)} \\ &= (a')' && \text{(since } a \in S\text{)}. \end{aligned}$$

Also,

$$\begin{aligned} a' + b' &= (a + b')' && \text{(by definition)} \\ &= ((a + b)')' && \text{(since } a \in S\text{)} \\ &= (a' + b)' && \text{(by definition)}. \end{aligned}$$

So $a' \in S$. Hence, by axiom (v), $S = \mathbf{N}$.

We now prove that there is exactly one binary operation in $\mathbf{N}$ satisfying (i) and (ii).

Suppose $\oplus$ is another binary operation in $\mathbf{N}$ satisfying (i) and (ii). For any fixed $a \in \mathbf{N}$ let

$$S = \{n \in \mathbf{N} | a + n = a \oplus n\}.$$

Then $1 \in S$ because

$$\begin{aligned} a + 1 &= a' && \text{[by (i)]} \\ &= a \oplus 1 && \text{[by (i)]}. \end{aligned}$$

Next, let $n \in S$. Then

$$\begin{aligned} a + n' &= (a + n)' && \text{[by (ii)]} \\ &= (a \oplus n)' && \text{(since } n \in S\text{)} \\ &= a \oplus n' && \text{[by (ii)]}. \end{aligned}$$

So $n' \in S$. Hence, by axiom (v), $S = \mathbf{N}$. This proves that "+" and "$\oplus$" are the same binary operations. □

1.4 Theorem. *The binary operation + in* $\mathbf{N}$ *satisfies the following laws:*

(i) *For all* $a,b,c \in \mathbf{N}$,

$$(a + b) + c = a + (b + c) \qquad \textit{(associative law of addition).}$$

(ii) For all $a,b \in \mathbf{N}$,

$$a + b = b + a \qquad \textit{(commutative law of addition).}$$

Proof. (i) Let a and b be fixed (but arbitrary) natural numbers, and let

$$S = \{c \in \mathbf{N} | (a + b) + c = a + (b + c)\}.$$

Then by definition of $+$,

$$(a + b) + 1 = (a + b)' = a + b' = a + (b + 1).$$

Thus, $1 \in S$. Now suppose $c \in S$. Then

$$\begin{aligned} (a + b) + c' &= ((a + b) + c)' && \text{(by definition of } +\text{)} \\ &= (a + (b + c))' && \text{(since } c \in S\text{)} \\ &= a + (b + c)' && \text{(by definition of } +\text{)} \\ &= a + (b + c') && \text{(again by definition of } +\text{).} \end{aligned}$$

Thus, $c' \in S$. Therefore, by axiom (v), it follows that $S = \mathbf{N}$. This proves the associative law of addition.

The proof of (ii) is similar and is left to the reader. □

The next theorem is the fundamental theorem for defining ordering in the set of natural numbers.

1.5 Theorem. *Let $a,b \in \mathbf{N}$. Then exactly one of the following statements holds:*

(i) $a = b$.
(ii) $a = b + u$ *for some* $u \in \mathbf{N}$.
(iii) $b = a + v$ *for some* $v \in \mathbf{N}$.

Proof. First we show that (i) and (ii) cannot simultaneously hold. If $a = b$ and $a = b + u$, then $a = a + u$. We show that this is not possible. Consider the set $S = \{n \in \mathbf{N} | n \neq n + u\}$. Now $1 + u = u + 1 = u'$ by the properties of the binary operation $+$. Further, by axiom (iii), $1 \neq u'$. So $1 \in S$. Let $n \in S$. Then $n \neq n + u$. Suppose $n' = n' + u$. Then $n' = (n + u)'$, by definition of $+$, so by axiom (iv), $n = n + u$, a contradiction. Thus, $n' \neq n' + u$, so $n' \in S$. Therefore, by axiom (v), $S = \mathbf{N}$. In particular, $a \neq a + u$. This proves that (i) and (ii) cannot hold simultaneously.

Similarly, (i) and (iii) cannot hold simultaneously.

The proof that (ii) and (iii) cannot hold together is a straightforward application of the associative and commutative laws of addition.

We now proceed to show that one of the three statements must hold.

Let $a \in \mathbf{N}$ be a fixed (but arbitrary) number. Let

$$T = \{b \in \mathbf{N} | \text{one of the three statements in the theorem is true}\}.$$

We assert that $1 \in T$, for either $a = 1$ or else if $a \neq 1$, then by Corollary 1.2,

$$\begin{aligned} a &= u' \qquad \text{(for some } u \in \mathbf{N}) \\ &= 1 + u \qquad \text{(by definition of } +). \end{aligned}$$

Hence, $a = 1$ or $a = 1 + u$, showing that $1 \in T$, as asserted. Now assume that $b \in T$. We show $b' \in T$. If $b = a$, then $b' = a' = a + 1$. So (iii) holds and, thus, $b' \in T$. If $b = a + u$, then $b' = (a + u)' = a + u'$. So again (iii) holds, and, thus, $b' \in T$. If $a = b + u$, then we have to consider two subcases:

Subcase (A). $u = 1$. Then $a = b + 1 = b'$, so (i) holds. Thus, $b' \in T$.

Subcase (B). $u \neq 1$. Then $u = v'$ for some $v \in \mathbf{N}$ (Corollary 1.2). Thus,

$$a = b + u = b + v' = b + (1 + v) = (b + 1) + v = b' + v,$$

showing that $b' \in T$. Hence, $T = \mathbf{N}$, which proves the theorem. □

1.6 Theorem (cancellation laws for addition)

(i) $a + u \neq a$.

(ii) $a + x = a + y \Rightarrow x = y$

for all $a, u, x, y \in \mathbf{N}$.

Proof. (i) Follows from the previous theorem.

(ii) *Case* 1. $x = y + v$ for some $v \in \mathbf{N}$. Then $a + x = a + y \Rightarrow a + y + v = a + y$, a contradiction by (i).

Case 2. $y = x + v'$ for some $v' \in \mathbf{N}$. Then, again, $a + x = a + y \Rightarrow a + x = a + x + v'$, a contradiction by (i). Hence, by the previous theorem $x = y$. □

Definition. *Let* $a, b \in \mathbf{N}$. a *is said to be* greater than b *if* $a = b + u$ *for some* $u \in \mathbf{N}$. *We denote this by* $a > b$.

Definition. *Let* $a, b \in \mathbf{N}$. a *is said to be* less than b *if* $b = a + v$ *for some* $v \in \mathbf{N}$. *We denote this by* $a < b$.

Theorem 1.5 can be restated as

1.7 Remark (trichotomy law of natural numbers). Given $a, b \in \mathbf{N}$, one and only one of the following statements holds:

(i) $a = b$.
(ii) $a > b$.
(iii) $a < b$.

1.8 Remark. $a > b$ if and only if $b < a$.

1.9 Remark. If $a \neq 1$, then $1 < a$.

Proof. $a \neq 1 \Rightarrow \exists b \in \mathbf{N}$ such that $a = b'$ by Corollary 1.2. Then $a = 1 + b$ by the properties of the binary operation $+$. Therefore, by definition of less than, $1 < a$. □

The notation $a \geq b$ shall mean that either $a = b$ or $a > b$. We give a similar meaning for $a \leq b$.

1.10 Theorem. *Let $a, b \in \mathbf{N}$. Then $a < b$ if and only if $a + 1 \leq b$.*

Proof. $a < b \Rightarrow b = a + u$ for some $u \in \mathbf{N}$. If $u = 1$, then $b = a + 1$. If $u \neq 1$, then, by Corollary 1.2, $u = v'$ for some $v \in \mathbf{N}$. Then

$$b = a + u \Rightarrow b = a + v' \Rightarrow b = a + (1 + v)$$
$$\Rightarrow b = (a + 1) + v \Rightarrow a + 1 < b.$$

Hence, in any case, $a + 1 \leq b$. The proof of the converse is obvious. □

We now prove a basic property of the set of natural numbers that is the basis of the second principle of mathematical induction.

1.11 Theorem (well-ordering property of natural numbers). *Every nonempty set of natural numbers possesses a least member.*

Proof. Let S be the given set, and let

$$T = \{n \in \mathbf{N} \mid n \leq a \text{ for all } a \in S\}.$$

Then by Remark 1.9, $1 \in T$. Let $n \in T$. Now, for each $a \in S$, $a' > a$, so $a' \notin T$. This implies $T \neq \mathbf{N}$. But then by Peano's axiom (v), there exists $t \in T$ such that $t' \notin T$. We now claim that $t \in S$ and is the required element. First, by definition of T, $t \leq a$ for all $a \in S$. For if $t \notin S$, then $t < a$ for all $a \in S$. But then $t + 1 \leq a$ by Theorem 1.10. This gives $t' =$

$t+1 \in T$, a contradiction. Therefore, $t \in S$ and $t \leq a$ for all $a \in S$. □

We now prove the second principle of induction.

1.12 Theorem (second principle of induction). *Let S be a subset of* $\mathbf{N}$ *such that*

(i) $1 \in S$.
(ii) $n \in S$ *whenever* $m \in S$ *for all positive integers* $m < n$.

Then $S = \mathbf{N}$.

Proof. Let $T = \{n \in \mathbf{N} | n \notin S\}$. If $T = \emptyset$, we are done, so assume $T \neq \emptyset$. Then by the well-ordering property of $\mathbf{N}$, T contains a least member, say t. By (i), $1 \notin T$, so $t \neq 1$. Thus, by Remark 1.9, $1 < t$. But then all natural numbers less than t belong to S. So by hypothesis (ii), $t \in S$, a contradiction. Hence, $T = \emptyset$; that is, $S = \mathbf{N}$. □

We now introduce a second binary operation in $\mathbf{N}$ to be called multiplication or product.

1.13 Theorem. *There exists one and only one binary operation* $\cdot$ *in* $\mathbf{N}$ *satisfying*

(i) $a \cdot 1 = a$,
(ii) $a \cdot b' = a \cdot b + a$

for all $a, b \in \mathbf{N}$.

We often write $a \cdot b$ simply as ab, which is called the product of a by b or the "number obtained from multiplication of a by b."

Proof. The proof is similar to the proof of Theorem 1.3, concerning existence and uniqueness of the binary operation $+$. □

1.14 Theorem. *The binary operation of multiplication in* $\mathbf{N}$ *satisfies the following laws:*

(i) $ab = ba$ *(commutative law)*,
(ii) $(ab)c = a(bc)$ *(associative law)*,
(iii) $a(b+c) = ab + ac$ *(distributive law)*,

for all $a, b, c \in \mathbf{N}$.

Proof. The proofs of (i) and (ii) are exactly the same as those for the corresponding laws for addition.

(iii) $a(b+1) = ab' = ab + a = ab + a \cdot 1$, by definition of multiplication. Assume that for a given $c \in \mathbf{N}$, we have $a(b+c) = ab + ac$ for all $a, b \in \mathbf{N}$. We prove $a(b+c') = ab + ac'$ for all $a, b \in \mathbf{N}$. Now

$$\begin{aligned} a(b+c') &= a((b+c)') && \text{(by definition of addition)} \\ &= a(b+c) + a && \text{(by definition of multiplication)} \\ &= (ab+ac) + a && \text{(by induction hypothesis)} \\ &= ab + (ac + a) && \text{(by associative law of addition)} \\ &= ab + ac' && \text{(by definition of multiplication).} \end{aligned}$$

Hence, by the principle of induction, $a(b+c) = ab + ac$ for all $a,b,c \in \mathbf{N}$. □

2 Integers

After having given a systematic development of the system of natural numbers $\mathbf{N}$, we now extend this to the set $\mathbf{Z}$ of integers. The necessity of extending the system of natural numbers arises from the fact that an equation of the type $a = x + b$, where $a, b, \in \mathbf{N}$, does not always possess a solution in $\mathbf{N}$ (see Theorem 1.5).

Consider an equivalence relation $\sim$ on $\mathbf{N} \times \mathbf{N}$ defined as follows:

$$(a,b) \sim (c,d) \quad \text{iff} \quad a + d = b + c.$$

Clearly, $\sim$ is an equivalence relation on $\mathbf{N} \times \mathbf{N}$. We denote the equivalence class of (a,b) by $\overline{(a,b)}$ and define binary operations + and · called, respectively, addition and multiplication, in the set $\mathbf{N} \times \mathbf{N}/\sim$ of equivalence classes by the following rules:

$$\begin{aligned} \overline{(a,b)} + \overline{(c,d)} &= \overline{(a+c, b+d)}, \\ \overline{(a,b)}\,\overline{(c,d)} &= \overline{(ac+bd, ad+bc)}. \end{aligned}$$

These definitions of addition and multiplication are well defined, as can be checked in a routine manner. We note that $\overline{(1,1)}$ is the identity element for addition, and $\overline{(1+1,1)}$ is the identity element for multiplication. For we have

$$\overline{(a,b)} + \overline{(1,1)} = \overline{(a+1, b+1)} = \overline{(a,b)}$$

and

$$\begin{aligned} \overline{(a,b)}\,\overline{(1+1,1)} &= \overline{(a(1+1) + b, a + b(1+1))} \\ &= \overline{(a+a+b, a+b+b)} = \overline{(a,b)}. \end{aligned}$$

The set $\mathbf{N} \times \mathbf{N}/\sim$ of equivalence classes is denoted by $\mathbf{Z}$.

2.1 Theorem. $(\mathbf{Z},+,\cdot)$ *is a commutative integral domain with unity.*

Proof. $(\mathbf{Z},+)$ is clearly a commutative semigroup with identity $\overline{(1,1)}$. Because $\overline{(a,b)} + \overline{(b,a)} = \overline{(a+b,a+b)} = \overline{(1,1)}$, it follows that each element of $(\mathbf{Z},+)$ has an additive inverse, which proves that $(\mathbf{Z},+)$ is an abelian group.

Further, $(\mathbf{Z},\cdot)$ is a commutative semigroup with identity $\overline{(1+1,1)}$. The two distributive laws follow directly. Hence, $(\mathbf{Z},+,\cdot)$ is a commutative ring with unity.

Let us denote, as usual, the additive identity $\overline{(1,1)}$ by 0. To prove $\mathbf{Z}$ is an integral domain, let $x,y \in \mathbf{Z}$ with $xy = 0$. Let $x = \overline{(a,b)}$ and $y = \overline{(c,d)}$. Then

$$\begin{aligned} xy = 0 &\Rightarrow \overline{(a,b)}\,\overline{(c,d)} = \overline{(1,1)} \\ &\Rightarrow \overline{(ac+bd,ad+bc)} = \overline{(1,1)} \\ &\Rightarrow (ac+bd)+1 = ad+bc+1. \end{aligned}$$

Suppose $x \neq 0$. Then $a \neq b$. So either $a = b+u$ or $b = a+v$.

Case 1. $a = b+u$. Then

$$\begin{aligned} ac+bd+1 &= ad+bc+1 \\ &\Rightarrow bc+uc+bd+1 \\ &= bd+ud+bc+1 \\ &\Rightarrow (bc+bd+1)+uc \\ &= (bd+bc+1)+ud, \end{aligned}$$

by commutativity and associativity of addition in $\mathbf{N}$.

But then by Theorem 1.6(ii), $uc = ud$. In case $c \neq d$, then either $c = d+s$ or $d = c+t$. If $c = d+s$, then $u(d+s) = ud \Rightarrow ud+us = ud$, a contradiction by Theorem 1.6(i). Similarly, $d = c+t$ leads to a contradiction. Hence, $uc = ud$ must imply $c = d$. So then $y = \overline{(c,d)} = 0$.

Case 2. $b = a+v$. An exactly similar computation yields that $c = d$, so $y = 0$.

Hence, $(\mathbf{Z},+,\cdot)$ is an integral domain. □

2.2 Theorem. $\mathbf{N}$ *embeds in* $\mathbf{Z}$ *under the mapping* $n \mapsto \overline{(n+1,1)}$, $n \in \mathbf{N}$, *that preserves addition and multiplication.*

Proof. Let $a,b \in \mathbf{N}$ and $\overline{(a+1,1)} = \overline{(b+1,1)}$. Then $(a+1)+1 = (b+1)+1$. But then by Theorem 1.6(ii), $a = b$. Thus $\mathbf{N}$ embeds in $\mathbf{Z}$. Now

$$\overline{(a+b+1,1)} = \overline{(a+1+b+1,1+1)} = \overline{(a+1,1)} + \overline{(b+1,1)}$$

and

$$\overline{(ab+1,1)} = \overline{(ab+1+a+b,1+a+b)} = \overline{(a+1,1)}\,\overline{(b+1,1)}.$$

This proves the theorem. □

2.3 Remark. By Theorem 2.2 we may thus identify each element n in $\mathbf{N}$ with its image $\overline{(n+1,1)}$ in $\mathbf{Z}$. Henceforth, we write n for its image $\overline{(n+1,1)}$, and $-n$ for the additive inverse $\overline{(1,n+1)}$. Further, $\overline{(a,b)} = \overline{(a+1,1)} + \overline{(1,b+1)} = a + (-b)$. So we identify $\overline{(a,b)}$ with $a-b$.

Also, the copy of $\mathbf{N}$ in $\mathbf{Z}$ under this embedding is denoted by $\mathbf{Z}^+$. The elements of $\mathbf{Z}$ are called *integers*, and those of the subset $\mathbf{Z}^+$ are called *positive integers*.

We also denote $\{-n \in \mathbf{Z} \mid n \in \mathbf{Z}^+\}$, called the set of *negative integers*, by $\mathbf{Z}^-$.

2.4 Theorem (trichotomy law of integers). *If $x \in \mathbf{Z}$, then one and only one of the following holds:*

(i) $x = 0$.
(ii) $x \in \mathbf{Z}^+$.
(iii) $-x \in \mathbf{Z}^+$.

Equivalently, $\mathbf{Z}$ is the disjoint union of its subsets $\{0\}$, $\mathbf{Z}^+$, and $\mathbf{Z}^-$.

Proof. Let $x \in \mathbf{Z}$, $x \neq 0$. Then

$$x = \overline{(a,b)}, \quad a \neq b, \quad a,b \in \mathbf{N}.$$

Then either $a = b + u$ for some $u \in \mathbf{N}$, or $b = a + v$ for some $v \in \mathbf{N}$. We show that

$$a = b + u \quad \text{iff} \quad x \in \mathbf{Z}^+,$$

and

$$b = a + v \quad \text{iff} \quad -x \in \mathbf{Z}^+.$$

Let $a = b + u$. Then

$$x = \overline{(a,b)} = \overline{(b+u,b)} = \overline{(u+1,1)} \in \mathbf{Z}^+.$$

Conversely, let $x \in \mathbf{Z}^+$. Then $x = \overline{(u+1,1)}$ for some $u \in \mathbf{N}$. But since $x = \overline{(a,b)}$, $u + 1 + b = a + 1$. Then by Theorems 1.4 and 1.6, $a = b + u$, as desired. Next, let $b = a + v$. Then

$$-x = -\overline{(a,b)} = \overline{(b,a)} = \overline{(a+v,a)} = \overline{(v+1,1)} \in \mathbf{Z}^+.$$

Conversely, let $-x \in \mathbf{Z}^+$. Then $-x = \overline{(v+1,1)}$ for some $v \in \mathbf{N}$. Again, because $x = \overline{(a,b)}$, $-x = \overline{(b,a)}$; therefore, $\overline{(b,a)} = \overline{(v+1,1)}$. Hence, $b+1 = a+v+1$; that is, $b = a+v$. This completes the proof. □

We have proved that the ring $(\mathbf{Z},+,\cdot)$ contains a subset $\mathbf{Z}^+$ that is (i) closed under $+$, (ii) closed under $\cdot$, and (iii) if $x \in \mathbf{Z}$, then one and only one of the following is true: $x = 0$, $x \in \mathbf{Z}^+$, $-x \in \mathbf{Z}^+$. The set $\mathbf{Z}^+$ is called the set of *positive elements* of $\mathbf{Z}$.

Definition. *Any ring R with a subset P of elements satisfying (i), (ii), and (iii) in the previous paragraph is called an* ordered domain.

Definition. *Let $a,b \in \mathbf{Z}$. Then a is said to be* greater than *b if $a - b \in \mathbf{Z}^+$. We denote this by $a > b$.*

Definition. *Let $a,b \in \mathbf{Z}$. Then a is said to be* less than *b if $b - a \in \mathbf{Z}^+$. We denote this by $a < b$.*

2.5 Remark (trichotomy law of integers). By Theorem 2.4, given $a,b \in \mathbf{Z}$, one and only one of the following statements holds:

(i) $a = b$.
(ii) $a > b$.
(iii) $a < b$.

2.6 Remark. $a > b$ if and only if $b < a$.

The notation $a \geq b$ shall mean that either $a = b$ or $a > b$. We give a similar meaning for $a \leq b$.

2.7 Theorem. *The order relations in $\mathbf{Z}$ satisfy the following laws:*

(i) Transitive law: If $a < b$ and $b < c$, then $a < c$.
(ii) Addition to an inequality: If $a < b$, then $a + c < b + c$.
(iii) Multiplication of an inequality: If $a < b$ and $0 < c$, then $ac < bc$.
(iv) Law of trichotomy: For any a and b in $\mathbf{Z}$, one and only one of the relations $a = b$. $a > b$. or $a < b$ holds.

Proof

(i) $a < b$ and $b < c \Rightarrow b - a, c - b \in Z^+ \Rightarrow (b - a) + (c - b) \in \mathbf{Z}^+$
$\Rightarrow c - a \in \mathbf{Z}^+ \Rightarrow a < c.$

(ii) and (iii) are left to the reader.
(iv) See Remark 2.5. □

2.8 Remark. For any ordered domain R, we can define in exactly the same manner the order relations $>$, $<$, $\geq$, and $\leq$, and we can prove the laws stated in Theorem 2.4 and Remark 2.5 for the domain $\mathbf{Z}$ of integers.

2.9 Theorem. *In any ordered ring R all squares of nonzero elements are positive.*

Proof. Let $0 \neq a \in R$. By condition (iii) of the definition of ordered ring, either a or $-a$ is positive. In the first case a^2 is positive by the multiplication property of positive elements. In the second case $-a$ is positive. But $a^2 = (-a)(-a)$, by the properties of a ring. Thus, a^2 is again positive. □

2.10 Corollary. *In any ordered ring R with unity e, we have* $e > 0$.

Proof. $e = ee$; hence, $e > 0$. □

2.11 Theorem. $(\mathbf{Z}, +, \cdot)$ *is an ordered domain with* $\mathbf{Z}^+$ *as positive elements. Moreover,* $\mathbf{Z}^+$ *is well ordered.*

Proof. Conditions (i) and (ii) in the definition of ordered ring for the set $\mathbf{Z}^+$ of positive elements follow from the fact that $\mathbf{Z}^+$ is the copy of $\mathbf{N}$ in $\mathbf{Z}$ (Theorem 2.2), and these conditions hold in $\mathbf{N}$. Condition (iii) follows from Theorem 2.4. Further, $\mathbf{Z}^+$ is well ordered by Theorem 1.11. □

We are now ready to prove that the set of integers $\mathbf{Z}$ can also be characterized as an ordered domain whose positive elements form a well-ordered set.

2.12 Theorem. *Any ordered domain D with unity whose positive elements are well ordered is isomorphic to the ring of integers* $\mathbf{Z}$.

Proof. Let e be the unity of D. Define a mapping $f: Z \to D$ by $f(n) = ne$, $n \in \mathbf{Z}$. Then f is a ring homomorphism, for

$$f(m+n) = (m+n)e = me + ne = f(m) + f(n),$$
$$f(mn) = (mn)e = (me)(ne) = f(m)f(n).$$

f is also 1-1, for let $n \neq 0$ and $ne = 0$. Choose $m \in \mathbf{Z}^+$ such that $m = n$ if $n > 0$, $m = -n$ if $n < 0$. Then $me = 0$, $m \in \mathbf{Z}^+$. Because, in an ordered ring, the sum of positive elements is positive, $me = e + e + \cdots + e$ is also positive. Hence, $ne = 0$ with $n \neq 0$ leads to a contradiction. Therefore, f is 1-1.

We now show $\text{Im} f = D$. If not, let $S = D - \text{Im} f$. We claim S contains positive elements. Let $x \in S$. Then $-x$ also belongs to S; otherwise if $-x \in \text{Im} f$, then $x \in \text{Im} f$, a contradiction.

Let m be the minimal element in the set of positive elements of S.

Case 1. $m = e$. This is not possible because $e \in \text{Im} f$.

Case 2. $m - e > 0$. We have $m - e \notin \text{Im} f$ because $m \notin \text{Im} f$. But $m - e < m$, a contradiction to the minimality of m.

Case 3. $m - e < 0$. Then $m^2 < m < e$. If $m^2 \in \text{Im} f$, then $\exists k \in \mathbf{Z}^+$ such that $f(k) = m^2$. But then $ke = m^2 < e$, a contradiction. Thus, $m^2 \notin \text{Im} f$; hence, $m^2 \in D - \text{Im} f$. Again, $m^2 < m$ yields a contradiction by the minimality of m.

Hence, $S = \varnothing$; therefore, f is an onto mapping. This proves $\mathbf{Z} \simeq D$. □

The theorem proved above shows that upto isomorphism $\mathbf{Z}$ is a unique ordered domain whose positive elements are well-ordered.

We have now completed our program to establish the properties of the natural numbers and to obtain the ring of integers by starting from Peano's axioms.

CHAPTER 14

Modules and vector spaces

1 Definition and examples

Let M be an additive abelian group, and let End(M) be the ring of endomorphisms of M (Section 3.1(c), Chapter 9). If $r \in \text{End}(M)$, $m \in M$, then rm will denote the image of m by r. Therefore

(i) $r(m_1 + m_2) = rm_1 + rm_2$,
(ii) $(r_1 + r_2)m = r_1m + r_2m$,
(iii) $(r_1r_2)m = r_1(r_2m)$,
(iv) $1m = m$,

where $r, r_1, r_2 \in R$, $m, m_1, m_2 \in M$.

We then say M is a *left module* over the ring $R = \text{End}(M)$ according to the following definition.

Definition. *Let R be a ring, M an additive abelian group, and $(r,m) \mapsto rm$, a mapping of $R \times M$ into M such that*

(i) $r(m_1 + m_2) = rm_1 + rm_2$,
(ii) $(r_1 + r_2)m = r_1m + r_2m$,
(iii) $(r_1r_2)m = r_1(r_2m)$,
(iv) $1m = m$ *if* $1 \in R$,

for all $r, r_1, r_2 \in R$ and $m, m_1, m_2 \in M$. Then M is called a left R-module, *often written as ${}_RM$.*

If R is a division ring, then a left R-module is called a left *vector space* over R.

Often rm is called the scalar multiplication or just multiplication of m by r on the left. We define right R-modules similarly. If R is a commutative ring and M is a left R-module, then M can be made into a right R-module by defining $mr = rm$. This definition of mr makes M a right R-module (verify!). In this case, abstractly speaking, a left R-module is the same as a right R-module. Thus, when R is commutative, we do not distinguish between left and right R-modules and simply call them R-modules.

We list some elementary properties of an R-module M:

(i) $0m = 0$, $m \in M$,
(ii) $a0 = 0$, $a \in R$,
(iii) $(-a)m = -(am) = a(-m)$, $a \in R$, $m \in M$,

where 0 on the right sides of (i) and (ii) is the zero of M, and 0 on the left side of (i) is the zero of R.

To prove (i), consider $am = (a + 0)m = am + 0m$. To prove (ii), consider $am = a(m + 0) = am + a0$. To prove (iii), consider $0 = 0m = (a + (-a))m = am + (-a)m$, and also consider $0 = a0 = a(m + (-m)) = am + a(-m)$.

Throughout, all modules are left modules unless otherwise stated.

1.1 Examples of modules

(a) Let A be any additive abelian group. Then A is a left (also right) $\mathbf{Z}$-module, because

$$\begin{aligned} k(a_1 + a_2) &= ka_1 + ka_2, \\ (k_1 + k_2)a &= k_1a + k_2a, \\ (k_1k_2)a &= k_1(k_2a), \\ 1a &= a \end{aligned}$$

for all integers $k, k_1, k_2 \in \mathbf{Z}$ and for all $a, a_1, a_2 \in A$.

(b) Let R be a ring. Then R itself can be regarded as a left R-module by defining am, $m \in R$, $a \in R$, to be the product of a and m as elements of the ring R.

Then the distributive laws and the associative law for multiplication in the ring R show that R is a left R-module. Similarly, R is also a right R-module.

(c) Let M be the set of $m \times n$ matrices over a ring R. Then M is a module over R, because M is an additive abelian group under the usual addition of matrices, and the usual scalar multiplication (ra_{ij}) of the matrix $(a_{ij}) \in M$ by the element $r \in R$ satisfies axioms (i)–(iv) for a module. In particular, the set of $1 \times n$ (or $n \times 1$) matrices – the set of n-tuples, denoted by R^n – is a module over R.

Further, by choosing $R = \mathbf{R}$ and $n = 2$ (or 3), we obtain that the set of vectors in a plane (or in space) forms a vector space over the field $\mathbf{R}$.

(d) *(Direct product of modules).* Let M and N be R-modules. In the cartesian product $M \times N$, define

$$(x,y) + (x',y') = (x + x', y + y'),$$
$$r(x,y) = (rx,ry)$$

for all $x,x' \in M$, $y,y' \in N$, and $r \in R$. Then $M \times N$ becomes an R-module, called the direct product of the R-modules M and N.

Problems

1. Show that the polynomial ring $R[x]$ over a ring R is an R-module.
2. Let R be a ring and let S denote the set of all sequences (a_i), $i \in \mathbf{N}$, $a_i \in R$. Define

   $(a_i) + (b_i) = (a_i + b_i), \qquad \alpha(a_i) = (\alpha a_i),$

   where $\alpha, a_i, b_i \in R$. Then S is a left R-module.
3. Let M be an additive abelian group. Show that there is only one way of making it a $\mathbf{Z}$-module.

2 Submodules and direct sums

Definition. *A nonempty subset N of an R-module M is called an* R-submodule *(or simply submodule) of M if*

(i) $a - b \in N$ for all $a,b \in N$.
(ii) $ra \in N$ for all $a \in N$, $r \in R$.

Clearly, (0) *or simply* 0 *and M are R-submodules, called* trivial submodules. *In case R is a field, N is called a* subspace *of M.*

We remark that if N is an R-submodule of M, then N is also an R-module in its own right.

2.1 Examples of submodules

(a) Each left ideal of a ring R is an R-submodule of the left R-module R, and conversely. This follows from the definition of a left ideal.

(b) The subset $W = \{(\alpha,0,0) | \alpha \in F\}$ of F^3 is a subspace of the vector space F^3.

(c) If M is an R-module and $x \in M$, then the set

$$Rx = \{rx | r \in R\}$$

is an R-submodule of M, for

$$r_1x - r_2x = (r_1 - r_2)x \in Rx,$$
$$r_1(r_2x) = (r_1r_2)x \in Rx, \qquad \text{for all } r_1, r_2 \in R.$$

(d) If M is an R-module and $x \in M$, then the set

$$K = \{rx + nx | r \in R, n \in \mathbf{Z}\}$$

is an R-submodule of M containing x. Further, if R has unity, then $K = Rx$. First, $(K,+)$ is clearly an abelian subgroup of $(M,+)$. Next, let $a \in R$, $rx + nx \in K$. Then

$$a(rx + nx) = a(rx) + a(nx)$$
$$= (ar)x + a(x + \cdots + x) \quad \text{or} \quad arx + a((-x) + \cdots + (-x)),$$

according to whether n is a positive or a negative integer. But then

$$a(rx + nx) = ((ar) + a + \cdots + a)x \quad \text{or} \quad ((ar) + (-a) + \cdots + (-a))x,$$

since $a(-x) = (-a)x$. Therefore,

$$a(rx + nx) = ux \qquad \text{for some } u \in R.$$

Hence, $a(rx + nx) \in K$, for all a,r in R and for all n (including 0) in $\mathbf{Z}$. Choosing $r = 0 \in R$ and $n = 1 \in \mathbf{Z}$ in $rx + nx$ gives $x \in K$. It is worth noting that if L is any other R-submodule of M containing x, then L contains all elements of the form $rx + nx$, $r \in R$, $n \in \mathbf{Z}$; hence, $K \subset L$. Thus, K is the smallest R-submodule of M containing x, usually denoted (x).

Suppose R has unity e. Then we show that $K = Rx$. Let $rx + nx \in K$. For $n > 0$,

$$rx + nx = rx + n(ex) = rx + (ex + \cdots + ex)$$
$$= (r + e + \cdots + e)x = ux$$

for some $u \in R$. Thus, $rx + nx \in Rx$. Similarly, if $n \leq 0$, then $rx + nx \in Rx$. So $K \subset Rx$. Trivially, $Rx \subset K$.

2.2 Theorem. *Let $(N_i)_{i\in\Lambda}$ be a family of R-submodules of an R-module M. Then $\bigcap_{i\in\Lambda} N_i$ is also an R-submodule.*

Proof. Let $x, y \in \bigcap_{i\in\Lambda} N_i$, $a \in R$. Then for all $i \in \Lambda$, $x - y \in N_i$ and $ax \in N_i$, because N_i are R-submodules. Thus, $x - y$, $ax \in \bigcap_{i\in\Lambda} N_i$, which proves that $\bigcap_{i\in\Lambda} N_i$ is an R-submodule. □

Let S be a subset of an R-module M. Let $\mathscr{A} = \{N | N$ is an R-submodule of M containing $S\}$. Then $\mathscr{A} \neq \emptyset$ because $M \in \mathscr{A}$. Let $K = \bigcap_{N \in \mathscr{A}} N$. Then K is the smallest R-submodule of M containing S and is denoted by (S). The smallest R-submodule of M containing a subset S is called the R-submodule *generated by S*. If $S = \{x_1,...,x_m\}$ is a finite set, then (S) is also written as $(x_1,...,x_m)$.

Definition. *An R-module M is called a* finitely generated *module if* $M = (x_1,...,x_k)$ *for some* $x_i \in M$, $1 \leq i \leq k$. *The elements* $x_1,...,x_k$ *are said to generate M.*

Definition. *An R-module M is called a* cyclic *module if* $M = (x)$ *for some* $x \in M$.

Example 2.1(d) shows that a cyclic module generated by x is precisely $\{rx + nx | r \in R, n \in \mathbf{Z}\}$, and if R has unity then it simplifies to $\{rx | r \in R\}$, that is, to Rx.

2.3 Theorem. *If an R-module M is generated by a set* $\{x_1,x_2,...,x_n\}$ *and* $1 \in R$, *then* $M = \{r_1x_1 + r_2x_2 + \cdots + r_nx_n | r_i \in R\}$. *The right side is symbolically written* $\Sigma_{i=1}^n Rx_i$.

Proof. Clearly, if $m,m_1,m_2 \in \Sigma_{i=1}^n Rx_i$ and $r \in R$, then $m_1 - m_2 \in \Sigma_{i=1}^n Rx_i$ and $rm \in \Sigma_{i=1}^n Rx_i$. Thus, $\Sigma_{i=1}^n Rx_i$ is an R-submodule of M. Also, $1x_i = x_i \in Rx_i \subset \Sigma_{i=1}^n Rx_i$. Thus, all $x_i \in \Sigma_{i=1}^n Rx_i$. But since M is the smallest submodule of M containing all x_i, the submodule $\Sigma_{i=1}^n Rx_i$ must be equal to M. □

If an element $m \in M$ can be expressed as $m = a_1x_1 + \cdots + a_nx_n, a_i \in R$ and $x_i \in M$, $i = 1,...,n$, then we say that m is a *linear combination* of the elements $x_1,...,x_n$ over R.

We remark that the set of generators of a module need not be unique. For example, let S be the set of all polynomials in x over a field F of degree $\leq n$. Then S is a vector space over F with $\{1,x,x^2,x^3,...,x^n\}$ and $\{1,1 + x,x^2,x^3,...,x^n\}$ as two distinct sets of generators.

Definition. *Let* (N_i), $1 \leq i \leq k$, *be a family of R-submodules of a module M. Then the submodule generated by* $\bigcup_{i=1}^k N_i$, *that is, the smallest submodule containing the submodules* N_i, $1 \leq i \leq k$, *is called the* sum *of* submodules N_i, $1 \leq i \leq k$, *and is denoted by* $\Sigma_{i=1}^k N_i$.

2.4 Theorem. *If (N_i), $1 \le i \le k$, is a family of R-submodules of a module M, then*

$$\sum_{i=1}^{k} N_i = \{x_1 + \cdots + x_k | x_i \in N_i\}.$$

Proof. Let $S = \{x_1 + \cdots + x_k | x_i \in N_i\}$. If $x_1 + \cdots + x_k$ and $y_1 + \cdots + y_k$ belong to S, then

$$(x_1 + \cdots + x_k) - (y_1 + \cdots + y_k) = (x_1 - y_1) + \cdots + (x_k - y_k)$$

also belong to S, because $x_i - y_i \in N_i$, $1 \le i \le k$. Also, if $r \in R$, then

$$r(x_1 + \cdots + x_k) = rx_1 + \cdots + rx_k \in S,$$

because each rx_i, $1 \le i \le k$, is in N_i. Thus, S is a left R-submodule.

Further, if K is any left R-submodule that contains each submodule N_i, then K contains all elements of the form $x_1 + \cdots + x_k$, $x_i \in N_i$. Thus, K contains S. Hence, S is the smallest submodule containing each N_i, $1 \le i \le k$. Therefore, by definition of $\Sigma_{i=1}^{k} N_i$,

$$S = \sum_{i=1}^{k} N_i. \quad \square$$

Remark. The sum $\Sigma_{i \in \Lambda} N_i$ of a family $(N_i)_{i \in \Lambda}$ of R-submodules of a module M is defined similarly as the submodule generated by $\bigcup_{i \in \Lambda} N_i$.

Following exactly the proof of Theorem 2.4, one obtains

$$\sum_{i \in \Lambda} N_i = \left\{ \sum_{\text{finite}} x_i | x_i \in N_i \right\},$$

where $\Sigma_{\text{finite}}\, x_i$ stands for any finite sum of elements of R-submodules N_i, $i \in \Lambda$.

Definition. *The sum $\Sigma_{i \in \Lambda} N_i$ of a family $(N_i)_{i \in \Lambda}$ of R-submodules of an R-module M is called a* direct sum *if each element x of $\Sigma_{i \in \Lambda} N_i$ can be* uniquely *written as $x = \Sigma_i x_i$, where $x_i \in N_i$* and *$x_i = 0$ for almost all i in Λ.*

When the sum $\Sigma_{i \in \Lambda} N_i$ is direct, we write it as $\oplus\, \Sigma_{i \in \Lambda} N_i$. Should it happen that Λ is a finite set $\{1,\ldots,k\}$, then the direct sum $\oplus\, \Sigma_{i \in \Lambda} N_i$ is also written as $N_1 \oplus \cdots \oplus N_k$.

Each N_i in the direct sum $\oplus\, \Sigma_{i \in \Lambda} N_i$ is called a *direct summand* of the direct sum.

2.5 Theorem. *Let $(N_i)_{i \in \Lambda}$ be a family of R-submodules of an R-module M. Then the following are equivalent:*

(i) $\Sigma_{i \in \Lambda} N_i$ *is a direct sum.*
(ii) $0 = \Sigma_i x_i \in \Sigma_{i \in \Lambda} N_i$ *implies* $x_i = 0$ *for all i.*
(iii) $N_i \cap \Sigma_{j \in \Lambda, j \neq i} N_j = (0)$, $i \in \Lambda$.

Proof. Exactly similar to the proof of Theorem 3.2 in Chapter 10. □

Problems

1. Let $V = \mathbf{R}^3$ be a vector space of 3-tuples over the real field **R**. Determine if W is a subspace of V, where W is the set of all $(\lambda_1, \lambda_2, \lambda_3)$ such that
   (a) $\lambda_1 = 0$. (d) $\lambda_1 + 1 = 2\lambda_2$.
   (b) $\lambda_1 = \lambda_2$. (e) $\lambda_1 + \lambda_2 \geq 0$.
   (c) $\lambda_1 \lambda_2 = 0$.
2. Show that the set of all functions f from the real field **R** to **R** can be made into a vector space by the usual operations of sum and scalar product.
3. Let V be the vector space of Problem 2. In each of the following determine if W is a subspace, where W is the set of all functions $f: \mathbf{R} \to \mathbf{R}$ satisfying
   (a) $f(1) = 0$. (d) $f(3) \geq 0$.
   (b) $f(0) = 1$. (e) $\lim_{t \to 1} f(t)$ exists.
   (c) $f(3) = 2f(2)$.
4. *Modular law.* Let A, B, and C be R-submodules of an R-module M such that $A \supset B$. Show that

   $A \cap (B + C) = B + (A \cap C)$.

   Give an example of three R-submodules A,B,C of an R-module M such that

   $A \cap (B + C) \neq (A \cap B) + (A \cap C)$.

   In other words the set of all R-submodules of an R-module M is a modular lattice which is not necessarily distributive.
5. Let M be an R-module. Show that the set $\{x \in R \mid xM = 0\}$ is an ideal of R. (M is called *faithful* if this ideal is zero.)
6. Let M be an R-module and $RM = \{\Sigma_i r_i m_i \mid r_i \in R, m_i \in M\}$. Show that RM is a submodule of M. (Here Σ denotes finite sum.)

7. Let $V = R^3$ be a vector space over a field R. Let

$$x_1 = (1,0,0), \qquad x_2 = (1,1,0), \qquad x_3 = (1,1,1).$$

Show that

$$V = Rx_1 + Rx_2 + Rx_3.$$

Also show that

$$V = \oplus \sum_{i=1}^{3} Rx_i.$$

8. Let $K \oplus K'$ and $L \oplus L'$ be direct sums of submodules of M such that $K \oplus K' = L \oplus L'$. Show that $K = L$ does not necessarily imply $K' = L'$.
9. Let R be the ring of all 2×2 upper triangular matrices over the field $\mathbf{Z}/(2)$.
   (a) List all direct summands of R as a left R-module; that is, all left ideals A of the ring R such that $A \oplus B = R$ for some left ideal B of R.
   (b) For each direct summand in (a) list all idempotents generating it as a left R-module.
10. Let $e_1,\ldots,e_n$ be pairwise orthogonal idempotents in a ring R with unity. Prove that
    (a) $\oplus \Sigma_{i=1}^{n} Re_i$ is a direct sum of left ideals.
    (b) $e = e_1 + \cdots + e_n$ is an idempotent of R.
    (c) $\{e_1,\ldots,e_n, 1 - e\}$ is a maximal set of pairwise orthogonal idempotents of R.
11. For any two idempotents e, f in R, prove that

$$Re + Rf = Re \oplus R(f - fe).$$

12. Let I be a left ideal in a ring R with unity. Show that $I^2 = I$ if I is a direct summand of R as a left R-module.
13. Let R be a commutative ring, and let I be a finitely generated ideal in R with $I = I^2$. Show that I is a direct summand of R. Give an example to show that if I is not finitely generated, then I need not be a direct summand.

3 *R*-homomorphisms and quotient modules

Definition. *Let f be a mapping of an R-module M to an R-module N such that*

(i) $f(x + y) = f(x) + f(y)$
(ii) $f(rx) = rf(x)$

for all $x,y \in M$ and $r \in R$. Then f is called an R-linear mapping *(or simply a* linear mapping*) or an* R-homomorphism *of M into N. $Hom_R(M,N)$ denotes the set of R-homomorphisms of M into N. If $M = N$, then f is called an* endomorphism *of M, and then the set $Hom_R(M,M)$ is also denoted by $End_R(M)$.*

If R is a field or a division ring, then f is also called a linear transformation *of the vector space M to the vector space N.*

Let $f\colon M \to N$ be an R-homomorphism of an R-module M into an R-module N. Then f, in particular, is a group homomorphism. Thus

(a) $f(0) = 0$.
(b) $f(-x) = -f(x)$, $x \in M$.
(c) $f(x - y) = f(x) - f(y)$, $x,y \in M$.

Definition. *Let $f\colon M \to N$ be an R-homomorphism of an R-module M into an R-module N. Then*

(a) The set $K = \{x \in M \mid f(x) = 0\}$ is called the kernel *of f and written* Ker f.
(b) The set $f(M) = \{f(x) \mid x \in M\}$ is called the homomorphic image *(or simply image) of M under f and is denoted by* Im f.

It can be easily proved that Ker f is an R-submodule of M, and that Im f is an R-submodule of N (Problem 1). Further, Ker $f = (0)$ if and only if f is 1-1. If f is 1-1, we say that M is isomorphic (or R-isomorphic) into N, or M is embeddable in N, or there is a copy of M in N, and we write it as $M \hookrightarrow N$.

If f is both 1-1 and onto, then we say that M is isomorphic (or R-isomorphic) onto N, and we write it as $M \simeq N$. It can be easily shown that $\simeq$ is an equivalence relation in the set of R-modules. Clearly, $M \simeq M$ by the identity mapping. Let $M \simeq N$ under a mapping f. We show that the inverse mapping $f^{-1}\colon N \to M$ is an R-homomorphism. Let $y, y_1, y_2 \in N$ and $a \in R$. Then there exist unique $x, x_1, x_2 \in M$ such that $f(x) = y$, $f(x_i) = y_i$, $i = 1,2$. Now,

$$\begin{aligned} f^{-1}(y_1 + y_2) &= f^{-1}(f(x_1) + f(x_2)) = f^{-1}(f(x_1 + x_2)) \\ &= x_1 + x_2 = f^{-1}(y_1) + f^{-1}(y_2). \end{aligned}$$

Further,

$$f^{-1}(ay) = f^{-1}(af(x)) = f^{-1}(f(ax)) = ax = af^{-1}(y).$$

Therefore, $N \simeq M$ under f^{-1}. Finally, if $M \simeq N$ under a mapping f, and $N \simeq K$ under a mapping g, then $M \simeq K$ under the composite mapping gf.

3.1 Examples of *R*-homomorphisms

(a) Let M be an R-module. Then the mapping $i: x \mapsto x$ of M onto M is clearly an R-homomorphism of M onto M. It is called the *identity endomorphism* of M. Similarly, the mapping $\bar{0}: x \mapsto 0$ of M into M is also an R-homomorphism of M into M; this is called the *zero endomorphism* of M.

(b) Let M be an R-module over a commutative ring R, and let a be some fixed element of R. Then the mapping

$$f: x \mapsto ax, \qquad x \in M,$$

of M into M is an R-homomorphism of M into M.

(c) Let $f: R^n \to R$ be the mapping defined by

$$f(x_1, \dots, x_n) = x_i \qquad (i \text{ fixed}).$$

Then f is an R-homomorphism of the R-module R^n onto the R-module R; this is called the *projection of* R^n *onto the ith component*.

(d) Let V be the vector space of real-valued functions of a real variable that have derivatives of all orders. Let D be the derivative operator. Then the mapping

$$f \mapsto D(f), \qquad f \in V,$$

of V into V is an **R**-homomorphism of V.

(e) Let A be any $m \times n$ matrix over F. Consider the mapping $T: v \mapsto vA$ of the vector space F^m to the vector space F^n, where the elements of F^m and F^n are written as $1 \times m$ and $1 \times n$ matrices, respectively. Then T is an F-homomorphism (i.e., linear transformation) since

$$T(v_1 + v_2) = (v_1 + v_2)A = v_1 A + v_2 A = T(v_1) + T(v_2),$$
$$T(av) = (av)A = a(vA) = aT(v),$$

for all $v, v_1, v_2 \in F^m$, and $a \in F$.

Theorems analogous to the fundamental theorems on homomorphisms and isomorphisms of groups and rings also hold for modules. But before we can obtain those theorems we need to define the concept of quotient modules (or factor modules).

Let N be an R-submodule of an R-module M. Let $a_1, a_2 \in M$. We say $a_1 \equiv a_2 \pmod{N}$ if $a_1 - a_2 \in N$. It follows immediately that $\equiv$ is an equivalence relation. Let $\bar{a}$ denote the equivalence class containing $a \in M$. Clearly, $\bar{a}$ is the set of elements $a + x$ with $x \in N$, and thus $\bar{a}$ is also written as $a + N$. We denote the set of equivalence classes by M/N (some authors denote it by $M - N$). In M/N we define the binary operations as follows:

$$\bar{a} + \bar{b} = \overline{a+b}, \quad \bar{a}, \bar{b} \in M/N,$$
$$r\bar{a} = \overline{ra}, \quad \bar{a} \in M/N, r \in R.$$

Following the construction of quotient groups, one can easily verify that these binary operations are well defined, and they make the set M/N an R-module. This module is called a *quotient module* (or a *factor module*) of M modulo N.

If R is a field, then M/N is called the *quotient space* of M modulo N.

It is quite natural to study the structure of the quotient module M/N: for example, how the submodules of M/N are related to the submodules of M.

3.2 Theorem. *The submodules of the quotient module M/N are of the form U/N, where U is a submodule of M containing N.*

Proof. Let $f: M \to M/N$ be the canonical mapping; that is, $f(x) = x + N$, $x \in M$. Let X be an R-submodule of M/N. Consider $U = \{x \in M \mid f(x) \in X\}$. We claim that U is an R-submodule of M. For if $x, y \in U$ and $r \in R$, then $f(x-y) = f(x) - f(y) \in X$ and $f(rx) = rf(x) \in X$, which shows that U is an R-submodule of M. Also, $N \subset U$, because for all $x \in N$, $f(x) = \bar{0} \in X$. Thus, N is an R-submodule of U. Also, if $x \in X$, then there exists $y \in M$ such that $f(y) = x$, because f is an onto mapping. So by definition of U, $y \in U$. Hence, $X \subset f(U)$. Clearly, $f(U) \subset X$. Thus, $X = f(U)$. But $f(U) = U/N$. Thus, $X = U/N$. □

The proof of the following theorem on R-homomorphisms is exactly similar to that of the corresponding theorem for groups or rings.

3.3 Theorem (fundamental theorem of R-homomorphisms). *Let f be an R-homomorphism of an R-module M into an R-module N. Then $M/\operatorname{Ker} f \simeq f(M)$.*

Proof. Consider the mapping $g: M/\operatorname{Ker} f \to f(M)$ given by $g(m + \operatorname{Ker} f) = f(m)$. Along exactly the same lines as the proof of the fundamental theorems of homomorphisms for groups or rings, we can show that g is an R-isomorphism of $M/\operatorname{Ker} f$ onto $f(M)$. □

3.4 Theorem. *Let A and B be R-submodules of R-modules M and N, respectively. Then*

$$\frac{M \times N}{A \times B} \simeq \frac{M}{A} \times \frac{N}{B}.$$

Proof. Define a mapping

$$f: M \times N \to \frac{M}{A} \times \frac{N}{B}$$

by $f(m,n) = (m + A, n + B)$, $m \in M$, $n \in N$. It is a straightforward verification that f is an onto R-homomorphism. Now

$$\begin{aligned} \operatorname{Ker} f &= \{(m,n) | (m + A, n + B) = (0 + A, 0 + B)\} \\ &= \{(m,n) | m \in A, n \in B\} \\ &= A \times B. \end{aligned}$$

Therefore, by the fundamental theorem of R-homomorphisms

$$\frac{M \times N}{A \times B} \simeq \frac{M}{A} \times \frac{N}{B}. \quad \square$$

3.5 Examples

(a) Let R be a ring with unity. An R-module M is cyclic if and only if $M \simeq R/I$ for some left ideal I of R.

Solution. Let $M = Rx$ be a cyclic module generated by x. Let $I = \{r \in R | rx = 0\}$. Then I is a left ideal of R. Define a mapping $f: R \to Rx$ by $f(r) = rx$, $r \in R$. It is obvious that f is an R-homomorphism that is also onto. Also, $\operatorname{Ker} f = \{r \in R | rx = 0\} = I$. Hence, by the fundamental theorem of R-homomorphisms, $R/I \simeq Rx$. For the converse, we note that the left R-module R/I is generated by $1 + I \in R/I$; that is, $R(1 + I) = R/I$. Hence, R/I is cyclic.

(b) Let M be an R-module. Then $\operatorname{Hom}_R(M,M)$ is a subring of $\operatorname{Hom}(M,M)$.

Solution. Recall that $\operatorname{Hom}(M,M)$, the set of endomorphisms of M, regarded as an abelian group, is a ring (Example 3.1(c), Chapter 9). Clearly, $\operatorname{Hom}_R(M,M) \subset \operatorname{Hom}(M,M)$. Let $f,g \in \operatorname{Hom}_R(M,M)$, $x \in M$, and $r \in$

R. Then

$$(f-g)(rx) = f(rx) - g(rx) = rf(x) - rg(x)$$
$$= r(f(x) - g(x)) = r((f-g)(x)).$$

Further,

$$(fg)(rx) = f(g(rx)) = f(rg(x)) = r(f(g(x))) = r((fg)(x)).$$

Therefore, $f-g, fg \in \mathrm{Hom}_R(M,M)$. Hence, $\mathrm{Hom}_R(M,M)$ is a subring of $\mathrm{Hom}(M,M)$.

(c) Let R be a ring with unity. Let $\mathrm{Hom}_R(R,R)$ denote the ring of endomorphisms of R regarded as a right R-module. Then $R \simeq \mathrm{Hom}_R(R,R)$ as rings.

Solution. Consider the mapping $f: R \to \mathrm{Hom}_R(R,R)$, given by $f(a) = a^*$, where $a^*(x) = ax$, $x \in R$. Let $x,y,r \in R$. Then

$$a^*(x+y) = a(x+y) = ax + ay = a^*(x) + a^*(y),$$

and

$$a^*(xr) = a(xr) = (ax)r = (a^*(x))r.$$

Thus, a^* is an R-homomorphism of the right R-module R into itself; that is, $a^* \in \mathrm{Hom}_R(R,R)$.

We now show that f is a ring homomorphism. Let $a,b \in R$. Then for any $x \in R$,

$$(a+b)^*(x) = (a+b)x = ax + bx$$
$$= a^*(x) + b^*(x) = (a^* + b^*)(x).$$

Thus, $(a+b)^* = a^* + b^*$. Similarly,

$$(ab)^*(x) = (ab)(x) = a(bx) = a(b^*(x))$$
$$= a^*(b^*((x))) = (a^*b^*)(x),$$

so $(ab)^* = a^*b^*$. Hence,

$$f(a+b) = (a+b)^* = a^* + b^* = f(a) + f(b)$$

and

$$f(ab) = (ab)^* = a^*b^* = f(a)f(b).$$

Finally, we show that f is both 1-1 and onto. Let $a,b \in R$ and $a^* = b^*$. Then $a^*(x) = b^*(x)$ for all $x \in R$. This gives $ax = bx$ for all $x \in R$, so $(a-b)R = 0$. Then $a = b$. Thus, f is 1-1. Now suppose $t \in \mathrm{Hom}_R(R,R)$. Let $t(1) = a$. We claim $t = a^*$; that is, a is a preimage of t.

Now for any $x \in R$,

$$t(x) = t(1x) = t(1)x = ax = a^*(x).$$

Hence, $t = a^*$ as claimed. This proves that f is also an onto mapping.

(d) Let R be a ring with unity. Let $\mathrm{Hom}_R(R,R)$ denote the ring of endomorphisms of R regarded as a left R-module. Then $R^{\mathrm{op}} \simeq \mathrm{Hom}_R(R,R)$ as rings.

Solution. As in (c) we consider the mapping

$$f: R^{\mathrm{op}} \to \mathrm{Hom}_R(R,R),$$

given by $f(a) = a^*$, where $a^*(x) = a \circ x = xa$. Then a^* is an R-homomorphism of the left R-module R into itself, and the mapping f is a ring isomorphism. The proof is exactly similar to the one for a right R-module R given in Example (c).

We call a sequence (finite or infinite) of R-modules and R-homomorphisms

$$\cdots \xrightarrow{f_{n-1}} M_{n-1} \xrightarrow{f_n} M_n \xrightarrow{f_{n+1}} M_{n+1} \longrightarrow \cdots$$

exact if $\mathrm{Im}\, f_n = \mathrm{Ker}\, f_{n+1}$ for all n.

(e) Suppose that the following diagram of R-modules and R-homomorphisms

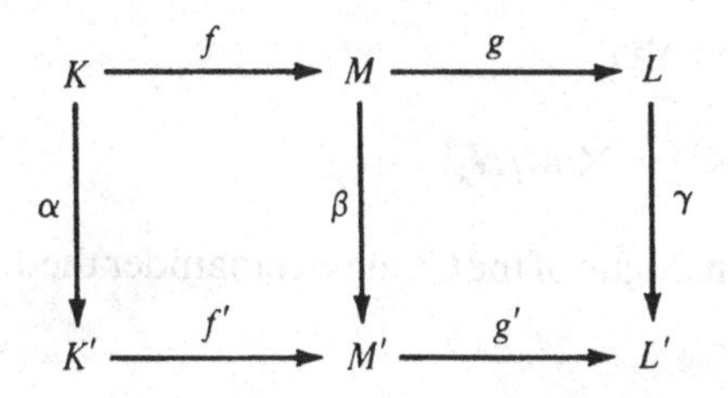

is commutative and has exact rows. Show that

(i) If α,γ, and f' are 1-1, then so is β.
(ii) If α,γ, and g are onto, then so is β.

Solution. Let $m \in \mathrm{Ker}\, \beta$. Because the diagram commutes, $\gamma g(m) = g'\beta(m) = 0$, so $g(m) = 0$ because γ is 1-1. Therefore, $m \in \mathrm{Ker}\, g = \mathrm{Im}\, f$ because the top row is exact. This implies $m = f(k)$ for some $k \in K$. Again, because the diagram commutes, $f'\alpha = \beta f$. Thus, $f'\alpha(k) = \beta f(k) =$

$\beta(m) = 0$. This implies $k = 0$, since both α and f' are 1-1. Then $m = f(k) = f(0) = 0$, proving (i). The solution of (ii) is similar.

Problems

1. If $f: M \to N$ is an R-homomorphism of an R-module M to an R-module N, show that
   (a) Ker f is an R-submodule of M.
   (b) Im f is an R-submodule of N.
2. If A and B are R-submodules of an R-module M, show that $(A + B)/A \simeq B/A \cap B$ as R-modules.
3. Prove that $\text{Hom}_{\mathbf{Z}}(\mathbf{Q},\mathbf{Q}) \simeq \mathbf{Q}$ as rings.
4. Let M be an R-submodule and $x \in M$ be such that $rx = 0, r \in R$, implies $r = 0$. Then show that $Rx \simeq R$ as R-modules.
5. Let V be a vector space over a field F generated by $x_1, x_2, \ldots, x_n$ and suppose that any relation $a_1x_1 + a_2x_2 + \cdots + a_nx_n = 0$, $a_i \in F$, implies $a_1 = 0 = a_2 = \cdots = a_n$. Show that $V \simeq F^n$.
6. Let $M = K \oplus K' = L \oplus L'$ be direct sums of submodules of M such that $K = L$. Show that $K' \simeq L'$.
7. Suppose I is a left ideal of a ring R with unity, and $R/I \simeq R$ regarded as R-modules. Prove that $I = Re$ for some idempotent $e \in R$.
8. Let $N_1, \ldots, N_k$ be a family of R-submodules of an R-module M. Suppose

   $$N_i + (N_1 \cap \cdots \cap N_{i-1} \cap N_{i+1} \cap \cdots \cap N_k) = M$$

   for all $i = 1, \ldots, k$. Show that

   $$M \Big/ \bigcap_{i=1}^{k} N_i \simeq M/N_1 \times \cdots \times M/N_k.$$

   (Note that this is an analogue of the Chinese remainder theorem.)

4 Completely reducible modules

Definition. *An R-module M is called* simple *or* irreducible *if $RM \neq (0)$ and (0), M are the only R-submodules of M. (Here RM is the set of finite sums $\Sigma r_i m_i$, $r_i \in R$, $m_i \in M$. If $1 \in R$, $RM = 0$ only if $M = 0$.)*

A trivial example of a simple module is a field or a division ring R regarded as a module over itself.

Let $R = F_n$ be the matrix ring over a field F. Then $M = Re_{kk}$ is a simple

R-module, where e_{kk} is a matrix unit; for, let $N \neq 0$ be an R-submodule of Re_{kk} and let $0 \neq \sum_{i=1}^{n} a_{ik}e_{ik} \in N$. Then for some l, $1 \leq l \leq n$, $a_{lk} \neq 0$; so a_{lk}^{-1} exists. This gives

$$e_{kk} = (a_{lk}^{-1}e_{kl})\left(\sum_{i=1}^{n} a_{ik}e_{ik}\right) \in N.$$

Hence, $N = Re_{kk}$, proving that M is simple. Note that M is the set of matrices all of whose entries are zero, except possibly in the kth column.

A curious reader might ask if a minimal left ideal in a ring R (defined in Section 3, Chapter 10) is a simple R-module. The answer, in general, is no. To see this, let A be an additive abelian group of order p, p prime. Make A into a ring by trivial multiplication. Then A is a minimal left ideal, but A is not a simple A-module because $A^2 = 0$. However, for rings with unity a minimal left ideal in a ring R is obviously a simple R-module.

We now prove the following characterization of simple modules.

4.1 Theorem. *Let R be a ring with unity, and let M be an R-module. Then the following statements are equivalent:*

(i) M is simple.
(ii) $M \neq (0)$, and M is generated by any $0 \neq x \in M$.
(iii) $M \simeq R/I$, where I is a maximal left ideal of R.

Proof. (i) $\Rightarrow$ (ii) Let $0 \neq x \in M$. Then $(x) = Rx$ is a nonzero R-submodule generated by x. So $M = (x)$.

(ii) $\Rightarrow$ (i) Let $0 \neq N$ be an R-submodule of M. Let $0 \neq x \in N$. Then by (ii) $M = (x) \subset N$. Hence, $N = M$, proving that M is simple.

We may note that, for rings without unity, (ii), in general, does not imply (i) (consider the earlier example of a ring A in which A as an A-module satisfies (ii), but A is not simple).

(i) $\Rightarrow$ (iii) Because $RM = M \neq 0$, there exists $0 \neq x \in M$ such that $Rx \neq 0$. But Rx is clearly an R-submodule of M. Because M is simple, $Rx = M$. Define a mapping $f: R \rightarrow Rx$ by $f(a) = ax$. f is an R-homomorphism of R regarded as an R-module to M. f is clearly surjective. Let $I = \text{Ker } f$. Then by the fundamental theorem of R-homomorphisms $R/I \simeq Rx$. Because R/I is simple, it follows by Theorem 3.2 that the only left ideal of R that contains I properly is R alone. Thus, I is a maximal left ideal.

(iii) $\Rightarrow$ (i) Follows from Theorem 3.2. $\square$

The most important fact about simple modules is the following basic result. called Schur's lemma.

4.2 Theorem (Schur's lemma). *Let M be a simple R-module. Then* $\text{Hom}_R(M,M)$ *is a division ring.*

Proof. Let $0 \neq \phi \in \text{Hom}_R(M,M)$. Consider the R-submodules $\text{Ker}\,\phi$ and $\text{Im}\,\phi$ of M. If $\text{Ker}\,\phi = M$, then $\phi = 0$, a contradiction. Thus, $\text{Ker}\,\phi = (0)$, so ϕ is injective. Further, if $\text{Im}\,\phi = (0)$, then $\phi = 0$, a contradiction. Therefore, $\text{Im}\,\phi = M$, which implies ϕ is surjective. Hence, ϕ is bijective, which proves that ϕ is invertible. □

Definition. *An R-module M is called* completely reducible *if* $M = \Sigma_{\alpha\in\Lambda} M_\alpha$, *where M_α are simple R-submodules.*

4.3 Theorem. *Let $M = \Sigma_{\alpha\in\Lambda} M_\alpha$ be a sum of simple R-submodules M_α. Let K be a submodule of M. Then there exists a subset Λ' of Λ such that $\Sigma_{\alpha\in\Lambda'} M_\alpha$ is a direct sum, and $M = K \oplus (\oplus\Sigma_{\alpha\in\Lambda'} M_\alpha)$.*

Proof. Let $S = \{A \subset \Lambda \mid \Sigma_{\alpha\in A} M_\alpha$ is a direct sum, and $K \cap \Sigma_{\alpha\in A} M_\alpha = (0)\}$. $S \neq \varnothing$ since $\varnothing \in S$. (Note, by $\Sigma_{\alpha\in A} M_\alpha$ we mean (0) when A is an empty set.) S is partially ordered by inclusion, and every chain (A_i) in S has an upper bound $\cup A_i$ in S. Thus, by Zorn's lemma, S has a maximal member, say A. Let $N = K \oplus (\oplus\Sigma_{\alpha\in A} M_\alpha)$. We claim $N = M$. So let $\beta \in \Lambda$. Because M_β is simple, either $M_\beta \cap N = (0)$ or $M_\beta \cap N = M_\beta$. Now $M_\beta \cap N = (0)$ implies $M_\beta \cap \oplus\Sigma_{\alpha\in A} M_\alpha = (0)$; so $\Sigma_{\alpha\in A\cup\{\beta\}} M_\alpha$ is a direct sum, and it has zero intersection with K. But this contradicts the maximality of A. Therefore, for all $\beta \in \Lambda$, $M_\beta \subset N$, proving that $N = M$. □

By choosing $K = (0)$ in Theorem 4.3, we obtain

4.4 Corollary. *If $M = \Sigma_{\alpha\in\Lambda} M_\alpha$ is the sum of a family of simple R-modules $(M_\alpha)_{\alpha\in\Lambda}$, then there exists a subfamily $(M_\alpha)_{\alpha\in\Lambda'\subset\Lambda}$ such that $M = \oplus\Sigma_{\alpha\in\Lambda'} M_\alpha$.*

As a consequence of the above results, we can show that every nonzero submodule and every nonzero quotient module of a completely reducible module is completely reducible (Problems 1 and 2).

Problems

1. Let M be a completely reducible module, and let K be a nonzero submodule of M. Show that K is completely reducible. Also show

that K is a direct summand of M. [*Hint:* $M = \oplus\Sigma_{\alpha\in\Lambda} M_\alpha$ and $\exists \Lambda' \subset \Lambda$ such that $M = K \oplus (\oplus\Sigma_{\alpha\in\Lambda'} M_\alpha)$. Then

$$K \simeq \frac{M}{\oplus\Sigma_{\alpha\in\Lambda'} M_\alpha} = \frac{\oplus\Sigma_{\alpha\in\Lambda''} M_\alpha \oplus\Sigma_{\alpha\in\Lambda'} M_\alpha}{\oplus\Sigma_{\alpha\in\Lambda'} M_\alpha} \simeq \sum_{\alpha\in\Lambda''} M_\alpha.]$$

2. Let M be a completely reducible module, and let $K \neq M$ be a submodule of M. Show that M/K is completely reducible.
3. Show that $\mathbf{Z}/(p_1 p_2)$ is a completely reducible $\mathbf{Z}$-module, where p_1 and p_2 are distinct primes.
4. Let $R = F_2$ be the 2×2 matrix ring over a field F. Show that R as an R-module is completely reducible. Prove the same if $R = F_n$ for any positive integer n. Show also that if A is any left ideal in R, then $A = Re$ for some idempotent e in R.
5. Let $R = \begin{bmatrix} F & F \\ 0 & F \end{bmatrix}$ and $M = \begin{bmatrix} 0 & F \\ 0 & 0 \end{bmatrix}$. Show that R/M is a completely reducible R- (as well as R/M-) module.
6. Let A and B be rings such that ${}_A A$ and ${}_B B$ are completely reducible modules. Let $R = A \oplus B$ be the ring direct sum of A and B. Show that ${}_R R$ is completely reducible.
7. Let R be a ring with unity. Show that R as an R-module is completely reducible if and only if each R-module M is completely reducible. [*Hint:* $M = \Sigma_{x\in M} Rx$, and Rx is a homomorphic image of R as a left R-module.]

5 Free modules

Throughout this section, unless otherwise stated, R is a nonzero ring with unity.

Definition. *A list – that is, a finite sequence –* $x_1,\dots,x_n$ *of elements of an* R*-module* M *is called* linearly independent *if, for any* $a_1,\dots,a_n \in R$, $\Sigma_{i=1}^n a_i x_i = 0$ *implies* $a_1 = a_2 = \cdots = a_n = 0$.

A finite sequence is called linearly dependent *if it is not linearly independent.*

A subset S of an R-module M is called linearly independent if every finite sequence of distinct elements of S is linearly independent. Otherwise S is called linearly dependent.

An example of a linearly independent set in the F-module $F[x]$ is $\{1,x,x^2,x^3,\dots\}$. The set $\{1,x,1+x,x^2\}$ is clearly a linearly dependent set.

In a module $M = R^n$ over a ring R with unity, consider the set $\{e_1,e_2,\dots,e_n\}$, where e_i is the n-tuple in which all components except the

ith are zero and the ith component is 1. This set $\{e_1,e_2,\dots,e_n\}$ is a linearly independent set in R^n and is indeed a basis according to the following definition.

Definition. *A subset B of an R-module M is called a* basis *if*

(i) M is generated by B.
(ii) B is a linearly independent set.

The basis $\{e_1,e_2,\dots,e_n\}$ of R^n that we have described is called the standard basis *of R^n.*

Of course, not every module has a basis. For example, consider a cyclic group G and regard it as a $\mathbf{Z}$-module. Then G has a basis if and only if G is infinite. For let $G=(a)$ and ma, $m\in\mathbf{Z}$, be some basis element. Then $\lambda(ma)=0$, $\lambda\in\mathbf{Z}$, must imply $\lambda=0$. However, if G is a finite group of order n, then $n(ma)=0$, a contradiction. Hence, G must be infinite if it has a basis. But if $G=(a)$ is an infinite cyclic group, then $\{a\}$ is clearly a basis of G as a $\mathbf{Z}$-module. This proves our assertion.

Definition. *An R-module M is called a* free module *if M admits a basis. In other words, M is free if there exists a subset S of M such that M is generated by S, and S is a linearly independent set.*

We regard (0) as a free module whose basis is the empty set.

5.1 Theorem. *Let M be a free R-module with a basis $\{e_1,\dots,e_n\}$. Then $M\simeq R^n$.*

Proof. Define a mapping $\phi: M\to R^n$ by

$$\phi\left(\sum_{i=1}^{n} r_ie_i\right)=\sum_{i=1}^{n} r_if_i,$$

where

$$f_i=(0,\dots,1,0,\dots,0)\in R^n.$$

Because $\Sigma_{i=1}^n r_ie_i=\Sigma_{i=1}^n r_i'e_i$ implies, by the linear independence of the e_i's, $r_i=r_i'$ for all i, ϕ is well defined. If $m=\Sigma_{i=1}^n r_ie_i$, $m'=\Sigma_{i=1}^n r_i'e_i$, and $r\in R$, then it immediately follows that

$$\phi(m+m')=\phi(m)+\phi(m') \qquad \text{and} \qquad \phi(rm)=r\phi(m).$$

Further, if $\phi(m)=0$, then $\Sigma_{i=1}^n r_if_i=0$. This implies $(r_1,\dots,r_n)=0$, and hence each $r_i = 0$, proving that ϕ is 1-1. It is clear that ϕ is onto, and, hence, ϕ is an isomorphism. □

5.2 Theorem. *Let V be a nonzero finitely generated vector space over a field F. Then V admits a finite basis.*

Proof. Let $\{x_1,x_2,...,x_n\}$ be a subset of V that generates V. If this set is not linearly independent, then there exist $\alpha_1,\alpha_2,...,\alpha_n$ in F, not all zero, such that $\alpha_1 x_1 + \alpha_2 x_2 + \cdots + \alpha_n x_n = 0$. For simplicity we may assume $\alpha_1 \neq 0$. Then $x_1 = \beta_2 x_2 + \cdots + \beta_n x_n$, where $\beta_2 = -\alpha^{-1}\alpha_2,...,\beta_n = -\alpha_1^{-1}\alpha_n$. This shows that the set $\{x_2,...,x_n\}$ also generates V. If $\{x_2,...,x_n\}$ is a linearly independent set, we are done, otherwise we proceed as before and omit one more element.

Continuing like this, we arrive at a linearly independent set that also generates V. Note that this process must end before we exhaust all elements; in the extreme case if we had come to a single element that generates V, then that single element will form a basis because each nonzero element in V forms a linearly independent set. Hence, V does admit a finite basis. □

Remark. Because we regard the empty set as the basis of a zero vector space, Theorem 5.2 shows that each finitely generated vector space has a basis with finite cardinality. Indeed, it can be shown that every vector space has a basis, finite or infinite – a consequence of Zorn's lemma.

5.3 Theorem. *Let M be a finitely generated free module over a commutative ring R. Then all bases of M are finite.*

Proof. Let (e_i), $i \in \Lambda$, be a basis of M, and let $\{x_1,x_2,...,x_n\}$ be a set of generators of M. Then each x_j can be written as

$$x_j = \sum_i \alpha_{ij} e_i, \qquad \alpha_{ij} \in R,$$

and all but finite number of α_{ij} are zero. Thus, the set S of those e_i's that occur in expressions of all the x_j's, $j = 1,2,...,n$, is finite. □

5.4 Theorem. *Let M be a finitely generated free module over a commutative ring R. Then all bases of M have the same number of elements.*

Proof. Equivalently, the theorem states that if $R^m \simeq R^n$, then $m = n$. Let $m < n$, let ϕ: $R^m \to R^n$ be an R-isomorphism, and let $\psi = \phi^{-1}$. Let $(e_1,...,e_m)$ and $(f_1,...,f_n)$ be ordered bases of R^m and R^n, respectively. Let us write

$$\phi(e_i) = a_{1i}f_1 + \cdots + a_{ni}f_n, \qquad 1 \le i \le m,$$
$$\psi(f_j) = b_{1j}e_1 + \cdots + b_{mj}e_m, \qquad 1 \le j \le n.$$

Let $A = (a_{ji})$ and $B = (b_{kj})$ be $n \times m$ and $m \times n$ matrices. Then

$$\psi\phi(e_i) = \sum_{k=1}^{m}\sum_{j=1}^{n} b_{kj}a_{ji}e_k, \qquad 1 \le i \le m.$$

Thus, by the linear independence of the e_i's and by the fact that $\psi = \phi^{-1}$, we have

$$\sum_{j=1}^{n} b_{kj}a_{ji} = \delta_{ki}.$$

This yields

$$BA = \begin{bmatrix} b_{11} & \cdots & b_{1n} \\ \vdots & & \vdots \\ b_{m1} & \cdots & b_{mn} \end{bmatrix} \begin{bmatrix} a_{11} & \cdots & a_{m1} \\ \vdots & & \vdots \\ a_{n1} & \cdots & a_{nm} \end{bmatrix} = I_m,$$

where I_m is the $m \times m$ identity matrix. Similarly, $AB = I_n$. Let

$$A' = [A\ 0] \qquad \text{and} \qquad B' = \begin{bmatrix} B \\ 0 \end{bmatrix}$$

be $n \times n$ augmented matrices, where each of the 0 blocks is a matrix of appropriate size. Then

$$A'B' = I_n, \qquad B'A' = \begin{bmatrix} I_m & 0 \\ 0 & 0 \end{bmatrix}.$$

This implies $\det(A'B') = 1$ and $\det(B'A') = 0$. But A' and B' are $n \times n$ matrices over a commutative ring. So $\det(A'B') = \det(B'A')$, which yields a contradiction. Hence, $m \ge n$. By symmetry $n \ge m$. This proves $m = n$. □

Definition. *The number of elements in any basis of a finitely generated free module M over a commutative ring R with unity is called the* rank *of M, written* rank M.

In particular, if R is a field then the rank of M is known as the dimension *of the vector space M and is denoted by* dim M.

We call a vector space V over a field F finite dimensional if it is finitely generated; otherwise we call it an infinite-dimensional vector space.

5.5 Examples

(a) Every finitely generated module is a homomorphic image of a finitely generated free module.

Solution. Let M be an R-module with generators $x_1,...,x_n$. Let e_i be the n-tuple with all entries 0 except at the ith place, where the entry is 1. Then $\{e_1,...,e_n\}$ are linearly independent over R and generate a free module R^n. Define a mapping $\phi: R^n \to M$ by

$$\phi\left(\sum_{i=1}^{n} r_i e_i\right) = \sum_{i=1}^{n} r_i x_i.$$

Because each element $x \in R^n$ has a unique representation as $\Sigma_{i=1}^{n} r_i e_i$, ϕ is well defined.

Further, if $x = \Sigma_{i=1}^{n} r_i e_i$, $y = \Sigma_{i=1}^{n} r_i' e_i$, and $r \in R$, then it is clear that

$$\phi(x+y) = \phi(x) + \phi(y), \qquad \phi(rx) = r\phi(x).$$

Hence, ϕ is an R-homomorphism of R^n onto M. If K is the kernel of ϕ, it follows by the fundamental theorem of homomorphisms that $R^n/K \simeq M$.

(b) Let V be a vector space over a field F with a basis $(e_i)_{i\in\Lambda}$. Prove the following:

(i) $V = \oplus\Sigma_{i\in\Lambda} Fe_i \simeq \oplus\Sigma_{i\in\Lambda} F_i$, $F_i = F$.
(ii) V is completely reducible.
(iii) If W is a subspace of V, then there exists a subspace W' such that $V = W \oplus W'$.
(iv) Suppose $|\Lambda| < \infty$, and let $\{e_1,...,e_k\}$ be a linearly independent subset of V. Show that there exists a basis of V containing $\{e_1,...,e_k\}$. (This is also true without the assumption that $|\Lambda| < \infty$.)

Solution. (i) $V = \oplus\Sigma_{i\in\Lambda} Fe_i$ follows from the fact that the set $(e_i)_{i\in\Lambda}$ generates V and is a linearly independent set. Next, it is clear that the mapping $\alpha \mapsto \alpha e_i$, $\alpha \in F$, is a bijective linear mapping from F to Fe_i regarded as vector spaces over F. Thus, $F \simeq Fe_i$, proving (i).

(ii) Because $V \simeq \oplus\Sigma_{i\in\Lambda} F_i$, $F_i = F$, and F is a simple F-module, it implies that V is completely reducible.

(iii) Because V is completely reducible, the result follows by Theorem 4.3.

(iv) Let $W = \oplus\Sigma_{i=1}^{k} Fe_i$. By (iii) $V = W \oplus W'$ for some subspace W' of V. Let $\{e_{k+1},...,e_n\}$ be a basis of W'. Then $\{e_1,...,e_k,e_{k+1},...,e_n\}$ is a basis of V.

Problems

1. Let R be a commutative ring with unity, and let $e \neq 0, 1$ be an idempotent. Prove that Re cannot be a free R-module.
2. Prove that the direct product $M_1 \times \cdots \times M_k$ of free R-modules M_i is again free.
3. Let $(x_i)_{i \in \Lambda}$ be a basis of a free R-module M. Prove that $M = \oplus\sum_{i \in \Lambda} Rx_i$.
4. Consider the diagram of R-modules and R-homomorphisms

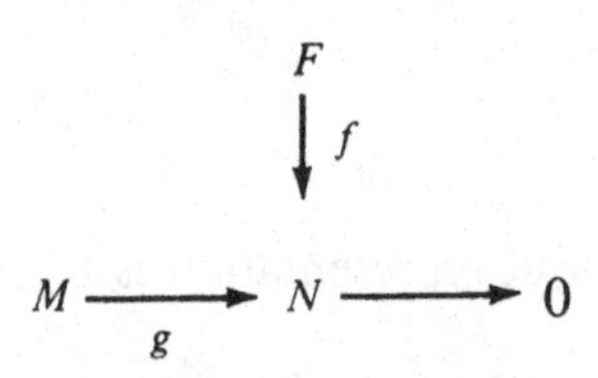

with exact row; that is, g is surjective. Suppose F is free. Show that there exists an R-homomorphism $h: F \to M$ such that $gh = f$.
5. Prove that ${}_{\mathbf{Z}}\mathbf{Q}$ is not a free module. Can you generalize this to ${}_D K$, where D is a commutative integral domain with unity and K is its field of fractions?
6. Show that every ideal of $\mathbf{Z}$ is free as $\mathbf{Z}$-module.
7. Show that every principal left ideal in an integral domain R with unity is free as a left R-module.
8. Show that every module is a homomorphic image of a free module.
9. An onto homomorphism $g: A \to B$ from an R-module A to an R-module B is said to split if Ker$g$ is a direct summand of A. Show that if B is free and $g: A \to B$ is an onto homomorphism, then g splits.

6 Representation of linear mappings

In this section F denotes a field. We show that to each linear mapping ϕ from a vector space V over F to a vector space U over F there corresponds a matrix A with respect to bases $\mathscr{B}$ and $\mathscr{C}$ of U and V, respectively. Further, if A' is the matrix of ϕ with respect to another pair of bases $\mathscr{B}'$ and $\mathscr{C}'$ of U and V, respectively, then $A' = P^{-1}AQ$ for some invertible matrices P and Q. Matrices A and $A' = P^{-1}AQ$ are called *equivalent*.

Let U and V be vector spaces over a field F with ordered bases $\mathscr{B} = (e_1, e_2, ..., e_m)$ and $\mathscr{C} = (f_1, f_2, ..., f_n)$, respectively. Let $\phi: V \to U$ be a linear mapping given by

$$\phi(f_j) = \sum_{i=1}^{m} a_{ij}e_i, \qquad j = 1,\ldots,n,$$

where $a_{ij} \in F$. Then the $m \times n$ matrix $A = (a_{ij})$ is called the matrix of the linear mapping ϕ with respect to the given bases $\mathscr{B}$ and $\mathscr{C}$. Conversely, any $m \times n$ matrix $A = (a_{ij})$ over F determines a unique linear mapping

$$\phi(f_j) = \sum_{i=1}^{m} a_{ij}e_i, \qquad j = 1,\ldots,n,$$

of V to U whose matrix is A.

Thus, $\phi \leftrightarrow A$ is a 1-1 correspondence between $\mathrm{Hom}_F(V,U)$ and $F^{m \times n}$. Indeed this correspondence $\phi \leftrightarrow A$ preserves addition and scalar multiplication; that is, if $\phi \leftrightarrow A$ and $\psi \leftrightarrow B$, then $\phi + \psi \leftrightarrow A + B$ and $\alpha\phi \leftrightarrow \alpha A$, where $\alpha \in F$. Therefore we have

6.1 Theorem. *$Hom_F(V,U) \simeq F^{m \times n}$ as vector spaces over F.*

If $V = U$ in Theorem 6.1, the correspondence $\phi \leftrightarrow A$ also preserves multiplication. This yields that the algebra of $n \times n$ matrices over a field F is, abstractly speaking, the same as the algebra of linear mappings of an n-dimensional vector space V. We record this in

6.2 Theorem. *$Hom_F(V,V) \simeq F_n$ as algebras over F, where $\dim V = n$.*

Our purpose now is to consider how a change of basis of a vector space affects the matrix of a linear mapping. Let $\mathscr{B} = (e_1, e_2,\ldots,e_m)$ and $\mathscr{B}' = (e'_1, e'_2,\ldots,e'_m)$ be two ordered bases of U. Then each e'_j can be written as a unique linear combination of e_i's, say

$$e'_j = \sum_{i=1}^{m} p_{ij}e_i, \qquad j = 1,\ldots,m, \tag{1}$$

where $p_{ij} \in F$. The $m \times m$ matrix $P = (p_{ij})$ is called the *matrix of transformation* from $\mathscr{B}'$ to $\mathscr{B}$. Similarly, each e_j is a unique linear combination of e'_i's, say

$$e_j = \sum_{i=1}^{m} p'_{ij}e'_i, \qquad j = 1,\ldots,m, \tag{2}$$

where $p'_{ij} \in F$. Then the $m \times m$ matrix $Q = (p'_{ij})$ is called the *matrix of* transformation from $\mathscr{B}$ to $\mathscr{B}'$.

We proceed to show that $PQ = I = QP$. Substituting (1) for e'_i in (2), we obtain

$$e_j = \sum_{i=1}^{m} p'_{ij} \sum_{k=1}^{m} p_{ki} e_k, \qquad j = 1,\dots,m. \tag{3}$$

Because the e_i are linearly independent, we equate coefficients in (3) and get

$$\sum_{i=1}^{m} p'_{ij} p_{ki} = \delta_{jk}.$$

Hence, $QP = I$. Similarly, $PQ = I$. On the other hand, let $\mathscr{B} = (e_1, e_2, \dots, e_m)$ be a basis of U, and let $P = (p_{ij})$ be an $m \times m$ invertible matrix. We claim that $\mathscr{B}' = (e'_1, e'_2, \dots, e'_m)$, where

$$e'_j = \sum_{i=1}^{m} p_{ij} e_i, \qquad j = 1,\dots,m, \tag{4}$$

is also a basis of U. For, let

$$\alpha_1 e'_1 + \alpha_2 e'_2 + \cdots + \alpha_m e'_m = 0, \qquad \alpha_i \in F.$$

Substituting for e'_j from (4), we obtain

$$\alpha_1 \sum_{i=1}^{m} p_{i1} e_i + \alpha_2 \sum_{i=1}^{m} p_{i2} e_i + \cdots + \alpha_m \sum_{i=1}^{m} p_{im} e_i = 0.$$

Because $e_1, e_2, \dots, e_m$ are linearly independent, we get

$$\alpha_1 p_{i1} + \alpha_2 p_{i2} + \cdots + \alpha_m p_{im} = 0, \qquad i = 1,\dots,m.$$

Therefore,

$$P \begin{bmatrix} \alpha_1 \\ \alpha_2 \\ \cdot \\ \cdot \\ \cdot \\ \alpha_n \end{bmatrix} = 0, \qquad \text{so} \qquad \begin{bmatrix} \alpha_1 \\ \alpha_2 \\ \cdot \\ \cdot \\ \cdot \\ \alpha_n \end{bmatrix} = 0,$$

because P is invertible. This proves our claim that $(e'_1, e'_2, \dots, e'_m)$ is also a basis of U.

We are now ready to prove the following theorem, which describes the effect of a change of a basis on the matrix of a linear mapping.

6.3 Theorem. *Let U and V be vector spaces over F of dimensions m and n, respectively. Let $A = (a_{ij})$ be the matrix of a linear mapping ϕ:*

$V \to U$ with respect to a given pair of ordered bases $\mathscr{B} = (e_1, e_2, \ldots, e_m)$ and $\mathscr{C} = (f_1, f_2, \ldots, f_n)$ of U and V, respectively. Then

(i) The matrix of ϕ with respect to a new pair of bases $\mathscr{B}' = (e'_1, e'_2, \ldots e'_m)$ and $\mathscr{C}' = (f'_1, f'_2, \ldots, f'_n)$ of U and V, respectively, is $P^{-1}AQ$, where P and Q are the matrices of transformations from $\mathscr{B}'$ to $\mathscr{B}$ and $\mathscr{C}'$ to $\mathscr{C}$, respectively.

(ii) For any given pair of invertible matrices P and Q of sizes $m \times m$ and $n \times n$, respectively, there is a basis $\mathscr{B}'$ of U and $\mathscr{C}'$ of V such that the matrix of ϕ with respect to $\mathscr{B}'$ and $\mathscr{C}'$ is $P^{-1}AQ$.

Proof. (i) We have

$$\phi(f_j) = \sum_{i=1}^{m} a_{ij} e_i, \qquad j = 1, \ldots, n.$$

Let $A' = (a'_{ij})$ be the matrix of ϕ with respect to bases $\mathscr{B}'$ and $\mathscr{C}'$. Then

$$\phi(f'_j) = \sum_{i=1}^{m} a'_{ij} e'_i, \qquad j = 1, \ldots, n. \tag{1}$$

Since $Q = (q_{ij})$ and $P^{-1} = (p'_{ij})$ are the matrices of transformations from $\mathscr{C}'$ to $\mathscr{C}$ and $\mathscr{B}$ to $\mathscr{B}'$, we have

$$f'_j = \sum_{k=1}^{n} q_{kj} f_k, \qquad j = 1, \ldots, n,$$

and

$$e_i = \sum_{l=1}^{m} p'_{li} e'_l, \qquad i = 1, \ldots, m.$$

Hence,

$$\begin{aligned}\phi(f'_j) = \phi\left(\sum_{k=1}^{n} q_{kj} f_k\right) &= \sum_{k=1}^{n} q_{kj} \phi(f_k) \\ &= \sum_{k=1}^{n} q_{kj} \sum_{i=1}^{m} a_{ik} e_i \\ &= \sum_{k=1}^{n} q_{kj} \sum_{i=1}^{m} a_{ik} \sum_{l=1}^{m} p'_{li} e'_l.\end{aligned}$$

That is,

$$\phi(f'_j) = \sum_{l=1}^{m} \left(\sum_{k=1}^{n} \sum_{i=1}^{m} p'_{li} a_{ik} q_{kj} \right) e'_l. \tag{2}$$

Because $e'_1, e'_2, \ldots, e'_m$ are linearly independent, (1) and (2) yield

$$a'_{lj} = \sum_{k=1}^{n} \sum_{i=1}^{m} p'_{li} a_{ik} q_{kj}.$$

Therefore, $A' = P^{-1}AQ$.

(ii) The proof follows from the discussion preceding the theorem and (i). □

6.4 Example

Let $\phi: F^3 \to F^2$ be the linear mapping given by $\phi(a,b,c) = (a+b+c, b+c)$. Find the matrix A of ϕ with respect to the standard bases of F^3 and F^2. Also, find the matrix A' of ϕ with respect to the bases $\mathscr{B}' = ((-1,0,2),(0,1,1),(3,-1,0))$ and $\mathscr{C}' = ((-1,1),(1,0))$ of F^3 and F^2, respectively. Verify that $A' = P^{-1}AQ$, where P and Q are as in Theorem 6.3.

Solution

$$\begin{aligned} \phi(1,0,0) &= (1,0) = 1(1,0) + 0(0,1), \\ \phi(0,1,0) &= (1,1) = 1(1,0) + 1(0,1), \\ \phi(0,0,1) &= (1,1) = 1(1,0) + 1(0,1). \end{aligned}$$

So $A = \left[\begin{smallmatrix} 1 & 1 & 1 \\ 0 & 1 & 1 \end{smallmatrix}\right]$. Further,

$$\begin{aligned} \phi(-1,0,2) &= (1,2) = 2(-1,1) + 3(1,0), \\ \phi(0,1,1) &= (2,2) = 2(-1,1) + 4(1,0), \\ \phi(3,-1,0) &= (2,-1) = -1(-1,1) + 1(1,0). \end{aligned}$$

Thus, $A' = \left[\begin{smallmatrix} 2 & 2 & -1 \\ 3 & 4 & 1 \end{smallmatrix}\right]$. We now find the matrix P^{-1}:

$$\begin{aligned} (1,0) &= 0(-1,1) + 1(1,0), \\ (0,1) &= 1(-1,1) + 1(1,0). \end{aligned}$$

So $P^{-1} = \left[\begin{smallmatrix} 0 & 1 \\ 1 & 1 \end{smallmatrix}\right]$. Similarly,

$$Q = \begin{bmatrix} -1 & 0 & 3 \\ 0 & 1 & -1 \\ 2 & 1 & 0 \end{bmatrix}.$$

By actual computation,

$$P^{-1}AQ = \begin{bmatrix} 2 & 2 & -1 \\ 3 & 4 & 1 \end{bmatrix} = A'.$$

Problems

1. Let V be the vector space of polynomials of degree less than or equal to 3 over a field F. Let D be the differentiation operator on

V. Choose bases of V

$$\mathscr{B}_1 = \{1,x,x^2,x^3\} \qquad \text{and} \qquad \mathscr{B}_2 = \{1,1+x,(1+x)^2,(1+x)^3\}.$$

(a) Find the matrix of D with respect to $\mathscr{B}_1$.
(b) Find the matrix of D with respect to $\mathscr{B}_2$.
(c) What is the matrix of transformation from $\mathscr{B}_1$ to $\mathscr{B}_2$?

2. Let V be the vector space generated by a set B of linearly independent differentiable functions from $\mathbf{R}$ to $\mathbf{R}$ and let D denote the differential operator $\dfrac{d}{dx}$. Find the matrix of the linear mapping D: $V \to V$ with respect to a basis B, where
   (a) $B = \{e^{-x}, e^{3x}\}$,
   (b) $B = \{e^x, xe^x\}$,
   (c) $B = \{1, x, x^2\}$,
   (d) $B = \{1, x, e^x, e^{2x}, e^{3x}\}$,
   (e) $B = \{\sin x, \cos x\}$.

3. Let $V = \mathbf{R}^2$ be the vector space over the reals. Let T be the linear mapping defined by $T(x,y) = (2x,2x+3y)$. Choose $\mathscr{B}_1$ and $\mathscr{B}_2$ to be the standard basis and $\{(\cos\alpha, \sin\alpha),(-\sin\alpha, \cos\alpha)\}$, respectively, of V.
   (a) Find the matrix of T with respect to $\mathscr{B}_1$.
   (b) Find the matrix of transformation from $\mathscr{B}_1$ to $\mathscr{B}_2$ with respect to the standard basis.
   (c) Find the matrix of T with respect to $\mathscr{B}_2$.

7 Rank of a linear mapping

In this section we introduce the concept of rank, which is of fundamental importance in linear algebra. It has wide applications: in particular, it gives the condition for consistency of a linear system of equations over a field, and it is a tool for reducing a matrix to canonical form.

Definition. *Let U and V be finite-dimensional vector spaces over a field F, and let ϕ: $V \to U$ be a linear mapping. The dimension of the subspace $Im\,\phi$ is called the* rank of the linear mapping ϕ.

Let F^n denote the set of n-tuples over F. The elements of the vector space F^n will be written as $1 \times n$ or sometimes as $n \times 1$ matrices interchangeably. The ambiguity of the meaning of the notation F^n should not cause any confusion because the context always makes it clear.

Let A be an $m \times n$ matrix over a field F. The rows of A can be looked upon as members of the vector space F^n. The subspace of F^n generated by the rows of A is called the *row space* of A, denoted by $R(A)$.

Definition. *The* row rank *of a matrix A is the dimension of the row space $R(A)$.*

We know that to each linear mapping there corresponds a class of equivalent matrices (Theorem 6.3). The next theorem gives a connection between the rank of a linear mapping and the row rank of the corresponding class of matrices.

7.1 Theorem. *Let $\phi: V \to U$ be a linear mapping. Then*

$$\text{rank } \phi = \text{row rank } A,$$

where A is any matrix of ϕ.

Before we prove the theorem, we prove a series of lemmas.

7.2 Lemma. *Let A be an $m \times n$ matrix. Then*

$$R(A) = \{xA \mid x \in F^m\}.$$

Proof. Let $A_1, \ldots, A_m$ denote the rows of A, and $x = (x_1, \ldots, x_m) \in F^m$. Then

$$xA = [x_1 \ldots x_m] \begin{bmatrix} A_1 \\ \cdot \\ \cdot \\ \cdot \\ A_m \end{bmatrix} = x_1 A_1 + \cdots + x_m A_m.$$

Thus, every element of $R(A)$ is a linear combination of the rows of A, and, conversely, any linear combination of the rows of A is of the form xA. □

7.3 Lemma. *Let A be an $m \times n$ matrix over F, and let P be an invertible $m \times m$ matrix over F. Then $R(A) = R(PA)$.*

Proof. Let $xA \in R(A)$. Then $xA = (xP^{-1})(PA) \in R(PA)$. Conversely, let $xPA \in R(PA)$. Then $xPA = (xP)A \in R(A)$. □

7.4 Lemma. *Let A be an $m \times n$ matrix over F, and let Q be an $n \times n$ invertible matrix over F. Then $\dim R(A) = \dim R(AQ)$.*

Proof. Let $\{u_1A,...,u_rA\}$ be a basis of $R(A)$. Because Q is invertible, the set $\{u_1AQ,...,u_rAQ\}$ is trivially a linearly independent set of vectors in the vector space $R(AQ)$. Further, if $uAQ \in R(AQ)$, then $uA \in R(A)$ and, hence, $uA = \alpha_1 u_1 A + \cdots + \alpha_r u_r A$, $\alpha_i \in F$. Therefore, $uAQ = \alpha_1 u_1 AQ + \cdots + \alpha_r u_r AQ$. This proves that $\{u_1AQ,...,u_rAQ\}$ is a basis of $R(AQ)$. □

7.5 Lemma. *Let A be an $m \times n$ matrix over F. Let P and Q, respectively, be $m \times m$ and $n \times n$ invertible matrices. Then*

$$\dim R(A) = \dim R(PAQ).$$

Proof. Follows by Lemmas 7.3 and 7.4. □

7.6 Lemma. *Let ϕ: $V \to U$ be a linear mapping with rank $\phi = r$. Then there exist bases $\mathscr{B}$ and $\mathscr{B}'$ of V and U, respectively, such that the matrix of ϕ with respect to $\mathscr{B}$ and $\mathscr{B}'$ is*

$$\begin{bmatrix} I_r & 0 \\ 0 & 0 \end{bmatrix}.$$

Proof. Choose a basis of Ker ϕ and complete it to a basis $\mathscr{B} = \{v_1,...,v_r, v_{r+1},...,v_m\}$ of V, where $\{v_{r+1},...,v_m\}$ is a basis of Ker ϕ [Example 5.5(b)]. We claim that the list $\phi(v_1),...,\phi(v_r)$ is linearly independent. For if $\alpha_1\phi(v_1) + \cdots + \alpha_r\phi(v_r) = 0$, then $\alpha_1 v_1 + \cdots + \alpha_r v_r \in \text{Ker}\ \phi$, so

$$\alpha_1 v_1 + \cdots + \alpha_r v_r = \beta_{r+1} v_{r+1} + \cdots + \beta_m v_m$$

for some $\beta_i \in F$. Then by the linear independence of $\{v_1,...,v_m\}$, we obtain that each α_i as well as each β_i is zero.

Further, since Im ϕ is generated by $\phi(v_1),...,\phi(v_m)$, out of which $\phi(v_{r+1}),...,\phi(v_m)$ are all zero, and the remaining r are linearly independent, it follows that $\{\phi(v_1),...,\phi(v_r)\}$ is a basis of Im $\phi \subset U$. Again complete the linearly independent set $\{\phi(v_1),...,\phi(v_r)\}$ to a basis $\mathscr{B}' = \{\phi(v_1),...,\phi(v_r),w_1,...,w_s\}$ of V.

It is clear that the matrix of ϕ with respect to bases $\mathscr{B}$ and $\mathscr{B}'$ is

$$\begin{bmatrix} I_r & 0 \\ 0 & 0 \end{bmatrix}. \quad \square$$

We remark that the proof of Lemma 7.6 contains the following useful result.

7.7 Rank nullity theorem. *If ϕ: $V \to U$ is a linear mapping from a vector space V to a vector space U, then*

$$\dim V = \dim \operatorname{Ker} \phi + \dim \operatorname{Im} \phi.$$

(dim Ker ϕ is called the nullity of ϕ.)

Proof of Theorem 7.1. Let

$$A = \begin{bmatrix} I_r & 0 \\ 0 & 0 \end{bmatrix}$$

be a matrix of ϕ. The assertion of the theorem is true for A. If B is another matrix of ϕ with respect to a different pair of bases of V and U, then by Theorem 6.3, $B = PAQ$ for some invertible matrices P,Q. This implies $r = \dim R(A) = \dim R(PAQ) = \dim R(B)$ (Lemma 7.5). □

Similarly, we can prove that the subspace of F^m generated by the columns of the matrix A is also of dimension r. This subspace is called the *column space* of A.

Definition. *The* rank *of a matrix is defined to be the common value of the dimension of the row space and that of the column space of the matrix.*

Thus, the rank of a linear mapping ϕ is the same as the rank of a matrix A of ϕ.

We conclude with the remark that, in practice, the computation of the rank of a matrix is done by performing elementary row and column operations on the matrix (see Chapter 20 for the computation of the rank of a matrix over a PID).

7.8 Example

Find the rank of the linear mapping ϕ: $\mathbf{R}^4 \to \mathbf{R}^3$, where

$\phi(a,b,c,d) = (a + 2b - c + d, -3a + b + 2c - d, -3a + 8b + c + d)$.

Solution. First we find the matrix of ϕ relative to standard bases of $\mathbf{R}^4$ and $\mathbf{R}^3$. Now

$$\begin{aligned}
\phi(1,0,0,0) &= (1,-3,-3) = 1(1,0,0) - 3(0,1,0) - 3(0,0,1),\\
\phi(0,1,0,0) &= (2,1,8) = 2(1,0,0) + 1(0,1,0) + 8(0,0,1),\\
\phi(0,0,1,0) &= (-1,2,1) = -1(1,0,0) + 2(0,1,0) + 1(0,0,1),\\
\phi(0,0,0,1) &= (1,-1,1) = 1(1,0,0) - 1(0,1,0) + 1(0,0,1).
\end{aligned}$$

Thus, the matrix of ϕ is

$$A = \begin{bmatrix} 1 & 2 & -1 & 1 \\ -3 & 1 & 2 & -1 \\ -3 & 8 & 1 & 1 \end{bmatrix}.$$

Let R_1, R_2, and R_3 denote the first, second, and third rows, respectively, of A. Because $3R_1 + 2R_2 - R_3 = 0$, it follows that R_1, R_2, and R_3 are linearly dependent. However, straightforward computations yield that if $\alpha R_1 + \beta R_2 = 0$, $\alpha, \beta \in \mathbf{R}$, then $\alpha = 0 = \beta$. Thus, R_1 and R_2 are linearly independent. Hence, the maximal number of linearly independent rows of the matrix A is 2, proving that rank $\phi = 2$.

Problems

1. Find the rank of linear mappings $\phi: \mathbf{R}^n \to \mathbf{R}^m$ in each of the following cases:
   (a) $n = 4$, $m = 3$;
   $$\phi(a,b,c,d) = (2a - b + 3c + d, a - 8b + 6c + 8d, a + 2b - 2d).$$
   (b) $n = 5$, $m = 4$;
   $$\phi(a,b,c,d,e) = (2a + 3b + c + 4e, 3a + b + 2c - d + e, 4a - b + 3c - 2d - 2e, 5a + 4b + 3c - d + 6e).$$
2. (a) Let A and B be two $m \times n$ matrices over a field F. Show that
   $$\text{rank}(A + B) \le \text{rank } A + \text{rank } B.$$
   (b) Let A and B be $m \times n$ and $n \times k$ matrices, respectively. Show that
   $$\text{rank}(AB) \le \min(\text{rank } A, \text{rank } B).$$
   [*Hint:* $\text{rank}(AB) = \text{row rank}(AB) = \dim(F^m AB) \le \dim(F^m B) = \text{rank } B.$
   Similarly, column $\text{rank}(AB) = \dim(ABF^k)$, etc.]
3. Let A and B be two $n \times n$ matrices over a field F. Show
   (a) $\text{rank}(AB) \ge \text{rank } A + \text{rank } B - n$.
   (b) If $A = A^2$, $\text{rank } A + \text{rank}(I - A) = n$.

4. Let $S_1, S_2, \ldots, S_k$ be subspaces of F^n. Show

$$\dim(S_1 \cap S_2 \cap \cdots \cap S_k) \geqslant \sum_{i=1}^{k} \dim S_i - (k-1)n.$$

5. Show rank $A = \operatorname{rank}(A^t A)$, where A is a matrix over reals. Generalize this result over the field of complex numbers.
6. Let A be an $n \times n$ nilpotent matrix over a field F. Show

$$\operatorname{rank} A + \operatorname{rank} A^m \leq n, \text{ where } A^{m+1} = 0.$$

PART IV

Field theory

CHAPTER 15

Algebraic extensions of fields

1 Irreducible polynomials and Eisenstein criterion

Let F be a field, and let $F[x]$ be the ring of polynomials in x over F. We know that $F[x]$ is an integral domain with unity and contains F as a proper subring. A polynomial $f(x)$ in $F[x]$ is called *irreducible* if the degree of $f(x) \geq 1$ and, whenever $f(x) = g(x)h(x)$, where $g(x), h(x) \in F[x]$, then $g(x) \in F$ or $h(x) \in F$. If a polynomial is not irreducible, it is called *reducible*.

We remark that irreducibility of a polynomial depends on the nature of the field. For example, $x^2 + 1$ is irreducible over **R** but reducible over **C**.

1.1 Properties of $F[x]$

We recall some of the basic properties of $F[x]$.

(i) The division algorithm holds in $F[x]$. This means that if $f(x) \in F[x]$ and $0 \neq g(x) \in F[x]$, then there exist unique $q(x)$, $r(x) \in F[x]$ such that $f(x) = g(x)q(x) + r(x)$, where $r(x) = 0$ or degree $r(x) <$ degree $g(x)$.
(ii) $F[x]$ is a PID (Theorem 3.2, Chapter 11).
(iii) $F[x]$ is a UFD (Example 1.2(b), Chapter 11).
(iv) The units of $F[x]$ are the nonzero elements of F.
(v) If $p(x)$ is irreducible in $F[x]$, then $F[x]/(p(x))$ is a field, and conversely.

We now proceed to prove some results for testing whether a polynomial is reducible or irreducible.

1.2 Proposition. *Let $f(x) \in F[x]$ be a polynomial of degree > 1. If $f(\alpha) = 0$ for some $\alpha \in F$, then $f(x)$ is reducible over F.*

Proof. By 1.1(i) we can write $f(x) = (x - \alpha)q(x) + r$, where $r \in F$. Then $0 = f(\alpha) = r$, so $f(x) = (x - \alpha)q(x)$. Because $f(x)$ has degree > 1, $q(x) \notin F$; so $f(x)$ is reducible over F. □

Definition. *Let E be a field containing the field F, and let $f(x) \in F[x]$. An element $\alpha \in E$ will be called a* root *or a* zero *of $f(x)$ if $f(\alpha) = 0$. (If $f(x) = a_0 + a_1x + \cdots + a_kx^k$, then $f(\alpha)$ stands for the element $a_0 + a_1\alpha + \cdots + a_k\alpha^k$ in E.)*

The converse of Proposition 1.2 holds for a certain class of polynomials, as given in the following proposition.

1.3 Proposition. *Let $f(x) \in F[x]$ be a polynomial of degree 2 or 3. Then $f(x)$ is reducible if and only if $f(x)$ has a root in F.*

Proof. If $f(x)$ is reducible, then $f(x) = f_1(x)f_2(x)$, where $f_1(x)$ and $f_2(x)$ are nonconstant polynomials, each of which has degree less than 3. But this implies that either $f_1(x)$ or $f_2(x)$ must be of degree 1. Let $f_1(x) = ax + b$, with $a \neq 0$. Then $f_1(-ba^{-1}) = 0$ and, hence, $f(-ba^{-1}) = 0$, which proves that $-ba^{-1}$ is a root of $f(x)$. □

Recall that a polynomial $f(x) \in \mathbf{Z}[x]$ is called *primitive* if the greatest common divisor of the coefficients of $f(x)$ is 1.

Definition. *A polynomial $a_0 + a_1x + \cdots + a_nx^n$ over a ring R is called* monic *if $a_n = 1$.*

Clearly, every monic polynomial $f(x) \in \mathbf{Z}[x]$ is primitive. The following lemma was proved earlier (Lemma 4.2, Chapter 11).

1.4 Lemma. *If $f(x), g(x) \in \mathbf{Z}[x]$ are primitive polynomials, then their product $f(x)g(x)$ is also primitive.*

A polynomial $f(x) \in \mathbf{Z}[x]$ is called irreducible over $\mathbf{Z}$ if $f(x)$ is an irreducible element in $\mathbf{Z}[x]$. An irreducible polynomial over $\mathbf{Z}$ must be primitive.

1.5 Lemma (Gauss). *Let $f(x) \in \mathbf{Z}[x]$ be primitive. Then $f(x)$ is reducible over $\mathbf{Q}$ if and only if $f(x)$ is reducible over $\mathbf{Z}$.*

Proof. If $f(x)$ is reducible over $\mathbf{Z}$, then, trivially, $f(x)$ is reducible over $\mathbf{Q}$. Conversely, suppose $f(x)$ is reducible over $\mathbf{Q}$. Let $f(x) = u(x)v(x)$ with $u(x), v(x) \in \mathbf{Q}[x]$ and $u(x) \notin \mathbf{Q}$, $v(x) \notin \mathbf{Q}$. Then $f(x) = (a/b)u'(x)v'(x)$, where $u'(x)$ and $v'(x)$ are primitive polynomials in $\mathbf{Z}[x]$. Then $bf(x) = a(u'(x)v'(x))$. The g.c.d of coefficients of $bf(x)$ is b, and the g.c.d. of the coefficients of $au'(x)v'(x)$ is a, by Lemma 1.4. Hence, $b = \pm a$, so $f(x) = \pm u'(x)v'(x)$. Therefore, $f(x)$ is reducible over $\mathbf{Z}$. □

Note. $\deg u'(x) = \deg u(x)$ and $\deg v'(x) = \deg v(x)$.

For polynomials $f(x) \in \mathbf{Z}[x]$ that are not necessarily primitive, we have

1.6 Lemma. *If $f(x) \in \mathbf{Z}[x]$ is reducible over $\mathbf{Q}$, then it is also reducible over $\mathbf{Z}$.*

Proof. The proof exactly parallels the proof of the corresponding result in Lemma 1.5. □

If $f(x) \in \mathbf{Z}[x]$ has a root in $\mathbf{Q}$, we can prove the stronger result

1.7 Theorem. *Let $f(x) = a_0 + a_1x + \cdots + a_{n-1}x^{n-1} + x^n \in \mathbf{Z}[x]$ be a monic polynomial. If $f(x)$ has a root $a \in \mathbf{Q}$, then $a \in \mathbf{Z}$ and $a | a_0$.*

Proof. Write $a = \alpha/\beta$, where $\alpha, \beta \in \mathbf{Z}$ and $(\alpha, \beta) = 1$. Then

$$a_0 + a_1\left(\frac{\alpha}{\beta}\right) + \cdots + a_{n-1}\left(\frac{\alpha^{n-1}}{\beta^{n-1}}\right) + \frac{\alpha^n}{\beta^n} = 0.$$

Multiply the above equation by β^{n-1} to obtain

$$a_0\beta^{n-1} + a_1\alpha\beta^{n-2} + \cdots + a_{n-1}\alpha^{n-1} = -\frac{\alpha^n}{\beta}.$$

Because $\alpha, \beta \in \mathbf{Z}$, it follows that $\alpha^n/\beta \in \mathbf{Z}$, so β must be ± 1. The last equation also shows $\alpha | a_0$. Hence, $a = \pm\alpha \in \mathbf{Z}$ and $a | a_0$. □

1.8 Theorem (Eisenstein criterion). *Let $f(x) = a_0 + a_1x + \cdots + a_nx^n \in \mathbf{Z}[x]$ $n \geq 1$. If there is a prime p such that $p^2 \nmid a_0$, $p|a_0$, $p|a_1,\ldots,p|a_{n-1}$, $p \nmid a_n$, then $f(x)$ is irreducible over $\mathbf{Q}$.*

Proof. Suppose

$$f(x) = (b_0 + b_1x + \cdots + b_rx^r)(c_0 + c_1x + \cdots + c_sx^s),$$

with $b_i, c_i \in \mathbf{Z}$, $b_r \neq 0 \neq c_s$, $r < n$, and $s < n$. Then $a_0 = b_0c_0$ and $a_n = b_rc_s$. Then since $p|a_0$ and $p^2 \nmid a_0$, either $p|b_0$ and $p \nmid c_0$, or $p|c_0$ and $p \nmid b_0$. Consider the case $p|c_0$ but $p \nmid b_0$. Because $p \nmid a_n$, it follows that $p \nmid b_r$ and $p \nmid c_s$. Let c_m be the first coefficient in $c_0 + \cdots + c_sx^s$ such that $p \nmid c_m$. Then note that $a_m = b_0c_m + b_1c_{m-1} + \cdots + b_mc_0$. From this we see that $p \nmid a_m$ (otherwise, $p|c_m$), so $m = n$. Then $n = m \leq s < n$, which is impossible. Similarly, if $p|b_0$ and $p \nmid c_0$, we arrive at an absurdity. Hence, by Lemma 1.6, $f(x)$ is irreducible over $\mathbf{Q}$. □

1.9 Examples

(a) $x^2 - 2$ is irreducible over $\mathbf{Q}$.

Solution. Follows from the Eisenstein criterion by setting $p = 2$.

(b) $\Phi_p(x) = 1 + x + \cdots + x^{p-1}$ is irreducible over $\mathbf{Q}$, where p is prime.

Solution. Write $\Phi_p(x) = (x^p - 1)/(x - 1)$. Let

$$\begin{aligned} g(x) &= \Phi_p(x+1) = ((x+1)^p - 1)/((x+1) - 1) \\ &= \frac{1}{x}\left(x^p + \binom{p}{1}x^{p-1} + \cdots + \binom{p}{p-1}x\right) \\ &= x^{p-1} + \binom{p}{1}x^{p-2} + \cdots + \binom{p}{p-1}. \end{aligned}$$

Note that p divides all the coefficients except that of x^{p-1}, and p^2 does not divide the constant term. Thus, $g(x)$ is irreducible over $\mathbf{Q}$. Hence, $\Phi_p(x)$ is also irreducible over $\mathbf{Q}$ [for reducibility of $\Phi_p(x)$ implies reducibility of $g(x)$].

Problems

1. Show that $x^3 + 3x + 2 \in \mathbf{Z}/(7)[x]$ is irreducible over the field $\mathbf{Z}/(7)$.
2. Show that $x^4 + 8 \in \mathbf{Q}[x]$ is irreducible over $\mathbf{Q}$.

3. Show that $x^3 - x - 1 \in \mathbf{Q}[x]$ is irreducible over $\mathbf{Q}$.
4. Show that $x^3 + ax^2 + bx + 1 \in \mathbf{Z}[x]$ is reducible over $\mathbf{Z}$ if and only if either $a = b$ or $a + b = -2$.
5. Determine all (a) quadratic, (b) cubic, and (c) biquadratic irreducible polynomials over $\mathbf{Z}/(2)$.
6. Determine which of the following polynomials are irreducible over $\mathbf{Q}$.
   (a) $x^3 - 5x + 10$.
   (b) $x^4 - 3x^2 + 9$.
   (c) $2x^5 - 5x^4 + 5$.

2 Adjunction of roots

Definition. *If F is a subfield of a field E, then E is called an* extension field *of F or simply an* extension *of F.*

If E is an extension of F, we denote this fact by the diagram

$$\begin{array}{c} E \\ | \\ F \end{array}$$

If E is an extension of F, then trivially E is a vector space over F. The dimension of E over F is usually written as $[E:F]$.

Definition. *Let E be an extension of F. Then the dimension of E considered as a vector space over F is called the* degree of E over F.

Therefore, the degree of E over F is written as $[E:F]$. If $[E:F] < \infty$, then E is called a finite extension of F (or simply finite over F); otherwise E is called an infinite extension of F.

2.1 Theorem. *Let $F \subseteq E \subseteq K$ be fields. If $[K:E] < \infty$ and $[E:F] < \infty$, then*

(i) $[K:F] < \infty$.
(ii) $[K:F] = [K:E][E:F]$.

Proof: Let $(v_1, \ldots, v_m)$ be a basis of K over E, and let $(w_1, \ldots, w_n)$ be a basis of E over F. If $u \in K$, then

$$u = \sum_{i=1}^{m} a_i v_i, \qquad a_i \in E. \tag{1}$$

Because $a_i \in E$,

$$a_i = \sum_{j=1}^{n} b_{ij} w_j, \qquad b_{ij} \in F. \tag{2}$$

Substituting the a_i's from (2) into (1), we get

$$u = \sum_{i=1}^{m} \sum_{j=1}^{n} b_{ij} v_i w_j, \qquad b_{ij} \in F.$$

This shows that the mn elements $v_i w_j$, $i = 1,\ldots,m, j = 1,\ldots,n$, form a set of generators of the vector space K over F.

This proves (i). To prove (ii), we show that the mn elements $v_i w_j$ are linearly independent over F. Suppose we have an equation

$$\sum_{i=1}^{m} \sum_{j=1}^{n} c_{ij} v_i w_j = 0, \qquad c_{ij} \in F.$$

We can rewrite this as

$$\sum_{i=1}^{m} \lambda_i v_i = 0, \quad \text{where } \lambda_i = \sum_{j=1}^{n} c_{ij} w_j \in E, \qquad i = 1,\ldots,m.$$

Because the v_i's are linearly independent over E, the last equation gives

$$\lambda_i = 0, \qquad i = 1,\ldots,m;$$

that is,

$$\sum_{j=1}^{n} c_{ij} w_j = 0, \qquad i = 1,2,\ldots,m.$$

Because $c_{ij} \in F$ and the w_j's are linearly independent over F, the last equation gives

$$c_{ij} = 0, \qquad \text{for } i = 1,\ldots,m \text{ and } j = 1,\ldots,n,$$

which shows that the mn elements $v_i w_j$, $i = 1,\ldots,m, j = 1,\ldots,n$, of K form a basis of K over F. This proves (ii). □

We recall that a 1-1 homomorphism of a field F into a field E is called an *embedding* of F into E.

2.2 Lemma. *Let E and F be fields, and let $\sigma: F \to E$ be an embedding of F into E. Then there exists a field K such that F is a subfield of K and σ can be extended to an isomorphism of K onto E.*

Proof. Let S be a set whose cardinality is the same as that of $E - \sigma(F)$ ($=$ the complement of $\sigma(F)$ in E) and that is disjoint from F. Let f be a 1-1 correspondence from S to $E - \sigma(F)$. Set $K = F \cup S$. Then we can extend the embedding $\sigma\colon F \to E$ to a mapping $\sigma^*\colon K \to E$ as follows: $\sigma^*(a) = \sigma(a)$ if $a \in F$, $\sigma^*(a) = f(a)$ if $a \in S$. Clearly, σ^* is a well-defined, 1-1, onto mapping. We now define a field structure on K. If $x, y \in K$, we define

$$x + y = (\sigma^*)^{-1}(\sigma^*(x) + \sigma^*(y)),$$
$$xy = (\sigma^*)^{-1}(\sigma^*(x)\sigma^*(y)).$$

Our definitions of addition and multiplication coincide with the given addition and multiplication of elements of the original field F, and it is clear that F is a subfield of K. Hence, K is the desired field. □

In view of the above lemma, if σ is an embedding of a field F in a field E, we can identify F with the corresponding image $\sigma(F)$ in E and can consider E as an extension of F. To express this identification of F with $\sigma(F)$, we write x in place of $\sigma(x)$ for each $x \in F$. *Henceforth, whenever there is an embedding of a field F into a field E, we say that E is an extension of F.*

2.3 Theorem. *Let $p(x)$ be an irreducible polynomial in $F[x]$. Then there exists an extension E of F in which $p(x)$ has a root.*

Proof. Because $p(x)$ is irreducible in $F[x]$, $(p(x))$ is a maximal ideal in $F[x]$; so $E = F[x]/(p(x))$ is a field. As explained in the previous paragraph, we identify each $a \in F$ with its coset $\bar{a} = a + (p(x))$ and regard E as an extension field of F.

Let

$$p(x) = a_0 + a_1 x + \cdots + a_n x^n.$$

Then $\overline{p(x)} = 0$ in E; that is, $\overline{a_0 + a_1 n + \cdots + a_n x^n} = 0$; that is, $\bar{a}_0 + \bar{a}_1 \bar{x} + \cdots + \bar{a}_n \bar{x}^n = 0$, where $\bar{x} = x + (p(x)) \in E$; that is, $a_0 + a_1 \bar{x} + \cdots + a_n \bar{x}^n = 0$, which shows that the element $\bar{x}$ in E is a root of $p(x) \in F[x]$. □

2.4 Corollary (Kronecker theorem). *Let $f(x) \in F[x]$ be a nonconstant polynomial. Then there exists an extension E of F in which $f(x)$ has a root.*

Proof. If $f(x)$ has a zero in F, then we take $E = F$. Suppose $f(x)$ has no zero in F. Let $p(x)$ be an irreducible factor of $f(x)$ in $F[x]$ and set $E = F[x]/(p(x))$. Then by Theorem 2.3, E is an extension field of F in which $p(x)$, and hence, $f(x)$ has a root. □

Let $p(x)$ be an irreducible polynomial in $F[x]$ having a root, say u, in an extension field E of F. We denote by $F(u)$ the subfield of E generated by F and u that is, the smallest subfield of E containing F and u. Consider the mapping $\phi: F[x] \to E$ defined by

$$\phi(b_0 + b_1 x + \cdots + b_m x^m) = b_0 + b_1 u + \cdots + b_m u^m,$$

where $b_0 + b_1 x + \cdots + b_m x^m \in F[x]$. Obviously, ϕ is a homomorphism whose kernel contains $p(x)$, because $p(u) = 0$. We show that Ker $\phi = (p(x))$.

Because $F[x]$ is a PID, $\text{Ker}(\phi) = (g(x))$ for some $g(x) \in F[x]$. Then $p(x) \in \text{Ker}\,\phi$ implies $p(x) = g(x)h(x)$, for some $h(x) \in F[x]$. Because $p(x)$ is irreducible over F, $h(x) \in F$. Thus, Ker $\phi = (g(x)) = (p(x))$.

By the fundamental theorem of homomorphisms,

$$\begin{aligned} F[x]/(p(x)) &\simeq \text{image of } \phi \\ &= \{b_0 + b_1 u + \cdots + b_m u^m \in E \mid b_0 \\ &\quad + b_1 x + \cdots + b_m x^m \in F[x]\} \\ &= F[u], \quad \text{say.} \end{aligned}$$

Because $F[x]/(p(x))$ is a field, the set $F[u]$ is a field. Obviously, $F[u]$ is the smallest subfield of E containing F and u, so $F(u) = F[u]$. If the degree of $p(x)$ is n, then u cannot satisfy any polynomial in $F[x]$ of degree less than n. This shows that the set

$$(1, u, u^2, \ldots, u^{n-1})$$

forms a basis of $F(u)$ over F, and $[F(u): F] = n$.

We summarize these results in

2.5 Theorem. *Let $p(x)$ be an irreducible polynomial in $F[x]$ and let u be a root of $p(x)$ in an extension E of F. Then*

(i) $F(u)$, the subfield of E generated by F and u, is the set

$$F[u] = \{b_0 + b_1 u + \cdots + b_m u^m \in E \mid b_0 + b_1 x + \cdots + b_m x^m \in F[x]\};$$

(ii) If the degree of $p(x)$ is n, the set $(1, u, \ldots, u^{n-1})$ forms a basis of $F(u)$ over F; that is, each element of $F(u)$ can be written uniquely as $c_0 + c_1 u + \cdots + c_n u^{n-1}$, where $c_i \in F$ and $[F(u): F] = n$. □

2.6 Examples

(a) Consider the irreducible polynomial $x^2 + 1 \in \mathbf{R}[x]$. If u is a root of $x^2 + 1$ in some extension K of $\mathbf{R}$, then the field $\mathbf{R}(u) = \{a + bu | a, b \in \mathbf{R}\}$ contains all the roots of $x^2 + 1$.

Solution. Note that the existence of K follows from Theorem 2.3. Further, by Theorem 2.5, $\mathbf{R}(u)$ is a vector space over $\mathbf{R}$ and has a basis $\{1, u\}$. Thus, $\mathbf{R}(u) = \{a + bu | a, b \in \mathbf{R}\}$. Clearly, if u is a root of $x^2 + 1$, then $-u$ is also a root. Hence, $\mathbf{R}(u)$ contains both the roots of $x^2 + 1$. (u is generally written as $\sqrt{-1}$, and $\mathbf{R}(u)$ is called the field of complex numbers denoted by $\mathbf{C}$.)

(b) Consider the irreducible polynomial $x^2 + x + 1 \in \mathbf{Z}/(2)[x]$. If u is a root of $p(x)$ in some extension K of $\mathbf{Z}/(2)$, show that the subfield $\mathbf{Z}/(2)[u]$ of K has four elements.

Solution. By Theorem 2.5, elements of $\mathbf{Z}/(2)[u]$ are $a + bu$, where $a, b \in \mathbf{Z}/(2)$. Thus, $\mathbf{Z}/(2)[u] = \{0, 1, u, 1 + u\}$, where $u^2 + u + 1 = 0$, is a field of four elements.

Problems

1. Show that $p(x) = x^2 - x - 1 \in \mathbf{Z}/(3)[x]$ is irreducible over $\mathbf{Z}/(3)$. Show that there exists an extension K of $\mathbf{Z}/(3)$ with nine elements having all roots of $p(x)$.
2. Show that $x^3 - 2 \in \mathbf{Q}[x]$ is irreducible over $\mathbf{Q}$. Find (if it exists) an extension K of $\mathbf{Q}$ having all roots of $x^3 - 2$ such that $[K:\mathbf{Q}] = 6$.
3. Find the smallest extension of $\mathbf{Q}$ having a root of $x^4 - 2 \in \mathbf{Q}[x]$.
4. Find the smallest extension of $\mathbf{Q}$ having a root of $x^2 + 4 \in \mathbf{Q}[x]$.
5. Let $f_1(x), \ldots, f_m(x)$ be a set of polynomials over a field F. Show that there exists an extension E of F in which each polynomial has a root.

3 Algebraic extensions

Definition. *Let E be an extension of F. An element $\alpha \in E$ is said to be* algebraic *over F if there exist elements $a_0, \ldots, a_n$ ($n \geq 1$) of F, not all equal to 0, such that*

$$a_0 + a_1\alpha + \cdots + a_n\alpha^n = 0.$$

In other words, an element $\alpha \in E$ is algebraic over F if there exists a non-constant polynomial $p(x) \in F[x]$ such that $p(\alpha) = 0$.

3.1 Theorem. *Let E be an extension field of F, and let $u \in E$ be algebraic over F. Let $p(x) \in F[x]$ be a polynomial of the least degree such that $p(u) = 0$. Then*

(i) $p(x)$ is irreducible over F.
(ii) If $g(x) \in F[x]$ is such that $g(u) = 0$, then $p(x)|g(x)$.
(iii) There is exactly one monic polynomial $p(x) \in F[x]$ of least degree such that $p(u) = 0$.

Proof. (i) Let $p(x) = p_1(x)p_2(x)$, and deg $p_1(x)$, deg $p_2(x)$ be less than deg $p(x)$. Then $0 = p(u) = p_1(u) \cdot p_2(u)$. This gives $p_1(u) = 0$ or $p_2(u) = 0$; that is, u satisfies a polynomial of degree less than that of $p(x)$, a contradiction. So $p(x)$ is irreducible over F.

(ii) By the division algorithm $g(x) = p(x)q(x) + r(x)$, where $r(x) = 0$ or deg $r(x) <$ deg $p(x)$. Then $g(u) = p(u)q(u) + r(u)$; that is, $r(u) = 0$. Because $p(x)$ is of the least degree among the polynomials satisfied by u, $r(x)$ must be 0. Thus, $p(x)|g(x)$.

(iii) Let $g(x)$ be a monic polynomial of least degree such that $g(u) = 0$. Then by (ii) $p(x)|g(x)$ and $g(x)|p(x)$, which gives $p(x) = g(x)$ since both are monic polynomials. □

Definition. *The monic irreducible polynomial in $F[x]$ of which u is a root will be called the* minimal polynomial *of u over F.*

Definition. *An extension field E of F is called* algebraic *if each element of E is algebraic over F.*

Extensions that are not algebraic are called transcendental *extensions.*

3.2 Theorem. *If E is a finite extension of F, then E is an algebraic extension of F.*

Proof. Let $u \in E$ and $[E:F] = n$. Then $\{1,\dots,u^n\}$ must be a linearly dependent set of elements of E over F. Thus, there exist $a_0, a_1, \dots, a_n$ (not all zero) in F such that $a_0 + a_1 u + \cdots + a_n u^n = 0$. Thus, u is algebraic over F. This proves the theorem. □

As a consequence of Theorems 3.2 and 2.5, we have

3.3 Theorem. *If E is an extension of F and $u \in E$ is algebraic over F, then $F(u)$ is an algebraic extension of F.*

Proof. If $p(x)$ is the minimal polynomial of u over F and degree $p(x)$ is n, then by Theorem 2.5, $[F(u):F] = n < \infty$. Therefore, by Theorem 3.2, $F(u)$ is an algebraic extension of F. □

Not every algebraic extension is finite, as the following example shows.

3.4 Example

Consider the field $E = \mathbf{Q}(\sqrt{2}, \sqrt{3}, \ldots, \sqrt{p}, \ldots)$, the smallest subfield of the field of real numbers containing the rationals $\mathbf{Q}$ and the square roots of all the positive primes. We claim that

$$\mathbf{Q} \subset \mathbf{Q}(\sqrt{2}) \subset \mathbf{Q}(\sqrt{2}, \sqrt{3}) \subset \cdots$$

is an infinite properly ascending chain of subspaces over $\mathbf{Q}$. In other words, each new adjunction of the square root of a prime is a proper extension. Let $p_1, \ldots, p_n$ be n distinct positive primes. Set $F = \mathbf{Q}(\sqrt{p_1}, \ldots, \sqrt{p_n})$. If q is any other prime not equal to any p_i, $i = 1, \ldots, n$, then we show $\sqrt{q} \notin F$. We prove this by induction on n. Let $n = 0$. Then $F = \mathbf{Q}$. Then by a proof analogous to the classical proof that $\sqrt{2} \notin \mathbf{Q}$, we can show $\sqrt{q} \notin \mathbf{Q}$, so the result holds for $n = 0$.

Assume now that the result is true for $n-1$ primes and suppose that $F = \mathbf{Q}(\sqrt{p_1}, \ldots, \sqrt{p_{n-1}}, \sqrt{p_n})$. Set $F_0 = \mathbf{Q}(\sqrt{p_1}, \ldots, \sqrt{p_{n-1}})$. By induction the result holds for F_0; in addition, $F = F_0(\sqrt{p_n})$ is an extension of F_0 of degree 2. If possible let $\sqrt{q} \in F$. Then

$$\sqrt{q} = a + b\sqrt{p_n}, \qquad a, b \in F_0.$$

This gives

$$q = a^2 + b^2 p_n + 2ab\sqrt{p_n},$$

which implies that $\sqrt{p_n} \in F_0$, a contradiction. Hence, $\sqrt{q} \notin F$, as asserted. Therefore,

$$\mathbf{Q} \subset \mathbf{Q}(\sqrt{2}) \subset \mathbf{Q}(\sqrt{2}, \sqrt{3}) \subset \cdots$$

is an infinite properly ascending chain.

Hence, $E = \mathbf{Q}(\sqrt{2}, \sqrt{3}, \ldots, \sqrt{p}, \ldots)$ is not a finite extension of $\mathbf{Q}$. Further, if $\alpha \in E$, then there exists a positive integer r such that $\alpha \in \mathbf{Q}(\sqrt{p_1}, \sqrt{p_2}, \ldots, \sqrt{p_r})$, which is clearly a finite extension (in fact of degree 2^r) of $\mathbf{Q}$. Thus, α is algebraic over $\mathbf{Q}$. Hence, E is an algebraic extension of $\mathbf{Q}$, and $[E:\mathbf{Q}]$ is infinite.

Definition. *An extension E of F is called* finitely generated *if there exists a finite number of elements $u_1, u_2, \ldots, u_r$ in E such that the smallest subfield of E containing F and $\{u_1, u_2, \ldots, u_r\}$ is E itself. We then write $E = F(u_1, \ldots, u_r)$.*

A finitely generated extension need not be algebraic.

3.5 Example

Let $F[x]$ be a polynomial ring over a field F in a variable x. Consider the field of quotients E of $F[x]$. The elements of E are of the form

$$(a_0 + a_1 x + \cdots + a_m x^m)/(b_0 + b_1 x + \cdots + b_n x^n),$$

where $a_i, b_i \in F$ and not all b_i are zero. Thus, E is generated by x over F; that is, $E = F(x)$. Clearly, by the definition of a polynomial ring, x cannot be algebraic over F. Hence, E is not an algebraic extension.

However, we have the following

3.6 Theorem. *Let $E = F(u_1, \ldots, u_r)$ be a finitely generated extension of F such that each u_i, $i = 1, \ldots, r$, is algebraic over F. Then E is finite over F and, hence, an algebraic extension of F. (The case for $r = 1$ was proved in Theorem 3.3.)*

Proof. Set $E_i = F(u_1, \ldots, u_i)$, $1 \leq i \leq r$. Observe that if an element in E is algebraic over a field F, then, trivially, it is algebraic over any field K such that $E \supset K \supset F$. Therefore, each u_i is algebraic over E_{i-1}, $i = 1, \ldots, r$, with $E_0 = F$. Also, $E_i = E_{i-1}(u_i)$. Therefore, by Theorem 2.5, $[E_i : E_{i-1}]$ is finite, say n_i. By Theorem 2.1,

$$[E:F] = [E:E_{r-1}][E_{r-1}:E_{r-2}] \cdots [E_1:F];$$

hence,

$$[E:F] = n_r n_{r-1} \cdots n_1.$$

Thus, E is a finite extension of F and therefore algebraic over F. $\square$

An application of Theorem 3.6 gives

3.7 Theorem. *Let E be an extension of F. If K is the subset of E consisting of all the elements that are algebraic over F, then K is a subfield of E and an algebraic extension of F.*

Proof. We need only show that if $a,b \in E$ and are algebraic over F, then $a \pm b$, ab, and a/b (if $b \neq 0$) are also algebraic over F. This follows from the fact that all these elements lie in $F(a,b)$, which, by Theorem 3.6, is an algebraic extension of F.

Thus, K is an algebraic extension of F in E. □

The subfield K in Theorem 3.7 is called the *algebraic closure of F in E*. In the next section we define an algebraically closed field as one that has no proper algebraic extensions, and we show that for any given field F there exists a unique (up to isomorphism) algebraic extension $\bar{F}$ that is algebraically closed. This $\bar{F}$ is also called the algebraic closure of F, without any reference to its being in another field such as the field K already defined.

We next prove an important result that we often use in the following sections.

Definition. *Let K and L be extension fields of a field F. Then a nonzero homomorphism (hence an embedding) $\sigma: K \to L$, such that $\sigma(a) = a$ for all $a \in F$, is called an* F-homomorphism of K into L *or an* embedding of K in L over F.

The above terminology is consistent with the earlier definition of an F-homomorphism of vector spaces, for we can regard K and L as vector spaces over F, and then if σ is a nonzero homomorphism of K into L (as fields) with $\sigma(a) = a$ for all $a \in F$, then $\sigma(ax) = \sigma(a)\sigma(x) = a\sigma(x)$.

3.8. Theorem. *Let E be an algebraic extension of F, and let $\sigma: E \to E$ be an embedding of E into itself over F. Then σ is onto and, hence, an automorphism of E.*

Proof. Let $a \in E$, and let $p(x)$ be its minimal polynomial over F. Let E' be the subfield of E containing F and generated over F by all roots of $p(x)$ that lie in E. Thus, E' is generated over F by a finite set of elements in E that are algebraic over F. So by Theorem 3.6, E' is a finite algebraic extension of F. Furthermore, σ must map a root of $p(x)$ onto a root of $p(x)$ and, hence, maps E' into itself. Now $\sigma(E') \simeq E'$, because σ is 1-1. Thus, $[\sigma(E'):F] = [E':F]$. But since $\sigma(E')$ is a subspace of E', it follows that $\sigma(E') = E'$. This implies that there exists $b \in E'$ such that $\sigma(b) = a$. Hence, σ is onto, which completes the proof of the theorem. □

Problems

1. Let $F \subset K \subset E$ be three fields such that K is an algebraic extension of F and $\alpha \in E$ is algebraic over K. Show that α is algebraic over F.
2. Prove that $\sqrt{2}$ and $\sqrt{3}$ are algebraic over **Q**. Find the degree of
   (a) $\mathbf{Q}(\sqrt{2})$ over **Q**.
   (b) $\mathbf{Q}(\sqrt{3})$ over **Q**.
   (c) $\mathbf{Q}(\sqrt{2}, \sqrt{3})$ over **Q**.
   (d) $\mathbf{Q}(\sqrt{2}+\sqrt{3})$ over **Q**.
3. Determine the minimal polynomials over **Q** of the following numbers:
   (a) $\sqrt{2}+5$. (b) $3\sqrt{2}+5$. (c) $\sqrt{-1+\sqrt{2}}$.
   (d) $\sqrt{2}-3\sqrt{3}$.
4. Find a suitable number a such that
   (a) $\mathbf{Q}(\sqrt{2},\sqrt{5}) = \mathbf{Q}(a)$. (b) $\mathbf{Q}(\sqrt{3}, i) = \mathbf{Q}(a)$.
5. Let E be an extension of F, and let $a,b \in E$ be algebraic over F. Suppose that the extensions $F(a)$ and $F(b)$ of F are of degrees m and n, respectively, where $(m,n) = 1$. Show that $[F(a,b):F] = mn$.
6. If E is an extension field of F and $[E:F]$ is prime, prove that there are no fields properly between E and F.
7. Let E be an extension field of F. If $a \in E$ has a minimal polynomial of odd degree over F, show that $F(a) = F(a^2)$.
8. Let $x^n - a \in F[x]$ be an irreducible polynomial over F, and let $b \in K$ be its root, where K is an extension field of F. If m is a positive integer such that $m|n$, find the degree of the minimal polynomial of b^m over F.
9. Let F be a field, and let $F(x)$ be a field of rational functions. Let $L \supsetneq F$ be a subfield of $F(x)$. Prove that $F(x)$ is an algebraic extension of L.
10. Give an example of a field E containing a proper subfield K such that E is embeddable in K and $[E:K]$ is finite.
11. Let $\omega = \cos\dfrac{2\pi}{n} + i\sin\dfrac{2\pi}{n}$, and $u = \cos\dfrac{2\pi}{n}$.
    Show that $[\mathbf{Q}(\omega):\mathbf{Q}(u)] = 2$.
12. Let D be an integral domain and F be a field in D such that $[D:F] < \infty$. Prove that D is a field.

4 Algebraically closed fields

Definition. *A field K is called* algebraically closed *if it possesses no proper algebraic extensions – that is, if every algebraic extension of K coincides with K.*

4.1 Theorem. *For any field K the following are equivalent:*

- *(i) K is algebraically closed.*
- *(ii) Every irreducible polynomial in $K[x]$ is of degree 1.*
- *(iii) Every polynomial in $K[x]$ of positive degree factors completely in $K[x]$ into linear factors.*
- *(iv) Every polynomial in $K[x]$ of positive degree has at least one root in K.*

Proof. (i) $\Rightarrow$ (ii) Let $p(x) \in K[x]$ be an irreducible polynomial of degree n. Then by Theorems 2.3 and 2.5, there exists a finite (hence algebraic) extension E of K such that $[E:K] = n$. But since K is algebraically closed, $E = K$, so $n = 1$.

(ii) $\Rightarrow$ (i) Let E be any algebraic extension of K, and let $a \in E$. Then the minimal polynomial of a over K is irreducible and, therefore, has degree 1. Therefore, $a \in K$ and, hence $E = K$. The equivalence of (ii), (iii), and (iv) is obvious. □

An example of an algebraically closed field is the field of complex numbers. Note that the algebraic closure of the field of real numbers **R** in the field of complex numbers **C** is **C** itself.

Definition. *If F is a subfield of a field E, then E is called an* algebraic closure *of F if*

- *(i) E is an algebraic extension of F.*
- *(ii) E is algebraically closed.*

The following is an easy fact, and its proof is left as an exercise.

If F is a subfield of an algebraically closed field K, then the algebraic closure $\overline{F}$ of F in K is also algebraically closed.

Next we show that any two algebraic closures of a field F (if they exist) are isomorphic under a mapping that keeps each element of F fixed. The proof of the existence of an algebraic closure $\overline{F}$ of F is given at the end.

We first prove a key lemma.

4.2 Lemma. *Let F be a field, and let $\sigma: F \to L$ be an embedding of F into an algebraically closed field L. Let $E = F(\alpha)$ be an algebraic extension of F. Then σ can be extended to an embedding $\eta: E \to L$, and the number of such extensions is equal to the number of distinct roots of the minimal polynomial of α.*

Proof. Let $p(x) = a_0 + a_1 x + \cdots + x^n$ be the minimal polynomial of α over F. Let

$$p^\sigma(x) = \sigma(a_0) + \sigma(a_1)x + \cdots + x^n \in L[x].$$

Let β be a root of $p^\sigma(x)$ in L. Recall that if α is algebraic over a field F, then a typical element of the field $F(\alpha)$ can be written uniquely as $b_0 + b_1\alpha + \cdots + b_k\alpha^k$, where $k <$ degree of the minimal polynomial of α over F, and $b_i \in F$, $i = 1,\ldots,k$.

Define $\eta: F(\alpha) \to L$ by the rule

$$\eta(b_0 + b_1\alpha + \cdots + b_k\alpha^k) = \sigma(b_0) + \sigma(b_1)\beta + \cdots + \sigma(b_k)\beta^k.$$

Then η is a well-defined mapping. Routine computation shows that η is a homomorphism. Thus, η is an embedding of $F(\alpha)$ into L, and it extends σ. Clearly, there is a 1-1 correspondence between the set of distinct roots of $p^\sigma(x)$ in L (hence, between the set of distinct roots of $p(x)$ in its splitting field over F) and the set of embeddings η of $F(\alpha)$ into L that extend σ. This proves the last assertion. □

4.3 Theorem. *Let E be an algebraic extension of a field F, and let $\sigma: F \to L$ be an embedding of F into an algebraically closed field L. Then σ can be extended to an embedding $\eta: E \to L$.*

Proof. Let S be the set of all pairs (K,θ), where K is a subfield of E containing F, and θ is an extension of σ to an embedding of K into L. If (K,θ) and (K',θ') are in S, we write $(K,\theta) \leq (K',\theta')$ if $K \subset K'$ and θ' restricted to K is θ. Because $(F,\sigma) \in S$, $S \neq \emptyset$. Also, if $\{(K_i,\theta_i)\}$ is a chain in S, we set $K = \cup K_i$ and define θ on K as follows. Let $a \in K$. Then $a \in K_i$ for some i, and we define $\theta(a) = \theta_i(a)$. θ is well defined. Let $a \in K_i$ and $a \in K_j$. Because either $K_i \subset K_j$ or $K_j \subset K_i$ by definition of a chain in S, we get $\theta_i(a) = \theta_j(a)$. Hence, θ is well defined. Then (K,θ) is an upper bound for the chain $\{(K_i,\theta_i)\}$. Using Zorn's lemma, let (K,η) be a maximal element in S. Then η is an extension of σ, and we contend that $K = E$. Otherwise, there exists $\alpha \in E$, $\alpha \notin K$. Then by Lemma 4.2 the embedding $\eta: K \to L$ has an extension $\eta^*: K(\alpha) \to L$, thereby contradicting the maximality of (K,η). Hence, $K = E$, which proves the theorem. □

We are now ready to show that any two algebraic closures of a field are isomorphic.

4.4 Theorem. *Let K and K' be algebraic closures of a field F. Then $K \simeq K'$ under an isomorphism that is an identity on F.*

Proof. Let $\lambda: F \to K$ be the injection; that is, $\lambda(a) = a$ for all $a \in F$. By Theorem 4.3, λ can be extended to an embedding $\lambda^*: K' \to K$. Now $K' \simeq \lambda^*(K')$. Hence, $\lambda^*(K')$ is also an algebraically closed field containing F. Because K is an algebraic extension of F, K is also an algebraic extension of $\lambda^*(K')$, which lies between F and K. But then $\lambda^*(K') = K$, so λ^* is an isomorphism of K' onto K, as desired. □

Theorem 4.4 shows that an algebraic closure of a field F is unique up to isomorphism. Henceforth, we denote by $\overline{F}$ the algebraic closure of F.

Now we want to show the existence of an algebraically closed field K containing a given field F and algebraic over it. The proof of this result, as given in the following theorem, is due to Artin and needs the concept of a polynomial ring in an infinite number of variables. There are other proofs, but all of them need some additional concepts that are not discussed in this book. For the sake of completeness we have chosen to give the proof by Artin. For those who find the proof involved, they may skip it at the first reading; this will not impede their understanding any of the following material.

We first outline the concept of a polynomial ring in an infinite number of variables over a field F.

Let F be a field, and let $S = (x_i)_{i \in \Lambda}$ be an infinite set of commuting indeterminates or variables. Then the elements of the form

$$\sum_{\text{finite}} a_i x_{i_1} x_{i_2} \cdots x_{i_n}, \qquad a_i \in F,\ x_{i_j} \in S,$$

with natural addition and multiplication form a ring $F[S]$ called the polynomial ring over F in S. Note that for a polynomial $\Sigma a_i x_{i_1} x_{i_2} \cdots x_{i_n}$ to be zero each coefficient a_i of each monomial $x_{i_n} \cdots x_{i_n}$ must be zero.

4.5 Theorem. *Let F be a field. Then there exists an algebraically closed field K containing F as a subfield.*

Proof. Let us first construct an extension K of F in which every polynomial in $F[x]$ of degree ≥ 1 has a root. Let S be a set in 1-1 correspondence with the set of all polynomials in $F[x]$ of degree ≥ 1, and suppose to each polynomial $f = f(x) \in F[x]$ of degree ≥ 1 we correspond the element $x_f \in S$.

Form the polynomial ring $F[S]$. We claim that if A is the ideal in $F[S]$ generated by all polynomials $f(x_f)$ of degree ≥ 1 in $F[S]$, then $A \neq F[S]$. Otherwise $1 \in A$, and, hence,

$$g_1 f_1(x_{f_1}) + \cdots + g_n f_n(x_{f_n}) = 1, \qquad \text{with } g_i \in F[S]. \tag{1}$$

Now $g_1,\ldots,g_n$ will involve only a finite number of variables. Write $x_{f_i} = x_i$ for each $f_i \in F[x]$. After reindexing, we can assume that $x_{f_1} = x_1,\ldots,x_{f_n} = x_n$, and the variables occurring in all g_i, $1 \leq i \leq n$, are in the set $\{x_1, x_2,\ldots,x_n,\ldots,x_m\}$. Thus, we can rewrite (1) as

$$\sum_{i=1}^{n} g_i(x_1,\ldots,x_m) f_i(x_i) = 1. \tag{2}$$

Let E be an extension of F in which each polynomial $f_1,\ldots,f_n$ has a root (Problem 5, Section 2). Let α_i be a root of f_i in E, $i = 1,\ldots,n$. Put $x_i = \alpha_i$ for $i = 1,\ldots,n$ and $x_i = 0$ for $i = n+1,\ldots,m$ in relation (2) and obtain $0 = 1$, which is absurd. Thus, $A \neq F[S]$. Then by Zorn's lemma, A can be embedded in a maximal ideal M of $F[S]$. It is easy to see that $F[S]/M$ is a field containing a copy of F. Hence, $F[S]/M$ can be regarded as an extension of F. Also, each polynomial $f \in F[x]$ of degree ≥ 1 has a root in $F[S]/M$. Therefore, we have constructed a field K_1, namely, $F[S]/M$, that is an extension of F and in which every polynomial $f(x) \in F[x]$ of degree ≥ 1 has a root.

Inductively, we can now form a chain of fields $K_1 \subset K_2 \subset K_3 \subset \ldots$ such that every polynomial in $K_n[x]$ of degree ≥ 1 has a root in K_{n+1}. Let $K = \bigcup_{n=1}^{\infty} K_n$. Then K is a field, and every polynomial in $K[x]$ has its coefficients in some subfield K_n and, hence, has a root in $K_{n+1} \subset K$. This completes the proof. □

4.6 Theorem. *Let F be a field. Then there exists an extension $\overline{F}$ that is algebraic over F and is algebraically closed; that is, each field has an algebraic closure.*

Proof. Let K be an algebraically closed field containing F. If $\overline{F}$ is the set of all elements of K that are algebraic over F, then, by Theorem 3.7, $\overline{F}$ is an algebraic extension of F. Let $f(x) \in \overline{F}[x]$. Then $f(x)$ has a root $a \in K$ because K is algebraically closed. But then $a \in K$ is algebraic over $\overline{F}$, and because $\overline{F}$ is algebraic over F, we obtain, by Problem 1 in Section 3, that a is algebraic over F. Hence, $a \in \overline{F}$. Thus, $\overline{F}$ is algebraically closed, which proves that $\overline{F}$ is an algebraic closure of F. □

It is proved in Chapter 18 that every polynomial in $\mathbf{C}[x]$ has a root in $\mathbf{C}$ (fundamental theorem of algebra). This is equivalent to saying that $\overline{\mathbf{C}} = \mathbf{C}$. Further, the tower

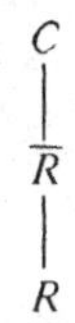

with $[\mathbf{C}:\mathbf{R}] = 2$ shows that $\overline{\mathbf{R}} = \mathbf{C}$. The algebraic closure $\overline{\mathbf{Q}}$ of $\mathbf{Q}$ in $\mathbf{C}$ is called the *field of algebraic numbers.* $\overline{\mathbf{Q}}$ is a countable set, so it is strictly contained in $\mathbf{C}$ (Theorem 5.3, Chapter 1).

CHAPTER 16

Normal and separable extensions

1 Splitting fields

Definition. *Let $f(x)$ be a polynomial in $F[x]$ of degree ≥ 1. Then an extension K of F is called a* splitting field *of $f(x)$ over F if*

(i) $f(x)$ factors into linear factors in $K[x]$; that is

$$f(x) = c(x - \alpha_1) \cdots (x - \alpha_n), \qquad \alpha_i \in K.$$

(ii) $K = F(\alpha_1, \ldots, \alpha_n)$; that is, K is generated over F by the roots $\alpha_1, \ldots, \alpha_n$ of $f(x)$ in K.

For example, (i) the field $\mathbf{Q}(\sqrt{2}) = \{a + b\sqrt{2} | a, b \in \mathbf{Q}\}$ is a splitting field of $x^2 - 2 \in \mathbf{Q}[x]$ over $\mathbf{Q}$; (ii) a splitting field of $x^2 + 1 \in \mathbf{R}[x]$ over $\mathbf{R}$ is the field $\mathbf{C}$.

We note that a polynomial $f(x) \in F[x]$ always has a splitting field, namely, the field generated by its roots in a given algebraic closure $\bar{F}$ of F.

1.1 Theorem. *If K is a splitting field of $f(x) \in F[x]$ over F, then K is a finite extension and, hence, an algebraic extension of F.*

Proof. Because $K = F(\alpha_1, \ldots, \alpha_n)$ and each α_i, $i = 1, \ldots, n$, is algebraic over F, the proof follows from Theorem 3.6 of Chapter 15. □

1.2 Theorem (uniqueness of splitting field). *Let K be a splitting field of the polynomial $f(x) \in F[x]$ over a field F. If E is another splitting field of*

$f(x)$ over F, then there exists an isomorphism $\sigma: E \to K$ that is identity on F.

Proof. Let $\overline{K}$ be an algebraic closure of K. Then $\overline{K}$ is algebraic over K. Because K is also algebraic over F, it follows that $\overline{K}$ is algebraic over F. Hence, $\overline{K} = \overline{F}$. Because E is an algebraic extension of F (Theorem 1.1), we get, by invoking Theorem 4.3 of Chapter 15, that the identity mapping $\lambda: F \to F$ can be extended to an embedding $\sigma: E \to \overline{K}$. Let $f(x) = a_0 + a_1x + \cdots + a_nx^n \in F[x]$. Set $f^\sigma(x) = \sigma(a_0) + \sigma(a_1)x + \cdots + \sigma(a_n)x^n$. Because σ is identity on F, $f^\sigma(x) = f(x)$. Now we have a factorization

$$f(x) = c(x - \alpha_1) \cdots (x - \alpha_n)$$

with $\alpha_i \in E$, $i = 1,\ldots,n$, $c \in F$. From the fact that $f(x) = f^\sigma(x)$ and $c \in F$, we obtain

$$f(x) = c(x - \sigma(\alpha_1)) \cdots (x - \sigma(\alpha_n)),$$

a unique factorization in $\overline{K}[x]$. But since $f(x)$ has a factorization in $K[x]$, say

$$f(x) = c(x - \beta_1) \cdots (x - \beta_n),$$

where $\beta_i \in K$, $i = 1,\ldots,n$, it follows that the sets $\{\sigma(\alpha_1),\ldots,\sigma(\alpha_n)\}$ and $\{\beta_1,\ldots,\beta_n\}$ are equal. Thus,

$$K = F(\beta_1,\ldots,\beta_n) = F(\sigma(\alpha_1),\ldots,\sigma(\alpha_n)) = \sigma(F(\alpha_1,\ldots,\alpha_n)) = \sigma(E).$$

Hence, σ is an isomorphism of E onto K. □

Theorem 1.2 proves that the splitting field of a polynomial over a given field is unique (up to isomorphism) if it exists. But recall that any field F has an algebraic closure $\overline{F}$ that contains roots of all polynomials over F. Thus, the intersection of all subfields of $\overline{F}$ containing all the roots of a given polynomial $f(x) \in F[x]$ is the splitting field of $f(x)$ over F.

1.3 Examples

(a) The degree of the extension of the splitting field of $x^3 - 2 \in \mathbf{Q}[x]$ is 6.

Solution. By Eisenstein's criterion $x^3 - 2 \in \mathbf{Q}[x]$ is irreducible over $\mathbf{Q}$, and indeed it is the minimal polynomial of $2^{1/3}$. Thus, $\mathbf{Q}[x]/(x^3 - 2) \simeq \mathbf{Q}(2^{1/3})$ with $[\mathbf{Q}(2^{1/3}):\mathbf{Q}] = 3$. Clearly, $\mathbf{Q}(2^{1/3})$ is not the splitting field of $x^3 - 2 \in \mathbf{Q}[x]$, since $x^3 - 2 = (x - 2^{1/3})(x^2 + 2^{1/3}x + 2^{2/3})$ implies $x^3 - 2$ has two complex roots, say α and $\overline{\alpha}$. Thus, $p(x) = x^2 + 2^{1/3}x +$

$2^{2/3} \in \mathbf{Q}(2^{1/3})[x]$ is irreducible over $\mathbf{Q}(2^{1/3})$. Hence,

$$\mathbf{Q}(2^{1/3})[x]/(p(x)) \simeq \mathbf{Q}(2^{1/3})(\alpha) = \mathbf{Q}(2^{1/3},\alpha),$$

and the degree of extension of $\mathbf{Q}(2^{1/3},\alpha)$ over $\mathbf{Q}(2^{1/3})$ is 2, the degree of $p(x)$. Because $\mathbf{Q}(2^{1/3},\alpha)$ contains one root α of $p(x)$, it will also contain the other root $\bar{\alpha}$. Hence, $\mathbf{Q}(2^{1/3},\alpha)$ is the splitting field of $x^3 - 2 \in \mathbf{Q}[x]$ over $\mathbf{Q}$. Finally,

$$\begin{aligned}[\mathbf{Q}(2^{1/3},\alpha): \mathbf{Q}] &= [\mathbf{Q}(2^{1/3},\alpha): \mathbf{Q}(2^{1/3})][\mathbf{Q}(2^{1/3}):\mathbf{Q}] \\ &= 3 \cdot 2 = 6.\end{aligned}$$

This completes the solution. We exhibit the bases of extensions in the following diagram:

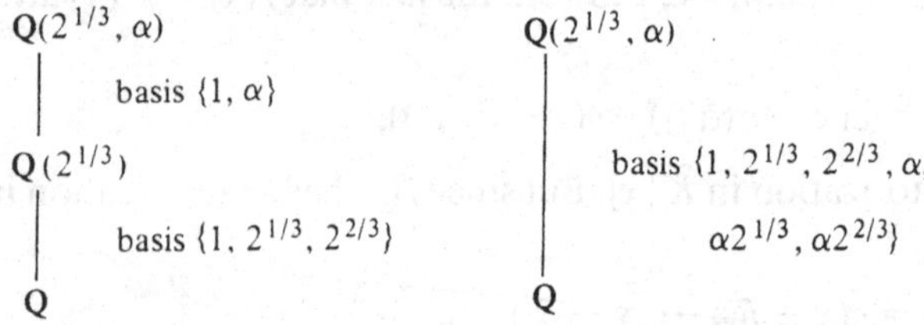

$$\begin{array}{ccc} \mathbf{Q}(2^{1/3},\alpha) & & \\ | & \text{basis } \{1, \alpha\} & \\ \mathbf{Q}(2^{1/3}) & & \\ | & \text{basis } \{1, 2^{1/3}, 2^{2/3}\} & \\ \mathbf{Q} & & \end{array} \qquad \begin{array}{cc} \mathbf{Q}(2^{1/3},\alpha) & \\ | & \text{basis } \{1, 2^{1/3}, 2^{2/3}, \alpha, \\ | & \alpha 2^{1/3}, \alpha 2^{2/3}\} \\ \mathbf{Q} & \end{array}$$

(b) Let p be prime. Then $f(x) = x^p - 1 \in \mathbf{Q}[x]$ has splitting field $\mathbf{Q}(\alpha)$, where $\alpha \neq 1$ and $\alpha^p = 1$. Also, $[\mathbf{Q}(\alpha): \mathbf{Q}] = p - 1$.

Solution

$$f(x) = (x - 1)(x^{p-1} + x^{p-2} + \cdots + x + 1).$$

We know

$$p(x) = x^{p-1} + \cdots + x + 1 \in \mathbf{Q}[x]$$

is irreducible over $\mathbf{Q}$. Let α be a root of $p(x)$ in the splitting field of $f(x)$ over $\mathbf{Q}$. Then, clearly, $\alpha^p = 1$ and $\alpha \neq 1$. We assert that $1,\alpha,\alpha^2,\ldots,\alpha^{p-1}$ are p distinct roots of $f(x)$. Clearly, $\alpha^p = 1$ implies $(\alpha^i)^p = 1$ for all positive integers i. Thus, we need to show that $1,\alpha,\alpha^2,\ldots,\alpha^{p-1}$ are distinct roots. Note that if m is the smallest positive integer such that $\alpha^m = 1$, then $m|p$. Thus, $m = p$. Hence, no two roots in the list $1,\alpha,\ldots,\alpha^{p-1}$ can be equal, whence these are all the p roots of $x^p - 1$. Hence, the splitting field of $x^p - 1 \in \mathbf{Q}[x]$ is $\mathbf{Q}(\alpha)$. Because the minimal polynomial of α is $p(x)$, which is of degree $p - 1$, we get $[\mathbf{Q}(\alpha): \mathbf{Q}] = p - 1$, which completes the solution.

We prove later that the degree of extension of the splitting field of $x^n - 1 \in \mathbf{Q}[x]$ is $\phi(n)$, where ϕ is Euler's function.

(c) Let $F = \mathbf{Z}/(2)$. The splitting field of $x^3 + x^2 + 1 \in F[x]$ is a finite field with eight elements.

Solution. Because $x^3 + x^2 + 1$ has degree 3, we know this is reducible if and only if it has a root in F. But by actual substitution neither 0 nor 1 is a root. Thus, $x^3 + x^2 + 1$ is irreducible over F. Let α be a root of this polynomial in its splitting field. Then we find

$$\begin{aligned} x^3 + x^2 + 1 &= (x + \alpha)(x^2 + (1 + \alpha)x + (\alpha + \alpha^2)) \\ &= (x + \alpha)(x + \alpha^2)(x + 1 + \alpha + \alpha^2). \end{aligned}$$

Therefore, $F(\alpha)$ is the splitting field of $x^3 + x^2 + 1$ over F, and $[F(\alpha): F] = 3$, the degree of the minimal polynomial $x^3 + x^2 + 1$ of α. Furthermore, $F(\alpha)$ has a basis $\{1, \alpha, \alpha^2\}$ over F. Therefore,

$$F(\alpha) = \{0, 1, \alpha, \alpha^2, \alpha + 1, \alpha^2 + 1, \alpha^2 + \alpha, \alpha^2 + \alpha + 1\},$$

where $\alpha^3 + \alpha^2 + 1 = 0$. This completes the solution.

(d) The splitting field of $f(x) = x^4 - 2 \in \mathbf{Q}[x]$ over $\mathbf{Q}$ is $\mathbf{Q}(2^{1/4}, i)$, and its degree of extension is 8.

Solution. $f(x) = x^4 - 2 \in \mathbf{Q}[x]$ is irreducible over $\mathbf{Q}$. We find $f(x)$ has a root $2^{1/4}$. So $f(x)$ is the minimal polynomial of $2^{1/4}$ over $\mathbf{Q}$. Thus,

$$[\mathbf{Q}(2^{1/4}): \mathbf{Q}] = 4, \qquad \text{the degree of } f(x).$$

Now $f(x) = (x - 2^{1/4})(x + 2^{1/4})(x^2 + 2^{1/2})$ and $x^2 + 2^{1/2}$ is irreducible over $\mathbf{Q}(2^{1/4})$. Thus, the root $2^{1/4}\, i$ has $x^2 + 2^{1/2}$ as its minimal polynomial over $\mathbf{Q}(2^{1/4})$, whence

$$[\mathbf{Q}(2^{1/4})(2^{1/4} i): \mathbf{Q}(2^{1/4})] = 2.$$

Because $\mathbf{Q}(2^{1/4})(2^{1/4} i) = \mathbf{Q}(2^{1/4}, i)$ is clearly the splitting field and

$$[\mathbf{Q}(2^{1/4}, i): \mathbf{Q}] = [\mathbf{Q}(2^{1/4}, i): \mathbf{Q}(2^{1/4})][\mathbf{Q}(2^{1/4}): \mathbf{Q}] = 4 \cdot 2 = 8,$$

we have completed the solution. We illustrate it in the following diagram:

$\mathbf{Q}(2^{1/4}, i)$

| basis $\{1, 2^{1/4}\, i\}$

$\mathbf{Q}(2^{1/4})$

| basis $\{1, 2^{1/4}, 2^{1/2}, 2^{3/4}\}$

$\mathbf{Q}$

Problems

1. Construct splitting fields over $\mathbf{Q}$ for the following polynomials.
(a) x^3-1. (b) x^4+1. (c) x^6-1. (d) $(x^2-2)(x^3-3)$.
Also find the degrees of extension over $\mathbf{Q}$.
2. Construct a splitting field for $x^3+x+1 \in \mathbf{Z}/(2)[x]$ and list all its elements.
3. Find conditions on a and b such that the splitting field of $x^3 + ax + b \in \mathbf{Q}[x]$ has degree of extension 3 over $\mathbf{Q}$.
4. In Problem 3 find conditions on a and b such that the degree of extension is 6 over $\mathbf{Q}$.
5. Let E be the splitting field of a polynomial of degree n over a field F. Show that $[E: F] \leq n!$
6. Let $f(x) \in F[x]$ be a polynomial of degree $n \geq 1$ over F. Show that $f(x)$ has exactly n roots in its splitting field. (A multiple root occurring with multiplicity m is counted m times.)
7. If an irreducible polynomial $p(x)$ over a field F has one root in a splitting field E of a polynomial $f(x) \in F[x]$, then $p(x)$ has all its roots in E.
8. Show that over any field $K \supset \mathbf{Q}$ the polynomial x^3-3x+1 is either irreducible or splits into linear factors.
9. Let $f(x) = x^4-2x^2-2 \in \mathbf{Q}[x]$. Find the roots α, β of $f(x)$ such that

$$\mathbf{Q}(\alpha) \simeq \mathbf{Q}(\beta).$$

What is the splitting field of $f(x)$?

2 Normal extensions

Let $(f_i(x))_{i \in \Lambda}$ be a family of polynomials of degree ≥ 1 over a field F. In the previous section we defined the splitting field of a polynomial over F. Now by a splitting field of a family $(f_i(x))_{i \in \Lambda}$ of polynomials we mean an extension E of F such that every $f_i(x)$ splits into linear factors in $E[x]$, and E is generated over F by all the roots of the polynomials $f_i(x)$, $i \in \Lambda$. If Λ is finite and our polynomials are $f_1(x),\dots, f_n(x)$, then their splitting field is a splitting field of the single polynomial $f(x) = f_1(x) \cdots f_n(x)$, obtained by taking the product. The proof of uniqueness (up to isomorphism) of a splitting field of a single polynomial can be extended to prove the uniqueness (up to isomorphism) of a splitting field of a family of polynomials over a given field.

The next theorem proves a set of equivalent statements for an extension E of F to be a splitting field of a family of polynomials over F.

2.1 Theorem. *Let E be an algebraic extension of a field F contained in an algebraic closure $\overline{F}$ of F. Then the following conditions are equivalent:*

(i) *Every irreducible polynomial in $F[x]$ that has a root in E splits into linear factors in E.*
(ii) *E is the splitting field of a family of polynomials in $F[x]$.*
(iii) *Every embedding σ of E in $\overline{F}$ that keeps each element of F fixed maps E onto E. (In other words, σ may be regarded as an automorphism of E.)*

Proof. (i) $\Rightarrow$ (ii) Let $\alpha \in E$, and let $p_\alpha(x)$ be its minimal polynomial over F. By (i), $p_\alpha(x)$ splits into linear factors in E. Thus, it follows immediately that E is a splitting field of the family $(p_\alpha(x))$, $\alpha \in E$.

(ii) $\Rightarrow$ (iii) Let $(f_i(x))$, $i \in \Lambda$, be a family of polynomials of which E is the splitting field. If α is a root of some $f_i(x)$ in E, then for any embedding σ of E into $\overline{F}$ that keeps each element of F fixed, we know $\sigma(\alpha)$ is a root of $f_i(x)$. Because E is generated by the roots of all the polynomials $f_i(x)$, it follows that σ maps E into itself. Thus, by Theorem 3.8 of Chapter 15, σ is an automorphism of E.

(iii) $\Rightarrow$ (i) Let $p(x) \in F[x]$ be an irreducible polynomial over F that has a root $\alpha \in E$. Let $\beta \in \overline{F}$ be another root of $p(x)$. We show that $\beta \in E$. Because α and β are roots of the same irreducible polynomial $p(x)$, we have F-isomorphisms

$$F(\alpha) \simeq F[x]/(p(x)) \simeq F(\beta).$$

Let σ: $F(\alpha) \to F(\beta)$ be the isomorphism given above. Then $\sigma(\alpha) = \beta$ and $\sigma(a) = a$ for all $a \in F$. By Theorem 4.3 of Chapter 15, σ can be extended to an embedding σ^*: $E \to \overline{F}$. But then, by (iii), σ^* is an automorphism of E; hence, $\sigma^*(\alpha) = \sigma(\alpha) = \beta \in E$. This completes the proof. □

Definition. *An extension E of a field F is called* normal *if E satisfies any one of the equivalent statements of Theorem* 2.1.

2.2 Examples of normal extensions

(a) **C** is a normal extension of **R**.

(b) **R** is not a normal extension of **Q**, for $x^3 - 2 \in \mathbf{Q}[x]$ is irreducible over **Q** and has a root $\sqrt[3]{2}$ in **R**, but it does not split into linear factors in **R** because it has complex roots.

(c) If $\alpha = \cos(\pi/4) + i\sin(\pi/4)$, then $\mathbf{Q}(\alpha)$ is a normal extension of **Q**. This follows from the fact that $\mathbf{Q}(\alpha)$ is the splitting field of $x^4 + 1 \in \mathbf{Q}[x]$.

(d) In general, any extension E of a field F, such that $[E:F] = 2$, is a normal extension. Let $\alpha \in E$, $\alpha \notin F$. Let $p(x)$ be the minimal polynomial of α over F. Then $[F(\alpha):F] =$ degree of $p(x)$. Because $[E:F(\alpha)][F(\alpha):F] = [E:F] = 2$, we must have $[E:F(\alpha)] = 1$ and $[F(\alpha):F] = 2$. Thus, $E = F(\alpha)$, and the degree of $p(x)$ is equal to 2. Because $p(x)$ has one root $\alpha \in E$, it must have its other root also in E. Hence, E is the splitting field of $p(x) \in F[x]$ and is thus a normal extension of F.

2.3 Example

Let E be a finite extension of F. Then E is a normal extension of F if and only if E is a splitting field of a polynomial $f(x) \in F[x]$.

Solution. By hypothesis $E = F(\alpha_1, \ldots, \alpha_n)$, where $\alpha_i \in E$ are algebraic over F. Let $p_i(x)$ be the minimal polynomial of α_i over F.

Assume first that E is a normal extension of F. Then $p_i(x)$ splits in E because it has one root $\alpha_i \in E$. Thus, $f(x) = p_1(x) \cdots p_n(x) \in F[x]$ has all roots in E. Because $E = F(\alpha_1, \ldots, \alpha_n)$ and $\alpha_1, \ldots, \alpha_n$ are some of the roots of $f(x)$, E must be the splitting field of $f(x)$. The converse follows from Theorem 2.1.

Problems

1. Which of the following extensions are normal over **Q**?
   (a) $\mathbf{Q}(\sqrt{-2})$. (b) $\mathbf{Q}(5\sqrt{7})$. (c) $\mathbf{Q}(\sqrt{-1})$.
   (d) $\mathbf{Q}(x)$, where x is not algebraic over **Q**.
2. Is $\mathbf{R}(\sqrt{-5})$ normal over **R**?
3. Let E be a normal extension of F and let K be a subfield of E containing F. Show that E is a normal extension over K. Give an example to show that K need not be a normal extension of F.
4. Let $F = \mathbf{Q}(\sqrt{2})$ and $E = \mathbf{Q}(\sqrt[4]{2})$. Show that E is a normal extension of F, F is a normal extension of **Q**, but E is not a normal extension of **Q**.
5. Show that every finite extension of a finite field is normal.
6. Let E_i, $i \in \Lambda$, be a family of normal extensions of a field F in some extension K of F. Show that $\bigcap_{i \in \Lambda} E_i$ is also a normal extension of F.
7. Show that the field generated by a root of $x^3 - x - 1$ over **Q** is not normal over **Q**.
8. Find the smallest normal extension (up to isomorphism) of $\mathbf{Q}(2^{1/4})$ in $\bar{\mathbf{Q}}$ (the algebraic closure of **Q**).

9. Find the smallest normal extension (up to isomorphism) of $\mathbf{Q}(2^{1/4}, 3^{1/4})$ in $\bar{\mathbf{Q}}$.
10. Let F, E, and K be fields such that $F \subseteq E \subseteq K$ and K is normal over F. Show that any F-homomorphism

    $\sigma: E \to K$

    can be extended to an F-automorphism of K.
11. Let $E \subset \bar{F}$ and $K \subset \bar{F}$ be normal extensions of a field F. Show that the subfield L generated by E and K is also normal over F.

3 Multiple roots

The purpose of this section is to discuss the multiplicity of roots of a polynomial over a field. For this purpose we introduce the concept of the derivative of a polynomial.

Let $f(x) = \Sigma_{i=0}^{n} a_i x^i$ be a polynomial over a field F. We define the *derivative* of $f(x)$ by $f'(x) = \Sigma_{i=1}^{n} i a_i x^{i-1}$. Then properties of derivatives that are familiar from calculus are not necessarily valid here. For example, $f'(x) = 0$ does not always imply that $f(x)$ is a constant: for example, if we set $f(x) = x^5$ in a field of characteristic 5 then $f'(x) = 5x^4 = 0$.

The ordinary rules for operating with derivatives, however, remain the same. It is easy to verify that taking the derivative is a linear operation; that is, we have

3.1 Theorem. $(af(x) + bg(x))' = af'(x) + bg'(x)$, *where* $a, b \in F$.

For the derivative of a product we have the usual rule.

3.2 Theorem. $(f(x)g(x))' = f'(x)g(x) + f(x)g'(x)$.

The proof follows easily from the definition of a derivative.

Let K be a splitting field of a polynomial $f(x) \in F[x]$. Let α be a root of $f(x)$. Then $(x - \alpha) | f(x)$ in $K[x]$. If $(x - \alpha)^s$ is the highest power of $(x - \alpha)$ that divides $f(x)$ in $K[x]$, then s is called the *multiplicity* of α. If $s = 1$, then α is called a *simple root*; if $s > 1$, then α is called a *multiple root*.

3.3 Theorem. *Let* $f(x) \in F[x]$ *be a polynomial of degree* ≥ 1 *with* α *as a root. Then* α *is a multiple root if and only if* $f'(\alpha) = 0$.

Proof. Because α is a root of $f(x)$, by the division algorithm we can write

$$f(x) = (x - \alpha)g(x).$$

Then $f'(x) = (x - \alpha)g'(x) + g(x)$. Clearly, α is a multiple root of $f(x)$ if and only if $g(\alpha) = 0$. Because $f'(\alpha) = g(\alpha)$, the proposition follows. □

3.4 Corollary. *Let $f(x)$ be an irreducible polynomial over F. Then $f(x)$ has a multiple root if and only if $f'(x) = 0$.*

Proof. By Theorem 3.3, α is a multiple root of $f(x)$ if and only if it is a root of $f'(x)$. Because $f(x)$ is irreducible, $a^{-1}f(x)$ is the minimal polynomial of α over F, where a is the leading coefficient of $f(x)$. Thus, if $f'(x) \neq 0$, degree of $f'(x) \geq$ degree of $a^{-1}f(x)$, a contradiction. Hence, $f'(x) = 0$. □

3.5 Corollary. *Any irreducible polynomial $f(x)$ over a field of characteristic 0 has simple roots. Also any irreducible polynomial $f(x)$ over a field F of characteristic $p \neq 0$ has multiple roots if and only if there exists $g(x) \in F[x]$ such that*

$$f(x) = g(x^p).$$

Proof. Let $f(x) = \Sigma_{i=0}^{n} a_i x^i$ be an irreducible polynomial over a field F. By Corollary 3.4, $f(x)$ has multiple roots if and only if $f'(x) = \Sigma_{i=1}^{n} ia_i x^{i-1} = 0$. This implies $ia_i = 0$, $1 \leq i \leq n$. In a field of characteristic 0 this gives $a_i = 0$, $1 \leq i \leq n$. But then $f(x) = a_0 \in F$, a contradiction. Thus, $f(x)$ has all roots simple if F is of characteristic zero. In case F is of characteristic $p \neq 0$, and if $a_i \neq 0$, we must have $p|i$. This implies $f(x) = \Sigma a_i x^i$, where either $a_i = 0$ or $p|i$. But then $f(x) = g(x^p)$ for a suitable polynomial $g(x) \in F[x]$. □

The following is an interesting fact regarding the multiplicity of roots of an irreducible polynomial over any field.

3.6 Theorem. *If $f(x) \in F[x]$ is irreducible over F, then all roots of $f(x)$ have the same multiplicity.*

Proof. Let $\overline{F}$ be the algebraic closure of F, and let α and β be roots of $f(x)$ in $\overline{F}$ with multiplicities k and k', respectively. We know that

$$F(\alpha) \simeq F[x]/(f(x)) \simeq F(\beta)$$

by

$$a_0 + a_1\alpha + \cdots + a_n\alpha^n \to a_0 + a_1 x + \cdots + a_n x^n + (f(x)) \mapsto a_0 + a_1\beta + \cdots + a_n\beta^n.$$

Let us denote this isomorphism of $F(\alpha)$ to $F(\beta)$ by σ. Clearly, $\sigma(\alpha) = \beta$. Consider the diagram

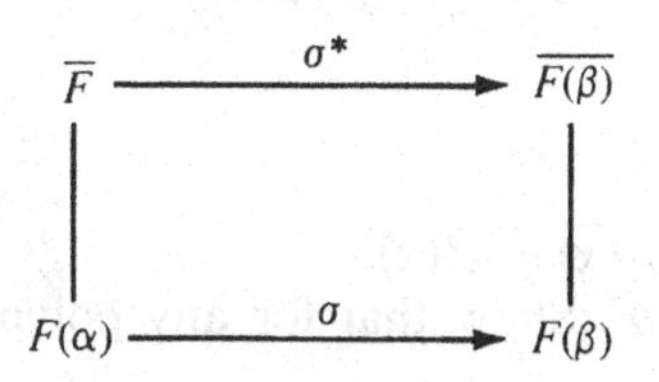

We know σ can be extended to an isomorphism σ^* from $\overline{F}$ to $\overline{F(\beta)} = \overline{F}$. Then σ^* induces a ring homomorphism

$$\eta: \overline{F}[x] \to \overline{F}[x]$$

given by

$$\eta(a_0 + a_1 x + \cdots + a_r x^r) = \sigma^*(a_0) + \sigma^*(a_1)x + \cdots + \sigma^*(a_r)x^r.$$

We note that $\eta(f(x)) = f(x)$. Because $\eta(x-\alpha)^k = (x-\beta)^k$, we get that $(x-\beta)^k$ is a factor of $f(x)$, so $k' \geq k$. The roles of α and β can be interchanged to show that $k \geq k'$. Hence, $k = k'$. □

3.7 Corollary. *If $f(x) \in F[x]$ is irreducible over F, then $f(x) = a\prod_{i=1}^{r}(x-\alpha_i)^k$, where α_i are the roots of $f(x)$ in its splitting field over F, and k is the multiplicity of each root.*

Proof. Obvious from Theorem 3.6. □

3.8 Example

Let $K = F(x)$ be the field of rational functions in one variable x over a field F of characteristic 3. (Indeed, $F(x)$ is the field of fractions of the polynomial ring $F[x]$.) Then the polynomial $y^3 - x$ in the polynomial ring $K[y]$ over K is irreducible over K and has multiple roots.

Solution. If $y^3 - x$ has a root in K, then there exists $g(x)/h(x)$ in K with $h(x) \neq 0$ such that $(g(x)/h(x))^3 = x$; that is, $g^3(x) = xh^3(x)$. But this implies that 3 (degree of $h(x)$) + 1 = 3 (degree of $g(x)$), which is impossible. Thus, $y^3 - x \in K[y]$ is irreducible over K. Now if β_1 and β_2 are two roots of $y^3 - x$ in its splitting field, then $\beta_1^3 = x = \beta_2^3$. But then $(\beta_1 - \beta_2)^3 = \beta_1^3 + (-1)^3\beta_2^3 = 0$, and, hence, $\beta_1 - \beta_2 = 0$. This shows that $y^3 - x$ has only one distinct root whose multiplicity is 3. This completes the solution.

In the next section we show that an irreducible polynomial over a finite field has only simple roots. Hence, it will follow then that the only fields over which an irreducible polynomial may have multiple roots are infinite fields of characteristic $p \neq 0$.

Problems

1. Verify that $(f(x) + g(x))' = f'(x) + g'(x)$.
2. Let F be a field and $a, h \in F$. Show that for any polynomial $f(x) \in F[x]$ of degree n,

$$f(a+h) = \sum_{k=0}^{n} \frac{1}{k!} f^{(k)}(a)h^k,$$

   where $f^{(k)}(x) = (f^{(k-1)}(x))'$ and $f^{(1)}(x) = f'(x)$, $k = 2,3,\ldots$. ($f^{(k)}(a)$ is called the kth derivative of $f(x)$ at $x = a$.)
3. Show that $f(x) \in F[x]$ has a root α of multiplicity $n > 1$ if and only if $f^{(k)}(\alpha) = 0$, $k = 1,\ldots,n-1$, and $f^{(n)}(\alpha) \neq 0$, where $f^{(i)}(\alpha)$ is the ith derivative of $f(x)$ at $x = \alpha$ as defined in Problem 2.
4. Let $f(x)$ be a polynomial of degree n over a field F of characteristic p. Suppose $f'(x) = 0$. Show that $p \mid n$ and that $f(x)$ has at most n/p distinct roots.

4 Finite fields

Definition. *A field is called* prime *if it has no proper subfield.*

Clearly, every field F contains a prime field – namely, the intersection of the family of its subfields, called the *prime field* of F.

We remark that $\mathbf{Q}$ and $\mathbf{Z}/(p)$, p prime, are prime fields. The following theorem shows that these are the only two kinds of prime fields.

4.1 Theorem. *The prime field of a field F is either isomorphic to $\mathbf{Q}$ or to $\mathbf{Z}/(p)$, p prime.*

Proof. Consider the mapping $f: \mathbf{Z} \to F$ given by $f(n) = ne$, e the unity of F. It is easily checked that f is a homomorphism.

Case 1. Ker $f = (0)$ (or, equivalently, char F is 0). Then f is an embedding of $\mathbf{Z}$ into F. This embedding can be extended to an embedding f^*:

$\mathbf{Q} \to F$ by defining $f^*(m/n) = me/ne$. Thus, $\mathbf{Q}$ embeds in F, and, hence, the prime field of F is isomorphic to $\mathbf{Q}$.

Case 2. $\operatorname{Ker} f \neq 0$. Because $\mathbf{Z}$ is a PID, $\operatorname{Ker} f = (m)$, m a positive integer. By the fundamental theorem of homomorphisms of rings $\mathbf{Z}/(m) \simeq \operatorname{Im} f$. This shows that $\mathbf{Z}/(m)$, being isomorphic to a subring of the field F, has no proper divisors of zero, so m must be a prime number p. Thus, $\mathbf{Z}/(p)$ embeds in F. Hence, the prime field of F is isomorphic to $\mathbf{Z}/(p)$. □

4.2 Theorem. *Let F be a finite field. Then*

(i) *The characteristic of F is a prime number p and F contains a subfield $F_p \simeq \mathbf{Z}/(p)$.*

(ii) *The number of elements of F is p^n for some positive integer n.*

Proof. The proof of (i) follows from Theorem 4.1.

To prove (ii), we regard F as a vector space over its prime field F_p. Let $(e_1,\ldots,e_n)$ be a basis of F over F_p. Then any element $x \in F$ can be written uniquely as

$$x = a_1e_1 + \cdots + a_ne_n, \qquad a_i \in F_p,\ i = 1,2,\ldots,n.$$

The number of elements of F is thus p^n, since each a_i in the previous expression for x can be chosen in p ways, p being the number of elements of F_p. □

A finite field is also called a *Galois field.* A Galois field with p^n elements is usually written as $GF(p^n)$.

We next prove that finite fields are splitting fields of suitable polynomials over F_p, the subfield with p elements.

4.3 Theorem. *Any finite field F with p^n elements is the splitting field of $x^{p^n} - x \in F_p[x]$. Consequently, any two finite fields with p^n elements are isomorphic.*

Proof. In the finite field F with p^n elements the nonzero elements form a multiplicative group of order $p^n - 1$. Thus, if $0 \neq \lambda \in F$, then $\lambda^{p^n-1} = 1$, so $\lambda^{p^n} = \lambda$. Also, if $\lambda = 0$, then $\lambda^{p^n} = \lambda$. Hence, all the p^n elements of F satisfy the equation $x^{p^n} - x = 0$. Because $x^{p^n} - x \in F_p[x]$ has only p^n roots, it follows that F coincides with the set of roots of $x^{p^n} - x$.

Let E and F be two finite fields with p^n elements. By Theorem 4.2, E and F contain subfields E_p and F_p, each of p elements. Also, E and F are splitting fields of $x^{p^n} - x$ over E_p and F_p, respectively. But since $E_p \simeq F_p$, it follows by uniqueness of splitting fields (up to isomorphism) that $E \simeq F$. This proves the theorem. □

The next theorem shows that there exists a field with p^n elements for any prime p and positive integer n. As a matter of notation, we denote $\mathbf{Z}/(p)$ by $\mathbf{Z}_p$.

4.4 Theorem. *For each prime p and each positive integer $n \geq 1$ the roots of $x^{p^n} - x \in \mathbf{Z}_p[x]$ in its splitting field over $\mathbf{Z}_p$ are all distinct and form a field F with p^n elements. Also, F is the splitting field of $x^{p^n} - x$ over $\mathbf{Z}_p$.*

Proof. Let $f(x) = x^{p^n} - x$. Because $f'(x) = p^n x^{p^n - 1} - 1 \neq 0$, by Theorem 3.3, $f(x)$ cannot have multiple roots. Thus, $f(x)$ has all its p^n roots distinct. We show that these roots form a field that is the splitting field of $f(x)$ over $\mathbf{Z}_p$. Let α and β be roots, where β is different from zero. Then

$$(\alpha \pm \beta)^{p^n} = \alpha^{p^n} \pm \beta^{p^n} = \alpha \pm \beta.$$
$$(\alpha\beta^{-1})^{p^n} = \alpha^{p^n}(\beta^{p^n})^{-1} = \alpha\beta^{-1}.$$

Thus, the set of roots forms a subfield of the splitting field and, therefore, coincides with it. □

We now show that any finite field has an extension of any given finite degree.

4.5 Theorem. *If F is a finite field with p^n elements and m is a positive integer, then there exists an extension field E of F such that $[E:F] = m$, and all such extensions are isomorphic.*

Proof. Let $\overline{F}$ be the algebraic closure of F. Consider the polynomial $f(x) = x^{p^{mn}} - x \in F[x]$. If $0 \neq u \in F$, then $u^{p^n - 1} = 1$, because the multiplicative group of F is of order $p^n - 1$. In addition, because $n | mn$, $(p^n - 1 | p^{mn} - 1)$, which gives $u^{p^{mn}-1} = 1$; that is, $u^{p^{mn}} = u$. This shows that each element of F satisfies $f(x)$.

By Theorem 4.5 the p^{mn} roots of $f(x)$ are distinct and form a field E. Thus, we have the tower

$$\begin{array}{l} \overline{F} \\ | \\ E \\ | \quad [E:F] = m \\ F \\ | \quad [F:F_p] = n \\ F_p \simeq \mathbf{Z}_p \end{array} \qquad \square$$

Next we prove an important result regarding the multiplicative group of nonzero elements of a finite field.

4.6 Theorem. *The multiplicative group of nonzero elements of a finite field is cyclic.*

Proof. We cite a result in group theory: Let a and b be elements of a finite abelian group G of orders m and n, respectively. Then there exists an element $c \in G$ whose order is the l.c.m. of m and n (Problem 14, Section 3, Chapter 4).

Let F^* be the multiplicative group of nonzero elements of F. By an application of the result cited, we can find an element $\alpha \in F^*$ whose order r is the l.c.m. of the orders of all the elements of F^*. Then the order of each element of F^* divides r. Hence, for all $a \in F^*$, $a^r = 1$. Because the polynomial $x^r - 1$ has at most r roots in F, it follows that the number of elements in $F^* \leq r$. However, $1, \alpha, \ldots, \alpha^{r-1}$ are all distinct and belong to F^*. Thus, F^* is generated by α. □

As an immediate consequence we have

4.7 Corollary. *Let E be a finite extension of a finite field F. Then $E = F(\alpha)$ for some $\alpha \in E$.*

Proof. Let the multiplicative group E^* of nonzero elements of E be generated by α. Then the smallest subfield $F(\alpha)$ of E containing F and α is E itself. □

Corollary 4.7 gives an important result contained in

4.8 Theorem. *Let F be a finite field. Then there exists an irreducible polynomial of any given degree n over F.*

Proof. Let E be an extension of F of degree n (Theorem 4.5). Then $E = F(\alpha)$ by Corollary 4.7. Because E is a finite extension of F, $\alpha \in E$ is algebraic over F. Let $p(x)$ be the minimal polynomial of α over F. Then $[F(\alpha):F] =$ degree of $p(x)$. But since $F(\alpha) = E$ and $[E:F] = n$, we have an irreducible polynomial $p(x)$ of degree n over F. □

4.9 Examples

(a) Show that a finite field F of p^n elements has exactly one subfield with p^m elements for each divisor m of n.

Solution. We quote a result in group theory: A cyclic group of order n has a unique subgroup of order d for each divisor d of n.

Now $F^* = F - \{0\}$ is a cyclic group of order $p^n - 1$. Because $m|n$, $p^m - 1$ divides $p^n - 1$. Thus, there exists a unique subgroup H of F^* of order $p^m - 1$. So for all $x \in H$, $x^{p^m-1} = 1$. Hence, $x^{p^m} = x$ for all $x \in H \cup \{0\}$. Because the roots of $x^{p^m} = x$ form a field, $H \cup \{0\}$ is our required subfield of the given field.

(b) If $f(x) \in F[x]$ is an irreducible polynomial over a finite field F, then all the roots of $f(x)$ are distinct.

Solution. Let F be a finite field with p^n elements. By Corollary 3.5, $f(x)$ has multiple roots if and only if $f(x) = \Sigma_{i=0}^{m} a_i(x^p)^i$. Because $a_i \in F$, $a_i^{p^n} = a_i$. Set $b_i = a_i^{p^{n-1}}$. Thus, $f(x)$ has multiple roots if and only if

$$f(x) = \sum_{i=0}^{m} (b_i x^i)^p = \left(\sum_{i=0}^{m} b_i x^i \right)^p,$$

a contradiction, because $f(x)$ is irreducible. Thus, $f(x)$ must have distinct roots.

The next problem is a converse of Theorem 4.6.

(c) If the multiplicative group F^* of nonzero elements of a field F is cyclic, then F is finite.

Solution. Let $F^* = (\alpha)$, where α generates F^*. If F^* is finite, then F is finite, and we are done. So assume F^* is an infinite cyclic group.

Case 1. The characteristic of F is $p > 0$. In this case $F = F_p(\alpha)$, where F_p is the subfield $\{0,1,2,...,p-1\}$ of F. Consider $1 + \alpha$. If $1 + \alpha = 0$, then $\alpha^2 = 1$, a contradiction, because F^* is infinite. If $1 + \alpha \neq 0$, then $1 + \alpha \in F^*$, so $1 + \alpha = \alpha^r$, where r is some positive or negative integer. In either case $1 + \alpha = \alpha^r$ yields a polynomial over F_p with α as its root. Thus, α is algebraic over F_p, so $[F_p(\alpha): F_p] =$ degree of the minimal polynomial of α over $F_p = r$, say. Then $F = F_p(\alpha)$ has p^r elements, a contradiction. So either the characteristic of F is 0, or F^* must be finite.

Case 2. The characteristic of F is 0. Here $0 \neq 1 \in F$. So $-1 = \alpha^r$, where r is some positive or negative integer. This implies $\alpha^{2r} = 1$; that is, $o(\alpha)$ is finite, a contradiction. Hence, F^* must be finite, so F must be a finite field.

(d) The group of automorphisms of a field F with p^n elements is cyclic of order n and generated by ϕ, where $\phi(x) = x^p$, $x \in F$. (ϕ is called the *Frobenius endomorphism*.)

Solution. Let F be a field with p^n elements. Let Aut(F) denote the group of automorphisms of F. Clearly, the mapping $\phi: F \to F$, defined by $\phi(x) = x^p$, is a homomorphism. Let $x^p = y^p$. Then $(x-y)^p = 0$. Thus, $x = y$. This shows that ϕ is 1-1 and, hence, onto. Thus, $\phi \in \text{Aut}(F)$. We note that $\phi^n = \text{identity}$ because $\phi^n(x) = x^{p^n} = x$ for all $x \in F$. Let d be the order of ϕ. We have $\phi^d(x) = x^{p^d}$ for all $x \in F$. Hence, each $x \in F$ is a root of the equation $t^{p^d} - t = 0$. This equation has p^d roots. It follows that $d \geq n$, whence $d = n$. Why?

Let α be a generator of the multiplicative cyclic group F^*. Then $F = F_p(\alpha)$, where F_p is the subfield of F with p elements. Let $f(x)$ be the minimal polynomial of α over F_p. Clearly, the degree of $f(x) = n$. We are interested in counting the number of extensions of the identity mapping $\lambda: F_p \to F$ to an automorphism $\lambda^*: F \to F$. This will then give us all the automorphisms of F, because, clearly, any automorphism of F keeps each element of F_p fixed.

By Lemma 4.2 of Chapter 15 it follows that the number of automorphisms of F is equal to the distinct roots of $f(x)$. However, by Example (b), $f(x)$ has all its roots distinct. Thus, the order of the group Aut(F) is n.

We showed in the beginning that there exists an element $\phi \in \text{Aut}(F)$ such that the order of ϕ is n. Hence, Aut(F) is a cyclic group generated by ϕ.

Problems

1. If F is a finite field of characteristic p, show that each element a of F has a unique pth root $\sqrt[p]{a}$ in F.
2. Construct fields with 4, 8, 9, and 16 elements.
3. Find generators for the multiplicative groups of fields with 8, 13, and 17 elements.
4. Find generators for the group of automorphisms of fields with 4, 8, 9, and 16 elements.
5. Let F be a field with four elements. Find irreducible polynomials over F of degrees 2, 3, and 4.
6. If F is a field and $f: F \to F$ is a mapping defined by

   $$f(x) = x^{-1}, \quad x \neq 0,$$
   $$f(x) = 0, \quad x = 0,$$

   show that f is an automorphism of F if and only if F has at most four elements.
7. Prove that in any finite field any element can be written as the sum of two squares.

8. If F is a finite field, then $H \cup \{0\}$ is a subfield of F for each subgroup H of the multiplicative group F^* if and only if $|F^*|$ is either 1 or a prime of the form $2^n - 1$, where n is a positive integer.
9. Show that $x^p - x - 1$ is irreducible over $\mathbf{Z}_p$. Let K be the splitting field of degree p over $\mathbf{Z}_p$. Show $K = \mathbf{Z}_p(\omega)$, where ω is a root of $x^p - x - 1$. For $p = 2, 3$, and 5, show that the order of ω in the group $K - \{0\}$ is $1 + p + p^2 + \cdots + p^{p-1}$.
10. Without actually computing, find the number of irreducible polynomials of (i) degree 2 and (ii) degree 3 over each of the fields $\mathbf{Z}_3$ and $\mathbf{Z}_5$.

5 Separable extensions

Definition. *An irreducible polynomial $f(x) \in F[x]$ is called a* separable polynomial *if all its roots are simple. Any polynomial $f(x) \in F[x]$ is called separable if all its irreducible factors are separable.*

A polynomial that is not separable is called inseparable.

Definition. *Let E be an extension of a field F. An element $\alpha \in E$ that is algebraic over F is called* separable *over F if its minimal polynomial over F is separable.*

An algebraic extension E of a field F is called a separable extension *if each element of E is separable over F.*

5.1 Remarks. (a) It follows by Corollary 3.5 that any polynomial over a field of characteristic zero is separable. Thus, if F is a field of characteristic 0, then any algebraic extension of F is separable.

(b) By Example 4.9(b), irreducible polynomials over finite fields have distinct roots. Hence, any algebraic extension of a finite field is separable.

(c) It was shown in Example 3.8 that if $K = F(x)$ is the field of rational functions over a field F of characteristic 3, then the polynomial $y^3 - x \in K[y]$ is irreducible over K. Also, $y^3 - x$ has all its roots equal, each being α, say. Hence, $K(\alpha)$ is not a separable extension of K.

Definition. *A field F is called* perfect *if each of its algebraic extensions is separable.*

Examples of perfect fields are fields of characteristic zero and finite fields.

We remarked that infinite fields of characteristic $p > 0$ have inseparable extensions. Thus, such fields are not, in general, perfect.

Definition. *An extension E of a field F is called a* simple extension *if $E = F(\alpha)$ for some $\alpha \in E$.*

5.2 Theorem. *If E is a finite separable extension of a field F, then E is a simple extension of F.*

Proof. If F is a finite field, then by Corollary 4.7, each finite extension E of F is simple. So suppose now that F is infinite. Because E is a finite extension of F, $E = F(a_1,...,a_n)$, where $a_i \in E$, $1 \le i \le n$, are algebraic over F. We first show that if $E = F(\alpha,\beta)$, then there exists an element $\theta \in E$ such that $E = F(\theta)$.Then the result will follow by induction. Let $p(x)$ and $q(x)$ be the minimal polynomials for α and β, respectively, over F. Let the roots of $p(x)$ be $\alpha = \alpha_1,...,\alpha_n$, and let those of $q(x)$ be $\beta = \beta_1,...,\beta_m$. Because E is a separable extension of F, all α_i, $1 \le i \le n$, and all β_j, $1 \le j \le m$, are distinct. Because F is infinite, there exists $a \in F$ such that $a \ne (\alpha_i - \alpha)/(\beta - \beta_j)$ for $1 \le i \le n$, $2 \le j \le m$. Then $a(\beta - \beta_j) \ne \alpha_i - \alpha$. So $a\beta + \alpha \ne \alpha_i + a\beta_j$ for $j \ne 1$. Set $\theta = a\beta + \alpha$. Then $\theta - a\beta_j \ne \alpha_i$ for all $i = 1,...,n$ and $j = 2,...,m$. Define $h(x) = p(\theta - ax) \in F(\theta)[x]$. Then $h(\beta) = p(\alpha) = 0$ and $h(\beta_j) = p(\theta - a\beta_j) \ne 0$ for $j \ne 1$. So β is a root of $h(x)$, but no β_j ($j \ne 1$) is a root of $h(x)$. Also, β is a root of $q(x)$. Regard $q(x) \in F(\theta)[x]$. Let $A(x) \in F(\theta)[x]$ be the minimal polynomial of β over $F(\theta)$. Then $A(x)|h(x)$ and $A(x)|q(x)$. Then any root of $A(x)$ is a root of $q(x)$ as well as a root of $h(x)$. But the only common root of $q(x)$ and $h(x)$ is β. Therefore, $A(x) = x - \beta$. This implies that $\beta \in F(\theta)$. Then since $\theta = a\beta + \alpha$, $\alpha \in F(\theta)$. Hence, $F(\alpha,\beta) = F(\theta)$. □

The next theorem gives a necessary and sufficient condition for a finite extension to be a simple extension.

5.3 Theorem. *Let E be a finite extension of a field F. Then the following are equivalent.*

(a) $E = F(\alpha)$ for some $\alpha \in E$.
(b) There are only a finite number of intermediate fields between F and E.

Proof. (a) ⇒ (b) Let $f(x) \in F[x]$ be the minimal polynomial of α over F. Let K be a subfield of E containing F, and let $g(x)$ be the minimal polynomial of α over K. Then since $g(x)$ is in $K[x]$, and $f(\alpha)=0$, $g(x)|f(x)$. If K' is the subfield of K containing F and the coefficients of the polynomial $g(x)$, then $g(x)\in K'[x]$, being irreducible over K, is also irreducible over K'. Also, $F(\alpha)=E$ implies $K(\alpha)=K'(\alpha)=E$. Thus, $[E:K]=$ degree of $g(x)=[E:K']$. Hence, $K=K'$.

Consider the mapping σ from the family of intermediate fields to the divisors of $f(x)$ in $E[x]$, given by $\sigma(K)=g(x)$, the minimal polynomial of α over K. By what has gone before, σ is 1-1. Because there are only finitely many divisors of $f(x)$, the family of intermediate fields between F and E is also finite.

(b) $\Rightarrow$ (a) If F is a finite field, then E is a finite field, and the result follows from Corollary 4.7. So assume F is infinite. We first prove that for any two elements $\alpha,\beta\in E$ there is an element $\gamma\in E$ such that $F(\alpha,\beta)=F(\gamma)$. For each $a\in F$ consider the linear combination $\gamma_a=\alpha+a\beta$ of α and β. The fields $F(\gamma_a)$ are intermediate fields between F and E. Because there are only a finite number of intermediate fields, there exist $a,b\in F, a\neq b$, such that $F(\gamma_a)=F(\gamma_b)$. But then $\gamma_a,\gamma_b\in F(\gamma_b)$ implies $\gamma_a-\gamma_b\in F(\gamma_b)$. Thus, $(a-b)\beta\in F(\gamma_b)$, and, hence, $\beta\in F(\gamma_b)$. Then $\gamma_b=\alpha+b\beta\in F(\gamma_b)$ implies $\alpha\in F(\gamma_b)$. Therefore, $F(\alpha,\beta)\subset F(\gamma_b)$. Because $F(\gamma_b)\subset F(\alpha,\beta)$, our assertion is proved.

We now choose $u\in E$ such that $[F(u):F]$ is as large as possible. Then we claim $E=F(u)$. Otherwise let $x\in E$, $x\notin F(u)$. We can find an element $t\in E$ such that $F(t)$ contains both u and x, with $F(t)\supsetneq F(u)$. This contradicts the choice of u. Hence, $E=F(u)$. □

5.4 Examples

(a) Let E be an extension of a field F, and let $\alpha\in E$ be algebraic over F. Then α is separable over F iff $F(\alpha)$ is a separable extension of F.

Solution. Let $\beta\in F(\alpha)$. We show that β is separable over F. We have $F\subseteq F(\beta)\subseteq F(\alpha)$. Let L be an algebraically closed field, and let $\sigma\colon F\to L$ be an embedding. Suppose $p_1(x)$ is the minimal polynomial of β over F that has m distinct roots. Then by Lemma 4.2 of Chapter 15, there are m distinct extensions, say $\sigma_1,\ldots,\sigma_m$, of σ to $F(\beta)$.

Further, let $p_2(x)$ be the minimal polynomial of α over $F(\beta)$, and suppose $p_2(x)$ has n distinct roots. Then again by the same lemma, for each σ_i, $1\leq i\leq m$, there are exactly n extensions σ_{ij}, $1\leq j\leq n$, to $F(\alpha)$.

It is clear that the set of mn embeddings (σ_{ij}), $1\leq i\leq m$, $1\leq j\leq n$, are the only possible embeddings from $F(\alpha)$ to L that extend $\sigma\colon F\to L$.

Now let $p_3(x)$ be the minimal polynomial of α over F. Then

$$\begin{aligned}[F(\alpha):F] &= \text{degree } p_3(x) \\ &= \text{number of distinct roots of } p_3(x), \text{ since } \alpha \text{ is separable over } F \\ &= \text{number of extensions of the embedding } \sigma \text{ to } F(\alpha).\end{aligned} \tag{1}$$

Moreover, α is separable over F implies α is separable over $F(\beta)$, and, hence, by the same reasoning as in the previous paragraph,

$$\begin{aligned}[F(\alpha):F(\beta)] &= \text{degree } p_2(x) \\ &= \text{number of distinct roots of } p_2(x) \\ &= \text{number of extensions of each } \sigma_i \text{ to } F(\alpha) \\ &= n.\end{aligned} \tag{2}$$

Also,

$$[F(\beta):F] = \text{degree } p_1(x);$$

and the number of distinct roots of $p_1(x)$

$$\begin{aligned}&= \text{the number of extensions of } \sigma \text{ to } F(\beta) \\ &= m.\end{aligned} \tag{3}$$

From (1)–(3),

$$\begin{aligned}mn = [F(\alpha):F] &= [F(\alpha):F(\beta)][F(\beta):F] \\ &= n \cdot \text{the degree of } p_1(x).\end{aligned}$$

Hence, $m = \text{degree } p_1(x) = $ the number of distinct roots of $p_1(x)$. Thus, $p_1(x)$ is a separable polynomial. Hence, β is separable over F.

The converse is clear.

(b) Let $F \subset E \subset K$ be three fields such that E is a finite separable extension of F, and K is a finite separable extension of E. Then K is a finite separable extension of F.

Solution. From Theorem 5.2 we know that $E = F(\alpha)$, $K = E(\beta)$ for some $\alpha \in E$, $\beta \in K$. Let $\gamma \in F(\alpha,\beta)$, $\gamma \notin F(\alpha)$. Then we have the diagram

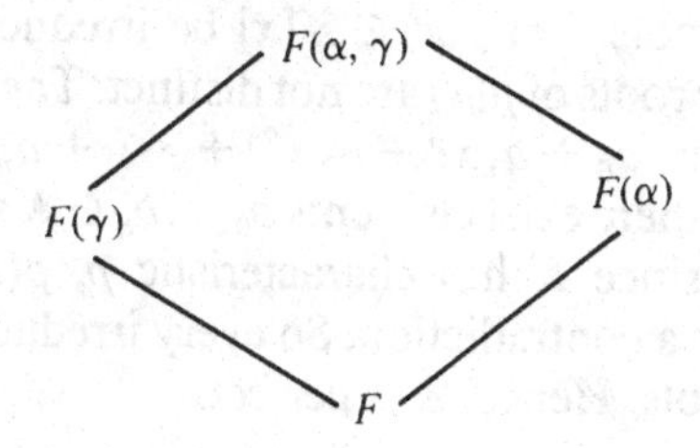

such that $F(\alpha)$ is a finite separable extension of F, and γ is a separable element over $F(\alpha)$. We prove that γ is separable over F. Let

$p_1(x)$ = the minimal polynomial of α over F with degree m,
$p_2(x)$ = the minimal polynomial of γ over $F(\alpha)$ with degree n,
$p_3(x)$ = the minimal polynomial of γ over F with degree s,
$p_4(x)$ = the minimal polynomial of α over $F(\gamma)$ with degree t.

Let $\sigma: F \rightarrow L$ be an embedding of F into an algebracially closed field L. Because α is separable over F, there are exactly m extensions (α_i), $1 \leq i \leq m$, of σ to $F(\alpha)$ (Lemma 4.2, Chapter 15). Also, since γ is separable over $F(\alpha)$, again by Lemma 4.2 of Chapter 15, there are exactly n extensions of each σ_i to $F(\alpha,\sigma)$. Let us call these n extensions $\sigma_{i1},...,\sigma_{in}$, where $1 \leq i \leq m$.

Therefore, there are precisely mn extensions of σ: $F \rightarrow L$ to $\sigma_{ij}: F(\alpha,\gamma) \rightarrow L$, $1 \leq i \leq m$, $1 \leq j \leq n$.

By considering extensions of $\sigma: F \rightarrow L$ to $F(\alpha,\gamma)$ via $F(\gamma)$, we obtain similarly that there are precisely st extensions to $F(\alpha,\gamma)$. Hence, $mn = st$.

Suppose γ is not separable over F. Then the number of extensions of σ to $F(\gamma)$ is $< s$ (Lemma 4.2, Chapter 15). This implies that the number of extensions of σ to $F(\alpha,\gamma)$ is $< st = mn$, a contradiction. Hence, γ is separable over F.

(c) If K is a field of characteristic $p \neq 0$, then K is perfect if and only if $K^p = K$ (i.e., if and only if every element of K has pth root in K).

Solution. Suppose K is perfect. Let a be any element of K. We claim that there is an element b in K such that $a = b^p$. We must show that the polynomial $f(x) = x^p - a$ has a root in K. Let b be a root of $f(x)$ in some extension field of K. Because K is perfect, b is separable over $K = K(a) = K(b^p)$. Let $p(x)$ be the minimal polynomial for b over K. Because b is a root of $x^p - b^p \in K[x]$, $p(x)$ is a factor of $x^p - b^p$ in $K[x]$. In $K[x]$ we have the decomposition $x^p - b^p = (x - b)^p$. So $p(x)$ is a power of $x - b$. But b is separable over K, so $p(x)$ has no multiple roots. Hence, $p(x) = x - b$. Because $p(x) \in K[x]$, it follows that $b \in K$.

Conversely, suppose that every element of K is the pth power of an element of K. To show that K is perfect, we show that every irreducible polynomial of $K[x]$ has distinct roots. Let $p(x) \in K[x]$ be irreducible. Assume, for contradiction, that the roots of $p(x)$ are not distinct. Then by Corollary 3.5, $p(x)$ has the form $a_0 + a_1x^p + a_2x^{2p} + \cdots + a_nx^{np}$, where $a_0,...,a_n \in K$. By hypothesis there exist elements $b_0,...,b_n \in K$ such that $a_i = b_i^p$ $(i = 0,1,...,n)$. Then since K has characteristic p, $p(x) = (b_0 + b_1x + \cdots + b_nx^n)^p$, which is a contradiction. So every irreducible polynomial of $K[x]$ has distinct roots. Hence, K is perfect.

Problems

1. Prove $\mathbf{Q}(\sqrt{2},\sqrt{3}) = \mathbf{Q}(\sqrt{2} + \sqrt{3})$.
2. Find $\theta \neq \sqrt{3} + \sqrt{5}$ such that $\mathbf{Q}(\sqrt{3},\sqrt{5}) = \mathbf{Q}(\theta)$.
3. Find θ such that $\mathbf{Q}(\sqrt{2},\omega) = \mathbf{Q}(\theta)$, where $\omega^3 = 1$, $\omega \neq 1$.
4. Let $F = \mathbf{Z}/(p)$, and let $E = F(x)$ be the field of rational functions, where p is prime. Let $K = F(x^p)$. Show that K is not perfect. (Consider $y^p - x^p \in K[y]$.)
5. Prove that a finite extension of a finite field is separable.
6. Prove that every extension of $\mathbf{Q}$ is separable.
7. Let α be a root of $x^p - x - 1$ over a field F of characteristic p. Show that $F(\alpha)$ is a separable extension of F.

CHAPTER 17

Galois theory

In this chapter we deal with the central results of Galois theory. The fundamental theorem on Galois theory establishes a one-to-one correspondence between the set of subfields of E, where E is a splitting field of a separable polynomial in $F[x]$, and the set of subgroups of the group of F-automorphisms of E. This correspondence transforms certain problems about subfields of fields into more amenable problems about subgroups of groups. Among the applications, this serves as the basis of Galois's criterion for solvability of an equation by radicals, as discussed in the next chapter, and provides a simple algebraic proof of the fundamental theorem of algebra.

1 Automorphism groups and fixed fields

Let E be an extension of a field F. We denote by $G(E/F)$ the *group of automorphisms* of E leaving each element of F fixed. The group $G(E/F)$ is also called the *group of F-automorphisms* of E. Throughout this section we confine ourselves to finite separable extensions and their groups of automorphisms. We recall that a finite separable extension E of F is simple. Thus, $E = F(\alpha)$ for some $\alpha \in E$. Let $p(x)$ be the minimal polynomial of α over F. Then $[F(\alpha):F] =$ degree of $p(x) = n$, say. Also by Lemma 4.2, Chapter 15, we get that the order of the group $G(E/F)$ is $\leq n$. Thus, we have

1.1 Theorem. *If E is a finite extension of a field F, then* $|G(E/F)| \leq [E:F]$.

1.2 Examples

(a) Consider $G = G(\mathbf{C}/\mathbf{R})$. If $a + ib \in \mathbf{C}$, $a,b \in \mathbf{R}$, and $\sigma \in G$, then $\sigma(a + ib) = \sigma(a) + \sigma(i)\sigma(b) = a + \sigma(i)b$. Also $-1 = \sigma(-1) = \sigma(i^2) = (\sigma(i))^2$. Thus, $\sigma(i) = \pm i$. Hence, $\sigma(a + ib) = a \pm ib$. Thus, there are only two possible **R**-automorphisms of **C**. Hence, $|G| = 2$. In this case $[\mathbf{C}:\mathbf{R}] = 2$ also.

(b) Consider $G = G(\mathbf{Q}(\sqrt[3]{2})/\mathbf{Q})$. $x^3 - 2$ is the minimal polynomial for $\sqrt[3]{2}$ over **Q**, so $[\mathbf{Q}(\sqrt[3]{2}):\mathbf{Q}] = 3$ and $\mathbf{Q}(\sqrt[3]{2})$ has a basis $\{1,\sqrt[3]{2},\sqrt[3]{4}\}$ over **Q**. Let $\sigma: \mathbf{Q}(\sqrt[3]{2}) \to \mathbf{Q}(\sqrt[3]{2})$ be an automorphism such that $\sigma(a) = a$ for all $a \in \mathbf{Q}$. Then

$$\begin{aligned}\sigma(a + b\sqrt[3]{2} + c\sqrt[3]{4}) &= \sigma(a) + \sigma(b)\sigma(\sqrt[3]{2}) + \sigma(c)\sigma(\sqrt[3]{4})\\ &= a + b\sigma(\sqrt[3]{2}) + c\sigma(\sqrt[3]{4}).\end{aligned}$$

Also

$$(\sigma(\sqrt[3]{2}))^3 = \sigma((\sqrt[3]{2})^3) = \sigma(2) = 2,$$

so $\sigma(\sqrt[3]{2})$ is a cube root of 2. Therefore, $\sigma(\sqrt[3]{2}) = \sqrt[3]{2}$, $\sqrt[3]{2}\omega$, or $\sqrt[3]{2}\omega^2$, where $\omega^3 = 1$, $\omega \neq 1$. Because $\sigma(\sqrt[3]{2})$ is real, the only possibility is $\sigma(\sqrt[3]{2}) = \sqrt[3]{2}$. Hence σ is the identity. Thus, $G(\mathbf{Q}(\sqrt[3]{2})/\mathbf{Q})$ is the trivial group.

Definition. *Let E be any field, and let H be a subgroup of the group of automorphisms of E. Then the set $E_H = \{x \in E \mid \sigma(x) = x \text{ for all } \sigma \in H\}$ is called the* fixed field *of the group of automorphisms H.*

It is easy to verify that E_H is a subfield of E. If E is an extension of a field F and $H < G(E/F)$ then $F \subset E_H \subset E$. For example, $\mathbf{C}_H = \mathbf{R}$, where $H = G(\mathbf{C}/\mathbf{R})$, and $\mathbf{Q}(\sqrt[3]{2})_H = \mathbf{Q}(\sqrt[3]{2})$, where $H = G(\mathbf{Q}(\sqrt[3]{2})/\mathbf{Q})$.

One of the main results in this section is

1.3 Theorem. *Let H be a finite subgroup of the group of automorphisms of a field E. Then*

$$[E:E_H] = |H|.$$

Before we prove this theorem we prove a lemma due to Dedekind.

1.4 Lemma (Dedekind). *Let F and E be fields, and let $\sigma_1,\sigma_2,\ldots,\sigma_n$ be distinct embeddings of F into E. Suppose that, for $a_1,a_2,\ldots,a_n \in E$, $\sum_{i=1}^{n} a_i\sigma_i(a) = 0$ for all $a \in F$. Then $a_i = 0$ for all $i = 1,2,\ldots,n$. (This is also* *expressed by saying that distinct embeddings of F into E are linearly independent over E.)*

Proof. Suppose there exist elements $a_1,...,a_n \in E$, not all zero, such that

$$a_1\sigma_1(x) + \cdots + a_n\sigma_n(x) = 0 \tag{1}$$

for all $x \in F$. Of all the equations of the form (1) there must be at least one for which the number of nonzero terms is least. Let

$$b_1\sigma_1(x) + \cdots + b_m\sigma_m(x) = 0, \tag{2}$$

for all $x \in F$, be such a relation, where $b_i \in E$, $i = 1,...,m$, and no $b_i = 0$. Clearly, $m > 1$.

Because $\sigma_1 \neq \sigma_m$ there exists $y \in F$ such that $\sigma_1(y) \neq \sigma_m(y)$. Replace x by yx in (2) and obtain

$$b_1\sigma_1(yx) + \cdots + b_m\sigma_m(yx) = 0$$

for all $x \in F$. But this implies

$$b_1\sigma_1(y)\sigma_1(x) + \cdots + b_m\sigma_m(y)\sigma_m(x) = 0 \tag{3}$$

for all $x \in F$. If we multiply (2) by $\sigma_1(y)$ and subtract from (3), we get

$$b_2(\sigma_1(y) - \sigma_2(y))\sigma_2(x) + \cdots + b_m(\sigma_1(y) - \sigma_m(y))\sigma_m(x) = 0.$$

The coefficient of $\sigma_m(x)$ is $b_m(\sigma_1(y) - \sigma_m(y)) \neq 0$. So we have an equation of the form (1) with fewer than m terms, a contradiction in view of the choice of relation (2). This proves the lemma. □

Proof of Theorem 1.3. Let $H = \{e = g_1,...,g_n\}$, and let $[E:E_H] = m$. If possible, suppose $m < n$. Let $\{a_1,...,a_m\}$ be a basis for E over E_H. Consider the system of m homogeneous linear equations

$$g_1(a_j)x_1 + \cdots + g_n(a_j)x_n = 0,$$

$j = 1,2,...,m$, in n unknowns $x_1,...,x_n$. Because $n > m$, this system has a nontrivial solution. So there exist $y_1,...,y_n \in E$, not all zero, such that

$$g_1(a_j)y_1 + \cdots + g_n(a_j)y_n = 0 \tag{1}$$

$j = 1,...,m$. Let a be any element of E. Then

$$a = \alpha_1 a_1 + \cdots + \alpha_m a_m,$$

where $\alpha_1,...,\alpha_m \in E_H$. So

$$g_1(a)y_1 + \cdots + g_n(a)y_n = g_1\left(\sum_{i=1}^{m} \alpha_i a_i\right)y_1 + \cdots + g_n\left(\sum_{i=1}^{m} \alpha_i a_i\right)y_n$$
$$= \sum_{i=1}^{m} \alpha_i g_1(a_i)y_1 + \cdots + \sum_{i=1}^{m} \alpha_i g_n(a_i)y_n$$
$$= \sum_{i=1}^{m} \alpha_i(g_1(a_i)y_1 + \cdots + g_n(a_i)y_n) = 0,$$

by (1).

Hence, the distinct embeddings $g_1,\ldots,g_n$ are linearly dependent, which contradicts the lemma. Therefore, $m \geq n$.

Suppose, if possible, $m > n$. Then there exists a set of $n+1$ elements of E that are linearly independent over E_H; let such a set be $\{a_1,\ldots,a_{n+1}\}$. Consider the family of n homogeneous linear equations

$$g_j(a_1)x_1 + \cdots + g_j(a_{n+1})x_{n+1} = 0, \qquad j = 1,\ldots,n,$$

in $n+1$ unknowns $x_1,\ldots,x_{n+1}$. So there exist $y_1,\ldots,y_{n+1} \in E$, not all zero, such that

$$g_j(a_1)y_1 + \cdots + g_j(a_{n+1})y_{n+1} = 0, \tag{2}$$

$j = 1,\ldots,n$. Choose $y_1,\ldots,y_{n+1}$ so that as few as possible are nonzero, and renumber so that

$$y_i \neq 0, \quad i = 1,\ldots,r \qquad \text{and} \qquad y_{r+1} = 0 = \cdots = y_{n+1}.$$

Equation (2) now becomes

$$g_j(a_1)y_1 + \cdots + g_j(a_r)y_r = 0. \tag{3}$$

Let $g \in H$ and operate on (3) with g. This gives a system of equations

$$gg_j(a_1)g(y_1) + \cdots + gg_j(a_r)g(y_r) = 0, \tag{4}$$

which is clearly equivalent to the system of equations

$$g_j(a_1)g(y_1) + \cdots + g_j(a_r)g(y_r) = 0. \tag{4$'$}$$

If we multiply relations (3) by $g(y_1)$, and (4′) by y_1, and subtract, we obtain

$$g_j(a_2)(y_2g(y_1) - g(y_2)y_1) + \cdots + g_j(a_r)(y_rg(y_1) - g(y_r)y_1) = 0, \tag{5}$$

$j = 1,\ldots,n$. (5) is a system of equations like (3) but with fewer terms, which gives a contradiction unless all the coefficients $y_ig(y_1) - y_1g(y_i)$ are equal to zero. If this happens, then $y_iy_1^{-1} = g(y_iy_1^{-1})$ for all $g \in H$. Thus, $y_iy_1^{-1} \in E_H$. Therefore, there exist $z_1,\ldots,z_r \in E_H$ such that $y_i = y_1z_i$,

$i = 1,\dots,r$. Then relation (3) with $j = 1$ becomes

$$g_1(a_1)y_1z_1 + \cdots + g_1(a_r)y_1z_r = 0.$$

This implies

$$g_1(a_1)z_1 + \cdots + g_1(a_r)z_r = 0,$$

because $y_1 \neq 0$. But then, because $z_i \in E_H$, $i = 1,\dots,r$, we get

$$g_1(a_1z_1) + \cdots + g_1(a_rz_r) = 0,$$

so

$$a_1z_1 + \cdots + a_rz_r = 0,$$

because g_1 is a nonzero embedding. But the linear independence of $a_1,\dots,a_r$ over E_H yields that $z_1 = 0 = \cdots = z_r$. This implies $y_1 = 0 = \cdots = y_r$, a contradiction. Hence, $[E:E_H] = n = |H|$, as desired. □

As an immediate consequence of Theorems 1.1 and 1.3, we have

1.5 Theorem. *Let E be a finite separable extension of a field F, and let $H < G(E/F)$. Then $G(E/E_H) = H$, and $[E:E_H] = |G(E/E_H)|$.*

Proof. Clearly, $H < G(E/E_H)$. By Theorem 1.3, $|H| = [E:E_H]$. Also by Theorem 1.1 we have

$$|H| = [E:E_H] \geq |G(E/E_H)| \geq |H|.$$

Hence, $H = G(E/E_H)$ and $[E:E_H] = |G(E/E_H)|$. This completes the proof. □

Another important consequence of Theorem 1.3 is the following characterization of normal separable extensions.

1.6 Theorem. *Let E be a finite separable extension of a field F. Then the following are equivalent:*

(i) E is a normal extension of F.
(ii) F is the fixed field of $G(E/F)$.
(iii) $[E:F] = |G(E/F)|$.

Proof. Because E is a finite separable extension of F, $E = F(\alpha)$ for some $\alpha \in E$ (Theorem 5.2, Chapter 16). Let $p(x)$ be the minimal polynomial of α over F, and let its degree be n. Then

$$[E:F] = [F(\alpha):F] = n. \tag{1}$$

Also, if E_0 is the fixed field of $G(E/F)$, then, by Theorem 1.3,

$$[E:E_0] = |G(E/F)|. \tag{2}$$

(i) $\Rightarrow$ (ii) By Lemma 4.2 in Chapter 15 the number of extensions of the inclusion mapping $F \to \overline{F}$ to the embedding $F(\alpha) \to \overline{F}$ is equal to the number of distinct roots of $p(x)$. Because E is a separable extension of F, $\alpha \in E$ is a separable element, so its minimal polynomial $p(x)$ over F must have distinct roots. So the number of distinct roots of $p(x)$ is equal to n.

In addition, because $E = F(\alpha)$ is a normal extension of F, any embedding σ: $F(\alpha) \to \overline{F}$ shall map $F(\alpha)$ onto $F(\alpha)$. Clearly, any member of $G(E/F)$ is an extension of the inclusion mapping $F \to \overline{F}$. Thus, it follows that

$$|G(E/F)| = \text{number of distinct roots of } p(x) = n. \tag{3}$$

Thus, (1)–(3) give

$$[E:F] = n = |G(E/F)| = [E:E_0].$$

Hence, $[E_0:F] = 1$, so $E_0 = F$ as desired. This proves (i) $\Rightarrow$ (ii).

(ii) $\Rightarrow$ (i) As explained earlier, $E = F(\alpha)$ for some $\alpha \in E$. Let $G(E/F) = \{\sigma_1 = \text{identity}, \sigma_2, \ldots, \sigma_n\}$. Consider the polynomial $f(x) = (x - \sigma_1(\alpha))(x - \sigma_2(\alpha)) \cdots (x - \sigma_n(\alpha))$. Now each $\sigma_i \in G(E/F)$ induces a natural homomorphism

$$\sigma_i^*: E[x] \to E[x],$$

where

$$\sigma_i^*(a_0 + a_1x + \cdots + a_mx^m) = \sigma_i(a_0) + \sigma_i(a_1)x + \cdots + \sigma_i(a_m)x^m.$$

So

$$\sigma_i^*(f(x)) = (x - (\sigma_i\sigma_1)(\alpha))(x - (\sigma_i\sigma_2)(\alpha)) \cdots (x - (\sigma_i\sigma_n)(\alpha)).$$

But since $\sigma_i\sigma_1, \sigma_i\sigma_2, \ldots, \sigma_i\sigma_n$ are distinct members of $G(E/F)$, these F-automorphisms are only a permutation of $\sigma_1, \sigma_2, \ldots, \sigma_n$. Hence, $\sigma_i^*(f(x)) = f(x)$ for all $i = 1, 2, \ldots, n$. Now by expanding $f(x)$ we have

$$f(x) = x^n - c_1x^{n-1} + c_2x^{n-2} + \cdots + (-1)^n c_n,$$

where $c_i \in E$. Therefore $\sigma_i^*(f(x)) = f(x)$ implies

$$\sigma_i(c_j) = c_j \qquad \text{for all } i, j = 1, 2, \ldots, n.$$

This gives that c_j is in the fixed field of $G(E/F)$. So, by hypothesis, $c_j \in F$, $j = 1, \ldots, n$. Hence, $f(x) \in F[x]$. Also, all the roots of $f(x)$ lie in E. Because

$E = F(\alpha)$ and α is one of the roots of $f(x)$, E is a splitting field of $f(x) \in F[x]$. This proves (ii) $\Rightarrow$ (i).

(ii) $\Rightarrow$ (iii) Follows from Theorem 1.5.

(iii) $\Rightarrow$ (ii) By equation (2) $[E:E_0] = |G(E/F)|$. Thus, by (iii), we get $[E:E_0] = [E:F]$. Hence, $E_0 = F$, proving (iii) $\Rightarrow$ (ii). This completes the proof of the theorem. □

1.7 Examples

(a) The group $G(\mathbf{Q}(\alpha)/\mathbf{Q})$, where $\alpha^5 = 1$ and $\alpha \neq 1$, is isomorphic to the cyclic group of order 4.

Solution. Clearly, $\alpha^5 - 1 = (\alpha - 1)(1 + \alpha + \alpha^2 + \alpha^3 + \alpha^4)$, so α is a root of a polynomial $p(x) = 1 + x + x^2 + x^3 + x^4 \in \mathbf{Q}[x]$. Because $p(x)$ is irreducible over $\mathbf{Q}$, $[\mathbf{Q}(\alpha):\mathbf{Q}] = 4$. Also, the roots of $x^5 - 1$ are $1,\alpha,\alpha^2,\alpha^3,\alpha^4$. So $\mathbf{Q}(\alpha)$ is the splitting field of $x^5 - 1 \in \mathbf{Q}[x]$ and, hence, a normal extension of $\mathbf{Q}$. Thus,

$$|G(\mathbf{Q}(\alpha)/\mathbf{Q})| = [\mathbf{Q}(\alpha):\mathbf{Q}] = 4.$$

Because $\{1,\alpha,\alpha^2,\alpha^3\}$ is a basis of $\mathbf{Q}(\alpha)$ over $\mathbf{Q}$, a typical element of $\mathbf{Q}(\alpha)$ is

$$a_0 + a_1\alpha + a_2\alpha^2 + a_3\alpha^3, \qquad a_i \in \mathbf{Q}.$$

The four $\mathbf{Q}$-automorphisms of $\mathbf{Q}(\alpha)$ are indeed the following:

$$\begin{aligned}
\sigma_1&: a_0 + a_1\alpha + a_2\alpha^2 + a_3\alpha^3 \mapsto a_0 + a_1\alpha + a_2\alpha^2 + a_3\alpha^3,\\
\sigma_2&: a_0 + a_1\alpha + a_2\alpha^2 + a_3\alpha^3 \mapsto a_0 + a_1\alpha^2 + a_2\alpha^4 + a_3\alpha^6\\
&\qquad = a_0 + a_1\alpha^2 + a_2\alpha^4 + a_3\alpha,\\
\sigma_3&: a_0 + a_1\alpha + a_2\alpha^2 + a_3\alpha^3 \mapsto a_0 + a_1\alpha^3 + a_2\alpha^6 + a_3\alpha^9\\
&\qquad = a_0 + a_1\alpha^3 + a_2\alpha + a_3\alpha^4,\\
\sigma_4&: a_0 + a_1\alpha + a_2\alpha^2 + a_3\alpha^3 \mapsto a_0 + a_1\alpha^4 + a_2\alpha^8 + a_3\alpha^{12}\\
&\qquad = a_0 + a_1\alpha^4 + a_2\alpha^3 + a_3\alpha^2.
\end{aligned}$$

These form a cyclic group of order 4 generated by σ_2 or σ_3.

(b) Let $E = \mathbf{Q}(\sqrt[3]{2},\omega)$, where $\omega^3 = 1$, $\omega \neq 1$. Let σ_1 be the identity automorphism of E, and let σ_2 be an automorphism of E such that $\sigma_2(\omega) = \omega^2$ and $\sigma_2(\sqrt[3]{2}) = \omega(\sqrt[3]{2})$. If $G = \{\sigma_1,\sigma_2\}$ then $E_G = \mathbf{Q}(\sqrt[3]{2}\omega^2)$.

Solution. Now $\alpha \in E_G$ if and only if $\sigma_2(\alpha) = \alpha$. Consider the following diagram of towers of fields:

$$\begin{array}{l} \mathbf{Q}(\sqrt[3]{2}, \omega) \\ \Big| \quad \text{basis } \{1, \omega\} \\ \mathbf{Q}(\sqrt[3]{2}) \\ \Big| \quad \text{basis } \{1, \sqrt[3]{2}, \sqrt[3]{4}\} \\ \mathbf{Q} \end{array}$$

Thus, $\{1,\sqrt[3]{2},\sqrt[3]{4},\omega,\omega\sqrt[3]{2},\omega\sqrt[3]{4}\}$ is a basis of $\mathbf{Q}(\sqrt[3]{2},\omega)$ over $\mathbf{Q}$. So let

$$\alpha = a + b\sqrt[3]{2} + c\sqrt[3]{4} + d\omega + e\omega\sqrt[3]{2} + f\omega\sqrt[3]{4}$$

be an element in $\mathbf{Q}(\sqrt[3]{2},\omega)$. Then

$$\begin{aligned} \sigma_2(\alpha) &= a + b\sqrt[3]{2}\omega + c\omega^2\sqrt[3]{4} + d\omega^2 + e\omega^2\omega\sqrt[3]{2} + f\omega^2\omega^2\sqrt[3]{4} \\ &= a + b\omega\sqrt[3]{2} + c(-1-\omega)\sqrt[3]{4} + d(-1-\omega) + e\sqrt[3]{2} + f\omega\sqrt[3]{4} \\ &= (a-d) + e\sqrt[3]{2} - c\sqrt[3]{4} - d\omega + b\omega\sqrt[3]{2} - (c-f)\omega\sqrt[3]{4}. \end{aligned}$$

So $\sigma_2(\alpha) = \alpha$ implies $a = a - d$, $b = e$, $c = -c$, $d = -d$, $e = b$, and $f = -c + f$. So we obtain $d = 0 = c$, $b = e$, and f, a arbitrary. Thus,

$$\begin{aligned} \alpha &= a + b\sqrt[3]{2}\,(1+\omega) + f\omega\sqrt[3]{4} \\ &= a - b\omega^2\sqrt[3]{2} + f\omega\sqrt[3]{4} \\ &= a - b\omega^2\sqrt[3]{2} + f(\omega^2\sqrt[3]{2})^2. \end{aligned}$$

Therefore, $E_G = \mathbf{Q}(\omega^2\sqrt[3]{2})$.

Problems

1. Let $E = \mathbf{Q}(\sqrt[3]{2},\omega)$ be an extension of a field $\mathbf{Q}$, where $\omega^3 = 1$, $\omega \neq 1$. For each of the following subgroups S_i of the group $G(E/\mathbf{Q})$ find E_{S_i}.

   (a) $S_1 = \{1,\sigma_2\}$, where $\sigma_2\colon \begin{cases} \sqrt[3]{2} \mapsto \sqrt[3]{2}\omega^2, \\ \omega \mapsto \omega^2. \end{cases}$

   (b) $S_2 = \{1,\sigma_3\}$, where $\sigma_3\colon \begin{cases} \sqrt[3]{2} \mapsto \sqrt[3]{2}\omega, \\ \omega \mapsto \omega^2. \end{cases}$

   (c) $S_3 = \{1,\sigma_4\}$, where $\sigma_4\colon \begin{cases} \sqrt[3]{2} \mapsto \sqrt[3]{2}, \\ \omega \mapsto \omega^2. \end{cases}$

   (d) $S_4 = \{1,\sigma_5,\sigma_6\}$, where

   $$\sigma_5\colon \begin{cases} \sqrt[3]{2} \mapsto \sqrt[3]{2}\omega, \\ \omega \mapsto \omega, \end{cases} \quad \text{and} \quad \sigma_6\colon \begin{cases} \sqrt[3]{2} \mapsto \sqrt[3]{2}\omega^2, \\ \omega \mapsto \omega. \end{cases}$$

2. Let E be a splitting field of $x^4 - 2 \in \mathbf{Q}[x]$ over $\mathbf{Q}$. Show that $G(E/\mathbf{Q})$ is isomorphic to the group of symmetries of a square.

Find subgroups of order 4 and their fixed fields. [*Hint*: $E = \mathbf{Q}(\sqrt[4]{2}, i)$. If σ is the automorphism of E given by $\sigma(\sqrt[4]{2}) = i\sqrt[4]{2}$ and $\sigma(i) = i$, then σ generates a subgroup of order 4. Also, if τ is the automorphism given by $\tau(\sqrt[4]{2}) = \sqrt[4]{2}$ and $\tau(i) = -i$, then we can describe all elements of $G(E/\mathbf{Q})$ in terms of σ and τ by means of the relation $\sigma^4 = 1 = \tau^2$, $\tau\sigma = \sigma^3\tau$. Also, there are three subgroups of order 4:

$C_4: \{1, \sigma, \sigma^2, \sigma^3\}$,
$C_{41}: \{1, \sigma^2, \tau, \sigma^2\tau\}$,
$C_{42}: \{1, \sigma^2, \sigma\tau, \sigma^3\tau\}$.

$\mathbf{Q}(i)$, $\mathbf{Q}(\sqrt{2})$, $\mathbf{Q}(i\sqrt{2})$ are respectively the fixed fields of C_4, C_{41}, C_{42}.]

2 Fundamental theorem of Galois theory

We are now ready to prove the fundamental theorem of Galois theory. First we give a couple of definitions.

Definition. *Let $f(x) \in F[x]$ be a polynomial, and let K be its splitting field over F. Then the group $G(K/F)$ of F-automorphisms of K is called the* Galois group of $f(x)$ over F.

Definition. *A finite, normal, and separable extension E of a field F is called a* Galois extension *of F.*

For example, if $f(x) \in F[x]$ is a polynomial over a field F of characteristic zero, then its splitting field E over F is a Galois extension of F.

2.1 Theorem (fundamental theorem of Galois theory). *Let E be a Galois extension of F. Let K be any subfield of E containing F. Then the mapping $K \mapsto G(E/K)$ sets up a one-to-one correspondence from the set of subfields of E containing F to the subgroups of $G(E/F)$ such that*

(i) *$K = E_{G(E/K)}$.*
(ii) *For any subgroup H of $G(E/F)$, $H = G(E/E_H)$.*
(iii) *$[E:K] = |G(E/K)|$, $[K:F] =$ index of $G(E/K)$ in $G(E/F)$.*
(iv) *K is a normal extension of F if and only if $G(E/K)$ is a normal subgroup of $G(E/F)$.*
(v) *If K is a normal extension of F, then $G(K/F) \simeq G(E/F)/G(E/K)$.*

Proof. By definition of normality, it follows that E is a normal extension of K. Thus, K is the fixed field of $G(E/K)$. This proves (i).

The proof of (ii) follows from Theorem 1.5. Note that the proof needs only the fact that E is a finite separable extension of F.

Because E is a normal extension of F and also of K, we have, by Theorem 1.6,

$$[E:F] = |G(E/F)| \quad \text{and} \quad [E:K] = |G(E/K)|.$$

Thus,

$$[E:F] = [E:K][K:F]$$

gives

$$|G(E/F)| = |G(E/K)|[K:F].$$

This proves $[K:F]$ = index of $G(E/K)$ in $G(E/F)$, as desired.

Now we proceed to prove (iv). Let $\overline{F}$ be an algebraic closure of F containing E. Recall that K is a normal extension of F if and only if each embedding $\sigma: K \to \overline{F}$, which keeps each element of F fixed, maps K onto K (Theorem 2.1, Chapter 16).

We assert that K is a normal extension of F if and only if for each $\sigma \in G(E/F)$, $\sigma(K) = K$. If K is a normal extension of F and $\sigma \in G(E/F)$, then σ restricted to K is an embedding of K into E and, hence, into $\overline{F}$. Thus, by Theorem 2.1 of Chapter 16, $\sigma(K) = K$. Conversely, let $\sigma: K \to \overline{F}$ be an embedding that keeps each element of F fixed. By Theorem 4.3, in Chapter 15, σ can be extended to $\sigma^*: E \to \overline{F}$. But then $\sigma^*(E) = E$, because E is a normal extension of F (Theorem 2.1, Chapter 16). Thus, $\sigma^* \in G(E/F)$. So, by hypothesis, $\sigma^*(K) = K$. Hence $\sigma(K) = K$ because σ^* is an extension of σ. Thus, we have shown that if $\sigma: K \to \overline{F}$ is an embedding that keeps each element of F fixed, then $\sigma(K) = K$. Therefore, by Theorem 2.1 of Chapter 16, K is a normal extension of F. This proves our assertion.

Therefore, K is a normal extension of F if and only if for all $\sigma \in G(E/F)$ and $k \in K$, $\sigma(k) \in K$. Then for all $\tau \in G(E/K)$, $\tau(\sigma(k)) = \sigma(k)$. This implies $(\sigma^{-1}\tau\sigma)(k) = k$ for all $k \in K$. Hence, $\sigma^{-1}\tau\sigma \in G(E/K)$. This proves $G(E/K)$ is a normal subgroup of $G(E/F)$. Retracing our steps, we see clearly that if $G(E/K)$ is a normal subgroup of $G(E/F)$, then $\tau(\sigma(k)) = \sigma(k)$ for all $\tau \in G(E/K)$, for all $\sigma \in G(E/F)$, and for all $k \in K$. We do know E is a normal extension of K. Hence, K is the fixed field of $G(E/K)$. Thus, $\tau(\sigma(k)) = \sigma(k)$ implies $\sigma(k) \in K$. Therefore, K is a normal extension of F, as desired.

Finally, we prove (v). Let K be a normal extension of F. By the preceding discussion, for all $\sigma \in G(E/F)$, $\sigma(K) = K$. Thus, σ induces an automorphism σ^* of K defined by $\sigma^*(k) = \sigma(k)$, $k \in K$. Clearly, $\sigma^* \in G(K/F)$. Consider the mapping $f: G(E/F) \to G(K/F)$ defined by $f(\sigma) = \sigma^*$. Let $\sigma_1, \sigma_2 \in G(E/F)$. Then

$$(\sigma_1^*\sigma_2^*)(k) = \sigma_1^*(\sigma_2^*(k)) = \sigma_1^*(\sigma_2(k)) = (\sigma_1\sigma_2)(k).$$

Therefore, $(\sigma_1\sigma_2)^* = \sigma_1^*\sigma_2^*$. Thus, f is a homomorphism of $G(E/F)$ into $G(K/F)$. Now $\operatorname{Ker} f = \{\sigma \in G(E/F) | \sigma^* = \text{identity}\}$. But $\sigma^* = \text{identity}$ if and only if $\sigma^*(k) = k$ for all $k \in K$. That is, $\sigma(k) = k$ for all $k \in K$, so $\sigma \in G(E/K)$. Hence, $\operatorname{Ker} f = G(E/K)$. Then by the fundamental theorem of homomorphisms

$$\frac{G(E/F)}{G(E/K)} \simeq \operatorname{Im} f \subset G(K/F). \tag{1}$$

Further, by (iii), we have

$$\left|\frac{G(E/F)}{G(E/K)}\right| = [K:F]. \tag{2}$$

Also, because K is normal over F,

$$|G(K/F)| = [K:F]. \tag{3}$$

Hence, by (1)–(3), we get

$$\frac{G(E/F)}{G(E/K)} \simeq G(K/F).$$

This completes the proof of the theorem. □

2.2 Examples

(a) If $f(x) \in F[x]$ has r distinct roots in its splitting field E over F, then the Galois group $G(E/F)$ of $f(x)$ is a subgroup of the symmetric group S_r.

Solution. Let $f(x) = a_0 + a_1x + \cdots + a_nx^n \in F[x]$ have r distinct roots $\alpha_1, \alpha_2, \ldots, \alpha_r$. Clearly, for each $\sigma \in G(E/F)$, $\sigma(\alpha_i)$ is again a root of $f(x)$. Also, if $\alpha_i \neq \alpha_j$, then $\sigma(\alpha_i) \neq \sigma(\alpha_j)$. Thus, $\sigma(\alpha_1), \sigma(\alpha_2), \ldots, \sigma(\alpha_r)$ is a permutation of $\alpha_1, \ldots, \alpha_r$. Let us set $\phi_\sigma(\alpha_i) = \sigma(\alpha_i)$, $i = 1, \ldots, r$. Then $\phi_\sigma \in S_r$, and we have a mapping $f: G(E/F) \to S_r$ given by $f(\sigma) = \phi_\sigma$. If $\sigma, \eta \in G(E/F)$, then

$$\phi_{\sigma\eta}(\alpha_i) = (\sigma\eta)(\alpha_i) = \sigma(\eta(\alpha_i)) = \phi_\sigma\phi_\eta(\alpha_i)$$

for all α_i. Hence, $\phi_{\sigma\eta} = \phi_\sigma\phi_\eta$. This shows that f is a homomorphism.

To show f is 1-1, let ϕ_σ = identity. Then $\phi_\sigma(\alpha_i) = \alpha_i$; that is, $\sigma(\alpha_i) = \alpha_i$ for all $i = 1,\dots,r$. Thus, σ = identity, since $E = F(\alpha_1,\dots,\alpha_r)$. Hence, f is an embedding of $G(E/F)$ into S_r.

(b) Let F be field of characteristic $\neq 2$. Let $x^2 - a \in F[x]$ be an irreducible polynomial over F. Then its Galois group is of order 2.

Solution. Clearly, if α is one root of $x^2 - a$, then $-\alpha$ is the other root. So $\alpha \neq -\alpha$ because char $F \neq 2$. Thus, $x^2 - a$ is separable over F. The splitting field $F(\alpha)$ of $x^2 - a$ over F is a finite, separable, and normal extension of degree 2 over F. Thus, $|G(F(\alpha)/F)| = 2$.

(c) Let F be a field of characteristic $\neq 2$ or 3. Let $f(x) = x^3 + bx + c$ be a separable polynomial over F. If $f(x)$ is irreducible over F, then the Galois group of $f(x)$ is of order 3 or 6. Also, the Galois group of $f(x)$ is S_3 if and only if $\Delta = -4b^3 - 27c^2$ is not a square in F; that is, there does not exist any element $\alpha \in F$ such that $\alpha^2 = \Delta$.

Solution. If $f(x)$ has a root $\alpha \in F$, then $f(x) = (x - \alpha)g(x)$, where $g(x) \in F[x]$. In case $g(x)$ has a root in F, then $f(x)$ splits into linear factors in F itself, so the Galois group of $f(x)$ is of order 1. In case $g(x)$ is irreducible over F, the splitting field E of $f(x)$ over F is an extension of degree 2, so $|G(E/F)| = [E:F] = 2$. Thus, if $f(x)$ is irreducible over F, then $|G(E/F)| \neq 1$ or 2.

Because $f(x)$ is also separable, all its roots are distinct. Hence, by Example (a), $G(E/F) < S_3$, where E is splitting field of $f(x)$ over F. Let $G = G(E/F)$. Then $|G| = 3$ or 6, for otherwise $f(x)$ is reducible, as discussed earlier.

Now we find necessary and sufficient conditions on the coefficients of $f(x)$ such that G is isomorphic to S_3. If α_1, α_2, and α_3 are the distinct roots of $f(x)$, we let $\delta = (\alpha_1 - \alpha_2)(\alpha_2 - \alpha_3)(\alpha_3 - \alpha_1)$ and $\Delta = \delta^2$.

Then for all $\sigma \in G$, $\sigma(\delta) = \pm\delta$. Hence, $\sigma(\Delta) = \Delta$. Thus, Δ is in the fixed field of G; that is, $\Delta \in F$. Now if $\delta \in F$, then $\sigma(\delta) = \delta$ for all $\sigma \in G$. Thus, σ cannot be an odd permutation. Hence, $\sigma \in A_3$, the subgroup of even permutations of S_3. Conversely, if $\sigma \in A_3$, then $\sigma(\delta) = \delta$. Hence, $G = A_3$ if and only if $\delta \in F$. Thus, $G = S_3$ if and only if $\delta \notin F$; that is, the polynomial $x^2 - \Delta \in F[x]$ is irreducible over F (since $\Delta = \delta^2 \in F$). By actual computation we have

$$\Delta = -4b^3 - 27c^2.$$

Thus, the Galois group of $f(x)$ is S_3 if and only if $-4b^3 - 27c^2$ is not a square in F.

(d) The Galois group of $x^4 - 2 \in \mathbf{Q}[x]$ is the octic group (= group of symmetries of a square).

Solution. $x^4 - 2$ is irreducible over $\mathbf{Q}$, and $x^4 - 2 = (x - \sqrt[4]{2})(x + \sqrt[4]{2})(x + i\sqrt[4]{2})(x - i\sqrt[4]{2}) \in \mathbf{Q}(\sqrt[4]{2},i)[x]$. Therefore, $E = \mathbf{Q}(\sqrt[4]{2},i)$ is the splitting field of $x^4 - 2$ over $\mathbf{Q}$. By Example 1.3(d) in Chapter 16, $[E:\mathbf{Q}] = 8$. Also, E is a normal separable extension of $\mathbf{Q}$. This gives $|G(E/\mathbf{Q})| = [E:\mathbf{Q}] = 8$. If $\sigma \in G(E/\mathbf{Q})$ and $\beta \in E$, then

$$\beta = a_0 + a_1\sqrt[4]{2} + a_2(\sqrt[4]{2})^2 + a_3(\sqrt[4]{2})^3 + a_4(i) + a_5(i\sqrt[4]{2}) + a_6(i(\sqrt[4]{2})^2) + a_7(i(\sqrt[4]{2})^3).$$

So we have

$$\sigma(\beta) = a_0 + a_1\sigma(\sqrt[4]{2}) + a_2\sigma(\sqrt[4]{2})^2 + a_3\sigma(\sqrt[4]{2})^3 + a_4\sigma(i) + a_5\sigma(i)\sigma(\sqrt[4]{2}) + a_6\sigma(i)\sigma(\sqrt[4]{2})^2 + a_7\sigma(i)\sigma(\sqrt[4]{2})^3.$$

Therefore, σ is determined by its effect on i and on $\sqrt[4]{2}$. Because $\sigma(i)$ must be i or $-i$; because $\sigma(\sqrt[4]{2})$ must be $\sqrt[4]{2}$, $-\sqrt[4]{2}$, $i\sqrt[4]{2}$, or $-i\sqrt[4]{2}$; and because there are eight elements in $G(E/\mathbf{Q})$, it follows that $G(E/\mathbf{Q}) = \{\sigma_1,\sigma_2,\ldots,\sigma_8\}$, where

$$\sigma_1: \begin{array}{l}\sqrt[4]{2}\mapsto\sqrt[4]{2}\\ i\mapsto i,\end{array} \quad \sigma_3: \begin{array}{l}\sqrt[4]{2}\mapsto-\sqrt[4]{2}\\ i\mapsto i,\end{array} \quad \sigma_5: \begin{array}{l}\sqrt[4]{2}\mapsto\sqrt[4]{2}\\ i\mapsto i,\end{array}$$

$$\sigma_7: \begin{array}{l}\sqrt[4]{2}\mapsto-\sqrt[4]{2},\\ i\mapsto -i,\end{array} \quad \sigma_2: \begin{array}{l}\sqrt[4]{2}\mapsto i\sqrt[4]{2}\\ i\mapsto i,\end{array} \quad \sigma_4: \begin{array}{l}\sqrt[4]{2}\mapsto-i\sqrt[4]{2}\\ i\mapsto i,\end{array}$$

$$\sigma_6: \begin{array}{l}\sqrt[4]{2}\mapsto i\sqrt[4]{2}\\ i\mapsto -i,\end{array} \quad \sigma_8: \begin{array}{l}\sqrt[4]{2}\mapsto-i\sqrt[4]{2}\\ i\mapsto -i,\end{array}$$

Let $\alpha_i = \sqrt[4]{2}$, $\alpha_2 = i\sqrt[4]{2}$, $\alpha_3 = -\sqrt[4]{2}$, $\alpha_4 = -i\sqrt[4]{2}$.

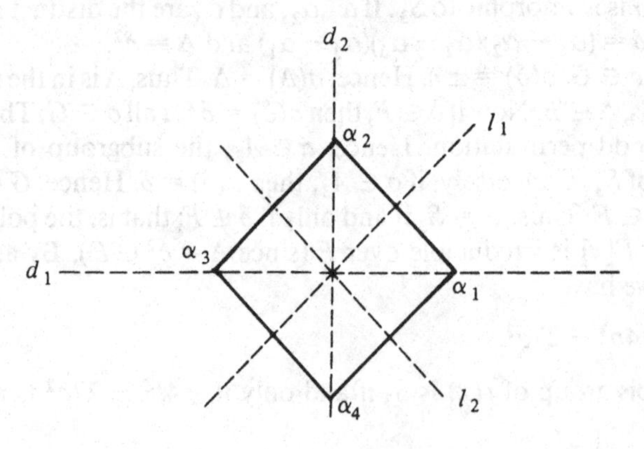

Then the elements of $G(E/\mathbf{Q})$ permute the roots $\alpha_1, \alpha_2, \alpha_3$, and α_4 of $x^4 - 2$ as follows (see the diagram):

σ_1: $0°$ rotation; σ_5: reflection about d_1;
σ_2: $90°$ rotation; σ_6: reflection about l_1;
σ_3: $180°$ rotation; σ_7: reflection about d_2;
σ_4: $270°$ rotation; σ_8: reflection about l_2.

(e) Illustrate the 1-1 correspondence between subgroups and subfields by using the Galois group of $x^4 - 2 \in \mathbf{Q}[x]$.

Solution. We saw in the solution to Example (d) that $E = \mathbf{Q}(\sqrt[4]{2}, i)$ is the splitting field of $x^4 - 2$ over $\mathbf{Q}$ and that the Galois group G of $x^4 - 2$ is the octic group. In the solution of that problem, σ_1 is the identity automorphism ϵ. Set $\sigma_2 = \sigma$; then $\sigma_3 = \sigma^2$, $\sigma_4 = \sigma^3$, and $\sigma^4 = \epsilon$. Next set $\sigma_5 = \tau$; then $\tau^2 = \epsilon$, $\sigma_6 = \sigma\tau$, $\sigma_7 = \sigma^2\tau$, and $\sigma_8 = \sigma^3\tau$. Finally, we have $\tau\sigma = \sigma^3\tau$.

The group G has four normal subgroups other than itself and the identity subgroup: $N_1 = \{\epsilon, \sigma\tau, \sigma^2, \sigma^3\tau\}$, $N_2 = \{\epsilon, \sigma, \sigma^2, \sigma^3\}$, $N_3 = \{\epsilon, \tau, \sigma^2, \sigma^2\tau\}$, and $N_4 = \{\epsilon, \sigma^2\}$, and four nonnormal subgroups: $H_1 = \{\epsilon, \sigma^3\tau\}$, $H_2 = \{\epsilon, \sigma\tau\}$, $H_3 = \{\epsilon, \sigma^2\tau\}$, and $H_4 = \{\epsilon, \tau\}$. The inclusion relations between these subgroups are shown in the following diagram:

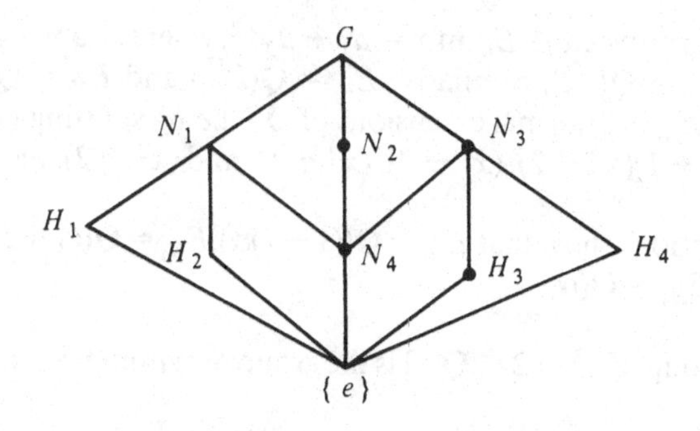

We now describe the fixed fields under these subgroups. We begin with E_{N_4}, the fixed field of the subgroup N_4. Let $\alpha = \sqrt[4]{2}$ and let

$$x = a_1 + a_2\alpha + a_3\alpha^2 + a_4\alpha^3 + a_5 i + a_6 i\alpha + a_7 i\alpha^2 + a_8 i\alpha^3$$

be any element of $\mathbf{Q}(\alpha, i)$. Then $x \in E_{N_4}$ iff $x = \sigma^2(x)$. Because $\sigma(\alpha) = i\alpha$ and $\sigma(i) = i$, we have

$$\begin{aligned}\sigma(x) &= a_1 + a_2 i\alpha - a_3\alpha^2 - a_4 i\alpha^3 + a_5 i - a_6\alpha - a_7 i\alpha^2 + a_8\alpha^3 \\ &= a_1 - a_6\alpha - a_3\alpha^2 + a_8\alpha^3 + a_5 i + a_2 i\alpha - a_7 i\alpha^2 - a_4 i\alpha^3, \\ \sigma^2(x) &= a_1 - a_2\alpha + a_3\alpha^2 - a_4\alpha^3 \\ &\quad + a_5 i - a_6 i\alpha + a_7 i\alpha^2 - a_8 i\alpha^3.\end{aligned}$$

Hence, $x = \sigma^2(x)$ if and only if $a_2 = -a_2$, $a_4 = -a_4$, $a_6 = -a_6$, $a_8 = -a_8$. So $a_2 = a_4 = a_6 = a_8 = 0$, and a_1, a_3, a_5, and a_7 are arbitrary. Therefore,

$$x = a_1 + a_3\alpha^2 + a_5 i + a_7 i\alpha^2.$$

Thus, $E_{N_4} = \mathbf{Q}(\sqrt{2}, i)$.

Next, to find E_{N_3}, we note that $x \in E_{N_3}$ iff $x = \tau(x) = \sigma^2(x) = \sigma^2\tau(x)$. Let

$$x = a_1 + a_2\alpha + a_3\alpha^2 + a_4\alpha^3 + a_5 i + a_6 i\alpha + a_7 i\alpha^2 + a_8 i\alpha^3.$$

Because $x = \sigma^2(x)$ implies $a_2 = a_4 = a_6 = a_8 = 0$, we have

$$x = a_1 + a_3\alpha^2 + a_5 i + a_7 i\alpha^2.$$

Further, $\tau(\alpha) = \alpha$ and $\tau(i) = -i$. So

$$\tau(x) = a_1 + a_3\alpha^2 - a_5 i - a_7 i\alpha^2.$$

Therefore, $x = \tau(x)$ implies $a_5 = 0 = a_7$, yielding

$$x = \tau(x) = a_1 + a_3\alpha^2.$$

Applying σ^2 we obtain

$$\sigma^2\tau(x) = a_1 + a_3\alpha^2,$$

since $\sigma(\alpha) = \alpha$. This proves $x \in E_{N_3}$ iff $x = a_1 + a_3\alpha^2$, where a_1 and a_3 are arbitrary. Thus, $E_{N_3} = \mathbf{Q}(\sqrt{2})$. Similarly, $E_{N_1} = \mathbf{Q}(i\sqrt{2})$ and $E_{N_2} = \mathbf{Q}(i)$. $E_{N_4}, E_{N_3}, E_{N_2}$, and E_{N_1} are normal extensions of $\mathbf{Q}$ – being splitting fields of polynomials $(x^2 + 1)(x^2 - 2)$, $(x^2 - 2)$, $(x^2 + 1)$, and $(x^2 + 2)$, respectively.

Similar computations show that $E_{H_1} = \mathbf{Q}((1 - i)\alpha)$, $E_{H_2} = \mathbf{Q}((1 + i)\alpha)$, $E_{H_3} = \mathbf{Q}(i\alpha)$, and $E_{H_4} = \mathbf{Q}(\alpha)$.

(f) The Galois group of $x^3 - 2 \in \mathbf{Q}[x]$ is the group of symmetries of the triangle.

Solution. The roots of equation $x^3 - 2 = 0$ are $\sqrt[3]{2}$, $\omega\sqrt[3]{2}$, and $\omega^2\sqrt[3]{2}$, where ω is a root of the irreducible polynomial $x^2 + x + 1$ over $\mathbf{Q}(\sqrt[3]{2})$. Thus, we write

$$x^3 - 2 = (x - \sqrt[3]{2})(x - \omega\sqrt[3]{2})(x - \omega^2\sqrt[3]{2}),$$

whence $E = \mathbf{Q}(\sqrt[3]{2}, \omega)$. It follows that $[E:\mathbf{Q}] = 6$. There are then exactly six automorphisms of E, for E is a splitting field and therefore normal. These automorphisms are determined by the manner in which they transform the roots of the above equation. The root $\sqrt[3]{2}$ can have only three images, ω only two. There are six possible combinations, and because there are six automorphisms, all combinations occur. The automorphism group is given by the table

	I	σ	σ^2	τ	$\sigma\tau$	$\sigma^2\tau$
$\sqrt[3]{2}$	$\sqrt[3]{2}$	$\omega\sqrt[3]{2}$	$\omega^2\sqrt[3]{2}$	$\sqrt[3]{2}$	$\omega\sqrt[3]{2}$	$\omega^2\sqrt[3]{2}$
ω	ω	ω	ω	ω^2	ω^2	ω^2

(g) The Galois group of $x^4+1 \in \mathbf{Q}[x]$ is the Klein four-group.

Solution. Let $E=\mathbf{Q}(\alpha)$, where $\alpha = e^{\pi i/4}$. Then because the roots of x^4+1 are α,α^3,α^5, and α^7, E is the splitting field of x^4+1 over $\mathbf{Q}$. Because x^4+1 is irreducible over $\mathbf{Q}$, $[E:\mathbf{Q}]=4$. Also the characteristic of $\mathbf{Q}$ is zero. So E is a normal separable extension of $\mathbf{Q}$. Thus, $|G(E/\mathbf{Q})| = [E:\mathbf{Q}]=4$. If $\sigma \in G(E/\mathbf{Q})$ and $\beta \in E$, then $\beta = a_0 + a_1\alpha + a_2\alpha^2 + a_3\alpha^3$ and $\sigma(\beta) = a_0 + a_1\sigma(\alpha) + a_2\sigma(\alpha)^2 + a_3\sigma(\alpha)^3$. Hence, σ is determined by its effect on α. Because $\sigma(\alpha)$ must be a root of x^4+1, and because there are four elements in $G(E/\mathbf{Q})$, it follows that $G(E/\mathbf{Q}) = \{\sigma_1,\sigma_3,\sigma_5,\sigma_7\}$, where $\sigma_1(\alpha)=\alpha$, $\sigma_3(\alpha)=\alpha^3$, $\sigma_5(\alpha)=\alpha^5$, and $\sigma_7(\alpha)=\alpha^7$. Note that $\sigma_3(\sigma_3(\alpha)) = \alpha^9 = \alpha$, $\sigma_5(\sigma_5(\alpha)) = \alpha^{25} = \alpha$, and $\sigma_7(\sigma_7(\alpha)) = \alpha^{49} = \alpha$; so $\sigma_3^2 = \sigma_1 =$ identity, $\sigma_5^2 = \sigma_1$, and $\sigma_7^2 = \sigma_1$. Therefore, $G(E/\mathbf{Q})$ is the Klein four-group (every element except the identity has order 2) $\simeq C_2 \times C_2$.

(h) Let n be a positive integer, and let F be a field containing all the nth roots of unity. Let K be the splitting field of $x^n - a \in F[x]$ over F. Then $K = F(\alpha)$, where α is any root of $x^n - a$, and the Galois group $G(K/F)$ is abelian.

Solution. If $\omega = \cos(2\pi/n) + i\sin(2\pi/n)$ and α is any root of $x^n - a$, then $\alpha,\alpha\omega,\ldots,\alpha\omega^{n-1}$ are all the roots of $x^n - a$. Thus, the splitting field K of $x^n - a$ over F is $F(\alpha)$. To show $G(K/F)$ is abelian, we let $\sigma_1,\sigma_2 \in G(K/F)$. Because α is root of $x^n - a$, $\sigma_1(\alpha)$ and $\sigma_2(\alpha)$ are also roots of $x^n - a$. So we may now write

$$\sigma_1(\alpha) = \alpha\omega^i \quad \text{and} \quad \sigma_2(\alpha) = \alpha\omega^j$$

for $0 \le i, j \le n-1$. Then

$$(\sigma_1\sigma_2)(\alpha) = \sigma_1(\sigma_2(\alpha)) = \sigma_1(\alpha\omega^j) = \sigma_1(\alpha)\omega^j = \alpha\omega^{i+j}.$$

Similarly, $(\sigma_2\sigma_1)(\alpha) = \alpha\omega^{i+j}$. Hence, $\sigma_1\sigma_2 = \sigma_2\sigma_1$, because $K = F(\alpha)$. Thus, $G(K/F)$ is abelian.

Problems

1. Find the Galois groups $G(K/\mathbf{Q})$ of the following extensions K of $\mathbf{Q}$:
   (a) $K = \mathbf{Q}(\sqrt{3}, \sqrt{5})$.
   (b) $K = \mathbf{Q}(\alpha)$, where $\alpha = \cos(2\pi/3) + i\sin(2\pi/3)$.
   (c) K is the splitting field of $x^4 - 3x^2 + 4 \in \mathbf{Q}[x]$.
2. In Examples 2.2(f) and 2.2(g) find the subgroups of $G(E/\mathbf{Q})$ and the corresponding fixed fields.
3. Let $u \in \mathbf{R}$ and let $\mathbf{Q}(u)$ be a normal extension of $\mathbf{Q}$ such that $[\mathbf{Q}(u):\mathbf{Q}] = 2^m$, where $m \geq 0$. Show that there exist intermediate fields K_i such that
   $$K_0 = \mathbf{Q} \subset K_1 \subset K_2 \subset \cdots \subset K_m = \mathbf{Q}(u),$$
   where $[K_i : K_{i-1}] = 2$.

3 Fundamental theorem of algebra

In this section we given an application of the fundamental theorem of Galois theory to prove that the field of complex numbers is algebraically closed.

3.1 Theorem (fundamental theorem of algebra). *Every polynomial $f(x) \in \mathbf{C}[x]$ factors into linear factors in $\mathbf{C}[x]$.*

Proof. Let $f(x) = a_0 + a_1x + \cdots + a_nx^n \in \mathbf{C}[x]$. Let

$$g(x) = (x^2+1)(a_0 + a_1x + \cdots + a_nx^n)(\bar{a}_0 + \bar{a}_1x + \cdots + \bar{a}_nx^n).$$

Then $g(x) \in \mathbf{R}[x]$. Let E be the splitting field of $g(x)$ over $\mathbf{R}$. Then $\mathbf{R} \subset \mathbf{C} \subset E$. We prove $E = \mathbf{C}$.

First, we assert that there does not exist a subfield K of E containing $\mathbf{C}$ such that $[K:\mathbf{C}] = 2$. Suppose such a subfield exists. Because K is a finite separable extension of $\mathbf{C}$, $K = \mathbf{C}(\alpha)$ for some $\alpha \in K$. If $p(x) \in \mathbf{C}[x]$ is the minimal polynomial of α over $\mathbf{C}$, then $\deg p(x) = 2$. Suppose $p(x) = x^2 + 2ax + b$. Then

$$p(x) = (x+a)^2 - (a^2 - b) = (x + a - \sqrt{a^2-b})(x + a + \sqrt{a^2-b}),$$

which is a contradiction, because $p(x)$ is irreducible over $\mathbf{C}$ (observe that if $c \in \mathbf{C}$, then $\sqrt{c} \in \mathbf{C}$). This proves our assertion.

Let $G = G(E/\mathbf{R})$ be the Galois group of $g(x)$ over $\mathbf{R}$. Let $|G| = 2^mq$, where q is an odd integer. Let H be the 2-Sylow subgroup of G, and let L be the corresponding subfield of E. We exhibit the foregoing considerations in the following diagram:

$$\begin{aligned} E &\longleftrightarrow \{e\} \\ L &\longleftrightarrow H \\ \mathbf{C} &\longleftrightarrow G(E/\mathbf{C}) \\ \mathbf{R} &\longleftrightarrow G(E/\mathbf{R}). \end{aligned}$$

Then $[E:L] = 2^m$, so $[L:\mathbf{R}] = q$. Also because L is a finite separable extension of $\mathbf{R}$, $L = \mathbf{R}(\beta)$ for some $\beta \in L$. Thus, the minimal polynomial

$$q(x) = b_0 + b_1x + \cdots + b_qx^q \in \mathbf{R}[x]$$

of β is of odd degree q. But this is impossible unless $q = 1$, because by the intermediate value theorem in calculus every equation of odd degree over the reals has a real root. Hence, $q = 1$, so $[E:\mathbf{R}] = 2^m$. This implies that $[E:\mathbf{C}] = 2^{m-1}$. Then the subgroup $G(E/\mathbf{C})$ of $G(E/\mathbf{R})$ is of order 2^{m-1}. If $m > 1$, let H' be the subgroup of $G(E/\mathbf{C})$ of order 2^{m-2}, and let L' be the corresponding subfield of E. Clearly, $L' \supseteq \mathbf{C}$. Then $[E:L'] = 2^{m-2}$, so $[L':\mathbf{C}] = 2$, a contradiction of the assertion proved in the beginning. Hence, $m = 1$. Then $[E:\mathbf{R}] = 2$ with $E \supseteq \mathbf{C}$ implies $E = \mathbf{C}$, as claimed. □

CHAPTER 18

Applications of Galois theory to classical problems

1 Roots of unity and cyclotomic polynomials

Definition. *Let E be a field, and let n be a positive integer. An element $\omega \in E$ is called a primitive nth root of unity in E if $\omega^n = 1$, but $\omega^m \neq 1$ for any positive integer $m < n$.*

Note that the complex numbers satisfying $x^n = 1$ form a finite subgroup H of the multiplicative group of the nonzero elements of the field $\mathbf{C}$ of complex numbers. Also H is a cyclic group generated by any primitive nth root ω of unity. There are exactly $\phi(n)$ primitive nth roots of unity for each positive integer n. These are $\cos(2k\pi)/n + i \sin(2k\pi)/n$, where k is a positive integer less than n and prime to n.

1.1 Theorem. *Let F be a field, and let U be a finite subgroup of the multiplicative group $F^* = F - \{0\}$. Then U is cyclic.*

In particular, the roots of $x^n - 1 \in F[x]$ form a cyclic group.

Proof. By Lemma 3.3 of Chapter 8, $U = S(p_1) \times \cdots \times S(p_k)$, where $|S(p_i)| = p_i^{r_i}$, and $p_1, \ldots, p_k$ are distinct primes. We proceed to show that each $S(p_i)$ is cyclic. Let $a \in S(p_i)$ be such that $o(a)$ is maximal, say $p_i^{s_i}$. Because $o(a) | p_i^{r_i}$, we have $s_i \leq r_i$. Further, for each $x \in S(p_i)$, $o(x) = p_i^{t_i} \leq p_i^{s_i}$. Therefore, $x^{p_i^{t_i}} = 1$ implies $x^{p_i^{s_i}} = 1$. Hence, for all $x \in S(p_i)$, $x^{p_i^{s_i}} = 1$. Because the equation $x^{p_i^{s_i}} = 1$ has at most $p_i^{s_i}$ roots, it follows that $p_i^{s_i} \geq p_i^{r_i}$. But this yields $s_i = r_i$, since $p_i^{r_i} \geq p_i^{s_i}$. Therefore, $o(a) = p_i^{r_i} = |S(p_i)|$, and, hence, $S(p_i)$ is a cyclic group generated by a. We also know that if A and B are cyclic groups of orders m and n, respectively, with $(m,n) = 1$, then $A \times B$ is again cyclic [Example 4.5(a), Chapter 4]. Hence, U is cyclic. □

1.2 Theorem. *Let F be a field and let n be a positive integer. Then there exists a primitive nth root of unity in some extension E of F if and only if either char $F = 0$ or char $F \nmid n$.*

Proof. Let $f(x) = x^n - 1 \in F[x]$, and let char $F = 0$ or char $F \nmid n$. Then $f'(x) = nx^{n-1} \neq 0$. Thus, $f(x)$ has n distinct roots (in its splitting field E over F), and they clearly form a group, say H. So by Theorem 1.2 this group H, consisting of the n distinct roots of $x^n - 1$, is a cyclic group. Now if $\omega \in H$ is a generator of H, then $\omega^n = 1$, but $\omega^m \neq 1$ for any positive integer $m < n$. Hence, ω is a primitive nth root of unity in an extension field E of F.

Conversely, let ω be a primitive nth root of unity in some extension field E of F. Then $1, \omega, \omega^2, \ldots, \omega^{n-1}$ are clearly n distinct roots of $f(x) = x^n - 1$. So $f(x)$ does not possess multiple roots. But then $f'(x) = nx^{n-1} \neq 0$. This implies either char $F = 0$ or char $F \nmid n$. □

Definition. *Let n be a positive integer, and let F be a field of characteristic zero or characteristic $p \nmid n$. Then the polynomial $\Phi_n(x) = \Pi_\omega(x - \omega)$, where the product runs over all the primitive nth roots ω of unity (i.e., the primitive nth root of $x^n - 1$ over F) is called the* nth cyclotomic polynomial.

For example, $\Phi_1(x) = x - 1$, $\Phi_2(x) = x + 1$, $\Phi_3(x) = x^2 + x + 1$, $\Phi_4(x) = x^2 + 1$, $\Phi_5(x) = x^4 + x^3 + x^2 + x + 1$, $\Phi_6(x) = x^2 - x + 1$.

1.3 Theorem. $\Phi_n(x) = \Pi_\omega(x - \omega)$, *$\omega$ primitive nth root in* **C**, *is an irreducible polynomial of degree $\phi(n)$ in* **Z**$[x]$.

Proof. Let E be the splitting field of $x^n - 1 \in \mathbf{Q}[x]$. Because the characteristic of **Q** is zero, E is a finite, separable, and normal extension of **Q**. So **Q** is the fixed field of $G(E/\mathbf{Q})$. Because for any primitive nth root ω of unity and any $\sigma \in G(E/\mathbf{Q})$, $\sigma(\omega)$ is again a primitive nth root of unity, the induced mapping σ^*: $E[x] \rightarrow E[x]$ keeps $\Phi_n(x)$ unaltered. Thus, the coefficients of $\Phi_n(x)$ lie in the fixed field of $G(E/\mathbf{Q})$; that is, $\Phi_n(x) \in \mathbf{Q}[x]$.

But then the facts that $\Phi_n(x)$ is a factor of $x^n - 1$ and $\Phi_n(x)$ is monic give $\Phi_n(x) \in \mathbf{Z}[x]$. Because the number of primitive nth roots of unity is $\phi(n)$, this also shows that $\Phi_n(x)$ is of degree $\phi(n)$.

We now show that $\Phi_n(x)$ is irreducible over $\mathbf{Z}$. Let $f(x) \in \mathbf{Z}[x]$ be an irreducible factor of $\Phi_n(x)$, and let ω be a root of $f(x)$. Of course, ω is a primitive nth root of unity. We shall prove that if p is a prime such that p does not divide n, then ω^p is also a root of $f(x)$. First note that ω^p is also a primitive nth root of unity. This follows from the fact that ω^p is also a generator of the cyclic group consisting of the roots of $x^n - 1$.

Because $f(x) \in \mathbf{Z}[x]$ is a factor of $\Phi_n(x)$, there exists $h(x) \in \mathbf{Z}[x]$ such that

$$\Phi_n(x) = f(x)h(x).$$

So if ω^p is not a root of $f(x)$, it must be a root of $h(x)$. Thus, ω is a root of $h(x^p)$. So $f(x)$ and $h(x^p)$ have a common factor over some extension of $\mathbf{Q}$. But this implies $f(x)$ and $h(x^p)$ have a common factor over $\mathbf{Q}$. (To prove this, use Euclid's division algorithm.) Because $f(x)$ is irreducible over $\mathbf{Z}$ and also over $\mathbf{Q}$, we get $f(x)$ divides $h(x^p)$. Write $h(x^p) = f(x)g(x)$. Because $f(x)$ and $h(x^p)$ are monic polynomials over $\mathbf{Z}$, $g(x)$ is also a monic polynomial over $\mathbf{Z}$. Let us denote by $\bar{f}(x)$ and $\bar{h}(x)$, respectively, the polynomials $f(x)$ (mod p) and $h(x)$ (mod p) – that is, polynomials obtained from $f(x)$ and $h(x)$ by replacing their coefficients $a \in \mathbf{Z}$ with $\bar{a} \in \mathbf{Z}/(p)$. Because $a^p \equiv a \pmod{p}$ for all integers a, we get

$$\bar{h}(x^p) = (\bar{h}(x))^p,$$

so $\overline{h(x^p)} = \overline{f(x)g(x)}$ gives that $\bar{h}(x)$ and $\bar{f}(x)$ have a common factor. Thus, from $\bar{\Phi}_n(x) = \bar{f}(x)\bar{h}(x)$ and $\Phi_n(x)|(x^n - 1)$, we get that $x^n - \bar{1}$ has multiple roots. But this is impossible. For if α is a multiple root, then the derivative of $x^n - \bar{1}$ should vanish at $x = \alpha$; that is, $\bar{n}\alpha^{n-1} = 0$. This implies $\alpha^{n-1} = 0$, because the characteristic p does not divide n; so $\alpha = 0$. Because $\alpha = 0$ is not a root of $x^n - 1$, we get a contradiction. Hence, we have shown that if ω is a root of $f(x)$, then ω^p is also a root of $f(x)$. Because any primitive nth root of unity can be obtained by raising ω to a succession of prime powers, with primes not dividing n, this implies that all primitive nth roots of unity are roots of $f(x)$. Hence, $f(x) = \Phi_n(x)$. Thus, $\Phi_n(x)$ is irreducible over $\mathbf{Z}$. □

1.4 Theorem. *Let ω be a primitive nth root of unity in* $\mathbf{C}$. *Then* $\mathbf{Q}(\omega)$ *is the splitting field of $\Phi_n(x)$ and also of $x^n - 1 \in \mathbf{Q}[x]$. Further, $[\mathbf{Q}(\omega):\mathbf{Q}] = \phi(n) = |G(\mathbf{Q}(\omega)/\mathbf{Q})|$ and $G(\mathbf{Q}(\omega)/\mathbf{Q}) \simeq (\mathbf{Z}/(n))^*$, the multiplicative group formed by the units of* $\mathbf{Z}/(n)$.

Proof. The minimal polynomial of ω is $\Phi_n(x)$ by Theorem 1.3, and because $\mathbf{Q}(\omega)$ contains a primitive nth root of unity, it contains all nth roots of unity; therefore, $\mathbf{Q}(\omega)$ is the splitting field of $\Phi_n(x)$ and of $x^n - 1$. Also, $[\mathbf{Q}(\omega):\mathbf{Q}] =$ degree of $\Phi_n(x) = \phi(n)$. Because $\mathbf{Q}(\omega)$ is a finite, separable, and normal extension of $\mathbf{Q}$,

$$|G(\mathbf{Q}(\omega)/\mathbf{Q})| = [\mathbf{Q}(\omega):\mathbf{Q}].$$

If $\sigma \in G(\mathbf{Q}(\omega)/\mathbf{Q})$, then $\sigma(\omega)$ is also a primitive nth root of unity; so $\sigma(\omega) = \omega^v$, where $v < n$ and $(v,n) = 1$. Denote σ by σ_v. We know there are $\phi(n)$ such v's, and they are precisely the members of the group $(\mathbf{Z}/(n))^*$. Let

$$f\colon (\mathbf{Z}/(n))^* \to G(\mathbf{Q}(\omega)/\mathbf{Q})$$

be a mapping defined by $f(v) = \sigma_v$. Then it is easy to see that f is both a 1-1 and an onto mapping. To check that f is a homomorphism, let v_1, $v_2 \in (\mathbf{Z}/(n))^*$. Write $v_1 v_2 = qn + r$ with $r < n$. By definition of multiplication in $(\mathbf{Z}/(n))^*$, $v_1 v_2 = r$. Now

$$\omega^{v_1 v_2} = \omega^{qn+r} = \omega^r.$$

So

$$f(v_1 v_2) = f(r) = \sigma_r = \sigma_{v_1 v_2} = \sigma_{v_1}\sigma_{v_2} = f(v_1)f(v_2).$$

Hence, $(\mathbf{Z}/(n))^* \simeq G(\mathbf{Q}(\omega)/\mathbf{Q})$. □

Remark. If p is an odd prime, then $(\mathbf{Z}/(p^e))^*$, the multiplicative group formed by the units of the ring $\mathbf{Z}/(p^e)$, is cyclic. The proof of this result is quite technical. The interested reader may find the proof in any advanced text. For $p = 2$ the following example [Example 1.5(a)] shows that the result is not necessarily true.

1.5 Examples

(a) $\Phi_8(x)$ and $x^8 - 1$ have the same Galois group, namely, $(\mathbf{Z}/(8))^* = \{1,3,5,7\}$, the Klein four-group.

Solution. Follows from Theorem 1.4.

(b) The Galois group of $x^4 + x^2 + 1$ is the same as that of $x^6 - 1$ and is of order 2.

Solution. Note $x^4 + x^2 + 1 = z^2 + z + 1$, where $z = x^2$. But $z^2 + z + 1$ is the minimal polynomial for a primitive third root of unity. So the splitting field of $x^4 + x^2 + 1$ will contain the square roots of $e^{2\pi i/3}$ and

$e^{4\pi i/3}$. Thus, we need to adjoin

$$e^{\pi i/3} = (e^{2\pi i/3})^{1/2}, \quad e^{4\pi i/3} = -(e^{2\pi i/3})^{1/2}, \quad e^{2\pi i/3} = (e^{4\pi i/3})^{1/2},$$
$$e^{5\pi i/3} = -(e^{4\pi i/3})^{1/2}.$$

So $E = \mathbf{Q}(\alpha)$, where $\alpha = e^{\pi i/3}$, is the splitting field of $x^4 + x^2 + 1 \in \mathbf{Q}[x]$. Because α is a primitive sixth root of unity, E is the splitting field of $x^6 - 1$. Then $G(E/\mathbf{Q}) = (\mathbf{Z}/(6))^* = \{\bar{1}, \bar{5}\}$ is the group of order 2.

2 Cyclic extensions

Definition. *Let E be a Galois extension of F. Then E is called a* cyclic extension *of F if $G(E/F)$ is a cyclic group.*

2.1 Examples of cyclic extensions

(a) If ω is a primitive pth root of unity and $E = \mathbf{Q}(\omega)$ is the splitting field of $x^p - 1 \in \mathbf{Q}[x]$, then E is a Galois extension of $\mathbf{Q}$ because $x^p - 1$ is a separable polynomial. Also, E is a cyclic extension.

(b) All finite extensions of finite fields are separable. Thus, the splitting field E of a polynomial $f(x)$ over a finite field F is a Galois extension. Furthermore, by Example 4.9(d) in Chapter 16, $G(E/F)$ is cyclic. Therefore, all splitting fields over finite fields are cyclic extensions.

The proof of the following useful result is easy and is left as an exercise for the reader.

2.2 Proposition. *Let F be a field of nonzero characteristic p. Then for every positive integer k the mapping π_k of F into itself, defined by $\pi_k(x) = x^{p^k}$ for all elements x of F, is an embedding of F into itself. (The mapping $\pi_1(x) = x^p$ is called the Frobenius endomorphism.)*

2.3 Lemma. *Let E be a finite extension of F. Suppose $f\colon G \to E^*$, $E^* = E - \{0\}$, has the property that $f(\sigma\eta) = \sigma(f(\eta))f(\sigma)$ for all $\sigma, \eta \in G$. Then there exists $\alpha \in E^*$ such that $f(\sigma) = \sigma(\alpha^{-1})\alpha$ for all $\sigma \in G$. (The mapping f in the hypothesis of the lemma is called a* crossed homomorphism.*)*

Proof. For all $\eta \in G$, $f(\eta) \in E^*$, so $f(\eta) \neq 0$. Thus, if

$$\sum_{\eta \in G} f(\eta)\eta(b) = 0$$

for all $b \in E^*$, then by the Dedekind lemma (Lemma 1.4, Chapter 17),

$f(\eta) = 0$, which is not true. Hence, there exists $b \in E^*$ such that

$$\sum_{\eta \in G} f(\eta)\eta(b) = \alpha \neq 0.$$

Then for any $\sigma \in G$, we get

$$\sum_{\eta \in G} \sigma(f(\eta)\eta(b)) = \sigma(\alpha),$$

which gives

$$\sum_{\eta \in G} \sigma(f(\eta))\sigma(\eta(b)) = \sigma(\alpha).$$

Thus, by using $\sigma(f(\eta)) = (f(\sigma))^{-1}f(\sigma\eta)$, we have

$$\sum_{\eta \in G} (f(\sigma))^{-1}f(\sigma\eta)\sigma\eta(b) = \sigma(\alpha).$$

But

$$\{\sigma\eta \mid \eta \in G\} = \{\eta \mid \eta \in G\}.$$

Hence,

$$(f(\sigma))^{-1} \sum_{\eta \in G} f(\eta)\eta(b) = \sigma(\alpha).$$

This implies

$$(f(\sigma))^{-1}\alpha = \sigma(\alpha).$$

Equivalently, we get

$$\alpha\sigma(\alpha^{-1}) = f(\sigma).$$

This completes the proof. □

2.4 Lemma (special case of Hilbert's problem 90). *Let E be a finite extension of F, and let $G = G(E/F)$ be a cyclic group of order n generated by σ. If $\omega \in E$ be such that $\omega\sigma(\omega)\sigma^2(\omega)\cdots\sigma^{n-1}(\omega) = 1$, then there exists $\alpha \in E^*$ such that $\omega = \sigma(\alpha)\alpha^{-1}$.*

Proof. By Lemma 2.3 we need only define $f\colon G \to E^*$ such that $f(\sigma) = \omega$ for some $\sigma \in G$ and f is a crossed homomorphism. Define $f\colon G \to E^*$ as follows: $f(1) = 1$, $f(\sigma) = \omega$, and $f(\sigma^i) = \sigma^{i-1}(\omega)\cdots\sigma(\omega)\omega$ for $2 \leq i \leq n-1$. We now check that f is a crossed homomorphism; that is, $f(\sigma\eta) = \sigma(f(\eta))f(\sigma)$. Take $\sigma^i, \sigma^j \in G$. If $i + j \equiv 0 \pmod{n}$, then $f(\sigma^i\sigma^j) = f(\sigma^n) = 1$; also

$$\begin{aligned}\sigma^i(f(\sigma^j))f(\sigma^i) &= \sigma^i(\sigma^{j-1}(\omega)\cdots\sigma(\omega)\omega)(\sigma^{i-1}(\omega)\cdots\sigma(\omega)\omega)\\ &= \sigma^{n-1}(\omega)\sigma^{n-2}(\omega)\cdots\sigma(\omega)\omega = 1, \text{ by hypothesis.}\end{aligned}$$

If $i+j$ is not a multiple of n, then

$$f(\sigma^i\sigma^j)=f(\sigma^{i+j})=f(\sigma^r)=\sigma^{r-1}(\omega)\sigma^{r-2}(\omega)\cdots\sigma(\omega)\omega,$$

where $i+j=qn+r$ and $r<n$. Next consider

$$\begin{aligned}\sigma^i f(\sigma^j)f(\sigma^i)&=\sigma^i(\sigma^{j-1}(\omega)\cdots\sigma(\omega)\omega)(\sigma^{i-1}(\omega)\cdots\sigma(\omega)\omega)\\&=\sigma^{(i+j)-1}(\omega)\sigma^{(i+j)-2}(\omega)\cdots\sigma(\omega)\omega\\&=\sigma^{r-1}(\omega)\sigma^{r-2}(\omega)\cdots\sigma(\omega)\omega.\end{aligned}$$

Hence, f is a crossed homomorphism. □

We may remark that the condition $\omega\sigma(\omega)\sigma^2(\omega)\cdots\sigma^{n-1}(\omega)=1$ is the usual condition that the norm of ω in E over F is 1, i.e., $N_{E/F}(\omega)=1$ (see Problem 4).

2.5 Theorem. *Let F contain a primitive nth root ω of unity. Then the following are equivalent:*

(i) E is a finite cyclic extension of degree n over F.
(ii) E is the splitting field of an irreducible polynomial $x^n-b\in F[x]$.

Furthermore note $E=F(\alpha)$, where α is a root of x^n-b.

Proof. (i) ⇒ (ii) Let σ be a generator of the finite cyclic group $G(E/F)$. By Lemma 2.4 there exists $\alpha\in E^*$ such that $\sigma(\alpha)=\omega\alpha$. Then it follows easily that, for all $i=1,2,\ldots,$

$$\sigma^i(\alpha)=\omega^i\alpha. \tag{1}$$

Then for all $i=1,2,\ldots,$

$$\sigma^i(\alpha^n)=(\sigma^i(\alpha))^n=\omega^{in}\alpha^n=\alpha^n.$$

Thus, $\alpha^n\in F$, and if $b=\alpha^n$, then $x^n-b\in F[x]$ and $x^n-b=\prod_{i=1}^n(x-\omega^i\alpha)$.

We show that $x^n-b\in F[x]$ is irreducible over F. Suppose $x^n-b=f(x)g(x)$, where $f(x)$ is a nonconstant irreducible monic polynomial over F. If $\omega^i\alpha$ is one root of $f(x)$, then for each positive integer j we have

$$\sigma^{j-i}(\omega^i\alpha)=\sigma^{j-i}(\omega^i)\sigma^{j-i}(\alpha).$$

But $\omega^i\in F$. So $\sigma^{j-i}(\omega^i)=\omega^i$. Also, from (1), $\sigma^{j-i}(\alpha)=\omega^{j-i}\alpha$. Thus, we have

$$\sigma^{j-i}(\omega^i\alpha)=\omega^i\omega^{j-i}\alpha=\omega^j\alpha.$$

Because any F-automorphism maps a root of a polynomial over F onto a root of that polynomial, we get that $\omega^j\alpha$ is also a root of $f(x)$. Hence, all the roots of $x^n - b$ are roots of $f(x)$. Thus, $f(x) = x^n - b$, which proves that $x^n - b$ is irreducible over F. Note that we have also shown that $E = F(\alpha)$, where α is a root of $x^n - b$, is a splitting field of $x^n - b$ over F. Thus, (i) $\Rightarrow$ (ii) and (iii).

We now show (ii) $\Rightarrow$ (i). Let $c \in E$ be a root of $x^n - b \in F[x]$. So $b = c^n$. Clearly, then, $c, c\omega, c\omega^2, \ldots, c\omega^{n-1}$ are n distinct roots of $x^n - b$, where $\omega \in F$ is a primitive nth root of unity. Thus, $x^n - b$ is a separable irreducible polynomial. Hence, $E = F(c)$ is a Galois extension of F. For each $\sigma \in G(E/F)$, let $\chi(\sigma)$ be defined by

$$\chi(\sigma) = \{k \in \mathbf{Z} \mid \sigma(c) = \omega^k c\}.$$

Then $\chi(\sigma) \neq \varnothing$ because $\sigma(c)$ is also a root of $x^n - b$. Moreover, for any $k \in \chi(\sigma)$, $\chi(\sigma) = \bar{k} \in \mathbf{Z}/(n)$, for $\omega^k c = \omega^j c$ if and only if $k \equiv j \pmod{n}$.

Further, if $\sigma, \tau \in G(E/F)$ and if $\sigma(c) = \omega^k c$ and $\tau(c) = \omega^j c$, then

$$(\sigma\tau)(c) = \sigma(\omega^j c) = \omega^j \sigma(c) = \omega^{j+k} c,$$

so

$$\chi(\sigma\tau) = \chi(\sigma) + \chi(\tau),$$

where the sum on the right is interpreted as the binary operation in the additive group $\mathbf{Z}/(n)$.

Finally, if $\chi(\sigma) = \bar{0}$, then $\sigma(c) = c$. So $\sigma =$ identity on E because $E = F(c)$. Consequently, χ is an isomorphism from $G(E/F)$ onto a subgroup of the additive group $\mathbf{Z}/(n)$. Also, $[F(c): F] =$ degree of the minimal polynomial of c over $F =$ degree of $x^n - b$, that is, n. Because $E = F(c)$ is a finite, separable, normal extension of F, $|G(E/F)| = [E:F] = n$. Because $G(E/F)$ is isomorphic to a subgroup of $\mathbf{Z}/(n)$, it follows that

$$G(E/F) \simeq \mathbf{Z}/(n). \quad \square$$

Problems

1. If a field F contains a primitive nth root of unity, then the characteristic of F is 0 or a prime p that does not divide n.
2. Let F contain a primitive nth root of unity, and let E be the splitting field of $x^m - b$ over F, where $m \mid n$ and m is prime. Then either $E = F$ or $x^m - b$ is irreducible over F. What can you say if m is not prime?
3. Let p be a prime and let F be a field. Prove that $x^p - b \in F[x]$ is reducible over F iff its splitting field is F or $F(\omega)$ according to whether char $F = p$ or char $F \neq p$, where ω is a primitive pth root of unity.

4. Let E be a finite separable normal extension over F and let $G(E/F) = \{\sigma_1 = 1, \sigma_2, \dots, \sigma_n\}$. If $a \in E$ we define

$$T_{E/F}(a) = \sum_{i=1}^{n} \sigma_i(a) \quad \text{and} \quad N_{E/F}(a) = \prod_{i=1}^{n} \sigma_i(a)$$

and call these respectively the *trace* and *norm* of a in E over F. Show:
   (a) $T_{E/F}(a) \in F$, $N_{E/F}(a) \in F$.
   (b) $T_{E/F}$ is an F-linear map of the vector space E over F.
   (c) $N_{E/F}$ is a group homomorphism from the group $E^* = E - \{0\}$ to the group $F^* = F - \{0\}$.
   (d) If $G(E/F)$ is a cyclic group generated by σ, then $N_{E/F}(a) = 1$ if and only if there exists $b \in E$ such that $a = (\sigma(b))^{-1}b$. (*Hint*: Generalize Lemma 2.4.)

3 Polynomials solvable by radicals

The object of this section is to use Galois theory to find necessary and sufficient conditions for a polynomial over a field F to be solvable by radicals. We also construct a polynomial of degree 5 that is not solvable by radicals. Before we prove these results, we need to explain the terms used.

Definition. *An extension E of a field F is an* extension by radicals *(or* radical extension*) if there are elements $\alpha_1, \dots, \alpha_r \in E$ and positive integers $n_1, \dots, n_r$ such that $E = F(\alpha_1, \dots, \alpha_r)$, $\alpha_1^{n_1} \in F$, and $\alpha_i^{n_i} \in F(\alpha_1, \dots, \alpha_{i-1})$, $1 < i \leq r$.*

For example, $\mathbf{Q}(\sqrt[3]{2})$ is a radical extension of $\mathbf{Q}$. Also, $\mathbf{Q}(\sqrt[3]{2}, \sqrt[5]{3})$ is a radical extension of $\mathbf{Q}$.

A useful fact about radical extension is

3.1 Remark. If E_r is a radical extension of $F = E_0$ with intermediate fields $E_1, \dots, E_{r-1}$ (written in ascending order), then there exists a radical extension E'_s of $F = E_0$ with intermediate fields $E'_1, \dots, E'_{s-1}$ (written in ascending order) such that

(i) $E'_s \supset E_r$.
(ii) E'_s is a normal extension of F.
(iii) E'_i is a splitting field of a polynomial of the form $x^{m_i} - b_i \in E'_{i-1}[x]$, $i = 1, \dots, s$.

To prove this remark, let us assume that we have an ascending chain of fields $F = E_0 \subset E_1 \subset \cdots \subset E_r$ starting with F such that $E_i = E_{i-1}(\alpha_i)$, where α_i is a root of $x^{n_i} - a_i \in E_{i-1}[x]$, $i = 1,...,r$. Let ω be a primitive nth root of unity, where $n = n_1 \cdots n_r$. Consider the tower

$$\begin{array}{l} E_1(\omega) \\ \quad | \\ E_0(\omega) = F(\omega) \\ \quad | \\ E_0 = F \end{array}$$

Clearly, $E_1(\omega)$ is a radical extension of F. Also $F(\omega)$, being a splitting field of $x^n - 1 \in F[x]$, is a normal extension of F. Thus, F is the fixed field of $G(F(\omega)/F)$, so the polynomial

$$f_1(x) = \prod_{\sigma \in G(F(\omega)/F)} (x^{n_1} - \sigma(a_1))$$

belongs to $F[x]$. (Here $f_1(x) = (x^{n_1} - a_1)^k$, $k = |G(F(\omega)/F)|$, since $a_1 \in F$.)

Let $g_1(x) = (x^n - 1)f_1(x)$. Then $g_1(x) \in F[x]$. Let K be the splitting field of $g_1(x)$ over F. Then K is a normal extension of F. Clearly, $\alpha_1 \in K$, $\omega \in K$, and $E_1 \subset K$. It is also clear that there is a finite ascending chain of fields between F and K such that each field is a splitting field of a polynomial of the form $x^m - b$ over the preceding field. Next we construct a field L such that L contains the fields K and E_2 and is a normal extension of F. So we consider a polynomial $g_2(x) = g_1(x)f_2(x)$, where $f_2(x) = \Pi_{\sigma \in G(K/F)} (x^{n_2} - \sigma(a_2))$. Because K is a normal extension over F, $f_2(x) \in F[x]$. Thus, $g_2(x) \in F[x]$. Let L be the splitting field of $g_2(x)$ over F. Then L contains α_2 and K, and, hence, $E_1(\alpha_2) = E_2 \subset L$. Therefore, L is a normal extension of F containing E_2. Further, in view of the nature of the polynomial $g_2(x)$, it is clear that there exists a finite ascending sequence of intermediate fields between K and L such that any member of the sequence is a splitting field of a polynomial of the form $x^m - b$ over the preceding member. Continuing like this, we can construct a radical extension E'_s of F having the desired properties. This proves the remark. □

Definition. *A polynomial* $f(x) \in F[x]$ *over a field* F *is said to be* solvable by radicals *if its splitting field* E *is contained in some radical extension of* F.

We note that a polynomial $f(x) \in F[x]$ is solvable by radicals if we can obtain every root of $f(x)$ by using a finite sequence of operations of addition, subtraction, multiplication, division, and taking nth roots, starting with elements of F.

Throughout this section we assume that all fields are of characteristic zero.

3.2 Theorem. *$f(x) \in F[x]$ is solvable by radicals over F if and only if its splitting field E over F has solvable Galois group $G(E/F)$.*

Proof. First suppose that $G(E/F)$ is solvable. Because the characteristic of F is zero, E is a normal separable extension. So $[E:F] = |G(E/F)| = n$, say. Assume first that F contains a primitive nth root of unity. Then F contains primitive mth roots of unity for all positive integers m that divide n. Let $G = G(E/F)$. Because G is solvable and finite, there is a chain $G = G_0 \supset G_1 \supset \cdots \supset G_r = (e)$ of subgroups of G such that $G_i \lhd G_{i-1}$ and G_{i-1}/G_i is cyclic. Let $F = F_0 \subset F_1 \subset \cdots \subset F_r = E$ be the corresponding subfields of E given by the fundamental theorem. Then $E_{G_i} = F_i$ and $G(E/F_i) = G_i$. Also, by the fundamental theorem, $G_1 = G(E/F_1) \lhd G(E/F) = G$ implies F_1 is a normal extension of F.

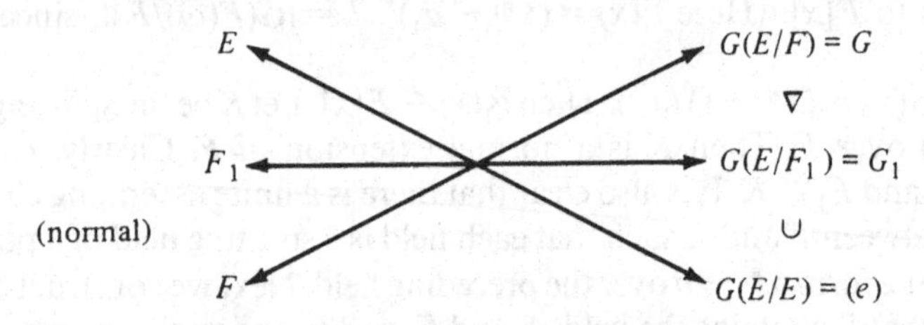

Now E can be regarded as the splitting field of $f(x)$ over F_1. So E is a finite normal extension of F_1. Then $G_2 \lhd G_1$ implies that F_2 is a normal extension of F_1.

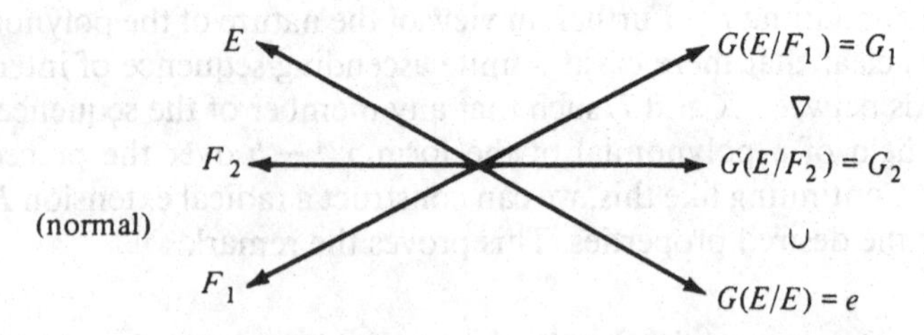

Continue in this way to show that F_i is a normal extension of F_{i-1}.

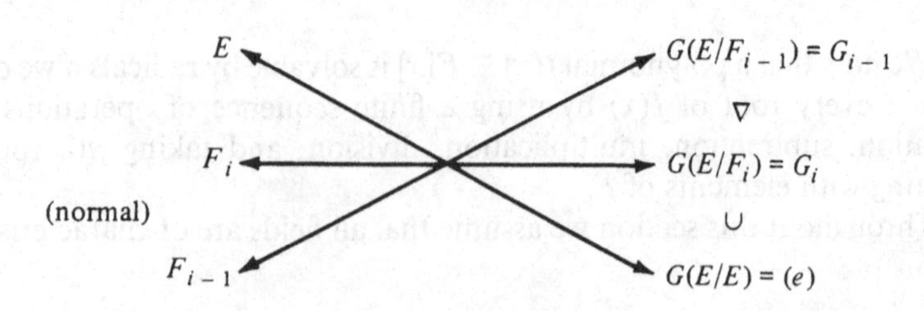

Furthermore, $G(F_i/F_{i-1}) = G(E/F_{i-1})/G(E/F_i) = G_{i-1}/G_i$ by the fundamental theorem. So F_i is a cyclic extension of F_{i-1}. Then by Theorem 2.5, F_i is the splitting field of an irreducible polynomial $x^{n_i} - b_i \in F_{i-1}[x]$ and $F_i = F_{i-1}(\alpha_i)$, where $\alpha_i^{n_i} = b_i \in F_{i-1}$. Then $E = F(\alpha_1,...,\alpha_r)$, $\alpha_1^{n_1} \in F$, and $\alpha_i^{n_i} \in F_{i-1} = F(\alpha_1,...,\alpha_{i-1})$ for $1 < i \leq r$. Thus, $f(x)$ is solvable by radicals over F.

Next we drop the assumption that F contains a primitive nth root of unity. The polynomial $x^n - 1 \in E[x]$ has roots in $\overline{E}$. Let ρ be a primitive nth root of unity lying in $\overline{E}$. Then $E(\rho)$ is the splitting field of $f(x)$ regarded as a polynomial over $F(\rho)$. Any $F(\rho)$-automorphism σ of $E(\rho)$ will leave the coefficients of the polynomial $f(x)$ unaltered. Now, for any automorphism $\sigma \in G(E(\rho)/F(\rho))$, we have $\sigma_0 = \sigma|_E \in G(E/F)$, since E is a normal extension of F. Further, the map $\sigma \mapsto \sigma_0$ is a 1-1 homomorphism of the group $G(E(\rho)/F(\rho))$ into $G(E/F)$. Then since a subgroup of a solvable group is solvable, $G(E(\rho)/F(\rho))$ is solvable. Now by the first part, $E(\rho)$ is a radical extension of $F(\rho)$; so $E(\rho)$ is a radical extension of F. Then the splitting field E of $f(x) \in F[x]$ is contained in the radical extension $E(\rho)$ of F, so $f(x)$ is solvable by radicals.

Before we prove the converse, we prove a lemma that deals with a particular case.

3.3 Lemma. *Let E be the splitting field of $x^n - a \in F[x]$. Then $G(E/F)$ is a solvable group.*

Proof. If F contains a primitive nth root of unity, then by Example 2.2(h) in Chapter 17 we know that $G(E/F)$ is abelian and, hence, solvable. Now suppose that F does not contain a primitive nth root of unity. Let $\rho \in \overline{F}$ be a generator of the cyclic group of the nth roots of unity. Let b be a root of $x^n - a$. Then $b\rho$ is also a root. So $\rho = b^{-1}(b\rho)$ is in the splitting field E of $x^n - a \in F[x]$. Consider $F \subset F(\rho) \subset E$. $F(\rho)$ is a normal extension of F, since $F(\rho)$ is the splitting field of $x^n - 1$; so $G(E/F(\rho))$ is a normal subgroup of $G(E/F)$ by the fundamental theorem of Galois theory.

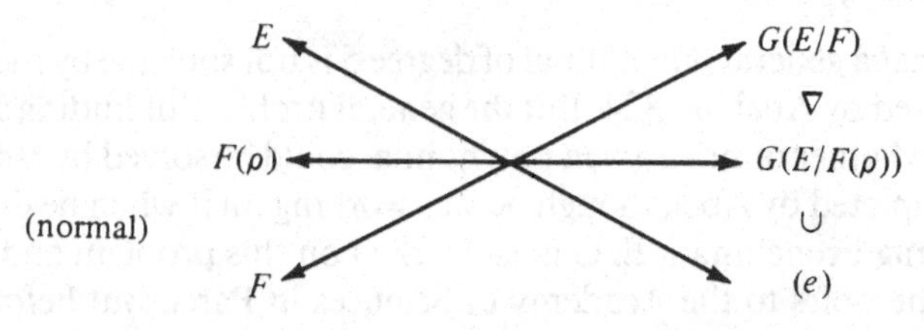

But $G(E/F(\rho))$ is abelian, because E is the splitting field of $x^n - a \in F[x]$. So $(e) \lhd G(E/F(\rho)) \lhd G(E/F)$ is a normal series. Again by the fundamental theorem of Galois theory, $G(E/F)/G(E/F(\rho)) \simeq G(F(\rho)/F)$, which is abelian (being isomorphic to $(\mathbf{Z}/(n))^*$ by Theorem 1.4 because $F(\rho)$ is the splitting field of $x^n - 1$). So $G(E/F)$ has a normal series with abelian factors whose last element is the trivial group. Therefore, $G(E/F)$ is solvable. □

We are now ready to complete the proof of the theorem.

In view of Remark 3.1, if a polynomial $f(x) \in F[x]$ is solvable by radicals, we may, without any loss of generality, assume that the splitting field E of $f(x)$ is contained in a radical extension E_r of F such that E_r is a normal extension of F and there exist intermediate fields $E_1, ..., E_{r-1}$ such that E_i is a splitting field of a polynomial of the form $x^{m_i} - b_i \in E_{i-1}[x]$. Thus, by the fundamental theorem of Galois theory

$$(e) \subset G(E_r/E_{r-1}) \subset G(E_r/E_{r-2}) \subset \cdots \subset G(E_r/F)$$

is a normal series. Also

$$G(E_r/E_{r-i})/G(E_r/E_{r-i+1}) \simeq G(E_{r-i+1}/E_{r-i})$$

is solvable by Lemma 3.3. Then

$$(e) \subset G(E_r/E_{r-1}) \subset G(E_r/E_{r-2}) \subset \cdots \subset G(E_r/F)$$

is a normal series with solvable quotient groups, so $G(E_r/F)$ is solvable. Further, since $G(E/F) \simeq G(E_r/F)/G(E_r/E)$, $G(E/F)$ is a homomorphic image of $G(E_r/F)$. Hence, it is solvable. □

Remark. We know that the symmetric group S_n is not solvable if $n \geq 5$. Thus, any polynomial whose Galois group is S_n, $n \geq 5$, is not solvable by radicals. In Problem 1(a) we give a polynomial of degree 5 whose Galois group is S_5. This shows that not all polynomials of degree 5 are solvable by radicals.

Historical background

The result that a general polynomial of degree 5 is not solvable by radicals was discovered by Abel in 1824. But the general problem of finding a way of deciding whether or not a given polynomial could be solved by radicals was not completed by Abel, though he was working on it when he died in 1829. A young Frenchman, E. Galois, worked on this problem and submitted his memoirs to the Academy of Sciences in Paris. But before he could get recognition, he died in a duel in 1832. It was Liouville who, while addressing the Academy in 1843, announced the profound work of Galois regarding the solvability of a polynomial by radicals.

Before we give applications of Theorem 3.2, we would like to recall the important fact [Example 2.2(a), Chapter 17] that the Galois group of a polynomial $f(x) \in F[x]$ having r distinct roots is embeddable in the symmetric group S_r, the group of all permutations of the r distinct roots $(\alpha_1,\dots,\alpha_r)$.

We call a subgroup H of S_n a *transitive permutation group* if, for all $i, j \in \{1,2,\dots,n\}$, there exists $\sigma \in H$ such that $\sigma(i) = j$.

We also recall the following result from group theory.

3.4 Theorem. *If p is a prime number and if a subgroup G of S_p is a transitive group of permutations containing a transposition $(a\ b)$, then $G = S_p$.*

We now prove

3.5 Theorem. *Let $f(x)$ be a polynomial over a field F with no multiple roots. Then $f(x)$ is irreducible over F if and only if the Galois group G of $f(x)$ is isomorphic to a transitive permutation group.*

Proof. Let $\alpha_1,\dots,\alpha_n$ be the roots of $f(x)$ in some splitting field E. Then for each $\sigma \in G$, $\sigma(\alpha_1),\dots,\sigma(\alpha_n)$ is a permutation of $\alpha_1,\dots,\alpha_n$. By Example 2.2(a) in Chapter 17, we may look upon G as a subgroup of S_n.

First assume $f(x)$ is irreducible over F. Then for each $i = 1,\dots,n$,

$$F(\alpha_i) \simeq F[x]/(f(x))$$

in which $\alpha_i \mapsto x + (f(x))$, $a \mapsto a + (f(x))$, $a \in F$. This isomorphism induces the isomorphism η: $F(\alpha_i) \to F(\alpha_j)$, where $\alpha_i \mapsto \alpha_j$ and $a \mapsto a$, $a \in F$. But since E is a normal extension of F, η can be extended to an F-homomorphism $\eta^*: E \to E$. Then $\eta^* \in G(E/F)$ and $\eta^*(\alpha_i) = \alpha_j$. Thus, G is a transitive permutation group.

Conversely, let G be transitive. Let $p(x)$ be the minimal polynomial for α_1 over F. Suppose α_i is any root. Because G is transitive, there exists $\sigma \in G$ such that $\sigma(\alpha_1) = \alpha_i$. Then $p(\alpha_i) = p(\sigma(\alpha_1)) = \sigma p(\alpha_1) = 0$. Hence, each α_i is a root of $p(x)$. Because $p(x) | f(x)$, it follows that $f(x) = cp(x)$, $c \in F$. Thus, $f(x)$ is irreducible over F. □

Next we prove

3.6 Theorem. *Let $f(x) \in \mathbf{Q}[x]$ be a monic irreducible polynomial over $\mathbf{Q}$ of degree p, where p is prime. If $f(x)$ has exactly two nonreal roots in $\mathbf{C}$, then the Galois group of $f(x)$ is isomorphic to S_p.*

Proof. Let $E \subset \mathbf{C}$ be a splitting field of $f(x)$ over $\mathbf{Q}$. Then by Theorem 3.5, $G(E/\mathbf{Q})$ is isomorphic to a transitive permutation group H, which is a subgroup of S_p. Let $\alpha_1,\ldots,\alpha_p$ be roots of $f(x)$, and let α_i be its complex root. Because $f(x) \in \mathbf{Q}[x]$, $\bar{\alpha}_i$ is also a root of $f(x)$. Hence, $\bar{\alpha}_i = \alpha_j$ for some $1 \leq j \leq p, j \neq i$. Consider the embedding $\sigma: z \mapsto \bar{z}$ from E to $\bar{\mathbf{Q}}$. Because E is a normal extension of $\mathbf{Q}$, σ maps E onto E. Thus, $\sigma \in G(E/\mathbf{Q})$. Then the permutation of the roots $\alpha_1,\ldots,\alpha_p$ of $f(x)$ corresponding to the element σ of the Galois group $G(E/\mathbf{Q})$ takes α_i to α_j and α_j to α_i, and keeps all α_k $(k \neq i,j)$ fixed. Hence, by Theorem 3.4, $H \simeq S_p$. Thus, $G(E/\mathbf{Q}) \simeq S_p$, as required. □

3.7 Examples

(a) Show that if an irreducible polynomial $p(x) \in F[x]$ over a field F has a root in a radical extension of F, then $p(x)$ is solvable by radicals over F.

Solution. Let E_r be a radical extension of F. By Remark 3.1 there exists a radical extension E'_s of F such that $E'_s \supset E_r$ and E'_s is a normal extension of F. Because $p(x)$ is irreducible over F and has a root in E_r, it has a root in E'_s. But because E'_s is a normal extension of F, it follows that E'_s contains a splitting field of $p(x)$. This shows that $p(x)$ is solvable by radicals.

(b) Show that the polynomial $x^7 - 10x^5 + 15x + 5$ is not solvable by radicals over $\mathbf{Q}$.

Solution. By Eisenstein's criterion $f(x) = x^7 - 10x^5 + 15x + 5$ is irreducible over $\mathbf{Q}$. Further, by Descartes's rule of signs it is known that

$$\text{the number of positive real roots} \leq \text{the number of changes in signs in } f(x) = 2,$$

and

$$\text{the number of negative real roots} \leq \text{the number of changes of signs in } f(-x) = 3.$$

Thus, the total number of real roots ≤ 5. Moreover, by the intermediate value theorem there are five real roots, one in each of the intervals $(-4,-3)$, $(-2,-1)$, $(-1,0)$, $(1,2)$, and $(3,4)$. So $f(x)$ has exactly two nonreal roots. By Theorem 3.6 the Galois group of $f(x)$ is S_7. Hence, by Theorem 3.2, $f(x)$ is not solvable by radicals.

Problems

1. Show that the following polynomials are not solvable by radicals over $\mathbf{Q}$:
   (a) $x^5 - 9x + 3$ (b) $2x^5 - 5x^4 + 5$
   (c) $x^5 - 8x + 6$ (d) $x^5 - 4x + 2$
2. Let $F[x_1, x_2, x_3]$ be a polynomial ring in x_1, x_2, x_3 over a field F. Let $K = F(x_1, x_2, x_3)$ be the field of rational functions (i.e., the field of fractions of the ring $F[x_1, x_2, x_3]$). Suppose
   $f(t) = t^3 - x_1 t^2 + x_2 t - x_3 \in K[t].$
   Prove that the Galois group of $f(t)$ over K is S_3. Generalize this result to a polynomial of degree n (see Theorem 4.1).

4 Symmetric functions

In this section we give an application of Galois theory to the symmetric functions. Let F be a field, and let $y_1, \ldots, y_n$ be n indeterminates. Consider the field of rational functions $F(y_1, \ldots, y_n)$ over F. If σ is a permutation of $\{1, \ldots, n\}$ – that is, $\sigma \in S_n$ – then σ gives rise to a natural map

$$\bar{\sigma}: F(y_1, \ldots, y_n) \to F(y_1, \ldots, y_n)$$

given by

$$\bar{\sigma}\left(\frac{f(y_1, \ldots, y_n)}{g(y_1, \ldots, y_n)}\right) = \frac{f(y_{\sigma(1)}, \ldots, y_{\sigma(n)})}{g(y_{\sigma(1)}, \ldots, y_{\sigma(n)})},$$

where $f(y_1, \ldots, y_n), g(y_1, \ldots, y_n) \in F[y_1, \ldots, y_n]$ and $g(y_1, \ldots, y_n) \neq 0$. It is immediate that $\bar{\sigma}$ is an automorphism of $F(y_1, \ldots, y_n)$ leaving each element of F fixed.

Definition. *An element* $f(y_1, \ldots, y_n)/g(y_1, \ldots, y_n)$ *of* $F(y_1, \ldots, y_n)$ *is called a* symmetric function *in* $y_1, \ldots, y_n$ *over* F *if it is left fixed by all permutations of* $1, \ldots, n$, *that is, for all* $\sigma \in S_n$,

$$\bar{\sigma}\left(\frac{f(y_1, \ldots, y_n)}{g(y_1, \ldots, y_n)}\right) = \frac{f(y_1, \ldots, y_n)}{g(y_1, \ldots, y_n)}.$$

Let $\bar{S}_n$ be the group of all F-automorphisms $\bar{\sigma}$ of $F(y_1, \ldots, y_n)$ corresponding to $\sigma \in S_n$. Obviously, $\bar{S}_n \simeq S_n$. Let K be the subfield of $F(y_1, \ldots, y_n)$ that is the fixed field of $\bar{S}_n$. Consider the polynomial

$$f(x) = \prod_{i=1}^{n} (x - y_i).$$

Now $f(x) \in F(y_1,...,y_n)[x]$. Clearly, the natural mapping

$$F(y_1,...,y_n)[x] \rightarrow F(y_1,...,y_n)[x]$$

induced by each $\bar{\sigma} \in \bar{S}_n$ leaves $f(x)$ unaltered. Thus, the coefficients of $f(x)$ are unaltered by each $\bar{\sigma} \in \bar{S}_n$. Hence, the coefficients lie in the fixed field K.

Let us write the polynomial $f(x)$ as $x^n + a_1x^{n-1} + a_2x^{-2} + \cdots + a_n$, where $a_i \in K$.

Definition. *If a_i is the coefficient of x^{n-i} in the polynomial $f(x) = \Pi_{i=1}^n (x-y_i)$, then $(-1)^i a_i$ is called the ith* elementary symmetric function *in $y_1,...,y_n$ and is denoted by s_i.*

Thus,

$$\begin{aligned}
s_1 &= y_1 + y_2 + \cdots + y_n, \\
s_2 &= y_1y_2 + y_1y_3 + \cdots + y_{n-1}y_n, \\
&\vdots \\
s_n &= y_1y_2 \cdots y_n.
\end{aligned}$$

We now prove the following theorem.

4.1 Theorem. *Let $s_1,...,s_n$ be the elementary symmetric functions in the indeterminates $y_1,...,y_n$. Then every symmetric function in $y_1,...,y_n$ over F is a rational function of the elementary symmetric functions. Also, $F(y_1,...,y_n)$ is a finite normal extension of $F(s_1,...,s_n)$ of degree $n!$, and the Galois group of this extension is isomorphic to S_n.*

Proof. Consider the field $E = F(s_1,...,s_n)$. Because K is the field of all symmetric functions in $y_1,...,y_n$ over F, $E \subset K$. Also, because $F(y_1,...,y_n)$ is a splitting field of the polynomial $f(x) = \Pi_{i=1}^n (x-y_i)$, of degree n over E, we have, by Theorem 3.6 in Chapter 15,

$$[F(y_1,...,y_n):E] \leq n!. \tag{1}$$

Further, as discussed in the beginning of this section,

$$[F(y_1,...,y_n):K] \geq |\bar{S}_n| = n!. \tag{2}$$

But since $E \subset K$, we obtain from (1) and (2) that $E = K$.

Now $f(x)$ is a separable polynomial over E, and $F(y_1,...,y_n)$ is its splitting field. Thus, $F(y_1,...,y_n)$ is a finite, separable, normal extension of E.

So

$$[F(y_1,...,y_n):E] = |G(F(y_1,...,y_n)/E)|. \tag{3}$$

Because $G(F(y_1,...,y_n)/E)$ is embeddable in S_n, and $[F(y_1,...,y_n):E] = n!$, we get from (3) that

$$G(F(y_1,...,y_n)/E) \simeq S_n.$$

Finally, the fact that $K = E$ shows that every symmetric function can be expressed as a rational function of the elementary symmetric functions $s_1,...,s_n$. □

4.2 Examples

We express the following symmetric polynomials as rational functions of the elementary symmetric functions:

(a) $x_1^2 + x_2^2 + x_3^2$.
(b) $(x_1 - x_2)^2(x_2 - x_3)^2(x_3 - x_1)^2$.

Solution. (a)

$$(x_1^2 + x_2^2 + x_3^2) = (x_1 + x_2 + x_3)^2 - 2(x_1x_2 + x_2x_3 + x_3x_1) = s_1^2 - 2s_2,$$

where s_1 and s_2 are elementary symmetric functions of x_1, x_2, and x_3.

(b) By simple computation it can be checked that

$$y_1 = x_1 - \frac{s_1}{3}, \quad y_2 = x_2 - \frac{s_1}{3}, \quad y_3 = x_3 - \frac{s_1}{3}$$

are the roots of $x^3 + 3\alpha x + \beta = 0$, where

$$\alpha = \frac{-s_1^2}{3} + s_2, \quad \beta = -s_3 - \frac{2s_1^3}{27} + \frac{s_1s_2}{3}.$$

Then the cubic equation whose roots are $(y_1 - y_2)^2$, $(y_2 - y_3)^2$, and $(y_3 - y_1)^2$ is

$$(3\alpha + y)^3 + 9\alpha(3\alpha + y)^2 + 27\beta^2 = 0. \tag{1}$$

Now

$$\begin{aligned}(x_1 - x_2)^2(x_2 - x_3)^2(x_3 - x_1)^2 &= (y_1 - y_2)^2(y_2 - y_3)^2(y_3 - y_1)^2 \\ &= \text{product of all the roots of (1)} \\ &= -27(\beta^2 + 4\alpha^3).\end{aligned}$$

Problem

1. Express the following symmetric functions as rational functions of elementary symmetric functions:
   (a) $x_1^3 + x_2^3 + x_3^3$.
   (b) $x_1^2 x_2^2 + x_2^2 x_3^2 + x_3^2 x_1^2$.
   (c) $(x_1^2 + x_2^2)(x_2^2 + x_3^2)(x_3^2 + x_1^2)$.
   (d) $(x_1 + x_2)^3(x_2 + x_3)^3(x_3 + x_1)^3$.

5 Ruler and compass constructions

The theory of fields provides solutions to many ancient geometric problems. Among such problems are the following:

1. To construct by ruler and compass a square having the same area as that of a circle.
2. To construct by ruler and compass a cube having twice the volume of a given cube.
3. To trisect a given angle by ruler and compass.
4. To construct by ruler and compass a regular polygon having n sides.

For these we must translate the geometric problem into an algebraic problem. We shall regard the plane as the coordinate plane $\mathbf{R}^2$ of analytic geometry. Let $P_0 \subset \mathbf{R}^2$. Assume P_0 has at least two points. We construct an ascending chain of subsets P_i of $\mathbf{R}^2$, $i = 0,1,2,\ldots$, inductively as follows: Let P_{i+1} be the union of P_i and the set of points obtained by intersection of (i) two distinct lines each passing through two distinct points in P_i, or (ii) two distinct circles each with its center in P_i and passing through another point in P_i, or (iii) a line and a circle of the types described in (i) and (ii).

Suppose that the coordinates of points in P_0 belong to a subfield K of $\mathbf{R}$. The equation of a line passing through two distinct points in P_0 and the equation of a circle whose center is in P_0 and that passes through another point in P_0 are

$$ax + by + c = 0, \qquad a,b,c \in K, \tag{1}$$
$$x^2 + y^2 + 2gx + 2fy + d = 0, \qquad g,f,d \in K, \tag{2}$$

respectively.

It follows then that the coordinates of the point of intersection of two such lines (1) lie in K. Also, the coordinates of the points of intersection of a line (1) and a circle (2), as well as the coordinates of the points of intersection of two distinct circles (2), lie in $K(\sqrt{\alpha_1})$, $\alpha_1 > 0$, $\alpha_1 \in K$. Likewise, we get that the coordinates of points in P_i lie in $K(\sqrt{\alpha_1},...,\sqrt{\alpha_i})$, $\alpha_1,...,\alpha_i > 0$, $\alpha_1 \in K$, $\alpha_2 \in K(\sqrt{\alpha_1}),..., \alpha_i \in K(\sqrt{\alpha_1},...,\sqrt{\alpha_{i-1}})$.

Definition. *(a) A point X is* constructible *from P_0 if $X \in P_i$ for some $i \in \{0, 1, 2, \ldots\}$.*

(b) A line l is constructible *from P_0 if it passes through two distinct points in some P_i, $i \in \{0,1,2,...\}$.*

(c) A circle C is constructible *from P_0 if its center is in some P_i, and it passes through another point in P_i, $i \in \{0,1,2,...\}$.*

From now on whenever a point X (a line l, a circle C) is constructible from $\mathbf{Q} \times \mathbf{Q}$, we shall also say that the point X (the line l, the circle C) is *constructible.*

Definition. *A real number u is* constructible *from $\mathbf{Q}$ if the point $(u,0)$ is constructible from $\mathbf{Q} \times \mathbf{Q}$, the subset of the plane $\mathbf{R}^2$.*

It then follows from all this that if $u \in \mathbf{R}$ is constructible from $\mathbf{Q}$, then there exists an ascending chain

$$\mathbf{Q} = K_0 \subset K_1 \subset K_2 \subset \cdots \subset K_n$$

of subfields $K_1, K_2, ..., K_n$ of $\mathbf{R}$ such that

(i) $u \in K_n$.
(ii) $K_i = K_{i-1}(\alpha_i)$, $1 \leq i \leq n$, where $\alpha_i^2 \in K_{i-1}$.

Thus, $[K_i : K_{i-1}] \leq 2$, and, hence, $[K_n : \mathbf{Q}] = 2^m$, $m \leq n$.

So we have shown

5.1 Theorem. *Let $u \in \mathbf{R}$ be constructible from $\mathbf{Q}$. Then there exists a subfield K of $\mathbf{R}$ containing u such that $[K:\mathbf{Q}] = 2^m$ for some positive integer m.*

5.2 Theorem. *Let K be the subset of $\mathbf{R}$ consisting of numbers constructible from $\mathbf{Q}$. Then K is a subfield containing square roots of all nonnegative numbers in K.*

Before we prove this theorem, we prove a series of lemmas.

5.3 Lemma. *The following are equivalent statements:*

(i) *$a \in \mathbf{R}$ is constructible from $\mathbf{Q}$.*
(ii) *$(a,0)$ is a constructible point from $\mathbf{Q} \times \mathbf{Q}$.*

(iii) *(a,a) is a constructible point from* $\mathbf{Q} \times \mathbf{Q}$.
(iv) *$(0,a)$ is a constructible point from* $\mathbf{Q} \times \mathbf{Q}$.

Proof. (i) $\Leftrightarrow$ (ii) Definition.

(ii) $\Rightarrow$ (iii) The circle $(x-a)^2 + y^2 = a^2$ is constructible because its center $(a,0)$ is a constructible point, and it passes through a constructible point $(0,0)$. Also, the line $x = y$ is constructible because it passes through constructible points $(0,0)$ and $(1,1)$. The point (a,a) is clearly a point of intersection of the circle and the line. Hence, (a,a) is constructible.

(iii) $\Rightarrow$ (iv) The circle $x^2 + y^2 = 2a^2$ is constructible because its center $(0,0)$ is constructible, and it passes through a constructible point (a,a). Also, the line $y = -x$ is constructible because it passes through two distinct constructible points $(0,0)$ and $(1,-1)$. One of the points of intersection of this circle and this line is $(-a,a)$. This implies that $(0,a)$ is a constructible point because it is the intersection of the constructible lines $y = a$ [which passes through two distinct constructible points $(-a,a)$ and (a,a)] and $x = 0$.

(iv) $\Rightarrow$ (ii) Follows by symmetry. □

Henceforth, whenever we say that a real number a is constructible, we mean that a is constructible from $\mathbf{Q}$.

5.4 Lemma. *If a is a constructible number, then $x = a$ and $y = a$ are constructible lines.*

Proof. If $a = 0$, then $x = 0$ is clearly constructible. So let $a \neq 0$. Then $x = a$ passes through two distinct constructible points $(a,0)$ and (a,a). Hence, $x = a$ is constructible. Similarly, $y = a$ is constructible. □

5.5 Lemma. *If a and b are constructible numbers, then (a,b) is a constructible point.*

Proof. (a,b) is the intersection of the constructible lines $x = a$ and $y = b$. □

5.6. Lemma. *If a and b are constructible numbers, then $a \pm b$ are also constructible.*

Proof. $(a \pm b, 0)$ are the points of intersection of the constructible line $y = 0$ and the constructible circle $(x-a)^2 + y^2 = b^2$ (the center $(a,0)$ is constructible; the point (a,b) through which the circle passes is constructible). □

5.7 Lemma. *If a and b are constructible numbers, then*

(i) ab is constructible.

(ii) a/b, $b \neq 0$, is constructible.

Proof. (i) The line $ay = -x + ab$ is constructible because it passes through constructible points $(0,b)$ and $(a,b-1)$. The intersection of this line with the constructible line $y = 0$ is $(ab,0)$. Hence, ab is constructible.

(Note that we have used the fact that if b is constructible, then $b - 1$ is also constructible – a consequence of Lemma 5.6.)

(ii) If $a = 0$, then it is clear. So let $a \neq 0$. Then the line $bx = a - y$ is constructible because it passes through two distinct constructible points: $(0,a)$ and $(a,a(1-b))$. The intersection of this line with the constructible line $y = 0$ is $(a/b,0)$. Hence, a/b is constructible. □

5.8 Lemma. *If $a > 0$ is constructible, then $\sqrt{a}$ is constructible.*

Proof. The point $(1,\sqrt{a})$ is a point of intersection of the constructible circle

$$\left(x - \frac{1+a}{2}\right)^2 + y^2 = \left(\frac{1+a}{2}\right)^2$$

[which passes through the constructible point (0,0) and which has constructible center $((1+a)/2,0)$] and the constructible line $x = 1$. Thus, $(1,\sqrt{a})$ is a constructible point.

Next, $(0,2\sqrt{a})$ is also a constructible point because it is a point of intersection of the constructible circle $(x-1)^2 + (y-\sqrt{a})^2 = a + 1$ and the constructible line $x = 0$. Therefore, $2\sqrt{a}$ is a constructible number. Then by Lemma 5.7, $\sqrt{a}$ is a constructible number. □

Proof of the theorem. Follows from Lemmas 5.6–5.8. □

5.9 Theorem. *If $u \in K_m$, where $K_0 = Q \subset K_1 \subset K_2 \subset \cdots \subset K_m$ is an ascending tower of fields K_i such that $[K_i:K_{i-1}] = 2$, then u is constructible.*

Equivalently, if $[\mathbf{Q}(u):\mathbf{Q}] = 2^t$ for some $t > 0$, then u is constructible.

Proof. Since rationals are constructible, the proof follows from Lemmas 5.6–5.8.

Definition. *An angle α is constructible by ruler and compass if the point* $(\cos\alpha, \sin\alpha)$ *is constructible from* $\mathbf{Q} \times \mathbf{Q}$.

5.10 Remark. The point $(\cos\alpha, \sin\alpha)$ is constructible from $\mathbf{Q}\times\mathbf{Q}$ if and only if $\cos\alpha$ is a constructible number (equivalently, if and only if $\sin\alpha$ is a constructible number).

Proof of the remark. Let $(\cos\alpha, \sin\alpha)$ be a constructible point. Then $(2\cos\alpha, 0)$ is a point of intersection of the constructible circle $(x-\cos\alpha)^2+(y-\sin\alpha)^2=1$ and the constructible line $y=0$, so $(2\cos\alpha, 0)$ is a constructible point. Thus, $2\cos\alpha$ is a constructible number. So by Theorem 5.2, $\cos\alpha$ is a constructible number. Conversely, assume that $\cos\alpha$ is a constructible number. Then by Theorem 5.2, $\sin\alpha$ is also a constructible number. This yields, by Lemma 5.3, that the points $(\cos\alpha, 0)$, $(\cos\alpha, \cos\alpha)$, $(0, \sin\alpha)$, and $(\sin\alpha, \sin\alpha)$ are constructible points. This means that the lines $x=\cos\alpha$ and $y=\sin\alpha$ are constructible lines. Hence, their intersection, namely, the point $(\cos\alpha, \sin\alpha)$, is a constructible point. The statement in parentheses can be proved the same way. □

5.11 Examples

(a) Problem of squaring a circle

It is impossible to construct a square equal in area to the area of a circle of radius 1.

Solution. Assume we have a circle of unit radius. If a is the side of the square whose area is equal to that of this circle, then $a^2=\pi$. But since π is not algebraic over $\mathbf{Q}$ (Hardy and Wright 1945), a^2 and, hence, a is not algebraic over $\mathbf{Q}$. So $[\mathbf{Q}(a):\mathbf{Q}]\neq 2^m$ for any positive integer m. Hence, by Theorem 5.1, a is not constructible by ruler and compass.

(b) Problem of duplicating a cube

It is impossible to construct a cube with a volume equal to twice the volume of a given cube by using ruler and compass only.

Solution. We can assume that the side of the given cube is 1. Let the side of the cube to be constructed be x. Then $x^3-2=0$. So we have to construct the number $2^{1/3}$ (the real cube root of 2). Because x^3-2 is irreducible over $\mathbf{Q}$,

$$[\mathbf{Q}(2^{1/3}):\mathbf{Q}]=3\neq \text{a power of } 2.$$

Thus, by Theorem 5.1, $2^{1/3}$ is not constructible from $\mathbf{Q}$ by ruler and compass.

(c) Problem of trisecting an angle

There exists an angle that cannot be trisected by using ruler and compass only.

Solution. We show that the angle $60°$ cannot be trisected by ruler and compass. Now if this angle can be trisected by ruler and compass, then the number $\cos 20°$ is constructible from $\mathbf{Q}$. This is equivalent to the constructibility of $2\cos 20°$ from $\mathbf{Q}$. Set $a = 2\cos 20°$. Then from $\cos 3\theta = 4\cos^3\theta - 3\cos\theta$, we deduce $a^3 - 3a - 1 = 0$. Because the polynomial $x^3 - 3x - 1 \in \mathbf{Q}[x]$ is irreducible over $\mathbf{Q}$ and has a root a, it follows that

$$[\mathbf{Q}(a):\mathbf{Q}] = 3 \neq \text{power of } 2.$$

Thus, by Theorem 5.1, $a = 2\cos 20°$ (or, equivalently, an angle of $20°$) cannot be constructed by ruler and compass from $\mathbf{Q}$. This completes the solution.

(d) Problem of constructing a regular n-gon

A regular n-gon is constructible (equivalently, the angle $\frac{2\pi}{n}$ is constructible) if and only if $\varphi(n)$ is a power of 2.

Solution. Let $\omega = \cos\frac{2\pi}{n} + i\sin\frac{2\pi}{n}$, where ω is a primitive nth root of unity. Then $\omega + \bar{\omega} = 2\cos\frac{2\pi}{n}$. Set $u = \cos\frac{2\pi}{n}$. To show that u is constructible, we need to prove that $[\mathbf{Q}(u):\mathbf{Q}] = 2^k$, $k \geq 0$. Consider the following tower:

$$\begin{array}{c} \mathbf{Q}(\omega) \\ | \\ \mathbf{Q}(u) \\ | \\ \mathbf{Q} \end{array}$$

Now $\left(\omega - \cos\frac{2\pi}{n}\right)^2 = -\sin^2\frac{2\pi}{n}$, and so $\omega^2 - 2\cos\frac{2\pi}{n}\omega + 1 = 0$. Thus ω satisfies

$$x^2 - \left(2\cos\frac{2\pi}{n}\right)x + 1 \in \mathbf{Q}(u),$$

which is clearly an irreducible polynomial over $\mathbf{Q}(u)$, proving that

$[\mathbf{Q}(\omega):\mathbf{Q}(u)] = 2$. Since by Theorem 1.4 $[\mathbf{Q}(\omega):\mathbf{Q}] = \varphi(n)$, $[\mathbf{Q}(u):\mathbf{Q}] = \frac{1}{2}\varphi(n)$. This shows that u is constructible if and only if $\varphi(n)$ is a power of 2.

Problems

1. Show that the angle $2\pi/5$ can be trisected using ruler and compass.
2. Show that it is impossible to construct a regular 9-gon or 7-gon using ruler and compass.
3. Show that it is possible to trisect $54°$ using ruler and compass.
4. Prove that the regular 17-gon is constructible with ruler and compass.
5. Find which of the following numbers are constructible:
   (i) $\sqrt{3}+1$. (ii) π^2+1.
   (iii) $\sqrt{\sqrt{3}-1}+1$. (iv) $\sqrt[3]{2}+1$.
   (v) $\sqrt[4]{\sqrt{2}+\sqrt{5}}$.

PART V

Additional topics

CHAPTER 19

Noetherian and artinian modules and rings

1 $\mathrm{Hom}_R(\oplus M_i, \oplus M_i)$

We begin this chapter by proving a result that is quite useful in studying the ring of endomorphisms of a direct sum of modules and that has applications in subsequent sections. Unless otherwise stated, all modules considered in this chapter are left modules.

1.1 Theorem. *Let $M = \oplus\sum_{i=1}^{k} M_i$ be a direct sum of R-modules M_i. Then*

$$\mathrm{Hom}_R(M,M) \simeq \begin{bmatrix} \mathrm{Hom}_R(M_1,M_1) & \mathrm{Hom}_R(M_2,M_1) & \cdots & \mathrm{Hom}_R(M_k,M_1) \\ \mathrm{Hom}_R(M_1,M_2) & \mathrm{Hom}_R(M_2,M_2) & \cdots & \mathrm{Hom}_R(M_k,M_2) \\ \vdots & \vdots & & \vdots \\ \mathrm{Hom}_R(M_1,M_k) & \mathrm{Hom}_R(M_2,M_k) & \cdots & \mathrm{Hom}_R(M_k,M_k) \end{bmatrix}$$

as rings. (The right side is a ring T, say, of $k \times k$ matrices $f = (f_{ij})$ under the usual matrix addition and multiplication, where $f_{ij} \in \mathrm{Hom}_R(M_j, M_i)$).

Proof. Let $\phi \in \mathrm{Hom}_R(M,M)$. Let $\lambda_j : M_j \to M$ and $\pi_i : M \to M_i$ be the natural inclusion and projection mappings, respectively. Then $\pi_i\phi\lambda_j \in \mathrm{Hom}_R(M_j,M_i)$. Define a mapping σ: $\mathrm{Hom}_R(M,M) \to T$ by setting $\sigma(\phi)$ to be the $k \times k$ matrix $(\pi_i\phi\lambda_j)$ whose (i, j) entry is $\pi_i\phi\lambda_j$, where $\phi \in \mathrm{Hom}_R(M,M)$. We proceed to show that σ is an isomorphism. So let $\phi,\psi \in \mathrm{Hom}_R(M,M)$. Then

$$\sigma(\phi + \psi) = (\pi_i(\phi + \psi)\lambda_j) = (\pi_i\phi\lambda_j) + (\pi_i\psi\lambda_j) = \sigma(\phi) + \sigma(\psi).$$

Further,

$$\sigma(\phi)\sigma(\psi) = (\pi_i\phi\lambda_l)(\pi_l\psi\lambda_j) = \left(\sum_{l=1}^{k} \pi_i\phi\lambda_l\pi_l\psi\lambda_j\right),$$

by definition of multiplication of matrices. But since $\Sigma_{l=1}^{k}\lambda_l\pi_l = 1$, it follows that

$$\sigma(\phi)\sigma(\psi) = (\pi_i\phi\psi\lambda_j) = \sigma(\phi\psi).$$

Therefore, σ is a homomorphism.

To prove that σ is injective, let $\sigma(\phi) = (\pi_i\phi\lambda_j) = 0$. Then $\pi_i\phi\lambda_j = 0$, $1 \leq i, j \leq k$. This implies $\Sigma_{i=1}^{k}\pi_i\phi\lambda_j = 0$. But since $\Sigma_{i=1}^{k}\pi_i = 1$, we obtain $\phi\lambda_j = 0$, $1 \leq j \leq k$. In a similar fashion we get $\phi = 0$, which proves that σ is injective. To prove that σ is surjective, let $f = (f_{ij}) \in T$, where $f_{ij}: M_j \to M_i$ is an R-homomorphism. Set $\phi = \Sigma_{i,j}\lambda_i f_{ij}\pi_j$. Then $\phi \in \mathrm{Hom}_R(M,M)$. By definition of σ, $\sigma(\phi)$ is the $k \times k$ matrix whose (s,t) entry is $\pi_s(\Sigma_{i,j}\lambda_i f_{ij}\pi_j)\lambda_t = f_{st}$, because $\pi_p\lambda_q = \delta_{pq}$. Hence, $\sigma(\phi) = (f_{st}) = f$. Thus, σ is also surjective. □

Problems

1. Let $M = M_1 \oplus M_2$ be the direct sum of simple modules M_1 and M_2 such that $M_1 \not\simeq M_2$. Show that the ring $\mathrm{End}_R(M)$ is a direct sum of division rings. [*Hint:* $\mathrm{Hom}_R(M_1,M_2) = 0$, etc.]
2. Let $M = M_1 \oplus M_2$ be the direct sum of isomorphic simple modules M_1, M_2. Show that $\mathrm{End}_R(M) \simeq D_2$, the 2×2 matrix ring over a division ring.

2 Noetherian and artinian modules

Recall that an R-module M is finitely generated if M is generated by a finite subset of M; that is, if there exist elements $x_1,\ldots,x_n \in M$ such that $M = (x_1,\ldots,x_n)$. This is equivalent to the statement: If $M = \Sigma_{\alpha\in\Lambda}M_\alpha$ is a sum of submodules M_α then there exists a finite subset Λ' of Λ such that $M = \Sigma_{\alpha\in\Lambda'}M_\alpha$. We now define a concept that is dual to that of a finitely generated module.

Definition. *An R-module M is said to be* finitely cogenerated *if, for each family* $(M_\alpha)_{\alpha\in\Lambda}$ *of submodules of M,*

$$\bigcap_{\alpha\in\Lambda} M_\alpha = 0 \Rightarrow \bigcap_{\alpha\in\Lambda'} M_\alpha = 0$$

for some finite subset Λ' *of* Λ.

We show that finitely generated and finitely cogenerated modules can be characterized as modules that have certain chain conditions on their submodules.

Definition. *An R-module M is called* noetherian (artinian) *if for every ascending (descending) sequence of R-submodules of M,*

$$M_1 \subset M_2 \subset M_3 \subset \cdots \qquad (M_1 \supset M_2 \supset M_3 \supset \cdots),$$

there exists a positive integer k such that

$$M_k = M_{k+1} = M_{k+2} = \cdots.$$

If M is noetherian (artinian), then we also say that the ascending (descending) chain condition for submodules holds in M, or M has acc (dcc) on submodules, or simply that M has acc (dcc).

Because the ring of integers **Z** is a principal ideal ring, any ascending chain of ideals of **Z** is of the form

$$(n) \subset (n_1) \subset (n_2) \subset \cdots,$$

where $n, n_1, n_2, \ldots$ are in **Z**. Because $(n_i) \subseteq (n_{i+1})$ implies $n_{i+1} | n_i$, any ascending chain of ideals in **Z** starting with n can have at most n distinct terms. This shows that **Z** as a **Z**-module is noetherian. But **Z** as a **Z**-module has an infinite properly descending chain

$$(n) \supset (n^2) \supset (n^3) \supset \cdots,$$

showing that **Z** is not artinian as a **Z**-module.

Before we give more examples, we prove two theorems providing us with criteria for a module to be noetherian or artinian.

2.1 Theorem. *For an R-module M the following are equivalent:*

(i) M is noetherian.
(ii) Every submodule of M is finitely generated.
(iii) Every nonempty set S of submodules of M has a maximal element (that is, a submodule M_0 in S such that for any submodule N_0 in S with $N_0 \supset M_0$ we have $N_0 = M_0$).

Proof. (i) $\Rightarrow$ (ii). Let N be a submodule of M. Assume that N is not finitely generated. For any positive integer k let $a_1, \ldots, a_k \in N$. Then $N \neq (a_1, \ldots, a_k)$. Choose $a_{k+1} \in N$ such that $a_{k+1} \notin (a_1, \ldots, a_k)$. We then obtain an infinite properly ascending chain

$$(a_1) \subsetneq (a_1, a_2) \subsetneq \cdots \subsetneq (a_1, \ldots, a_k) \subsetneq (a_1, \ldots, a_{k+1}) \subsetneq \cdots$$

of submodules of M, which is a contradiction to the hypothesis. Hence, N is finitely generated.

(ii) $\Rightarrow$ (iii) Let N_0 be an element of S. If N_0 is not maximal, it is properly contained in a submodule $N_1 \in S$. If N_1 is not maximal, then N_1 is properly contained in a submodule $N_2 \in S$. In case S has no maximal elements, we obtain an infinite properly ascending chain of submodules $N_0 \subset N_1 \subset N_2 \subset \cdots$ of M. Let $N = \bigcup_i N_i$. N is also a submodule of M. For let $x, y \in \bigcup_i N_i$ and $r \in R$. Then $x \in N_u$, $y \in N_v$. Because either $N_u \subset N_v$ or $N_v \subset N_u$, both x and y lie in one submodule N_u or N_v, and, hence, $x - y$ and rx lie in the same submodule. This implies $x - y \in N$ and $rx \in N$, and, hence, N is a submodule of M. By (ii) N is finitely generated. So there exist elements $a_1, a_2, \ldots, a_n \in N$ such that $N = (a_1, a_2, \ldots, a_n)$. Now $a_1, a_2, \ldots, a_n$ belong to a finite number ($\leq n$) of submodules N_i, $i = 1, 2, \ldots$. Hence, there exists N_k such that all a_i, $1 \leq i \leq n$, lie in N_k. Because $N_k \subset N$ and N is the smallest submodule containing all a_i, $1 \leq i \leq n$, it follows that $N_k = N$. But then $N_k = N_{k+1} = \cdots$, a contradiction. Thus, S must have a maximal element.

(iii) $\Rightarrow$ (i) Suppose we have an ascending sequence of submodules of M,

$$M_1 \subset M_2 \subset M_3 \cdots.$$

By (iii) the sequence $M_1, M_2, M_3, \ldots$ has a maximal element say M_k. But then $M_k = M_{k+1} = \cdots$. Hence, M is noetherian. □

The next theorem is dual to Theorem 2.1.

2.2 Theorem. *For an R-module M the following are equivalent:*

(i) M is artinian.

(ii) Every quotient module of M is finitely cogenerated.

(iii) Every nonempty set S of submodules of M has a minimal element (that is, a submodule M_0 in S such that for any submodule N_0 in S with $N_0 \subset M_0$, we have $N_0 = M_0$).

Proof. The proof is similar (indeed dual) to the proof of Theorem 2.1 and is thus left as an exercise. □

Definition. *A ring R is called a* left noetherian (artinian) *ring if R regarded as a left R-module is noetherian (artinian).*

Similarly, we define right noetherian (artinian) rings.

Throughout, unless otherwise stated, by a noetherian (artinian) ring we mean a left noetherian (artinian) ring.

In view of the importance of noetherian and artinian rings in themselves, we rewrite Theorems 2.1 and 2.2 for rings as follows:

2.3 Theorem. *Let R be a ring. Then the following are equivalent:*

(i) R is noetherian (artinian).
(ii) Let A be any left ideal of R. Then A (R/A) is finitely generated (cogenerated).
(iii) Every nonempty set S of left ideals of R has a maximal (minimal) element.

In particular, every principal left ideal ring is a noetherian ring.

2.4 Examples

(a) Let V be an n-dimensional vector space over a field F. Then V is both noetherian and artinian. For, if W is a proper subspace of V, then dim $W <$ dim $V = n$. Thus any properly ascending (or descending) chain of subspaces cannot have more than $n + 1$ terms.

(b) Let A be a finite-dimensional algebra with unity over a field F. Then A as a ring is both left and right noetherian as well as artinian. To see this, let $[A:F] = n$. If we observe that each left or right ideal is a subspace of A over F, it follows that any properly ascending (or descending) chain cannot contain more than $n + 1$ terms.

In particular, (i) if G is a finite group and F a field, then the group algebra $F(G)$ is both a noetherian and an artinian ring; (ii) the $m \times m$ matrix ring F_m over a field F is also a noetherian and artinian ring; (iii) the ring of upper (as well as lower) triangular matrices over a field F is both noetherian and artinian.

(c) Let $R = F[x]$ be a polynomial ring over a field F in x. Because $F[x]$ is a principal ideal domain, it follows by Theorem 2.3 that $F[x]$ is a noetherian ring. But $F[x]$ is not artinian, for there exists a properly descending chain of ideals in R, namely,

$$R \supset Rx \supset Rx^2 \supset \cdots.$$

However, every proper homomorphic image R/A, where A is a nonzero ideal in R, is artinian, because we know that R is a PID. Hence, $A = (p(x))$, $p(x) \in F[x]$. Let $p(x) = a_0 + a_1x + \cdots + a_nx^n$. Then $F[x]/(p(x))$ is a finite-dimensional algebra over F with a basis $\{1, \bar{x}, \ldots, \bar{x}^{n-1}\}$. Hence, by Example (b), $R/A = F[x]/(p(x))$ is an artinian ring.

(d) Let D_n be the $n \times n$ matrix ring over a division ring D. Then D_n is an n^2-dimensional vector space over D, and each left ideal as well as each right ideal of D_n is a subspace over D. Thus, any ascending or descending chain of left (as well as right) ideals cannot contain more than n^2+1 terms. Thus, D_n is both noetherian and artinian ring.

(e) Let p be a prime number, and let

$$R = \mathbf{Z}(p^\infty) = \left\{ \frac{m}{p^n} \in \mathbf{Q} \,\middle|\, 0 \le \frac{m}{p^n} < 1 \right\}$$

be the ring where addition is modulo positive integers, and multiplication is trivial; that is, $ab = 0$ for all $a, b \in R$. Then

(i) Each ideal in R is of the form

$$A_k = \left\{ \frac{1}{p^k}, \frac{2}{p^k}, \ldots, \frac{p^k - 1}{p^k}, 0 \right\},$$

where k is some positive integer.

(ii) R is artinian but not noetherian.

Solution. (i) Let $A \neq (0)$ be any ideal of R, and let k be the smallest positive integer such that for some positive integer m, $m/p^k \notin A$. Consider n/p^i, with $i \ge k$ and $(n, p) = 1$. We assert that $n/p^i \notin A$. Now $n/p^i \in A$ implies $np^{i-k}/p^i = n/p^k \in A$. Also, by choice of k, $1/p^{k-1} \in A$. Because $(n, p) = 1$, we can find integers a and b such that $an + bp = 1$. Then from n/p^k, $1/p^{k-1} \in A$, we have that na/p^k (reduced modulo whole numbers) and bp/p^k (reduced modulo whole numbers) lie in A. Hence, $1/p^k \in A$, a contradiction. Thus, no n/p^i, $i \ge k$, $(n, p) = 1$ can lie in A. Hence,

$$A = \left\{ \frac{1}{p^{k-1}}, \frac{2}{p^{k-1}}, \ldots, \frac{p^{k-1} - 1}{p^{k-1}}, 0 \right\}.$$

This ideal is denoted by A_{k-1}.

(iii) Because each ideal contains a finite number of elements, each descending chain of ideals must be finite. Hence, R is artinian.

Clearly, the chain

$$A_1 \subset A_2 \subset A_3 \subset \cdots$$

is an infinite properly ascending chain of left ideals, showing that R is not noetherian. Note that although each ideal $A \neq R$ is finite and, hence, finitely generated, R itself is not finitely generated.

2.5 Theorem. *Every submodule and every homomorphic image of a noetherian (artinian) module is noetherian (artinian).*

Proof. Follows at once from Theorem 2.1 (Theorem 2.2). □

2.6 Theorem. *Let M be an R-module, and let N be an R-submodule of M. Then M is noetherian (artinian) if and only if both N and M/N are noetherian (artinian).*

Proof. Let N and M/N be noetherian, and let K be any submodule of M. Then $(K+N)/N$ is a submodule of M/N, and, hence, it is finitely generated (Theorem 2.1). But then $(K+N)/N \simeq K/(N \cap K)$ implies $K/(N \cap K)$ is finitely generated, say

$$\frac{K}{N \cap K} = (\bar{x}_1) + \cdots + (\bar{x}_m), \qquad \bar{x}_i \in \frac{K}{N \cap K}.$$

Then

$$K = (x_1) + \cdots + (x_m) + N \cap K, \qquad x_i \in K.$$

Further, because N is noetherian, its submodule $N \cap K$ is finitely generated, say by $y_1, \ldots, y_n \in N \cap K$. This implies

$$K = (x_1) + \cdots + (x_m) + (y_1) + \cdots + (y_n).$$

Hence, K is finitely generated, so M is noetherian. The converse is Theorem 2.5. The proof for the artinian case is similar. □

An equivalent statement of Theorem 2.6 in the terminology of exact sequences is as follows.

Let $0 \to M_1 \to M \to M_2 \to 0$ be an exact sequence of R-modules. Then M is noetherian (artinian) if and only if both M_1 and M_2 are noetherian (artinian).

2.7 Theorem. *A subring of a noetherian (artinian) ring need not be noetherian (artinian).*

Proof. For the artinian case the ring of rational numbers **Q** is an artinian ring, but its subring **Z** is not an artinian ring.

For the noetherian case, the ring of 2×2 matrices over the rational numbers **Q** is a noetherian ring, but its subring $\left[\begin{smallmatrix} \mathbf{Z} & \mathbf{Q} \\ 0 & \mathbf{Q} \end{smallmatrix}\right]$ is not noetherian – that is, not left noetherian [see Example 2.15(e)]. □

2.8 Theorem. *Let R_i, $1 \leq i \leq n$, be a family of noetherian (artinian) rings each with a unity element. Then their direct sum $R = \oplus \sum_{i=1}^{n} R_i$ is again noetherian (artinian).*

Proof. We know that each left ideal A of R is of the form $A_1 \oplus \ldots \oplus A_n$, where A_i are left ideals in R_i. So if a left ideal $B = B_1 \oplus \ldots \oplus B_n$ of R is such that $A \subset B$, then it is clear that $A_i \subset B_i$, $1 \leq i \leq n$. Hence, any properly ascending (descending) chain of left ideals in R must be finite because each R_i is noetherian (artinian). □

2.9 Theorem. *If J is a nil left ideal in an artinian ring R, then J is nilpotent.*

Proof. Suppose $J^k \neq (0)$ for any positive integer k. Consider a family $\{J, J^2, J^3, \ldots\}$. Because R is artinian, this family has a minimal element, say $B = J^m$. Then $B^2 = J^{2m} \subset J^m = B$ implies $B^2 = B$. Consider another family

$$\mathcal{F} = \{A \mid A \text{ is a left ideal contained in } B \text{ with } BA \neq (0)\}.$$

Then $\mathcal{F} \neq \emptyset$ because $B \in \mathcal{F}$. Let A be a minimal element in $\mathcal{F}$. Then $BA \neq (0)$. This implies there exists an element $a \in A$ such that $Ba \neq 0$. But $Ba \subset A$ and $B(Ba) = B^2a = Ba \neq 0$. Thus, $Ba \in \mathcal{F}$. Hence, by minimality of A, $Ba = A$. This gives that there exists an element $b \in B$ such that $ba = a$. This implies $b^i a = a$ for all positive integers i. But because b is a nilpotent element, this implies $a = 0$, a contradiction. Therefore, for some positive integer k, $J^k = (0)$. □

2.10 Lemma. *Let R be a noetherian ring. Then the sum of nilpotent ideals in R is a nilpotent ideal.*

Proof. Let $B = \Sigma_{i \in \Lambda} A_i$ be the sum of nilpotent ideals in R. Because R is noetherian (i.e., left noetherian), B is finitely generated as a left ideal. Suppose $B = (x_1, \ldots, x_m)$. Then each x_i lies in the sum of finitely many A_i's. Hence, B is contained in the sum of a finite number of A_i's, say (after reindexing if necessary) $A_1, \ldots, A_n$. Thus, $B = A_1 + \cdots + A_n$. Then by Problem 1 of Section 5 in Chapter 10, B is nilpotent. □

Recall that if S is any nonempty subset of a ring R, then $l(S) = \{x \in R \mid xS = 0\}$ is a left ideal of R called the *left annihilator* of S in R.

2.11 Theorem. *Let R be a noetherian ring having no nonzero nilpotent ideals. Then R has no nonzero nil ideals.*

Proof. Let N be a nonzero nil ideal in R. Let $\mathcal{F} = \{l(n) \mid n \in N, n \neq 0\}$ be a family of left annihilator ideals. Because R is noetherian, $\mathcal{F}$ has a maximal member, say $l(n)$. Let $x \in R$. Then $nx \in N$, so there exists a smallest positive integer k such that $(nx)^k = 0$. Now, clearly, $l(n) \subset l((nx)^{k-1})$. Because $(nx)^{k-1} \neq 0$, $l((nx)^{k-1}) \in \mathcal{F}$. But then by maximality of $l(n)$, $l(n) = l((nx)^{k-1})$. Now

$$(nx)^k = 0 \Rightarrow nx \in l((nx)^{k-1}) = l(n) \Rightarrow nxn = 0.$$

Now $(RnR)^2 = RnRRnR = 0$. Therefore, by hypothesis, $RnR = 0$. If $1 \in R$, then $n = 0$, a contradiction. So in this case we are done. Otherwise, consider the ideal $(n) = nR + Rn + RnR + n\mathbf{Z}$ generated by n. Set $A = nR + Rn$. Because $nxn = 0$, for all $x \in R$, $A^2 = 0$. Thus, $(n) = A + n\mathbf{Z}$. But then if $n^k = 0$, we have $(A + n\mathbf{Z})^k = 0$. Therefore, by hypothesis, $A + n\mathbf{Z} = 0$, which gives $n = 0$, a contradiction. Hence, R has no nonzero nil ideals. □

Remark. Indeed, one can similarly show that R has no nonzero right or left nil ideals.

Next we show that a nil ideal in a noetherian ring is nilpotent.

2.12 Theorem. *Let N be a nil ideal in a noetherian ring R. Then N is nilpotent.*

Proof. Let T be the sum of nilpotent ideals in R. Then R/T has no nonzero nilpotent ideals, for if A/T is nilpotent, then $(A/T)^m = (0)$ implies $A^m/T = (0)$; so $A^m \subset T$. But since, by Lemma 2.10, T is nilpotent, there exists a positive integer k such that $(A^m)^k = (0)$. Hence, A itself is nilpotent, so $A \subset T$. This implies $A/T = (\bar{0})$.

Consider the nil ideal $(N + T)/T$ in R/T. By Theorem 2.11, $(N + T)/T = (0)$. This implies $N \subset T$, which is a nilpotent ideal. Hence, N is nilpotent. □

2.13 Remark. *If R is an artinian ring with identity, then it is known that R is noetherian.*

2.14 Theorem (Hilbert basis theorem). *Let R be a noetherian ring. Then the polynomial ring $R[x]$ is also a noetherian ring.*

Proof. Let $\mathcal{F}$ and $\mathcal{F}'$ be the families of left ideals of R and $R[x]$, respectively. Let n be a nonnegative integer. Define a mapping $\phi_n\colon \mathcal{F}' \to \mathcal{F}$, where

$$\phi_n(I) = \{a \in R \mid \exists ax^n + bx^{n-1} + \cdots \in I, a \neq 0\} \cup \{0\}.$$

It is easy to verify that $\phi_n(I) \in \mathcal{F}$. We claim that if $I, J \in \mathcal{F}'$ with $I \subset J$ and

$\phi_n(I) = \phi_n(J)$ for all $n \geq 0$, then $I = J$. Let $0 \neq f(x) \in J$ be of degree m. Because $\phi_m(I) = \phi_m(J)$, there exists $g_m(x) \in I$ with leading coefficient the same as that of $f(x)$, and $f(x) - g_m(x)$ is either 0 or of degree at most $m - 1$. Suppose $f(x) - g_m(x) \neq 0$. Because $f(x) - g_m(x) \in J$, we can similarly find $g_{m-1}(x) \in I$ such that $f(x) - g_m(x) - g_{m-1}(x) \in J$ and is either 0 or of degree at most $m - 2$. Continuing like this, we arrive, after at most m steps, at

$$f(x) - g_m(x) - g_{m-1}(x) - \cdots - g_i(x) = 0,$$

where $g_m(x), g_{m-1}(x), \ldots \in I$. But then $f(x) \in I$, which proves that $I = J$, as claimed.

Let $A_1 \subset A_2 \subset A_3 \subset \cdots$ be an ascending sequence of left ideals of $R[x]$. Then for each nonnegative integer n,

$$\phi_n(A_1) \subset \phi_n(A_2) \subset \phi_n(A_3) \subset \cdots$$

is an ascending sequence of left ideals of R; hence, there exists a positive integer $k(n)$ such that

$$\phi_n(A_{k(n)}) = \phi_n(A_{k(n)+1}) = \cdots. \tag{1}$$

Further, because R is noetherian, the collection of left ideals $(\phi_n(A_i))$, $n \in \mathbf{N}$, $i \in \mathbf{N}$, has a maximal element, say $\phi_p(A_q)$ (Theorem 2.3). Then

$$\begin{aligned} \phi_p(A_q) &= \phi_n(A_q) \qquad \text{(for all n} \geq p) \\ &= \phi_n(A_j) \qquad \text{(for all n} \geq p, j \geq q). \end{aligned}$$

Therefore, we may choose $k(n) = q$ for all $n \geq p$ in (1). Moreover, if $s = k(1) \cdots k(p-1)q$, then $\phi_n(A_s) = \phi_n(A_{s+1}) = \cdots$ for all $n \in \mathbf{N}$. Hence, by the result proved in the first paragraph, $A_s = A_{s+1} = \cdots$. Therefore, $R[x]$ is noetherian. □

2.15 Examples

(a) If R is noetherian, then each ideal contains a finite product of prime ideals.

Solution. Suppose that the family $\mathcal{F}$ of ideals in R that do not contain any product of prime ideals is nonempty. Then by Theorem 2.3, $\mathcal{F}$ has a maximal element, say A. Because $A \in \mathcal{F}$, A is not a prime ideal. Hence, there exist ideals B and C of R such that $BC \subset A$, but $B \not\subset A$, $C \not\subset A$. Consider

$$(B + A)(C + A) \subset BC + BA + AC + A^2 \subset A.$$

Because, $B + A \supsetneq A$, and $C + A \supsetneq A$, both $B + A$ and $C + A$ contain a product of prime ideals, and, hence, $(B+A)(C+A)$ contains a product of prime ideals. This implies that A contains a product of prime ideals, a contradiction. Hence, $\mathcal{F}=\varnothing$.

(b) A Boolean noetherian ring is finite and is indeed a finite direct product of fields with two elements.

Solution. Let R be a Boolean ring that is noetherian, and let $P \neq R$ be a prime ideal in R. Consider the homomorphic image R/P of R. Clearly, for all $\bar{x}=x+P \in R/P$, $\bar{x}^2=\bar{x}$. Because R/P is an integral domain and the only idempotents of an integral domain are zero and (possibly) the unity element, it follows that if $P \neq R$, then R/P has only two elements. Thus, R/P is a field isomorphic to $\mathbf{Z}/(2)$, so P is a maximal ideal. Now by Example 2.15(a), there exist prime ideals $P_1,\ldots,P_m$ such that $P_1P_2\cdots P_m=(0)$. But then $P_1 \cap P_2 \cap \cdots \cap P_m \subset P_i$, $1 \leq i \leq m$, gives $(P_1 \cap P_2 \cap \cdots \cap P_m)^m \subset P_1P_2 \cdots P_m=(0)$. Thus, $(P_1 \cap P_2 \cap \cdots \cap P_m)^m=(0)$, and because R is Boolean, $P_1 \cap P_2 \cap \cdots \cap P_m=(0)$. Further, because P_i are also maximal, $P_i+P_j=R$, $i \neq j$. Then by Example 3.4(a) of Chapter 10, we get

$$R \simeq R/P_1 \times R/P_2 \times \cdots \times R/P_m \simeq \mathbf{Z}/(2) \times \mathbf{Z}/(2) \times \cdots \times \mathbf{Z}/(2).$$

This completes the solution. Incidently, this also gives that the cardinality of R is 2^k for some positive integer k.

Remark. The above problem shows that any commutative noetherian ring R with no nonzero nilpotent elements in which each prime ideal $P \neq R$ is maximal is a finite direct sum of fields.

(c) If R is a noetherian ring, then $ab=1$, $a,b \in R$, if and only if $ba=1$.

Solution. Let $ab=1$ and $ba \neq 1$. Define

$$e_{ij}=b^{i-1}a^{j-1}-b^ia^j, \qquad i,j=1,2,3,\ldots.$$

It is easy to check that

$$e_{ij}e_{kl}=\delta_{jk}e_{il}.$$

This gives rise to an infinite set $\{e_{ii} | i=1,2,3,\ldots\}$ of orthogonal idempotents in R. Set $e_{ii}=f_i$ for all $i=1,2,3,\ldots$. Thus, we get an infinite properly ascending chain of left ideals

$$Rf_1 \subsetneq Rf_1 \oplus Rf_2 \subsetneq Rf_1 \oplus Rf_2 \oplus Rf_3 \subsetneq \cdots,$$

which is a contradiction. Hence, $ba=1$, which completes the solution.

(d) Any right ideal of the ring $R = \begin{pmatrix} \mathbf{Z} & \mathbf{Q} \\ 0 & \mathbf{Q} \end{pmatrix}$ can be generated by at most two elements; hence, R is right noetherian.

Solution. Let A be a nonzero right ideal of R. Let

$$X = \left\{ n \in \mathbf{Z} \,\middle|\, \begin{pmatrix} n & x \\ 0 & y \end{pmatrix} \in A \text{ for some } x, y \in \mathbf{Q} \right\}$$

Then it is clear that X is an ideal in $\mathbf{Z}$. Hence, $X = (n_0)$ for some $n_0 \in \mathbf{Z}$, because $\mathbf{Z}$ is a principal ideal ring.

Case 1. $X \neq (0)$. We claim

$$A = \begin{pmatrix} n_0 & 1 \\ 0 & 1 \end{pmatrix} R \quad \text{or} \quad A = \begin{pmatrix} n_0 & 1 \\ 0 & 0 \end{pmatrix} R;$$

that is, A is a principal right ideal of R generated by

$$\begin{pmatrix} n_0 & 1 \\ 0 & 1 \end{pmatrix}$$

or by

$$\begin{pmatrix} n_0 & 1 \\ 0 & 0 \end{pmatrix}.$$

First, let

$$\begin{pmatrix} n_0 & a \\ 0 & b \end{pmatrix} \in A, \qquad b \neq 0.$$

Then

$$\begin{pmatrix} n_0 & a \\ 0 & b \end{pmatrix} \begin{pmatrix} k & x \\ 0 & y \end{pmatrix} = \begin{pmatrix} n_0 k & n_0 x + ay \\ 0 & by \end{pmatrix} \in A$$

for all $k \in \mathbf{Z}$, $x, y \in \mathbf{Q}$. Taking $k = 1$, $y = 1/b$, $x = (1 - a/b)/n_0$, we see that

$$\begin{pmatrix} n_0 & 1 \\ 0 & 1 \end{pmatrix} \in A.$$

Next, let

$$\begin{pmatrix} n_0 m & c \\ 0 & d \end{pmatrix}$$

be an arbitrary element of A. Then

$$\begin{pmatrix} n_0 m & c \\ 0 & d \end{pmatrix} = \begin{pmatrix} n_0 & 1 \\ 0 & 1 \end{pmatrix} \begin{pmatrix} m & (c-d)/n_0 \\ 0 & d \end{pmatrix}.$$

Hence,

$$A = \begin{pmatrix} n_0 & 1 \\ 0 & 1 \end{pmatrix} R.$$

In case the (2,2) entry of each element of A is 0, then, exactly as before, it follows that

$$A = \begin{pmatrix} n_0 & 1 \\ 0 & 0 \end{pmatrix} R.$$

Case 2. $X = (0)$. Then either (i) A is a principal right ideal, or (ii) A is not a principal right ideal.

(i) Suppose A is a principal right ideal generated by some nonzero element of the form $\left(\begin{smallmatrix} 0 & \beta \\ 0 & \gamma \end{smallmatrix}\right)$, $\beta,\gamma \in \mathbf{Q}$. Then $A = \left(\begin{smallmatrix} 0 & \beta \\ 0 & \gamma \end{smallmatrix}\right)R$; that is,

$$A = \left\{ \begin{pmatrix} 0 & \beta y \\ 0 & \gamma y \end{pmatrix} \middle| \, y \in \mathbf{Q} \right\},$$

which includes the particular cases $A = \left(\begin{smallmatrix} 0 & 0 \\ 0 & \mathbf{Q} \end{smallmatrix}\right)$, when $\beta = 0$, and $A = \left(\begin{smallmatrix} 0 & \mathbf{Q} \\ 0 & 0 \end{smallmatrix}\right)$, when $\gamma = 0$.

(ii) A is not a principal right ideal. In this situation there must exist an element $\left(\begin{smallmatrix} 0 & \beta \\ 0 & \gamma \end{smallmatrix}\right) \in A$ such that $\beta \neq 0$, $\gamma \neq 0$. Otherwise, A becomes a principal right ideal as shown next. Suppose for all $\left(\begin{smallmatrix} 0 & \beta \\ 0 & \gamma \end{smallmatrix}\right) \in A$, $\gamma = 0$. Then

$$\begin{pmatrix} 0 & \beta \\ 0 & 0 \end{pmatrix} \in A \Rightarrow \begin{pmatrix} 0 & \beta \\ 0 & 0 \end{pmatrix} \begin{pmatrix} 0 & 0 \\ 0 & 1/\beta \end{pmatrix} = \begin{pmatrix} 0 & 1 \\ 0 & 0 \end{pmatrix} \in A;$$

so

$$\begin{pmatrix} 0 & 1 \\ 0 & 0 \end{pmatrix} \begin{pmatrix} n & x \\ 0 & y \end{pmatrix} = \begin{pmatrix} 0 & y \\ 0 & 0 \end{pmatrix} \in A \qquad \text{for all } y \in \mathbf{Q}.$$

So $A = \left(\begin{smallmatrix} 0 & \mathbf{Q} \\ 0 & 0 \end{smallmatrix}\right)$, the principal right ideal of R generated by $\left(\begin{smallmatrix} 0 & 1 \\ 0 & 0 \end{smallmatrix}\right)$. Similarly, if for all $\left(\begin{smallmatrix} 0 & \beta \\ 0 & \gamma \end{smallmatrix}\right) \in A$, $\beta = 0$, then

$$A = \begin{pmatrix} 0 & 0 \\ 0 & 1 \end{pmatrix} R = \begin{pmatrix} 0 & 0 \\ 0 & \mathbf{Q} \end{pmatrix}.$$

Hence, in possibility (ii) there exists $\left(\begin{smallmatrix} 0 & \beta \\ 0 & \gamma \end{smallmatrix}\right) \in A$, $\beta \neq 0$, $\gamma \neq 0$, such that

$$A \supsetneq \begin{pmatrix} 0 & \beta \\ 0 & \gamma \end{pmatrix} R = \left\{ \begin{pmatrix} 0 & \beta y \\ 0 & \gamma y \end{pmatrix} \middle| y \in \mathbf{Q} \right\}.$$

So we can choose

$$\begin{bmatrix} 0 & \beta' \\ 0 & \gamma' \end{bmatrix} \in A$$

such that $\beta'/\beta \neq \gamma'/\gamma$; that is, $\beta'\gamma - \beta\gamma' \neq 0$. Hence, given arbitrary $p,q \in \mathbf{Q}$, there exist $x,y \in \mathbf{Q}$ such that $p = \beta x + \beta' y$, $q = \gamma x + \gamma' y$. Therefore,

$$\begin{pmatrix} 0 & p \\ 0 & q \end{pmatrix} = \begin{pmatrix} 0 & \beta \\ 0 & \gamma \end{pmatrix}\begin{pmatrix} 0 & x \\ 0 & x \end{pmatrix} + \begin{pmatrix} 0 & \beta' \\ 0 & \gamma' \end{pmatrix}\begin{pmatrix} 0 & y \\ 0 & y \end{pmatrix} \in A.$$

Hence, $A = \left(\begin{smallmatrix} 0 & \mathbf{Q} \\ 0 & \mathbf{Q} \end{smallmatrix}\right)$.

Indeed, A is a right ideal generated by e_{12} and e_{22}. Summarizing, if A is a nontrivial right ideal in $R = \left(\begin{smallmatrix} \mathbf{Z} & \mathbf{Q} \\ 0 & \mathbf{Q} \end{smallmatrix}\right)$, then A is one of the following:

(i) $A = \begin{pmatrix} n_0 & 1 \\ 0 & 1 \end{pmatrix} R = \begin{pmatrix} n_0\mathbf{Z} & \mathbf{Q} \\ 0 & \mathbf{Q} \end{pmatrix}$ (n_0 a nonzero fixed element of $\mathbf{Z}$).

(ii) $A = \begin{pmatrix} n_0 & 1 \\ 0 & 0 \end{pmatrix} R = \begin{pmatrix} n_0\mathbf{Z} & \mathbf{Q} \\ 0 & 0 \end{pmatrix}$.

(iii) $A = \begin{pmatrix} 0 & \beta \\ 0 & \gamma \end{pmatrix} R = \left\{ \begin{pmatrix} 0 & \beta y \\ 0 & \gamma y \end{pmatrix} \middle| \, y \in \mathbf{Q} \right\}$ (where β,γ are fixed elements of $\mathbf{Q}$).

In particular, $A = \left(\begin{smallmatrix} 0 & \mathbf{Q} \\ 0 & 0 \end{smallmatrix}\right)$ or $\left(\begin{smallmatrix} 0 & 0 \\ 0 & \mathbf{Q} \end{smallmatrix}\right)$ according to whether $\gamma = 0$ or $\beta = 0$.

(iv) $A = \begin{pmatrix} 0 & \mathbf{Q} \\ 0 & \mathbf{Q} \end{pmatrix}$.

The first three are principal right ideals, the last one is not principal but can be generated by two elements. Thus, by Theorem 2.3, R is right noetherian.

The following is an example of a ring which is right noetherian but not left noetherian.

(e) The ring

$$R = \begin{bmatrix} \mathbf{Z} & \mathbf{Q} \\ 0 & \mathbf{Q} \end{bmatrix} = \left\{ \begin{bmatrix} a & b \\ 0 & c \end{bmatrix} \middle| \, a \in \mathbf{Z}, b,c \in \mathbf{Q} \right\}$$

is not left noetherian but right noetherian.

Solution. It is easy to verify that for each positive integer k,

$$A_k = \left\{ \begin{bmatrix} 0 & m/2^k \\ 0 & 0 \end{bmatrix} \middle| \, m \in \mathbf{Z} \right\}$$

is a left ideal of R and $A_k \subsetneq A_{k+1}$. Thus, there exists an infinite properly ascending chain of left ideals of R, namely,

$$A_1 \subset A_2 \subset A_3 \subset \cdots.$$

Hence, R is not left noetherian, but R is right noetherian, as shown in Example (d). We may invoke Theorem 2.6 to give the following alternative argument (perhaps a nice one too) to prove that R is right noetherian.

Let $A = \left[\begin{smallmatrix} 0 & 0 \\ 0 & \mathbf{Q} \end{smallmatrix}\right]$ be a right ideal of R. Then $R/A \simeq \left[\begin{smallmatrix} \mathbf{Z} & \mathbf{Q} \\ 0 & 0 \end{smallmatrix}\right]$ as right R-modules. Now it can be easily verified that nonzero R-submodules of $\left[\begin{smallmatrix} \mathbf{Z} & \mathbf{Q} \\ 0 & 0 \end{smallmatrix}\right]$are $\left[\begin{smallmatrix} (n) & \mathbf{Q} \\ 0 & 0 \end{smallmatrix}\right]$, when n is a fixed integer [see Example 1.5(e), Chapter 10]. Clearly, any ascending chain of R-submodules of the R-module $\left[\begin{smallmatrix} \mathbf{Z} & \mathbf{Q} \\ 0 & 0 \end{smallmatrix}\right]$ gives rise to an ascending chain of ideals in $\mathbf{Z}$. It then follows that $R/A \simeq \left[\begin{smallmatrix} \mathbf{Z} & \mathbf{Q} \\ 0 & 0 \end{smallmatrix}\right]$ is noetherian. Also, $A = \left[\begin{smallmatrix} 0 & 0 \\ 0 & \mathbf{Q} \end{smallmatrix}\right]$ as an R-module has no proper submodules. Thus, A is a noetherian R-module. But then by Theorem 2.6, R is noetherian.

Problems

1. Let $R_1, \ldots, R_n$ be a family of noetherian rings. Show that their direct sum $R = R_1 \oplus \cdots \oplus R_n$ is again noetherian. Prove the analogous result for modules.
2. Prove that the intersection of all prime ideals in a noetherian ring is nilpotent.
3. Show that every principal left ideal ring is noetherian.
4. Show that if R_i, $i = 1,2,\ldots$, is an infinite family of nonzero rings and if R is the direct product (or sum) of this family (R_i), $i = 1,2,\ldots$, then R cannot be noetherian.
5. Let R be a commutative noetherian ring, and let S be a multiplicative subset of R. Show that the ring of fractions R_S is also noetherian.
6. Let R be a noetherian ring. Show that the ring of $n \times n$ matrices R_n over R is also noetherian.
7. Let R be a noetherian integral domain. Then show that for all $0 \neq c \in R$, Rc is *large* [in the sense that $Rc \cap A \neq (0)$ for all nonzero left ideals A in R]. [*Hint:* If $Rc \cap A = (0)$, then $\Sigma_{n=1}^{\infty} Ac^n$ is a direct sum.]

8. Let R be a left artinian integral domain with more than one element. Show that R is a division ring.
9. If a ring R contains a division ring D as a subring such that R as a left vector space over D is finitely generated, show that R is left artinian.
10. (a) If K is any left ideal of a ring $R = A_n$, the $n \times n$ matrix ring over A with unity, show that $e_{11}K$ is a submodule of the free A-module $e_{11}R$. If L is any A-submodule of $e_{11}R$, show that $\Sigma_i e_{i1} L$ is a left ideal of R. Verify that these correspondences give a 1-1 onto mapping that preserves inclusions.
(b) Show that the matrix ring over an artinian (noetherian) ring is also artinian (noetherian).
11. Show that a left artinian ring cannot possess an infinite direct sum $\oplus \Sigma A_i$ of left ideals A_i of R.

3 Wedderburn–Artin theorem

In this section our aim is to prove that any left artinian ring R with unity without nonzero nilpotent ideals is a finite direct sum of matrix rings over division rings. Because any left artinian ring must contain minimal left ideals, it is appropriate to know their form. The following lemma is a useful fact.

3.1 Lemma. *Let A be a minimal left ideal in a ring R. Then either $A^2 = (0)$ or $A = Re$, where e is an idempotent in R.*

Proof. Suppose $A^2 \neq (0)$. Then there exists $a \in A$ such that $Aa \neq (0)$. But $Aa \subset A$, and the minimality of A shows $Aa = A$. From this it follows that there exists $e \in A$ such that $ea = a$, and, clearly, $e \neq 0$ because $a \neq 0$. Moreover, $e^2 a = ea$ or $(e^2 - e)a = 0$. If $B = \{c \in A \mid ca = 0\}$, then B is a left ideal of R such that $B \subset A$ and $B \neq A$. Therefore, we must have $B = (0)$. But then $(e^2 - e)a = 0$ implies $e^2 = e$. Now $Re \subset A$ and $Re \neq (0)$, because $0 \neq e = e^2 \in Re$. Accordingly, $Re = A$, and the proof is completed. □

We now prove the celebrated Wedderburn–Artin theorem.

3.2 Theorem (Wedderburn–Artin). *Let R be a left (or right) artinian ring with unity and no nonzero nilpotent ideals. Then R is isomorphic to a finite direct sum of matrix rings over division rings.*

Proof. We first establish that each nonzero left ideal in R is of the form Re for some idempotent e. So let A be any nonzero left ideal. By virtue of the dcc on left ideals in R, A contains a minimal left ideal M. By Lemma 3.1 either $M^2 = (0)$ or $M = Re$ for some idempotent e. If $M^2 = (0)$, then $(MR)^2 = (0)$; so, by hypothesis, $MR = (0)$, which gives $M = (0)$, a contradiction. Hence, $M = Re$. This yields that each nonzero left ideal contains a nonzero idempotent. Consider now a family $\mathcal{F}$ of left ideals, namely,

$$\mathcal{F} = \{R(1-e) \cap A \mid e \text{ is a nonzero idempotent in } A\}.$$

Clearly, $\mathcal{F}$ is nonempty. Because R is left artinian, $\mathcal{F}$ has a minimal member, say $R(1-e) \cap A$. We claim $R(1-e) \cap A = (0)$. Otherwise, there exists a nonzero idempotent $e_1 \in R(1-e) \cap A$. Clearly, $e_1 e = 0$. Set $e' = e + e_1 - ee_1$. It is easy to verify that $e'e' = e'$ and $e_1 e' \neq 0$. It is also obvious that $R(1-e') \cap A \subset R(1-e) \cap A$. But $e_1 e' \neq 0$ gives $e_1 \notin R(1-e') \cap A$ and $e_1 \in R(1-e) \cap A$. Hence, $R(1-e') \cap A \subsetneq R(1-e) \cap A$, a contradiction to the minimality of $R(1-e) \cap A$ in $\mathcal{F}$. This establishes our claim that $R(1-e) \cap A = (0)$. Next, let $a \in A$. Then $a(1-e) \in R(1-e) \cap A = (0)$. Thus, $a = ae$. Then $A \supset Re \supset Ae = A$ proves that $A = Re$, as asserted.

Let S be the sum of all minimal left ideals in R. Then $S = Re$ for some idempotent e. If $R(1-e) \neq (0)$, then there exists a minimal left ideal A contained in $R(1-e)$. But then $A \subset Re \cap R(1-e) = (0)$, a contradiction. Hence, $R(1-e) = (0)$, which proves that $R = S = \Sigma_{i \in \Lambda} A_i$, where (A_i), $i \in \Lambda$, is the family of all minimal left ideals in R. By Theorem 4.3 of Chapter 14 there exists a subfamily (A_i), $i \in \Lambda'$, of the family of the minimal left ideals such that $R = \oplus \Sigma_{i \in \Lambda'} A_i$. Let $1 = e_{i_1} + \cdots + e_{i_n}, 0 \neq e_{i_j} \in A_{i_j}$, $i_j \in \Lambda'$. Then $R = Re_{i_1} \oplus \cdots \oplus Re_{i_n}$. After reindexing if necessary, we may write $R = Re_1 \oplus \cdots \oplus Re_n$, a direct sum of minimal left ideals.

In the family of minimal left ideals $Re_1, \ldots, Re_n$, choose a largest subfamily consisting of all minimal left ideals that are not isomorphic to each other as left R-modules. After renumbering if necessary, let this subfamily be $Re_1, \ldots, Re_k$.

Suppose the number of left ideals in the family (Re_i), $1 \leq i \leq n$, that are isomorphic to Re_i is n_i. Then

$$R = \overbrace{[Re_1 \oplus \cdots]}^{n_1 \text{ summands}} \oplus \overbrace{[Re_2 \oplus \cdots]}^{n_2 \text{ summands}} \oplus \cdots \oplus \overbrace{[Re_k \oplus \cdots]}^{n_k \text{ summands}},$$

where each set of brackets contains pairwise isomorphic minimal left ideals, and no minimal left ideal in any pair of brackets is isomorphic to a minimal left ideal in another pair. By observing that $\mathrm{Hom}_R(Re_i, Re_j) = 0$, $i \neq j$, $1 \leq i, j \leq k$, and recalling Schur's lemma (Theorem 4.2, Chapter 14) that $\mathrm{Hom}_R(Re_i, Re_i) = D_i$, a division ring, we get, by invoking Theorem 1.1,

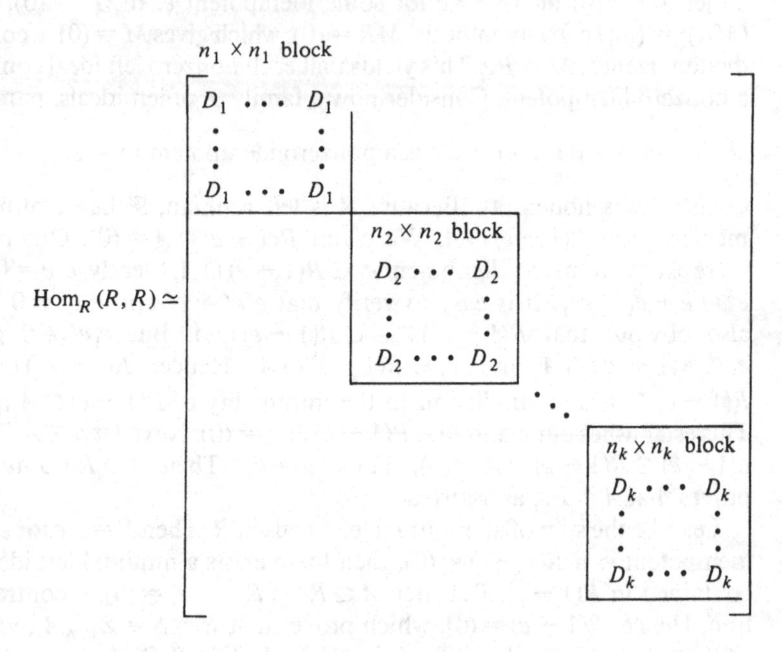

$$= \begin{bmatrix} (D_1)_{n_1} & & & \\ & (D_2)_{n_2} & & \\ & & \ddots & \\ & & & (D_k)_{n_k} \end{bmatrix}$$

$$\simeq (D_1)_{n_1} \oplus (D_2)_{n_2} \oplus \cdots \oplus (D_k)_{n_k}.$$

But since $\mathrm{Hom}_R(R,R) \simeq R^{op}$ as rings [Example 3.5(d), Chapter 14], and the opposite ring of a division ring is a division ring, R is a finite direct sum of matrix rings over division rings. □

Because the matrix rings over division rings [Example 2.4(d)] are both right and left noetherian and artinian, and a finite direct sum of noetherian and artinian rings is again noetherian and artinian (prove these), we get

3.3 Theorem. *Let R be a left (or right) artinian ring with unity and no nonzero nilpotent ideals. Then R is also right and left artinian (noetherian).*

Not every ring that is artinian on one side is artinian on the other side. The following example gives a ring that is left artinian but not right artinian.

3.4 Example

Let $R = \left[\begin{smallmatrix} \mathbf{Q} & \mathbf{Q} \\ 0 & 0 \end{smallmatrix}\right]$. If $A = \left[\begin{smallmatrix} 0 & \mathbf{Q} \\ 0 & 0 \end{smallmatrix}\right]$, then A is an ideal of R, and as a left R-module it is simple. Thus, A is left artinian. Also, the quotient ring $R/A \simeq \left[\begin{smallmatrix} \mathbf{Q} & 0 \\ 0 & 0 \end{smallmatrix}\right]$, which is a field. Therefore, R/A as an R/A-module (or as an R-module) is artinian. Because A is an artinian left R-module, we obtain that R as a left R-module is artinian (Theorem 2.6); that is, R is a left artinian ring, but R is not right artinian. For there exists a strictly descending chain

$$\begin{bmatrix} 0 & 2\mathbf{Z} \\ 0 & 0 \end{bmatrix} \supset \begin{bmatrix} 0 & 2^2\mathbf{Z} \\ 0 & 0 \end{bmatrix} \supset \begin{bmatrix} 0 & 2^3\mathbf{Z} \\ 0 & 0 \end{bmatrix} \supset \cdots$$

of right ideals of R.

As an application of the Wedderburn–Artin theorem, we prove the following useful result for a certain class of group algebras.

3.5 Theorem (Maschke). *If F is the field of complex numbers and G is a finite group, then*

$$F(G) \simeq F_{n_1} \oplus \cdots \oplus F_{n_k}$$

for some positive integers $n_1, \dots, n_k$.

Proof. We first prove that $F(G)$ has no nonzero nilpotent ideals. Let $G = \{g_1 = e, g_2, \dots, g_n\}$, and $x = \Sigma \alpha_i g_i \in F(G)$. Set $x^* = \Sigma \bar{\alpha}_i g_i^{-1}$, where $\bar{\alpha}_i$ denotes the complex conjugate of α_i. Then

$$xx^* = \sum_{i=1}^{n} |\alpha_i|^2 + \sum_{i=2}^{n} \beta_i g_i$$

for some $\beta_i \in F$. Hence, $xx^* = 0$ implies $\Sigma_{i=1}^{n} |\alpha_i|^2 = 0$, so each $\alpha_i = 0$; that is, $x = 0$. Thus, $xx^* = 0$ implies $x = 0$. Let A be a nilpotent ideal in $F(G)$. Let $a \in A$. Then $aa^* \in A$, so aa^* is nilpotent, say $(aa^*)^r = 0$. (We may assume r is even.) Set $b = (aa^*)^{r/2}$. Then $b^2 = 0$ and $b = b^*$. Thus, $bb^* = 0$, which gives $(aa^*)^{r/2} = b = 0$. Proceeding like this, we get $aa^* =$

0. Hence, $a = 0$, which proves that $A = (0)$. Hence, $F(G)$ has no nonzero nilpotent ideals.

Further, $F(G)$ is a finite-dimensional algebra with unity over the field F. Therefore, $F(G)$ is an artinian ring. Then by the Wedderburn–Artin theorem,

$$F(G) \simeq D^{(1)}_{n_1} \oplus \cdots \oplus D^{(k)}_{n_k},$$

where $D^{(i)}$, $1 \leq i \leq k$, are division rings. Now each $D^{(i)}_{n_i}$ contains a copy K of F in its center. In this way each $D^{(i)}_{n_i}$ is a finite-dimensional algebra over K (How?). Let $[D^{(i)}:K] = n$, and $a \in D^{(i)}$. Then $1, a, a^2, \ldots, a^n$ are linearly dependent over K. Thus, there exist $\alpha_0, \alpha_1, \ldots, \alpha_n$ (not all zero) in K such that $\alpha_0 + \alpha_1 a + \cdots + \alpha_n a^n = 0$. But since K is algebraically closed, $\alpha_0 + \alpha_1 x + \cdots + \alpha_n x^n \in K[x]$ has all its roots in K. Hence, $a \in K$, which shows that $D^{(i)} = K \simeq F$ and completes the proof. □

The next theorem gives some equivalent statements for left artinian rings that have no nonzero nilpotent ideals.

3.6 Theorem. *Let R be a ring with unity. Then the following are equivalent:*

(i) R is left artinian with no nonzero nilpotent ideals.
(ii) R is left artinian with no nonzero nil ideals.
(iii) R is a finite direct sum of minimal left ideals.
(iv) Each left ideal of R is of the form Re, e an idempotent.
(v) R is a finite direct sum of matrix rings over division rings.

(A ring satisfying any of these conditions is called a semisimple artinian ring.*)*

Proof. (i) ⇒ (ii) Follows from the fact that nil ideals in an artinian ring are nilpotent.

(i) ⇒ (iii) Follows from the proof given for the Wedderburn–Artin theorem.

(iii) ⇒ (iv) Follows from Problems 1 or 4 of Section 4 in Chapter 14.

(iv) ⇒ (v) Let $M = Re$ be a maximal left ideal of R with $e = e^2$. Then $R(1-e)$ is a minimal left ideal because $\dfrac{R}{Re} \simeq R(1-e)$ is a simple R-module. Thus if S is the sum of all minimal left ideals of R then $S \neq (0)$. We claim $S = R$. Otherwise, S is contained in a maximal left ideal Rf, say, where $f = f^2$. But then $R(1-f)$ is a minimal left ideal which is not contained in S because $S \cap R(1-f) \subset Rf \cap R(1-f) = (0)$, a contradiction. Thus $S = R$. Then, as shown in the latter part of Theorem 3.2, R is a finite-direct-sum of matrix rings over division rings.

(v) $\Rightarrow$ (i) Because the matrix rings over division rings are right and left artinian, and the finite direct sum of left (or right) artinian rings is again left (or right) artinian, we get R is left and right artinian. Further, the matrix rings over division rings are simple rings with unity and, hence, cannot possess nonzero nilpotent ideals. Now if $R = A_1 \oplus \cdots \oplus A_k$ is a finite direct sum of simple rings each with unity, then it is easy to verify that any nonzero ideal I is of the form $A_{i_1} \oplus \cdots \oplus A_{i_r}$, where $\{i_1,...,i_r\} \subset \{1,...,k\}$; hence, I cannot be nilpotent. □

3.7 Remarks on uniqueness

Exactly as in groups, we define the composition series of a module M as a sequence $(M_0, M_1, ..., M_k)$ such that $(0) = M_0 \subset M_1 \subset \cdots \subset M_k = M$, where each quotient module M_i/M_{i-1}, $i = 1,...,k$, is simple. Then the Jordan–Hölder theorem holds for modules possessing composition series. The proof carries over verbatim to modules.

3.8 Theorem. *Let D_n and D'_m be $n \times n$ and $m \times m$ matrix rings over division rings D and D', respectively, such that $D_n \simeq D'_m$. Then $n = m$ and $D \simeq D'$.*

Proof. Let (e_{ij}), $1 \le i,j \le n$, and (e'_{kl}), $1 \le k,l \le m$, denote the matrix units of the rings D_n and D'_m, respectively. Then

$$D_n = \oplus \sum_{i=1}^{n} D_n e_{ii} \qquad \text{and} \qquad D'_m = \oplus \sum_{k=1}^{m} D'_m e'_{kk}.$$

Because $D_n e_{ii}$ and $D'_m e'_{kk}$ are minimal left ideals of D_n and D'_m, respectively, we get two composition series

$$(0) \subset D_n e_{11} \subset D_n e_{11} \oplus D_n e_{22} \subset \cdots \subset D_n$$

and

$$(0) \subset D'_m e'_{11} \subset D'_m e'_{11} \oplus D'_m e'_{22} \subset \cdots \subset D'_m$$

of D_n and D'_m regarded as modules over themselves. Because $D_n \simeq D'_m$, it follows by the Jordan–Hölder theorem that $n = m$ and $D_n e_{ii} \simeq D'_m e'_{jj}$. This implies that $\mathrm{Hom}_{D_n}(D_n e_{ii}, D_n e_{ii}) \simeq \mathrm{Hom}_{D'_m}(D'_m e'_{jj}, D'_m e'_{jj})$. But $\mathrm{Hom}_{D_n}(D_n e_{ii}, D_n e_{ii}) \simeq D$. (Prove this.) Hence, $D \simeq D'$. □

The uniqueness of the decomposition of any artinian ring with no nonzero nilpotent ideals can similarly be proved. The interested reader is referred to any advanced text for studying the uniqueness of the direct sum decomposition of a module as given by Krull, Schmidt, Remak, Azumaya, and others.

Problems

1. Let R be a left artinian ring with unity and no nonzero nilpotent ideals. Then show that for each ideal I of R, R/I is also left artinian with no nonzero nilpotent ideals.
2. Let R be a prime left artinian ring with unity. Show that R is isomorphic to the $n \times n$ matrix ring over a division ring. Hence, show that a prime ideal in an artinian ring is maximal.
3. Let R be an artinian ring. Then show that the following sets are ideals and are equal:
   $N =$ sum of nil ideals,
   $U =$ sum of all nilpotent left ideals,
   $V =$ sum of all nilpotent right ideals.
   Also show that R/N has no nonzero nil ideals.
4. Let $R = \mathbf{Z}[x, y]/I$, where $I = (x^2 - 1, xy - y, y^2 - 2y)$. Show that
   (a) R has no nonzero nilpotent elements.
   (b) R is not the direct sum of minimal ideals.
5. Let R be a finite-dimensional algebra over an algebraically closed field F. Suppose R has no nonzero nil ideals. Show that R is isomorphic to the direct sum of matrix rings over F.

4 Uniform modules, primary modules, and Noether–Lasker theorem

Definition. *A nonzero module M is called* uniform *if any two nonzero submodules of M have nonzero intersection.*

Definition. *If U and V are uniform modules, we say U is subisomorphic to V and write $U \sim V$ provided U and V contain nonzero isomorphic submodules.*

Note $\sim$ is an equivalence relation and $[U]$ denotes the equivalence class of U.

Definition. *A module M is called* primary *if each nonzero submodule of M has uniform submodule and any two uniform submodules of M are subisomorphic.*

We note that **Z** as a **Z**-module is uniform and primary. Indeed, any uniform module must be primary. Another example of a uniform module is a commutative integral domain regarded as a module over itself.

4.1 Theorem. *Let M be a noetherian module or any module over a noetherian ring. Then each nonzero submodule contains a uniform module.*

Proof. Let $0 \neq x \in M$. It is enough to show xR contains a uniform submodule. If M is noetherian, the submodule xR is also noetherian. But if R is noetherian, then xR, the homomorphic image of R, is noetherian.

For convenience, call, as usual, a nonzero submodule N of M large if $N \cap K \neq 0$ for all nonzero submodules K of M.

Consider now the family F of all submodules of xR which are not large. Clearly $0 \in F$. Since xR is noetherian, F has a maximal member K, say. Because K is not large, $K \cap U = 0$ for some nonzero submodule U of xR. We claim U is uniform. Otherwise, there exist submodules A,B of U such that $A \cap B = (0)$. But then $(K \oplus A) \cap B = (0)$. For if $x \in (K \oplus A) \cap B$, then $x = k + a = b$ for some $k \in K$, $a \in A$, $b \in B$. This gives $k = b - a \in K \cap U = (0)$. Hence $k = 0$ and $b - a = 0$, which further yields $b = a \in A \cap B = (0)$. Thus $b = 0 = a$, proving $x = 0$; that is, $(K \oplus A) \cap B = (0)$. However, this yields a contradiction to the maximality of K. This shows U is uniform, completing the proof. □

Definition. *If R is a commutative noetherian ring and P is a prime ideal of R, then P is said to be* associated with the module M *if R/P embeds in M, or equivalently $P = r(x)$ for some $x \in M$, where $r(x) = \{a \in R \mid xa = 0\}$ denotes the annihilator of x.*

Definition. *A module M is called P-primary for some prime ideal P if P is the only prime ideal associated with M.*

4.2 Remark. If R is a commutative noetherian ring and P is a prime ideal of R, then an R-module is P-primary if and only if each nonzero submodule of M is subisomorphic to R/P.

4.3 Theorem. *Let U be a uniform module over a commutative noetherian ring R. Then U contains a submodule isomorphic to R/P for precisely one prime ideal P, that is, U is subisomorphic to R/P for exactly one prime ideal P.*

Proof. Consider the family F of annihilator ideals $r(x)$, $0 \neq x \in U$. Since R is noetherian, there exists a maximal member, say $r(x)$ in F. We claim $P = r(x)$ is prime. For, let $ab \in r(x)$, $a \notin r(x)$. Then $xab = 0$. Now $xa \neq 0$ and $r(xa) \supseteq r(x)$. By maximality of $r(x)$, $r(xa) = r(x)$. Thus $xab = 0$ implies $xb = 0$ and so $b \in r(x)$. This proves P is prime. Clearly $xR \simeq R/r(x) = R/P$. Therefore, R/P is embeddable in U. If for any other prime ideal Q, R/Q is embeddable in U, then $[R/P] = [R/Q]$. This implies there exist cyclic submodules xR and yR of R/P and R/Q, respectively, such that $xR \simeq yR$. But $xR \simeq R/P$ and $yR \simeq R/Q$. So $R/P \simeq R/Q$, which yields $P = Q$, proving the theorem. □

4.4 Remark. The ideal P in the above theorem is usually called the prime ideal associated with uniform module U.

4.5 Theorem. *Let M be a nonzero finitely generated module over a commutative noetherian ring R. Then there are only a finite number of primes associated with M.*

Proof. Consider the family F consisting of the direct sums of cyclic uniform submodules of M. F is not empty, as proved in Theorem 4.1. We partial order F by $\oplus \sum_{i \in I} x_i R \leq \oplus \sum_{j \in J} y_j R$ if and only if $I \subset J$ and $x_i R \subset y_i R$ for $i \in I$. By Zorn's Lemma, F has a maximal member $K = \oplus \sum_{j \in J} x_j R$, say. Since M is noetherian, K is finitely generated and let $K = \oplus \sum_{j=1}^{l} x_j R$. There exist $x_j a_j \in x_j R$ such that $r(x_j a_j) = P_j$, the prime ideal associated with $x_j R$. Set $x'_j = x_j a_j$ and $K' = \oplus \sum_{j=1}^{l} x'_j R$. Let $Q = r(x)$ be an associated prime ideal of M. We shall show $Q = P_j$ for some j, $1 \leq j \leq l$.

Since K is a maximal member of F, K as well as K' have the property that each intersects nontrivially with any nonzero submodule L of M. Now let $0 \neq y \in xR \cap K'$. Write $y = \sum_{j=1}^{l} x'_j b_j = xb$. We claim $r(x'_j b_j) = r(x'_j)$ whenever $x'_j b_j \neq 0$. Clearly $r(x'_j) \subset r(x'_j b_j)$. Let $x'_j b_j c = 0$. This implies $b_j c \in r(x'_j) = P_j$ and so $c \in P_j$ since $b_j \notin P_j$. Hence, $c \in r(x'_j)$.

Furthermore, we note $Q = r(x) = r(y) = \bigcap_{j=1}^{l} r(x'_j b_j) = \bigcap_{j \in \Lambda} P_j$, omitting those arising from $x'_j b_j = 0$, where $\Lambda \subset \{1, \ldots, l\}$. Therefore, $Q \subseteq P_j$ for all $j \in \Lambda$. Also, $\prod_{j \in \Lambda} P_j \subset \cap_{j \in \Lambda} P_j = Q$. Since Q is a prime ideal, at least one P_j appearing in the product $\prod_{j \in \Lambda} P_j$ must be contained in Q. Hence, $Q = P_j$ for some j. □

We conclude with Noether–Lasker theorem whose proof is a consequence of the above results (see Remarks 4.7).

4.6 Theorem (Noether–Lasker). *Let M be a finitely generated module over a commutative noetherian ring R. Then there exists a finite* *family $N_1,\ldots,N_l$ of submodules of M such that*

(a) $\bigcap_{i=1}^{l} N_i = 0$ *and* $\bigcap_{\substack{i=1\\ i\neq i_0}}^{l} N_i \neq 0$ *for all* $1 \le i_0 \le l$.

(b) *Each quotient M/N_i is a P_i-primary module for some prime ideal P_i.*

(c) *The P_i are all distinct,* $1 \le i \le l$.

(d) *The primary component N_i is unique if and only if P_i does not contain P_j for any $j \neq i$.*

4.7 Remarks. Let $\{U_i\}$, $1 \le i \le l$ be uniform submodules obtained as in the proof of Theorem 4.5. Choose N_i to be a maximal member in the family $\{K \subset M \mid K$ contains no submodule subisomorphic to $U_i\}$. With this choice of $N_1,\ldots,N_l$, (a), (b), and (c) follow directly.

This form of the Noether–Lasker theorem and the results leading to the theorem are taken from Barbara Osofsky, Noether–Lasker Primary Decomposition Revisited, preprint (1994), to appear in *American Mathematical Monthly.* The reader may refer to this article for additional details and to see how the statement given in Theorem 4.6 is indeed equivalent to the usual classical statement.

4.8 Exercise (Osofsky). Let $R = \mathbf{R}[X, Y, Z]$ be the ring of polynomials over the reals. Let I be the ideal of R generated by $\{X^2, XY, XZ^2\}$. Find the primes associated to the module $M = R/I$ and a set of primary components.

CHAPTER 20

Smith normal form over a PID and rank

1 Preliminaries

Throughout this chapter R is a principal ideal domain with unity. Thus, in particular, R may be a field. Recall that the rank of a finitely generated free R-module is defined to be the number of elements in any basis of the free module (Chapter 14, Section 5). For an $m \times n$ matrix A with entries in R we define the column rank and the row rank of A in terms of the rank of a certain free R-module, and we show in Section 3 that these two ranks are equal – the common value being known as the *rank* of A. But first we need a key lemma.

1.1 Lemma. *Let R be a principal ideal domain, and let F be a free R-module with a basis consisting of n elements. Then any submodule K of F is also free with a basis consisting of m elements, such that $m \leq n$.*

Proof. We know $F \simeq R^n$ as R-modules (Theorem 5.1, Chapter 14). To prove the theorem, we use induction on n. R^0 is interpreted as a (0) module, and this is free on the empty set. Therefore, we may assume that $n > 0$, and let us identify the copy of K in R^n (under the isomorphism $F \simeq R^n$) with K itself. Let $\pi: K \to R$ be the mapping defined by

$$\pi(x_1, \dots, x_n) = x_1.$$

If $\pi = 0$, then $K = \operatorname{Ker} \pi \subset R^{n-1}$, and the theorem follows by induction on n. If $\pi \neq 0$, its image is a nonzero ideal Ra in R; that is, $\pi(K) = Ra$, where $a \neq 0$. Choose $k \in K$ such that $\pi(k) = a$. We assert

$$K = Rk \oplus \operatorname{Ker} \pi.$$

For let $x \in K$. Write $\pi(x) = ba$, $b \in R$. Then $\pi(x - bk) = ba - ba = 0$. Hence,

$$x = bk + (x - bk)$$

implies that

$$x \in Rk + \text{Ker}\ \pi.$$

Thus,

$$K = Rk + \text{Ker}\ \pi.$$

To prove that the sum is direct, let $ck \in Rk \cap \text{Ker}\ \pi$, $c \in R$. Then $0 = \pi(ck) = ca$; hence, $c = 0$ because $a \neq 0$. This proves our assertion that $K = Rk \oplus \text{Ker}\ \pi$.

It is easy to check that the mapping $r \mapsto rk$ of R onto Rk is an R-isomorphism.

Further, $\text{Ker}\ \pi = \{(0, x_2, \ldots, x_n) | x_i \in R\}$. Thus, $\text{Ker}\ \pi \subsetneq R^{n-1}$. Hence, by induction, Ker π is free, with a basis consisting of at most $n - 1$ generators. Therefore,

$$K = \text{Ker}\ \pi \oplus Rk \simeq R^m \oplus R, \qquad m \leq n - 1,$$

so

$$K \simeq R^{m+1}, \qquad m + 1 \leq n. \quad \square$$

2 Row module, column module, and rank

Let A be an $m \times n$ matrix over R. The rows (columns) of A are the elements of the R-module $R^{1 \times n}$ ($R^{m \times 1}$) consisting of the $1 \times n$ ($m \times 1$) matrices over R. Generally, we write $R^{1 \times n}$ (and $R^{m \times 1}$) as R^n (and R^m). Using the notation R^n to denote rows as well as columns never creates any confusion because context always makes the meaning clear.

Definition. *Let A be an $m \times n$ matrix over R. The submodule of R^n generated by the m rows of A is called the* row module *of A; and the submodule of R^m generated by the n columns of A is called the* column module *of A.*

$R(A)$ and $C(A)$, respectively, denote the row module and the column module of the matrix A. If the ring R is a field, $R(A)$ and $C(A)$ are, respectively, called the *row space* and *column space* of matrix A.

Note that $R(A)$ and $C(A)$ are finitely generated submodules of free modules R^n and R^m, respectively. Thus by Lemma 1.1, both $R(A)$ and $C(A)$ are free modules.

Definition. *Let A be an $m \times n$ matrix over R. The rank of the module $R(A)$ $[C(A)]$ is called the* row rank (column rank) *of A.*

The proof of the following theorem is exactly the same as the proof of the corresponding theorem for matrices over a field (Lemma 7.5, Section 7, Chapter 14).

2.1 Theorem. *Let A be an $m \times n$ matrix over R. Let P and Q, respectively, be $m \times m$ and $n \times n$ invertible matrices over R. Then* row (column) rank (PAQ) = row (column) rank (A).

In order to compute the row rank or the column rank of a given matrix, we reduce the matrix into "normal" form by performing certain operations on rows and columns defined in the next section. Eventually, we obtain by this process the result that the row rank of a matrix A is equal to the column rank of the matrix (Theorem 3.3).

3 Smith normal form

We begin with a

Definition. *Let A be an $m \times n$ matrix over R. The following three types of operations on the rows (columns) of A are called* elementary row (column) operations.

(i) *Interchanging two rows (columns). We denote by $R_i \leftrightarrow R_j$ ($C_i \leftrightarrow C_j$) the operation of interchanging the ith and jth rows (columns).*

(ii) *Multiplying the elements of one row (column) by a nonzero element of R. We denote by αR_i (αC_i) the operation of multiplying the ith row (column) by α.*

(iii) *Adding to the elements of one row (column) α times the corresponding elements of a different row (column), where $\alpha \in R$. We denote by $R_i + \alpha R_j$ ($C_i + \alpha C_j$) the operation of adding to the elements of the ith row (column) α times the corresponding elements of the jth row (column).*

Recall that e_{ij}, $1 \leq i,j \leq n$, denote $n \times n$ matrix units, and $e_{ij}e_{kl} = \delta_{jk}e_{il}$, where δ_{jk} is the Kronecker delta.

3.1 **Theorem.** *Let A be an $m \times n$ matrix over R.*

(i) *If $E_{ij} = 1 - e_{ii} - e_{jj} + e_{ij} + e_{ji}$, then $E_{ij}A$ (AE_{ij}) is the matrix obtained from A by interchanging the ith and the jth rows (columns). Also, $E_{ij}^{-1} = E_{ij}$.*

(ii) *If $L_i(\alpha) = 1 + (\alpha - 1)e_{ii}$, and α is an invertible element in R, then $L_i(\alpha)A$ [$AL_i(\alpha)$] is the matrix obtained from A by multiplying the ith row (column) by α. Also, $L_i^{-1}(\alpha) = 1 + (\alpha^{-1} - 1)e_{ii}$.*

(iii) *If $M_{ij}(\alpha) = 1 + \alpha e_{ij}$, then $M_{ij}(\alpha)A$ [$AM_{ji}(\alpha)$] is the matrix obtained from A by multiplying the jth row (column) by α and adding it to the ith row (column). Also, $M_{ij}^{-1}(\alpha) = 1 - \alpha e_{ij}$ $(i \neq j)$.*

Proof. (i) $E_{ij}A = (1 - e_{ii} - e_{jj} + e_{ij} + e_{ji})A$, or

$$E_{ij}A = A - e_{ii}A - e_{jj}A + e_{ij}A + e_{ji}A. \tag{1}$$

We note that

(a) $e_{ii}A$ is the matrix whose ith row is the same as that of A, and all other rows are zero.

(b) $e_{ij}A$ is the matrix whose ith row is the jth row of A, and all other rows are zero.

Thus, (1) shows that $E_{ij}A$ is the matrix obtained from A by interchanging its ith and jth rows. Because $E_{ij}^2 = 1$, it follows that $E_{ij}^{-1} = E_{ij}$.

(ii) $L_i(\alpha)A = A - e_{ii}A + \alpha e_{ii}A$. Because

$$e_{ii}A = \begin{bmatrix} & 0 & \\ a_{i1} & \cdots & a_{in} \\ & 0 & \end{bmatrix} i\text{th row},$$

$L_i(\alpha)A$ is the matrix obtained from A by multiplying the ith row by α. Further, by actual multiplication

$$(1 + (\alpha - 1)e_{ii})(1 + (\alpha^{-1} - 1)e_{ii}) = 1.$$

This proves $L_i^{-1}(\alpha) = (1 + (\alpha^{-1} - 1))e_{ii}$.

(iii) $M_{ij}(\alpha)A = A + \alpha e_{ij}A$. Because

$$e_{ij}A = \begin{bmatrix} & 0 & \\ a_{j1} & \cdots & a_{jn} \\ & 0 & \end{bmatrix} i\text{th row},$$

$M_{ij}(\alpha)A$ is the matrix obtained from A by multiplying the jth row of A by α and adding it to the ith row. Finally,

$$(1 + \alpha e_{ij})(1 - \alpha e_{ij}) = 1 - \alpha e_{ij} + \alpha e_{ij} = 1,$$

so $M_{ij}^{-1}(\alpha) = 1 - \alpha e_{ij}$. □

Definition. *The matrices E_{ij}, $L_i(\alpha)$, and $M_{ij}(\alpha)$ are known as* elementary matrices.

We note that an elementary matrix is the result of performing a single elementary row or column operation on an identity matrix. Precisely,

E_{ij} = the matrix obtained from the identity matrix by interchanging the ith and jth rows (equivalently, interchanging the ith and jth columns).

$L_i(\alpha)$ = the matrix obtained from the identity matrix by multiplying the ith row by a nonzero $\alpha \in R$ (equivalently, by multiplying the ith column by α).

$M_{ij}(\alpha)$ = the matrix obtained from the identity matrix by adding to the elements of the ith row α times the corresponding elements of the jth row, where $\alpha \in R$ [$M_{ij}(\alpha)$ is also the matrix obtained from the identity matrix by adding to the elements of the jth column α times the corresponding elements of the ith column].

In addition to these three elementary row (column) operations, we apply a nonelementary operation to the rows and columns of A: multiplication by matrices of the form

$$\text{(iv)} \quad \begin{bmatrix} 1 & & & & & & & & \\ & 1 & & & & & & & \\ & & \ddots & & & & & & \\ & & & \begin{bmatrix} u & s \\ v & t \end{bmatrix} & & & & & \\ & & & & 1 & & & & \\ & & & & & 1 & & & \\ & & & & & & \ddots & & \\ & & & & & & & & 1 \end{bmatrix},$$

where $\left[\begin{smallmatrix} u & s \\ v & t \end{smallmatrix}\right]$ is invertible in R_2, the 2×2 matrix ring over R.

Multiplying A on the right (left) by a suitable matrix of the above form has the effect of replacing two of the entries on a given row (column) by their greatest common divisor and 0, respectively.

Definition. *Two $m \times n$ matrices A and B over R are said to be* equivalent *if there exists an invertible matrix $P \in R_m$ and an invertible matrix $Q \in R_n$ such that $B = PAQ$.*

It is clear that "being equivalent" defines an equivalence relation in the set of $m \times n$ matrices with entries in R.

We are now ready to state and prove that every matrix is equivalent to a "diagonal" matrix, known as the *Smith normal form.*

3.2 Theorem. *If A is an $m \times n$ matrix over a principal ideal domain R, then A is equivalent to a matrix that has the "diagonal" form*

$$\begin{bmatrix} a_1 & & & & & & & \\ & a_2 & & & & & & \\ & & \ddots & & & & & \\ & & & a_r & & & & \\ & & & & 0 & & & \\ & & & & & \ddots & & \\ & & & & & & & 0 \end{bmatrix},$$

where the $a_i \neq 0$ and $a_1 | a_2 | \cdots | a_r$.

Proof. We define the length $l(a)$ of $a \neq 0$ to be the number of prime factors occurring in the factorization, $a = p_1 p_2 \cdots p_r$, p_i primes (not necessarily distinct). We use the convention that $l(u) = 0$ if u is a unit.

If $A = 0$, then there is nothing to prove. Otherwise, let a_{ij} be a nonzero element of A with minimal length $l(a_{ij})$. Elementary row and column operations bring this element to the $(1,1)$ position. We may then assume that the nonzero element of A with smallest length is at the $(1,1)$ position. Let $a_{11} \nmid a_{1k}$. Interchanging the second and the kth columns, we may assume $a_{11} \nmid a_{12}$. Let $d = (a_{11}, a_{12})$ be the greatest common divisor of a_{11} and a_{12}. Then $l(d) < l(a_{11})$. There exist elements $u,v \in R$ such that $a_{11}u + a_{12}v = d$. Because $d = (a_{11}, a_{12})$, there exist $s,t \in R$ such that $a_{11} = ds$ and $a_{12} = dt$. Thus, $d(us + vt) = d$; so $us + vt = 1$. It can be verified that

$$\begin{bmatrix} u & t \\ v & -s \end{bmatrix} \begin{bmatrix} s & t \\ v & -u \end{bmatrix} = \begin{bmatrix} 1 & 0 \\ 0 & 1 \end{bmatrix},$$

which implies $\begin{bmatrix} u & t \\ v & -s \end{bmatrix}$ is invertible. Multiplying A on the right by

$$\begin{bmatrix} \begin{bmatrix} u & t \\ v & -s \end{bmatrix} & & & & 0 \\ & 1 & & & \\ & & 1 & & \\ 0 & & & \ddots & \\ & & & & 1 \end{bmatrix},$$

we obtain the matrix whose first row is

$$(d \quad 0 \quad b_{13} \quad \cdots \quad b_{1n}),$$

where $l(d) < l(a_{11})$.

If $a_{11} | a_{12}$, then by performing elementary row (column) operation (iii) on the first two columns (or, equivalently, by multiplying on the right by a suitable elementary matrix), we can reduce A to a matrix whose first row is again of the form $(d \ 0 \ b_{13} \ \cdots \ b_{1n})$, where $l(d) < l(a_{11})$.

Continuing this process yields an equivalent matrix whose first row has all entries 0, except the (1,1) entry.

Similarly, appropriate elementary row operations (i)–(iii) and the nonelementary operation of multiplying on the left by the matrix of the form given in (iv) reduce the elements in the first column after the (1,1) position to 0 and either keep the elements in the first row unaltered (i.e., all apart from (1,1) entry are zero) or reduce the length of the (1,1) entry. In the second case we repeat the process by which all the elements in the first row except the one at the (1,1) position are reduced to 0. Because $l(a_{11})$ is finite, this process (of alternately reducing the first row and the first column) must come to an end. When it does, we have reduced A to the form

$$P_1AQ_1 = \begin{bmatrix} a_1 & 0 & 0 & \cdots & 0 \\ 0 & & & & \\ \cdot & & & & \\ \cdot & & & A_1 & \\ \cdot & & & & \\ 0 & & & & \end{bmatrix},$$

where A_1 is an $(m-1) \times (n-1)$ matrix, and P_1 and Q_1 are $m \times m$ and $n \times n$ invertible matrices, respectively.

Similarly, there exist invertible matrices P_2' and Q_2' of orders

$(m-1)\times(m-1)$ and $(n-1)\times(n-1)$, respectively, such that

$$P_2'A_1Q_2' = \begin{bmatrix} a_2 & 0 & \cdots & 0 \\ 0 & & & \\ \cdot & & A_2 & \\ \cdot & & & \\ \cdot & & & \\ 0 & & & \end{bmatrix}$$

where A_2 is an $(m-2)\times(n-2)$ matrix. Let

$$P_2 = \begin{bmatrix} 1 & 0 \\ 0 & P_2' \end{bmatrix} \quad \text{and} \quad Q_2 = \begin{bmatrix} 1 & 0 \\ 0 & Q_2' \end{bmatrix}$$

be, respectively, $m\times m$ and $n\times n$ invertible matrices. Then

$$P_2P_1AQ_1Q_2 = \begin{bmatrix} a_1 & 0 & \cdots & 0 \\ 0 & a_2 & \cdots & 0 \\ \cdot & \cdot & & \\ \cdot & \cdot & A_2 & \\ \cdot & \cdot & & \\ 0 & 0 & & \end{bmatrix}.$$

Continuing like this (or by induction on $m+n$), we obtain

$$PAQ = \text{diag}(a_1, a_2, \ldots, a_r, 0, \ldots, 0).$$

Finally, we show that we can reduce PAQ further such that $a_1|a_2|\cdots|a_r$. Assume $a_1 \nmid a_2$. Add the second row to the first row. The first row then becomes

$$(a_1 \quad a_2 \quad 0 \quad \cdots \quad 0).$$

By performing these operations, we can reduce the length of a_1. Thus, by further reduction we may assume $a_1|a_2$ and, similarly, $a_1|a_i$, $i = 3,\ldots,r$. By repeating this procedure for a_2 in place of a_1, and so on, we finally reach a situation where $a_i|a_{i+1}$, $i = 1,\ldots,r-1$. □

Definition. *The nonzero diagonal elements of the matrix having the diagonal form given in Theorem* 3.2 *are called the* invariant factors *of* A.

It can be shown that the invariant factors are unique up to unit multipliers, and two $m\times n$ matrices are equivalent if and only if they have the same invariant factors. We do not intend to discuss these questions. The interested reader may refer to any graduate level text in abstract algebra.

3.3 Theorem. *Let A be an $m \times n$ matrix over a PID. Then* row rank A = column rank A.

Proof. Choose P and Q invertible matrices of suitable sizes such that PAQ is in Smith normal form. Then by Theorem 2.4

$$\begin{aligned}\text{row rank } A = \text{row rank } PAQ = r = \text{column rank } PAQ \\ = \text{column rank } A. \quad \square\end{aligned}$$

Definition. *Let A be an $m \times n$ matrix over a PID. The common value of the row rank of A and the column rank of A is called the* rank *of A.*

3.4 Examples

(a) Obtain the Smith normal form and rank for the matrix with integral entries

$$\begin{bmatrix} 1 & 2 & 3 \\ 4 & 5 & 0 \end{bmatrix}.$$

Solution:

$$\begin{bmatrix} 1 & 2 & 3 \\ 4 & 5 & 0 \end{bmatrix} \overset{R_2-4R_1}{\sim} \begin{bmatrix} 1 & 2 & 3 \\ 0 & -3 & -12 \end{bmatrix} \overset{C_2-2C_1,\, C_3-3C_1}{\sim} \begin{bmatrix} 1 & 0 & 0 \\ 0 & -3 & -12 \end{bmatrix}$$

$$\overset{C_3-4C_2}{\sim} \begin{bmatrix} 1 & 0 & 0 \\ 0 & -3 & 0 \end{bmatrix} \overset{(-1)R_2}{\sim} \begin{bmatrix} 1 & 0 & 0 \\ 0 & 3 & 0 \end{bmatrix}.$$

Clearly, rank $A = 2$.

Note. Sometimes one may end up with a "diagonal" matrix $\text{diag}(d_1, d_2, \ldots)$, where d_i does not necessarily divide d_{i+1} for some i. In such a situation one must continue performing suitable operations until arriving at the desired form.

(b) Find the invariant factors of the matrix

$$\begin{bmatrix} -x & 4 & -2 \\ -3 & 8-x & 3 \\ 4 & -8 & -2-x \end{bmatrix}$$

over the ring $\mathbf{Q}[x]$. Also find the rank.

Solution. The matrix can be proved to be equivalent to

$$\begin{bmatrix} 1 & 0 & 0 \\ 0 & 1 & 0 \\ 0 & 0 & (x-2)(x^2-4x+20) \end{bmatrix}$$

[we performed (i) $R_1 \leftrightarrow R_3$, (ii) $R_3 + (x/4)R_1$, (iii) $R_2 + \frac{3}{4}R_1$, (iv) $R_3 - 2R_2$, (v) $C_2 + 2C_1$, (vi) $C_3 - \frac{3}{4}C_2$, (vii) $\frac{1}{4}C_1$, (viii) $C_3 + (2+x)C_1$, (ix) $-4C_3$, (x) $R_3 + R_2$, (xi) $C_3 + (2-x)C_2$, (xii) $\frac{1}{16}R_3$, (xiii) $C_2 + [(2-x)/16]C_3$, (xiv) $R_2 \leftrightarrow R_3$ and $C_3 \leftrightarrow C_2$ in succession]. It will be a good exercise for the reader to obtain this form in fewer steps. Clearly, rank $A = 3$.

Problems

1. Obtain the Smith normal form and rank for the following matrices over a principal ideal domain R:

   (a) $\begin{bmatrix} 0 & 2 & -1 \\ -3 & 8 & 3 \\ 2 & -4 & -1 \end{bmatrix}$, where $R = \mathbf{Z}$.

   (b) $\begin{bmatrix} -x-3 & 2 & 0 \\ 1 & -x & 1 \\ 1 & -3 & -x-2 \end{bmatrix}$, where $R = \mathbf{Q}[x]$.

2. Find the invariant factors of the following matrix over $\mathbf{Q}[x]$:

$$\begin{bmatrix} 5-x & 1 & -2 & 4 \\ 0 & 5-x & 2 & 2 \\ 0 & 0 & 5-x & 3 \\ 0 & 0 & 0 & 4 \end{bmatrix}.$$

3. Find the rank of the subgroup of $\mathbf{Z}^4$ generated by each of the following lists of elements:

   (a) (3,6,9,0), (−4,−8,−12,0).
   (b) (2,3,1,4), (1,2,3,0), (1,1,1,4).
   (c) (−1,2,0,0), (2,−3,1,0), (1,1,1,1).

CHAPTER 21

Finitely generated modules over a PID

Our object in this chapter is to prove the fundamental structure theorem for finitely generated modules over a principal ideal domain. This theorem provides a decomposition of such a module into a direct sum of cyclic modules. In particular, it gives an independent proof of the fundamental theorem of finitely generated abelian groups proved earlier in Chapter 8. Applications of this theorem to linear algebra are given in Sections 4 and 5.

1 Decomposition theorem

1.1 Theorem. *Let R be a principal ideal domain, and let M be any finitely generated R-module. Then*

$$M \simeq R^s \oplus R/Ra_1 \oplus \cdots \oplus R/Ra_r,$$

a direct sum of cyclic modules, where the a_i are nonzero nonunits and $a_i | a_{i+1}$, $i = 1,\dots,r-1$.

Proof. Because M is a finitely generated R-module, $M \simeq R^n/K$ [Example 5.5(a), Chapter 14]. Further, by Lemma 1.1 of Chapter 20, $K \simeq R^m$, $m \leq n$. Let ϕ be this isomorphism from R^m to K. Thus, $K = \phi(R^m)$. Let $(e_1,\dots,e_m)$ be a basis of R^m. Let us write

$$\phi(e_1) = \begin{bmatrix} a_{11} \\ a_{21} \\ \cdot \\ \cdot \\ \cdot \\ a_{n1} \end{bmatrix} \in R^n$$

$$\vdots \qquad \vdots$$

$$\phi(e_m) = \begin{bmatrix} a_{1m} \\ a_{2m} \\ \cdot \\ \cdot \\ \cdot \\ a_{nm} \end{bmatrix} \in R^n.$$

Then $\phi(R^m) = AR^m$, where $A = (a_{ij})$ is an $n \times m$ matrix.

Choose invertible matrices P and Q of order $n \times n$ and $m \times m$, respectively, such that $PAQ = \text{diag}(a_1, a_2, \ldots, a_k, 0, \ldots, 0)$, where $a_1 | a_2 | \cdots | a_k$. Then

$$M \simeq \frac{R^n}{K} = \frac{R^n}{\phi(R^m)} = \frac{R^n}{AR^m} \simeq \frac{PR^n}{PAQR^m} = \frac{R^n}{PAQR^m}$$

$$= R^n / \begin{bmatrix} a_1 & & & & & & & \\ & a_2 & & & & 0 & & \\ & & \cdot & & & & & \\ & & & \cdot & & & & \\ & & & & \cdot & & & \\ & & & & & a_k & & \\ & & & & & & 0 & \\ & 0 & & & & & & \cdot \\ & & & & & & & & \cdot \\ & & & & & & & & & \cdot \\ & & & & & & & & & & 0 \end{bmatrix} \begin{bmatrix} R \\ R \\ \cdot \\ \cdot \\ \cdot \\ R \end{bmatrix}$$

$$= \frac{R \oplus R \oplus \cdots \oplus R}{Ra_1 \oplus Ra_2 \oplus \cdots \oplus Ra_k}$$

$$\simeq \frac{R}{Ra_1} \oplus \frac{R}{Ra_2} \oplus \cdots$$

$$\oplus \frac{R}{Ra_k} \oplus \overbrace{R \oplus \cdots \oplus R}^{n-k \text{ copies}} \qquad \text{(by Theorem 3.4, Chapter 14)}$$

$$= \frac{R}{Ra_u} \oplus \cdots \oplus \frac{R}{Ra_k} \oplus R^s$$

[by deleting the zero terms if any (corresponding to those a_i's that are units)]

$$= \frac{R}{Ra_1} \oplus \cdots \oplus \frac{R}{Ra_r} \oplus R^s \qquad \text{(by renumbering if necessary).}$$

Because for any ideal I [including (0)], R/I is a cyclic R-module, the proof is completed. □

2 Uniqueness of the decomposition

First, we give a

Definition. *An element x of an R-module M is called a* torsion element *if there exists a nonzero element $r \in R$ such that $rx = 0$.*

Definition. *A nonzero element x of an R-module M is called a* torsion-free element *if $rx = 0$, $r \in R$, implies $r = 0$.*

2.1 Theorem. *Let R be a principal ideal domain, and let M be an R-module. Then*

$$\text{Tor } M = \{x \in M \mid x \text{ is torsion}\}$$

is a submodule of M.

Proof. Left to the reader as an exercise. □

2.2 Theorem. *Let M be a finitely generated module over a principal ideal domain R. Then*

$$M = F \oplus \text{Tor } M,$$

where (i) $F \simeq R^s$ for some nonnegative integer s, and (ii) Tor $M \simeq$

$R/Ra_1 \oplus \cdots \oplus R/Ra_r$, where a_i are nonzero nonunit elements in R such that $a_1|a_2| \cdots |a_r$.

Proof. By the structure theorem for finitely generated modules over a PID, $M \simeq R^s \oplus R/Ra_1 \oplus \cdots \oplus R/Ra_r$, where a_i are nonzero nonunit elements in R such that $a_1|a_2| \cdots |a_r$. It then follows that

$$M = F \oplus T,$$

where $F \simeq R^s$, and $T \simeq R/Ra_1 \oplus \cdots \oplus R/Ra_r$. Clearly, $a_rT = 0$. Thus, $T \subset \text{Tor } M$.

Next, let $x \in \text{Tor } M$. Then $x = x_1 + x_2$, $x_1 \in F$, $x_2 \in T$. So $x_1 = x - x_2 \in \text{Tor } M$, and, hence, $rx_1 = 0$ for some $0 \neq r \in R$. If $(y_1,...,y_s) \in R^s$ is the image of x_1 under the isomorphism $F \simeq R^s$, then $r(y_1,...,y_s) = 0$. Hence, $ry_i = 0$, $1 \leq i \leq s$. Therefore, each $y_i = 0$, because $r \neq 0$. This yields $x_1 = 0$ and proves that $x \in T$, and, thus, $T = \text{Tor } M$. This completes the proof. □

We are now ready to prove the uniqueness of a decomposition of a finitely generated module over a principal ideal domain.

2.3 Theorem. *Let R be a principal ideal domain, and let M be a finitely generated R-module. Suppose*

$$M \simeq R^s \oplus R/Ra_1 \oplus \cdots \oplus R/Ra_u, \tag{1}$$

where a_i are nonzero nonunit elements in R such that $a_1|a_2| \cdots |a_u$.

$$M \simeq R^t \oplus R/Rb_1 \oplus \cdots \oplus R/Rb_v, \tag{2}$$

where a_i are nonzero nonunit elements in R such that $a_1|a_2|\cdots|a_u$, and Then $s = t$, $u = v$, and $Ra_i = Rb_i$, $1 \leq i \leq u$.

Proof. By Theorem 2.2, (1) and (2), respectively, give $M = F \oplus \text{Tor } M$, and $M = F' \oplus \text{Tor } M$, where $F \simeq R^s$, $F' \simeq R^t$. But then

$$\frac{M}{\text{Tor } M} = \frac{F \oplus \text{Tor } M}{\text{Tor } M} \simeq F.$$

Thus, $F \simeq F'$. This yields $s = t$ (Theorem 5.4, Chapter 14).

We now proceed to show that if

$$T = \text{Tor } M \simeq R/Ra_1 \oplus \cdots \oplus R/Ra_u \tag{3}$$

and

$$T = \text{Tor } M \simeq R/Rb_1 \oplus \cdots \oplus R/Rb_v, \tag{4}$$

where the a_i's and b_i's satisfy conditions as in (1) and (2), respectively, then $u = v$ and $Ra_i = Rb_i$. For any R-module X and a prime $p \in R$, we define

$$X_p = \{x \in X \mid px = 0\}.$$

Then from (3),

$$T_p \simeq \oplus \sum_{i=1}^{u} \left(\frac{R}{Ra_i}\right)_p.$$

Clearly,

$$\left(\frac{R}{Ra_i}\right)_p = \begin{cases} \dfrac{R(a_i/p)}{Ra_i} & \text{if } p \mid a_i, \\ (0) & \text{if } p \nmid a_i, \end{cases}$$

and $R(a_i/p)/Ra_i$ is a one-dimensional vector space over R/Rp generated by the element $a_i/p + Ra_i$.

Therefore, T_p is a vector space over R/Rp, and its dimension is equal to the number of terms R/Ra_i such that $p \mid a_i$.

Suppose that p is a prime dividing a_1, and hence a_i, for each $i = 1,\ldots,u$. Let T have a direct sum decomposition into v summands as in (2). Then p must divide at least u of the elements b_j, whence $u \leq v$. By symmetry, $u = v$. Therefore, any two decompositions of types (1) and (2) of any given module must have an equal number of summands. This fact plays an important role in proving the rest of the theorem.

From (3) we have $a_u T = 0$. But then (4) gives $Ra_u \subset Rb_u$. By symmetry, $Rb_u \subset Ra_u$. Hence, $Ra_u = Rb_u$.

Now assume $(a_i) = (b_i)$, $k \leq i \leq u$. We show $(a_{k-1}) = (b_{k-1})$. This is done by contradiction, obtained by multiplying (3) and (4) by a suitable element in R to make the first m summands zero in one of the decompositions (3) or (4), and the first $k-1$ summands zero in the other decomposition, for some $m < k-1$.

Let p be a prime in R such that $p^\alpha \mid a_{k-1}$, $p^{\alpha+1} \nmid a_{k-1}$ and $p^\beta \mid b_{k-1}$, $p^{\beta+1} \nmid b_{k-1}$. If $\alpha = \beta$ for each prime p, then we are fine because this means $(a_{k-1}) = (b_{k-1})$. Otherwise, for definiteness, let $\alpha > \beta$. Put

$$x = p^{\alpha-1}\frac{a_u}{p^\theta} \quad \left(= p^{\alpha-1}\frac{b_u}{p^\theta}\right) \in R, \qquad \text{where } p^\theta \mid a_u,\ p^{\theta+1} \nmid a_u.$$

Then (3) gives

$$xT = x\left[\frac{R}{Ra_1} \oplus \cdots \oplus \frac{R}{Ra_u}\right] = \frac{Rx + Ra_1}{Ra_1} \oplus \cdots \oplus \frac{Rx + Ra_u}{Ra_u}, \tag{5}$$

and its $(k-1)$th summand is

$$\frac{Rx + Ra_{k-1}}{Ra_{k-1}} = \frac{Rp^{\alpha-1}(a_u/p^\theta) + Rp^\alpha a'_{k-1}}{Rp^\alpha a'_{k-1}} \qquad \text{(where } (a'_{k-1}, p) = 1\text{)}$$
$$= \frac{Rd}{Rp^\alpha a'_{k-1}},$$

where $d = (p^{\alpha-1}(a_u/p^\theta), p^\alpha a'_{k-1})$, the g.c.d. of $p^{\alpha-1}(a_u/p^\theta)$ and $p^\alpha a'_{k-1}$. But

$$a_{k-1} | a_u \Rightarrow a'_{k-1} = \frac{a_{k-1}}{p^\alpha} \,\bigg|\, \frac{a_u}{p^\theta}$$
$$\Rightarrow \left(p^{\alpha-1}\frac{a_u}{p^\theta}, p^\alpha a'_{k-1}\right) = p^{\alpha-1} a'_{k-1}.$$

Therefore, $d = p^{\alpha-1}a'_{k-1}$. Thus, the $(k-1)$th summand in (5) is

$$\frac{Rp^{\alpha-1}a'_{k-1}}{Rp^\alpha a'_{k-1}} \simeq \frac{R}{Rp},$$

because in any integral domain $Ra/Rab \simeq R/Rb$, $0 \neq a, b \in R$.

It can similarly be shown that any summand preceding the $(k-1)$th summand is either $R/(p)$ or (0); and indeed if any summand is (0) then all the preceding ones are also zero.

Therefore, (5) can be rewritten as

$$xT \simeq \overbrace{(0) \oplus \cdots \oplus (0)}^{s \text{ terms}} \oplus \overbrace{R/(p) \oplus \cdots \oplus R/(p) \oplus R/(p)}^{t \text{ terms}} \oplus R/(p^{\lambda_k}) \oplus \cdots \oplus R/(p^{\lambda_u}), \tag{6}$$

where $s, t \geq 0$, and $s + t = k - 2$. Further, from (4),

$$xT = \frac{Rx + Rb_1}{Rb_1} \oplus \cdots \oplus \frac{Rx + Rb_u}{Rb_u}, \tag{7}$$

and its $(k-1)$th summand is

$$\frac{Rx + Rb_{k-1}}{Rb_{k-1}} = \frac{Rp^{\alpha-1}(b_u/p^\theta) + Rp^\beta b'_{k-1}}{Rp^\beta b'_{k-1}} \text{ (where } (b'_{k-1}, p) = 1\text{)}$$
$$= \frac{Rp^\beta b'_{k-1}}{Rp^\beta b'_{k-1}} \qquad \left(\text{since } \alpha > \beta, \text{ and } b_{k-1} | b_u \Rightarrow b'_{k-1} = \frac{b_{k-1}}{p^\beta} \,\bigg|\, \frac{b_u}{p^\theta} \Rightarrow p^\beta b'_{k-1} | p^{\alpha-1}\frac{b_u}{p^\theta}\right).$$

Therefore, $((Rx + Rb_{k-1})/Rb_{k-1}) = (0)$. Thus, $Rx \subset Rb_{k-1}$. Because $Rb_{k-1} \subset Rb_{k-2} \subset \cdots \subset Rb_1$, it follows that the first $k-1$ summands in (7) are all zero. Therefore, the decomposition (7) of xT may be rewritten as

$$xT = \overbrace{(0) \oplus \cdots \oplus (0)}^{(k-1)\text{ summands}} \oplus R/(p^{\lambda_k}) \oplus \cdots \oplus R/(p^{\lambda_u}). \tag{8}$$

Comparing (6) and (8), we obtain a contradiction, because, as remarked earlier, any two such decompositions of a module over a PID must have an equal number of nonzero summands. Hence, $\alpha = \beta$. This implies $Ra_{k-1} = Rb_{k-1}$. It then follows that $Ra_i = Rb_i$ for all $i = 1,\dots,u$. □

3 Application to finitely generated abelian groups

Because the ring of integers $\mathbf{Z}$ is a PID and any abelian group is a $\mathbf{Z}$-module, an immediate application of Theorems 1.1 and 2.3 gives an alternative proof of the decomposition theorem for a finitely generated abelian group (Theorem 2.1, Chapter 8).

3.1 Theorem. *Let A be a finitely generated abelian group. Then*

$$A \simeq \mathbf{Z}^s \oplus \mathbf{Z}/a_1\mathbf{Z} \oplus \cdots \oplus \mathbf{Z}/a_r\mathbf{Z}, \tag{1}$$

where s is a nonnegative integer and a_i are nonzero nonunits in $\mathbf{Z}$, such that

$$a_1 | a_2 | \cdots | a_r. \tag{2}$$

Further, the decomposition (1) of A subject to the condition (2) is unique. ($\mathbf{Z}^0$ is interpreted as the trivial group (0).)

In particular, if A is generated by $(x_1,\dots,x_n)$ subject to

$$\sum_{j=1}^{n} a_{ij}x_j = 0, \ 1 \le i \le m, \text{ then}$$

$$A \simeq \overbrace{\mathbf{Z} \times \cdots \times \mathbf{Z}}^{(n-r)\text{ copies}} \times \mathbf{Z}/a_1\mathbf{Z} \times \cdots \times \mathbf{Z}/a_r\mathbf{Z},$$

where $a_1,\dots,a_r$ are the invariant factors of the $m \times n$ matrix $A = (a_{ij})$.

3.2 Examples

(a) The abelian group generated by x_1 and x_2 subject to $2x_1 = 0$, $3x_2 = 0$ is isomorphic to $\mathbf{Z}/(6)$, because the matrix (of constraints) is $\begin{pmatrix} 2 & 0 \\ 0 & 3 \end{pmatrix}$, which is equivalent to $\begin{pmatrix} 1 & 0 \\ 0 & 6 \end{pmatrix}$.

(b) Find the abelian group generated by $\{x_1, x_2, x_3\}$ subject to

$$5x_1 + 9x_2 + 5x_3 = 0,$$
$$2x_1 + 4x_2 + 2x_3 = 0,$$
$$x_1 + x_2 - 3x_3 = 0.$$

Let

$$A = \begin{bmatrix} 5 & 9 & 5 \\ 2 & 4 & 2 \\ 1 & 1 & -3 \end{bmatrix}.$$

Perform $C_2 - C_1$, $R_1 - 2R_2$, $C_3 - C_1$, $C_2 - C_1, R_3 - R_1$, $(-1)R_3$, $(-1)C_2$, $C_1 \leftrightarrow C_2$, and $C_2 - C_1$ in succession to obtain that A is equivalent to

$$\begin{bmatrix} 1 & 0 & 0 \\ 0 & 2 & 0 \\ 0 & 0 & 4 \end{bmatrix}.$$

Hence, the desired abelian group is isomorphic to $\mathbf{Z}/1\mathbf{Z} \times \mathbf{Z}/2\mathbf{Z} \times \mathbf{Z}/4\mathbf{Z} = \mathbf{Z}/(2) \times \mathbf{Z}/(4)$.

Problem

Compute the invariants and write down the structures of the abelian groups with generators $x_1, \ldots, x_n$ subject to the following relations:

(a) $n = 2$; $x_1 + x_2 = 0$.
(b) $n = 3$; $3x_1 - 2x_2 = 0$, $x_1 + x_3 = 0$, $-x_1 + 3x_2 + 2x_3 = 0$.
(c) $n = 3$; $2x_2 - x_3 = 0$, $-3x_1 + 8x_2 + 3x_3 = 0$, and $2x_1 - 4x_2 - x_3 = 0$.

4 Rational canonical form

The fundamental theorem of finitely generated modules over a PID has interesting applications in obtaining important canonical forms for square matrices over a field. Given an $n \times n$ matrix A over a field F, our effort will be to choose an invertible matrix P such that $P^{-1}AP$ is in a desired canonical form. This problem is equivalent to finding a suitable basis of a vector space F^n such that the matrix of the linear mapping $x \mapsto Ax$ with respect to the new basis is in the required canonical form (Corollary 6.4, Chapter 14).

Not every matrix is similar to a diagonal matrix. For example, let $A = \left[\begin{smallmatrix} 1 & a \\ 0 & 1 \end{smallmatrix}\right]$, $a \neq 0$, be a matrix over $\mathbf{R}$. Suppose there exists an invertible matrix P such that

$$P^{-1}AP = \begin{bmatrix} d_1 & 0 \\ 0 & d_2 \end{bmatrix}.$$

Clearly,

$$\det(P^{-1}AP - xI) = \det(A - xI).$$

Thus,

$$\begin{vmatrix} d_1 - x & 0 \\ 0 & d_2 - x \end{vmatrix} = \begin{vmatrix} 1-x & a \\ 0 & 1-x \end{vmatrix}.$$

This implies $d_1 d_2 = 1$ and $d_1 + d_2 = 2$, so $d_1 = 1 = d_2$. Hence, $P^{-1}AP = I$; that is, $A = I$, a contradiction.

However, simple canonical forms are highly important in linear algebra because they provide an adequate foundation for the study of deeper properties of matrices. As an application of the fundamental theorem for finitely generated modules over a PID, we show that every matrix A over a field is similar to a matrix

$$\begin{bmatrix} B_1 & & & & \\ & B_2 & & & \\ & & \cdot & & \\ & & & \cdot & \\ & & & & \cdot \\ & & & & & B_s \end{bmatrix}$$

where B_i are matrices of the form

$$\begin{bmatrix} 0 & 0 & \cdots & 0 & * \\ 1 & 0 & \cdots & 0 & * \\ 0 & 1 & \cdots & 0 & * \\ \cdot & \cdot & & \cdot & \cdot \\ \cdot & \cdot & & \cdot & \cdot \\ \cdot & \cdot & & \cdot & \cdot \\ 0 & 0 & \cdots & 1 & * \end{bmatrix}.$$

Let V be a vector space over a field F, and let $T: V \to V$ be a linear mapping. We can make V an $F[x]$-module by defining the action of any polynomial $f(x) = a_0 + a_1 x + \cdots + a_m x^m$ on any vector $v \in V$ as

$$f(x)v = a_0 v + a_1(Tv) + \cdots + a_m(T^m v),$$

where $T^i v$ stands for $T^i(v)$. Clearly, this action of $F[x]$ on V is the extension of the action of F such that $xv = Tv$. First we note the following simple fact.

If V is a finite-dimensional vector space over F, then V is a finitely generated $F[x]$-module.

For if $\{v_1,...,v_m\}$ is a basis of V over F, then $\{v_1,...,v_m\}$ is a set of generators of V as $F[x]$-module.

Although it follows by Theorem 1.1 that the finitely generated $F[x]$-module V can be decomposed as a finite direct sum of cyclic modules, we provide a proof that gives an explicit algorithm for obtaining the decomposition in this particular situation.

4.1 Theorem. *Let $T \in \mathrm{Hom}_F(V,V)$, and let $f_1(x),...,f_n(x)$ be the invariant factors of $A - xI$, where A is a matrix of T. Then*

$$V \simeq \frac{F[x]}{(f_1(x))} \oplus \cdots \oplus \frac{F[x]}{(f_n(x))}.$$

Proof. Let $(v_1,...,v_n)$ be a basis of V, and let $A = (a_{ij})$ be the matrix of T relative to this basis. Then

$$Tv_j = \sum_{i=1}^{n} a_{ij}v_i, \qquad 1 \leq j \leq n.$$

Let

$$e_i = (0,...,0,\overset{i\text{th}}{1},0,...,0), \qquad i = 1,2,...,n,$$

be the standard basis of the $F[x]$-module $F[x]^n = F[x] \times \cdots \times F[x]$ (n times). As we remarked earlier, $v_1,...,v_n$ is a set of generators of the $F[x]$-module V. Let

$$\phi: F[x]^n \to V$$

be an $F[x]$-homomorphism sending $e_i \mapsto v_i$. We claim

$$\mathrm{Ker}\, \phi = (A - xI)F[x]^n.$$

For let

$$\begin{bmatrix} g_1(x) \\ \cdot \\ \cdot \\ \cdot \\ g_n(x) \end{bmatrix} \in F[x]^n,$$

and let $(A - xI)^j$ denote the jth column of $A - xI$. Then

$$\begin{aligned}(A - xI)\begin{bmatrix} g_1(x) \\ \cdot \\ \cdot \\ \cdot \\ g_n(x) \end{bmatrix} &= [(A - xI)^1 \cdots (A - xI)^n]\begin{bmatrix} g_1(x) \\ \cdot \\ \cdot \\ \cdot \\ g_n(x) \end{bmatrix} \\ &= (A - xI)^1 g_1(x) + \cdots + (A - xI)^n g_n(x) \\ &= g_1(x)[(a_{11} - x)e_1 + \cdots + a_{n1}e_n] \\ &\quad + \cdots + g_n(x)[a_{1n}e_1 + \cdots + (a_{nn} - x)e_n] \\ &\mapsto g_1(x)[(a_{11} - x)v_1 + \cdots + a_{n1}v_n] \\ &\quad + \cdots + g_n(x)[a_{1n}v_1 + \cdots + (a_{nn} - x)v_n] \\ &= g_1(x)[a_{11}v_1 + \cdots + a_{n1}v_n - Tv_1] \\ &\quad + \cdots + g_n(x)[a_{1n}v_1 + \cdots + a_{nn}v_n - Tv_n] \\ &= 0.\end{aligned}$$

So $(A - xI)F[x]^n \subset \operatorname{Ker} \phi$. In particular, for all $1 \leq j \leq n$,

$$f_j = (A - xI)^j = \sum_{i=1}^{n} a_{ij}e_i - xe_j \in \operatorname{Ker} \phi;$$

hence,

$$xe_j = \sum_{i=1}^{n} a_{ij}e_i - f_j. \tag{1}$$

Now suppose

$$\sum_{j=1}^{n} g_j(x)e_j \in \operatorname{Ker} \phi. \tag{2}$$

Repeated substitution of xe_j, $1 \leq j \leq n$, in (2) gives

$$\sum_{j=1}^{n} g_j(x)e_j = \sum_{j=1}^{n} h_j(x)f_j + \sum_{j=1}^{n} c_je_j. \tag{3}$$

Because f_j and, hence, $h_j(x)f_j \in \operatorname{Ker} \phi$, $\Sigma_{j=1}^{n} c_je_j \in \operatorname{Ker} \phi$. Thus, $\Sigma_{j=1}^{n} c_jv_j = 0$, which implies each $c_j = 0$. This proves

$$\begin{aligned}\sum_{j=1}^{n} g_j(x)e_j &= \sum_{j=1}^{n} h_j(x)(A - xI)^j \\ &= (A - xI)\begin{bmatrix} h_1(x) \\ \cdot \\ \cdot \\ \cdot \\ h_n(x) \end{bmatrix} \in (A - xI)F[x]^n.\end{aligned}$$

Hence, Ker $\phi = (A - xI)F[x]^n$, as claimed. Therefore,

$$V \simeq \frac{F[x]^n}{(A - xI)F[x]^n}.$$

Let P and Q be $n \times n$ invertible matrices over $F[x]$. Note that for any $n \times n$ invertible matrix T over $F[x]$, $TF[x]^n = F[x]^n$. So

$$\frac{F[x]^n}{P(A - xI)Q\,F[x]^n} = \frac{F[x]^n}{P(A - xI)\,F[x]^n}$$
$$= \frac{P\,F[x]^n}{P(A - xI)\,F[x]^n} \simeq \frac{F[x]^n}{(A - xI)\,F[x]^n},$$

by a canonical isomorphism sending

$$P\begin{bmatrix} g_1(x) \\ \cdot \\ \cdot \\ \cdot \\ g_n(x) \end{bmatrix} + P(A - xI)\,F[x]^n \mapsto \begin{bmatrix} g_1(x) \\ \cdot \\ \cdot \\ \cdot \\ g_n(x) \end{bmatrix} + (A - xI)F[x]^n.$$

Therefore,

$$V \simeq \frac{F[x]^n}{(A - xI)F[x]^n} \simeq \frac{F[x]^n}{P(A - xI)Q\,F[x]^n},$$

when P and Q are any $n \times n$ invertible matrices. Choose P and Q such that

$$P(A - xI)Q = \text{diag}(f_1(x),\dots,f_n(x)),$$

where $f_1(x)|f_2(x)| \cdots |f_n(x)$. Then

$$V \simeq \frac{F[x]^n}{P(A - xI)Q\,F[x]^n} = \frac{F[x] \oplus \cdots \oplus F[x]}{f_1(x)F[x] \oplus \cdots \oplus f_n(x)F[x]}$$
$$\simeq \frac{F[x]}{(f_1(x))} \oplus \cdots \oplus \frac{F[x]}{(f_n(x))}. \quad \square$$

Note. By abuse of terminology $f_1(x),\dots,f_n(x)$ are also called invariant factors of A or of T. The sequence of factors $f_k(x), f_{k+1}(x),\dots,f_n(x)$ obtained from the invariant factors of $A - xI$ by discarding units, if any, are known as the *invariant factors* of $F[x]$-module V.

The proof of Theorem 4.1 shows that $\det(A - xI)$ is the product of the invariant factors of the matrix $A - xI$.

Definition. *The polynomial* $f(x) = \det(A - xI)$ *is called the* characteristic polynomial *of A, and its roots are called* characteristic roots *(or* eigenvalues*) of A.*

Some authors define the characteristic polynomial of A to be $f(x) = \det(xI - A)$, which differs from our definition by only the factor $(-1)^n$, where n is the order of A.

Definition. *Let* $T \in \mathrm{Hom}_F(V, V)$. *If V as an* $F[x]$*-module relative to T is cyclic, then V is called a T-*cyclic space. *In this case T is also called a* cyclic endomorphism *of V.*

4.2 Theorem. *Let W be a subspace of V. Let* $T \in \mathrm{Hom}_F(V, V)$ *such that* $TW \subseteq W$. *Then W is a T-cyclic subspace iff there exists an element* $w \in W$ *such that* $\{w, Tw, \ldots, T^{k-1}w\}$ *is a basis of W for some* $k \geq 1$.

Proof. Let $\dim V = n$, and let W be a T-cyclic subspace generated by w. Since $\dim V = n$, the vectors $w, Tw, T^2w, \ldots, T^n w$ must be linearly dependent over F. Let $T^r w$ be the first of these vectors linearly dependent on the preceding ones. Then $w, Tw, \ldots, T^{r-1}w$ are linearly independent, while

$$T^r w = a_0 w + \cdots + a_{r-1}(Tw)^{r-1}, \qquad a_i \in F. \tag{1}$$

We claim that every $T^i w$ is linearly dependent on $w, Tw, \ldots, T^{r-1}w$. For $i \leq r - 1$ this is clear, and for $i = r$ it follows by (1). Thus, let us take $i > r$ and use induction on i. Applying T^{i-r} to (1), we obtain

$$T^i w = a_0(T^{i-r}w) + \cdots + a_{r-1}(T^{i-1}w),$$

and by induction hypothesis each term on the right side is linearly dependent on $w, Tw, \ldots, T^{r-1}w$. Hence, $T^i w$ is also. Because $W = F[x]w$, every element of W is of the form

$$(b_0 + b_1 x + \cdots + b_m x^m)w = (b_0 I + b_1 T + \cdots + b_m T^m)w,$$

where $b_i \in F$. It follows then from these considerations that $\{w, Tw, \ldots, T^{r-1}w\}$ is a basis of W.

Conversely, let $\{w, Tw, \ldots, T^{r-1}w\}$ be a basis of W.

This implies that for any element $c_0 w + c_1(Tw) + \cdots + c_{r-1}(T^{r-1}w)$ of W, there exists a polynomial $f(x) = c_0 + c_1 x + \cdots + c_{r-1}x^{r-1}$ such that

$$c_0 w + c_1(Tw) + \cdots + c_{r-1}(T^{r-1}w) = (c_0 + c_1 x + \cdots + c_{r-1}x^{r-1})w.$$

Hence, $W \subseteq F[x]w$. Because $F[x]w \subseteq W$, it follows that $W = (w)$; that is, W is a T-cyclic subspace. □

Definition. *Let* $f(x) = a_0 + a_1 x + \cdots + a_{k-1}x^{k-1} + x^k$ *be a monic polynomial over F. The* companion matrix *of* $f(x)$ *is the* $k \times k$ *matrix*

$$\begin{bmatrix} 0 & 0 & \cdots & 0 & -a_0 \\ 1 & 0 & \cdots & 0 & -a_1 \\ 0 & 1 & \cdots & 0 & -a_2 \\ \cdot & \cdot & & \cdot & \cdot \\ \cdot & \cdot & & \cdot & \cdot \\ \cdot & \cdot & & \cdot & \cdot \\ 0 & 0 & \cdots & 1 & -a_{k-1} \end{bmatrix}$$

4.3 Theorem. *Let* $T \in \operatorname{Hom}_F(V,V)$, *and let W be a T-cyclic subspace of V with a basis* $\{w, Tw, \ldots, T^{k-1}w\}$. *Then the matrix of T on W is the* $k \times k$ *companion matrix of a monic polynomial* $f(x)$, *where* $(f(x)) = \operatorname{ann} w$, *the annihilator of w in the ring* $F[x]$ *(also called the order ideal of w).*

Proof. Because $\{w, Tw, \ldots, T^{k-1}w\}$ is a basis of W,

$$T^k w = a_0 w + a_1(Tw) + \cdots + a_{k-1}(T^{k-1}w),$$

where a_i are unique elements of F. Thus,

$$(a_0 + a_1 x + \cdots + a_{k-1}x^{k-1} - x^k)w = 0.$$

If $\operatorname{ann} w = (f(x))$, then

$$a_0 + a_1 x + \cdots + a_{k-1}x^{k-1} - x^k \in (f(x)),$$

so $k \geq \operatorname{degree} f(x) = s$, say. So if $f(x) = b_0 + b_1 x + \cdots + b_{s-1}x^{s-1} + x^s$, then

$$\begin{aligned} 0 &= (b_0 + b_1 x + \cdots + b_{s-1}x^{s-1} + x^s)w \\ &= b_0 w + b_1 T(w) + \cdots + b_{s-1}T^{s-1}(w) + T^s w. \end{aligned}$$

Hence, $T^s w$ is a linear combination of $w, Tw, \ldots, T^{s-1}w$, whence $s \geq k$. This gives $s = k$. Therefore, if $g(x) = a_0 + a_1 x + \cdots + a_{k-1}x^{k-1} - x^k$, then $g(x) = cf(x)$, where $c \in F$. By comparing the coefficients of x^k on each side, we get $c = -1$. Hence, $f(x) = -g(x)$. Because

$$T^k w = a_0 w + a_1(Tw) + \cdots + a_{k-1}(T^{k-1}w),$$

we obtain that the matrix of T on W is

$$\begin{bmatrix} 0 & 0 & \cdots & 0 & a_0 \\ 1 & 0 & \cdots & 0 & a_1 \\ 0 & 1 & \cdots & 0 & a_2 \\ \cdot & \cdot & & \cdot & \cdot \\ \cdot & \cdot & & \cdot & \cdot \\ \cdot & \cdot & & \cdot & \cdot \\ 0 & 0 & \cdots & 1 & a_{k-1} \end{bmatrix}$$

which is the companion matrix of $f(x)$. This completes the proof. □

Remark. In Theorem 4.3, $f(x) = x^k - a_{k-1}x^{k-1} - \cdots - a_1 x - a_0$ is the minimal polynomial of T on W.

Let V be an n-dimensional vector space over F. Then dim $\text{Hom}_F(V,V), = n^2$, so if $0 \neq T \in \text{Hom}_F(V,V)$, the collection (T^i), $i = 0,1,2,\ldots,n^2$, must be linearly dependent over F. Let T^r be the first in this collection linearly dependent on the preceding ones. Thus,

$$T^r = a_0 I + a_1 T + \cdots + a_{r-1} T^{r-1}$$

for some $a_i \in F$. Hence, V as an $F[x]$-module relative to T has the property that

$$\begin{aligned} 0 &= (a_0 I + a_1 T + \cdots + a_{r-1} T^{r-1} - T^r)v \\ &= (a_0 + a_1 x + \cdots + a_{r-1} x^{r-1} - x^r)v. \end{aligned}$$

Because $a_0 + a_1 x + \cdots + a_{r-1}x^{r-1} - x^r \neq 0$, every element of the $F[x]$-module V is a torsion element. Because V is a finitely generated $F[x]$-module, by the structure theorem of finitely generated modules over a principal ideal domain (Theorem 1.1), we have

$$V = V_1 \oplus \cdots \oplus V_r,$$

where V_i $[\simeq F[x]/(f_i(x))]$ are cyclic $F[x]$-modules, and $f_1(x)|f_2(x)| \cdots |f_r(x)$; $(f_i(x))$ is the annihilator ideal (order ideal) of V_i in $F[x]$.

Because V_i is a T-cyclic space, the linear mapping T on V_i is represented by the companion matrix B_i of $f_i(x)$ with respect to a suitable basis $\mathcal{B}_i$ of V_i as shown in Theorem 4.3. So the linear transformation T on V is represented by the matrix A, which is the direct sum of the matrices $B_1,\ldots,B_r$ with respect to the basis $\mathcal{B} = \bigcup_{i=1}^r \mathcal{B}_i$. Thus, we have

4.4 Theorem. *Let $T \in \mathrm{Hom}_F(V,V)$. Then there exists a basis of V with respect to which the matrix of T is*

$$A = \begin{bmatrix} B_1 & & & & 0 \\ & B_2 & & & \\ & & \cdot & & \\ & & & \cdot & \\ 0 & & & \cdot & \\ & & & & B_r \end{bmatrix},$$

where B_i is the companion matrix of a certain unique polynomial $f_i(x)$, $i = 1,\dots,r$, such that $f_1(x)|f_2(x)| \cdots |f_r(x)$. □

This form of A is called the *rational canonical form* of the matrix of T. The uniqueness of the decomposition of V subject to $f_1(x)|f_2(x)| \cdots |f_r(x)$ shows that T has a unique representation by a matrix in a rational canonical form.

The polynomials $f_1(x),\dots,f_r(x)$ are the invariant factors of $A - xI$, but they are also called the invariant factors of A (see note following Theorem 4.1).

4.5 Examples

(a) Reduce the following matrix A to rational canonical form:

$$A = \begin{bmatrix} -3 & 2 & 0 \\ 1 & 0 & 1 \\ 1 & -3 & -2 \end{bmatrix}.$$

Solution. The invariant factors of $A - xI$ are obtained in Problem 1 (b) in Section 3 of Chapter 20. These are $1, 1, (x+1)^2(x+3) = 3 + 7x + 5x^2 + x^3$. Therefore, the rational canonical form is

$$\begin{bmatrix} 0 & 0 & -3 \\ 1 & 0 & -7 \\ 0 & 1 & -5 \end{bmatrix}.$$

(b) The rational canonical form of the 6×6 matrix A whose invariant factors are $(x-3),(x-3)(x-1),(x-3)(x-1)^2$ is

$$\begin{bmatrix} 3 & & & & & \\ & 0 & -3 & & & \\ & 1 & 4 & & & \\ & & & 0 & 0 & 3 \\ & & & 1 & 0 & -7 \\ & & & 0 & 1 & 5 \end{bmatrix}$$

Problem

1. Find rational canonical forms of the following matrices over **Q**:

$$\text{(a)} \begin{bmatrix} 1 & 5 & 7 \\ 0 & 4 & 3 \\ 0 & 0 & 1 \end{bmatrix}. \quad \text{(b)} \begin{bmatrix} 2 & 4 & 0 \\ 1 & 4 & 1 \\ 3 & 8 & 3 \end{bmatrix}. \quad \text{(c)} \begin{bmatrix} 1 & 1 & -2 & 4 \\ 0 & 1 & 2 & 2 \\ 0 & 0 & 1 & 3 \\ 0 & 0 & 0 & 0 \end{bmatrix}.$$

Some authors further decompose each B_i as the direct sum of matrices, each of which is a companion matrix of a polynomial that is a power of an irreducible polynomial, and call the resulting form rational canonical form. The powers of the irreducible polynomials corresponding to each block in the form thus obtained are then called the *elementary divisors of A*.

5 Generalized Jordan form over any field

We now give another application of the fundamental theorem of finitely generated modules over a PID. We shall prove that every matrix over **C** is similar to an "almost diagonal" matrix of the form

$$\begin{bmatrix} J_1 & & & & \\ & \cdot & & & \\ & & \cdot & & \\ & & & \cdot & \\ & & & & J_s \end{bmatrix},$$

where

$$J_i = \begin{bmatrix} \lambda_i & 0 & \cdots & 0 \\ 1 & \lambda_i & \cdots & 0 \\ 0 & 1 & & \\ \vdots & \vdots & & \\ 0 & 0 & \cdots & \lambda_i \end{bmatrix}.$$

But first we obtain a canonical form over any field, of which the Jordan canonical form is a special case.

Let V be a vector space of dimension n over F. Let $T \in \operatorname{Hom}_F(V,V)$. Because $\operatorname{Hom}_F(V,V)$ is an n^2-dimensional vector space over F, the list

$$1, T, T^2, \ldots, T^{n^2}$$

is linearly dependent, and, hence, there exist $\alpha_0, \alpha_1, \ldots, \alpha_{n^2} \in F$ (not all zero) such that

$$\alpha_0 + \alpha_1 T + \cdots + \alpha_{n^2} T^{n^2} = 0.$$

Therefore, T satisfies a nonzero polynomial over F. We begin with

5.1 Theorem. *Let V be a finite-dimensional vector space over a field F, and let $T \in \operatorname{Hom}_F(V,V)$. Suppose $f(x) = g(x)h(x)$ is a factorization of $f(x)$ in $F[x]$ such that $(g(x), h(x)) = 1$. Then $f(T) = 0$ if and only if $V =$ $\operatorname{Ker} g(T) \oplus \operatorname{Ker} h(T)$.*

Proof. Suppose $f(T) = 0$. Because $(g(x),h(x)) = 1$, there exist $a(x),b(x) \in F[x]$ such that $a(x)g(x) + b(x)h(x) = 1$. Hence,

$$a(T)g(T) + b(T)h(T) = 1. \tag{1}$$

Then for each $v \in V$,

$$a(T)g(T)v + b(T)h(T)v = v.$$

Because $g(T)h(T) = 0$, $a(T)g(T)v \in \operatorname{Ker} h(T)$, and $b(T)h(T)v \in \operatorname{Ker} g(T)$. Thus,

$$V = \operatorname{Ker} g(T) + \operatorname{Ker} h(T).$$

Further, if $w \in \operatorname{Ker} g(T) \cap \operatorname{Ker} h(T)$, then $g(T)w = 0 = h(T)w$. Thus, by (1) $w = 0$. Hence, $V = \operatorname{Ker} g(T) \oplus \operatorname{Ker} h(T)$. Conversely, let $V =$ $\operatorname{Ker} g(T) \oplus \operatorname{Ker} h(T)$, and let $v \in V$. Then $v = v_1 + v_2$, $v_1 \in \operatorname{Ker} g(T)$,

$v_2 \in \text{Ker } h(T)$. This gives $g(T)h(T)(v_1 + v_2) = 0$, because $g(T)$ and $h(T)$ commute. Therefore, $f(T)v = 0$ for all $v \in V$; so $f(T) = 0$. □

More generally, we can similarly prove

5.2 Theorem. *Let V be a finite-dimensional vector space over a field F, and let $T \in \text{Hom}_F(V,V)$. Suppose $f(x) = p_1^{e_1}(x) \dots p_k^{e_k}(x)$ is a factorization of $f(x)$ into irreducible polynomials $p_i(x)$ over F. Then $f(T) = 0$ if and only if*

$$V = \text{Ker } p_1^{e_1}(T) \oplus \cdots \oplus \text{Ker } p_k^{e_k}(T).$$

In the notation of Theorem 4.2, let us set $W_i = \text{Ker } p_i^{e_i}(T)$, $i = 1,\dots,k$. It is clear that $TW_i \subseteq W_i$. Let T_i be the restriction of T on W_i, so $T_i \in \text{Hom}_F(W_i, W_i)$. Let us make W_i an $F[x]$-module relative to T_i in the usual way. Then W_i is a finitely generated $F[x]$-module, and, hence, W_i is the finite direct sum of cyclic modules (w_{ij}), $1 \le j \le s$. Because every element of W_i is annihilated by $p_i^{e_i}(T)$, $p_i^{e_i}(T)w_{ij} = 0$. Our object is to find a basis of each cyclic module (w_{ij}) so that the matrix of T_i relative to the union of these bases is in generalized Jordan canonical form.

For clarity we drop all subscripts and proceed to find a basis of a T-cyclic subspace (v), where $p^s(x)v = 0$, $p(x)$ is an irreducible polynomial over F, and s is the smallest positive integer such that $p^s(x)v = 0$; that is, $p^s(x)$ is the minimal polynomial of T [or, equivalently, ann v is the ideal $(p^s(x))$].

5.3 Lemma. *Let $V = (v)$ be a T-cyclic space, and let the annihilator of v (i.e., the order ideal of v) be (p^s), where $p = p(x) = x^t + a_{t-1}x^{t-1} + \cdots + a_0$ is an irreducible polynomial over F. Then*

$$\begin{aligned} B = \{v, Tv, \dots, T^{t-1}v\} &\cup \{pv, pTv, \dots, pT^{t-1}v\} \\ &\cup \cdots \cup \{p^{s-1}v, p^{s-1}Tv, \dots, p^{s-1}T^{t-1}v\} \end{aligned}$$

is a basis of V, and the matrix of T relative to this basis is

$$\begin{bmatrix} P & 0 & \cdots & 0 \\ N & P & \cdots & 0 \\ 0 & N & \cdots & 0 \\ \vdots & \vdots & & \vdots \\ 0 & 0 & \cdots & P \end{bmatrix},$$

where P is the $t \times t$ companion matrix of p and

$$N = \begin{bmatrix} 0 & 0 & \cdots & 1 \\ 0 & 0 & \cdots & 0 \\ \cdot & \cdot & & \cdot \\ \cdot & \cdot & & \cdot \\ \cdot & \cdot & & \cdot \\ 0 & 0 & \cdots & 0 \end{bmatrix}$$

is a $t \times t$ matrix all of whose entries, except at the $(t,1)$ position, are zero.

Proof. Because $V = (v)$ is a T-cyclic space, V has a basis of the form $\{v, Tv, \ldots, T^{k-1}v\}$ for some positive integer k (Theorem 4.2). Further, $p^s(x)v = 0$ implies $p^s(T) = 0$. Therefore, T satisfies a polynomial of degree $n = st$. Thus, $T^n v$ is a linear combination of $v, Tv, \ldots, T^{n-1}v$. But then $n \geq k$. If possible let $k < n$. Then, because $\{v, Tv, \ldots, T^{k-1}v\}$ is a basis of V, we get that $T^k v$ is a linear combination of $v, Tv, \ldots, T^{k-1}v$, so there exists a polynomial $g(x)$ of degree $k < n$ such that $g(x)v = 0$, a contradiction. Hence, $n = k = \dim V$.

The set B of $n = st$ elements in the theorem must be linearly independent. Otherwise there will be a nonzero polynomial $h(x)$ over F of degree $< n$ such that $h(x)v = 0$, a contradiction. Hence, the set B is linearly independent and, therefore, a basis of V.

We now proceed to find the matrix of T relative to the basis B. But first we express each of

$$T^t v, pT^t v, \ldots, p^{s-1}T^t v$$

as a linear combination of the basis elements. Now

$$0 = p - (x^t + a_{t-1}x^{t-1} + \cdots + a_0),$$

so

$$0 = pv - (T^t v + a_{t-1}T^{t-1}v + \cdots + a_0 v),$$

which gives

$$T^t v = -a_0 v - a_1 Tv - \cdots - a_{t-1}T^{t-1}v + pv. \tag{1}$$

Further,

$$0 = p^2 - p(x^t + a_{t-1}x^{t-1} + \cdots + a_0)$$

gives

$$0 = p^2 v - pT^t v - a_{t-1}pT^{t-1}v - \cdots - a_0 pv.$$

Thus,

$$pT^tv = -a_0pv - a_1pTv - \cdots - a_{t-1}pT^{t-1}v + p^2v. \tag{2}$$

Similarly, by considering

$$0 = p^{s-1} - p^{s-2}(x^t + a_{t-1}x^{t-1} + \cdots + a_0),$$

we get

$$p^{s-2}T^tv = -a_0p^{s-2}v - a_1p^{s-2}Tv - \cdots - a_{t-1}p^{s-2}T^{t-1}v + p^{s-1}v. \tag{3}$$

Finally, to obtain the expression for $p^{s-1}T^tv$, we write

$$0 = p^sv = p^{s-1}(x^t + a_{t-1}x^{t-1} + \cdots + a_0)v,$$

so

$$p^{s-1}T^tv = -a_0p^{s-1}v - a_1p^{s-1}Tv - \cdots - a_{t-1}p^{s-1}T^{t-1}v. \tag{4}$$

Consider now

$$\begin{aligned}
Tv &= Tv,\\
T\,Tv &= T^2v,\\
&\vdots\\
T\,T^{t-2}v &= T^{t-1}v,\\
T\,T^{t-1}v &= T^tv = -a_0v - a_1Tv - \cdots - a_{t-1}T^{t-1}v + pv \quad \text{[from (1)]},\\
Tpv &= pTv,\\
TpTv &= pT^2v,\\
&\vdots\\
TpT^{t-2}v &= pT^{t-1}v,\\
TpT^{t-1}v &= pT^tv = -a_0pv - a_1pTv - \cdots - a_{t-1}pT^{t-1}v + p^2v \quad \text{[from (2)]},\\
&\vdots\\
Tp^{s-1}v &= p^{s-1}Tv,\\
Tp^{s-1}Tv &= p^{s-1}T^2v,\\
&\vdots\\
Tp^{s-1}T^{t-2}v &= p^{s-1}T^{t-1}v,\\
Tp^{s-1}T^{t-1}v &= p^{s-1}T^tv = -a_0p^{s-1}v - a_1p^{s-1}Tv - \cdots\\
&\quad - a_{t-1}p^{s-1}T^{t-1}v \quad \text{[from (4)]}.
\end{aligned}$$

Thus, the matrix of T relative to B is of the form

$$\begin{bmatrix} P & 0 & \cdots & 0 \\ N & P & \cdots & 0 \\ 0 & N & \cdots & \\ \vdots & \vdots & & \vdots \\ 0 & 0 & \cdots & P \end{bmatrix},$$

where P is the $t \times t$ companion matrix

$$\begin{bmatrix} 0 & 0 & \cdots & 0 & -a_0 \\ 1 & 0 & \cdots & 0 & -a_1 \\ 0 & 1 & \cdots & 0 & -a_2 \\ \cdot & \cdot & & \cdot & \cdot \\ \cdot & \cdot & & \cdot & \cdot \\ \cdot & \cdot & & \cdot & \cdot \\ 0 & 0 & \cdots & 1 & -a_{t-1} \end{bmatrix}$$

of the polynomial $p(x)$, and N is the $t \times t$ matrix all of whose entries are zero, except at the $(1, t)$ position, where the entry is 1. This completes the proof of the lemma. □

Let us call a matrix of the form

$$\begin{bmatrix} P & 0 & \cdots & 0 \\ N & P & \cdots & 0 \\ 0 & N & \cdots & \\ \vdots & \vdots & & \vdots \\ 0 & 0 & \cdots & P \end{bmatrix},$$

where P is the companion matrix of a certain irreducible polynomial over F of degree t, and N is a matrix all of whose entries are zero except at the $(1, t)$ position, where the entry is 1, a *generalized Jordan block.* It is noteworthy that if F is the field of complex numbers, then P is a 1×1 matrix, (λ), and N is a 1×1 matrix (1), so the generalized Jordan block becomes the usual *Jordan block.*

5.4 Theorem. *Let $T \in \operatorname{Hom}_F(V, V)$. Then there exists a basis of V such that the matrix of T relative to this basis is a direct sum of generalized Jordan blocks.*

Proof. Follows from Lemma 5.3 and the discussion preceding the lemma. □

Remark. The generalized Jordan representation (and, in particular, the usual canonical representation in Jordan form) of a linear mapping T is unique up to the ordering of the blocks. This is a consequence of the uniqueness of the decomposition of the $F[x]$-module V into the direct sum of cyclic modules, each of which is annihilated by a power of an irreducible polynomial over F.

The polynomials $p^s(x)$, where $p(x)$ is irreducible over F, corresponding to each generalized Jordan block are the elementary divisors of A.

The matrix version of Theorem 5.4 is

5.5 Theorem. *Let A be an $n \times n$ matrix. Then there exists an invertible matrix P such that $P^{-1}AP$ is a direct sum of generalized Jordan blocks J_i, where J_i are unique except for ordering.*

Note that the Jordan canonical form is determined by the set of elementary divisors.

5.6 Examples

(a) For a 3×3 matrix with invariant factors $(x-2)$ and (x^2-4), the elementary divisors are $(x-2)$,$(x-2)$, and $(x+2)$, so the Jordan canonical form is

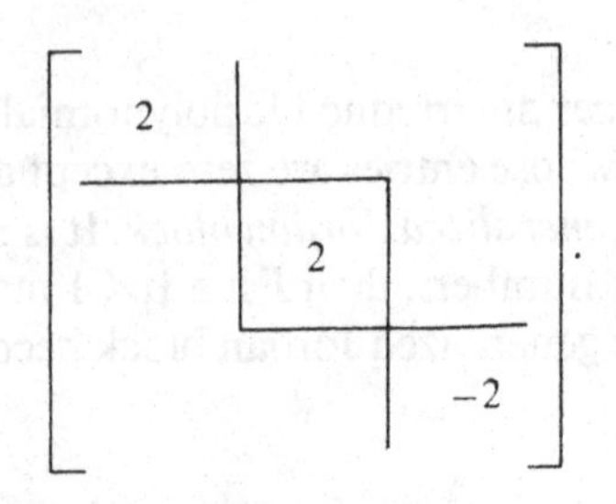

(b) For a 6×6 matrix with invariant factors $(x+2)^2$, $(x+3)^2(x+2)^2$ the elementary divisors are $(x+2)^2$,$(x+2)^2$, and $(x+3)^2$, so the Jordan canonical form is

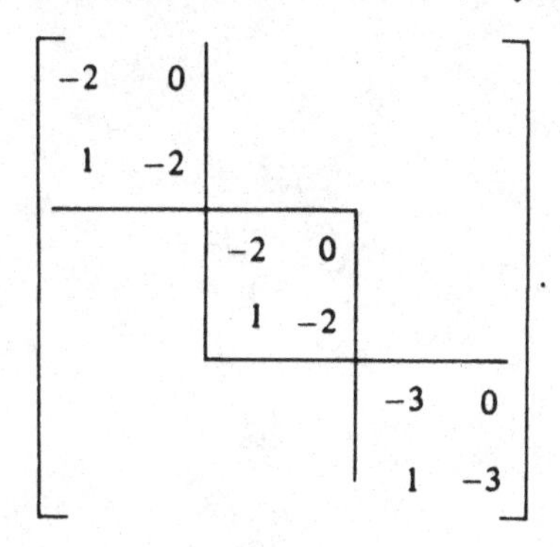

Problems

1. Find invariant factors, elementary divisors, and the Jordan canonical form of the matrices

$$\text{(a)} \begin{bmatrix} 0 & 4 & 2 \\ -3 & 8 & 3 \\ 4 & -8 & -2 \end{bmatrix}. \quad \text{(b)} \begin{bmatrix} 5 & \frac{1}{2} & -2 & 4 \\ 0 & 5 & 4 & 4 \\ 0 & 0 & 5 & 3 \\ 0 & 0 & 0 & 4 \end{bmatrix}.$$

2. Find all possible Jordan canonical forms of a matrix with characteristic polynomial $p(x)$ over $\mathbf{C}$ in each of the following cases:

(a) $p(x) = (x-1)^2(x+1)$.
(b) $p(x) = (x-2)^3(x+5)^2$.
(c) $p(x) = (x+1)(x+2)^2(x^2+1)$.

CHAPTER 22

Tensor products

1 Categories and functors

Let **C** be a class of objects $A, B, C, \ldots$, together with a family of disjoint sets $\mathrm{Hom}(A, B)$, one for each ordered pair A, B of objects. The elements of $\mathrm{Hom}(A, B)$ are called morphisms. If $\alpha \in \mathrm{Hom}(A, B)$, write $\alpha: A \to B$ and call A the domain of α and B the codomain of α. Suppose for each $\alpha \in \mathrm{Hom}(A, B)$ and $\beta \in \mathrm{Hom}(B, C)$ there is a unique morphism $\beta\alpha \in \mathrm{Hom}(A, C)$.

Definition. *We call* **C** *a* category *if the following axioms hold*:

(i) *Associativity.* $\forall \alpha \in Hom(A, B)$, $\forall \beta \in Hom(B, C)$, *and* $\forall \gamma \in Hom(C, D)$, $\gamma(\beta\alpha) = (\gamma\beta)\alpha$.

(ii) *Existence of identity morphisms.* $\forall A \in \mathbf{C}$, $\exists I_A \in Hom(A, A)$ *such that* $I_A f = f$ *and* $g I_A = g$ *whenever* $I_A f$ *and* $g I_A$ *are defined.*

It is clear that $\forall A \in \mathbf{C}$, I_A is unique.

1.1 Examples

(a) Let **C** be the family of abelian groups and let the homomorphisms between them be morphisms. Then **C** is a category, denoted by **Ab** or mod-**Z**.

(b) Let **C** be the family of right (or left) R-modules and let the R-homomorphisms between them be morphisms. Then **C** is a category denoted by mod-R (or R-mod).

(c) The family of groups also forms a category with homomorphisms as morphisms.

(d) With objects as sets and morphisms as maps between them, we obtain a category called the category of sets.

We now proceed to define a functor. Roughly speaking a functor is a map between two categories.

Definition. *T is called a covariant (contravariant) functor from a category $\mathbf{C}_1$ to a category $\mathbf{C}_2$ if for each object $A \in \mathbf{C}_1$ there is a unique object $T(A) \in \mathbf{C}_2$, and for each morphism $\alpha \in Hom(A, B)$ there is a unique morphism $T(\alpha) \in Hom(T(A), T(B))$ $[Hom(T(B), T(A))]$ such that the following axioms hold:*

(i) If $I: A \to A$ is an identity morphism in $\mathbf{C}_1$, then $T(I): T(A) \to T(A)$ is an identity morphism in $\mathbf{C}_2$.

(ii) If α, β are morphisms and $\beta\alpha$ is defined then $T(\beta\alpha) = T(\beta)T(\alpha)$ $[T(\beta\alpha) = T(\alpha)T(\beta)]$.

Note that if $\beta\alpha$ is defined then $T(\beta)T(\alpha)$ $[T(\alpha)T(\beta)]$ is also defined whenever T is a covariant (contravariant) functor.

1.2 Examples

(a) Let $\mathbf{C}$ be a category with objects as groups and morphisms as homomorphisms. Then T is a functor from $\mathbf{C}$ to $\mathbf{C}$ if $T(G) = G'$, the derived group of G, for if $\alpha: A \to B$ is a homomorphism then $\alpha|_{A'} = T(\alpha)$ maps $T(A) = A'$ to $T(B) = B'$ and $T(I) = I$, where I is the identity map. Furthermore, if β is another homomorphism then $T(\alpha\beta) = T(\alpha)T(\beta)$ (follows by definition of $T(\alpha)$). Thus T is a covariant functor.

(b) Let $\mathbf{C}$ be the category with objects as free groups and morphisms as homomorphisms. Let S be the category with objects as sets and morphisms as functions. Let A be a set in S and $F(A)$ denote the free group generated by A. For each function $\alpha: A \to B$, where A, B are objects (= sets) in S, let $F(\alpha): F(A) \to F(B)$ denote the homomorphism of the free group $F(A)$ to $F(B)$ obtained by extending the map α in a natural way. Then F is a covariant functor from S to $\mathbf{C}$.

(c) Let $\mathbf{C}$ be the category as in (a) with objects as groups and morphisms as homomorphisms, and let $\mathbf{Ab}$ be the category of abelian groups. Let $T: \mathbf{C} \to \mathbf{Ab}$ be the functor defined by $T(G) = G/G'$. Then T is a covariant functor.

(d) Let $\mathbf{C}$ be the category mod-F where F is a field, that is, the category of vector spaces over F. For a vector space V, let $V^* = \text{Hom}(V, F)$, the dual space of V. Then $T: \mathbf{C} \to \mathbf{C}$ defined by $T(V) = V^*$ is a contravariant functor. We may note that if $\alpha: V \to V$ is a morphism (= linear mapping) in $\mathbf{C}$ then $T(\alpha)$: $V^* \to V^*$ is defined by $T(\alpha)(\phi) = \phi\alpha$. Thus for all linear mapings α, β, $T(\alpha\beta) = T(\beta)T(\alpha)$.

Remark. Let R be a ring and mod-R be the category of right R-modules. Let $\text{Hom}_R(-, A)$ [$\text{Hom}_R(A, -)$] denote the functor from mod-R to mod-$\mathbf{Z}$ that sends X to $\text{Hom}_R(X, A)$ [$\text{Hom}_R(A, X)$]. Then $\text{Hom}_R(-, A)$ is a contravariant functor and $\text{Hom}_R(A, -)$ is a covariant functor.

In the next section we define the tensor product $M \otimes_R N$ of a right module M_R with a left module ${}_RN$, which is an abelian group in general. It can be shown that both $T_1 = (\cdot) \otimes_R N$ and $T_2 = M \otimes_R (\cdot)$ are covariant functors from mod-R to mod-$\mathbf{Z}$.

2 Tensor products

In this section and the others which follow, we will briefly describe the concept of the tensor product of modules. Tensor products occur often in all branches of mathematics. The use of tensor product is indeed a fundamental tool for scientists.

Let M, N, and P be abelian groups. A map f: $M \times N \to P$ is called *bi-additive* if

$$f(m_1 + m_2, n) = f(m_1, n) + f(m_2, n)$$

and

$$f(m, n_1 + n_2) = f(m, n_1) + f(m, n_2)$$

for all $m, m_1, m_2 \in M$ and $n, n_1, n_2 \in N$. This implies for a fixed $n \in N$, the map $m \to f(m, n)$ from M to P and for a fixed $m \in M$ the map $n \to f(m, n)$ from N to P are both homomorphisms. It follows that $f(m, 0) = 0 = f(0, n)$ and $f(m, -n) = -f(m, n) = f(-m, n)$ for all $m \in M$ and $n \in N$.

Suppose now M is a right R-module and N is a left R-module. Then the map $f: M \times N \to P$ is called *balanced* if it is bi-additive and satisfies

$$f(mr, n) = f(m, rn)$$

for all $m \in M$, $n \in N$, and $r \in R$.

Throughout this chapter all rings have unity unless otherwise stated.

Definition. *A tensor product of a right R-module M and a left R-module N is a pair (T, ϕ), where T is an abelian group and ϕ is a balanced map from $M \times N$ to T satisfying the following condition: Given any abelian* *group P and a balanced map f: $M \times N \to P$, there exists a unique homomorphism $\bar{f}$: $T \to P$ such that the diagram*

$$\begin{array}{ccc} M \times N & & \\ \downarrow \phi & \searrow f & \\ T & \xrightarrow[\bar{f}]{} & P \end{array}$$

is commutative, that is, $\bar{f}\phi = f$.

Remark. It is easy to check that if (T', ϕ') is another tensor product of M and N, then there exists an isomorphism α: $T \to T'$ such that $\phi' = \alpha\phi$. Thus the tensor product is unique, up to isomorphism.

2.1 Existence and construction of tensor product

We will now study a construction which yields an abelian group when we are given both a right R-module M and a left R-module N. In this construction, which is one of the most important in abstract algebra, the resulting group is indeed the tensor product of the given modules M and N.

Let $Z(M,N)$ denote the free abelian group with the set $M \times N$ as a basis. Let $Y(M,N)$ denote the subgroup of $Z(M,N)$ generated by the elements of the form

$$(m_1 + m_2, n) - (m_1, n) - (m_2, n),$$
$$(m, n_1 + n_2) - (m, n_1) - (m, n_2),$$
$$(mr, n) - (m, rn),$$

for all $m, m_1, m_2 \in M, n, n_1, n_2 \in N$, and $r \in R$. Let $T(M,N) = Z(M,N)/Y(M,N)$ be the quotient group. We proceed to show that $T(M,N)$ is the tensor product of M and N. Let P be any abelian group and f: $M \times N \to P$ be a balanced map. We need to produce a balanced map ϕ: $M \times N \to T(M,N)$ and a unique group homomorphism $\bar{f}$: $T(M,N) \to P$ such that the diagram

$$\begin{array}{ccc} M \times N & & \\ \downarrow \phi & \searrow f & \\ T(M,N) & \xrightarrow[\bar{f}]{} & P \end{array}$$

commutes. We choose ϕ to be the restriction of the canonical homomorphism

$$\pi: Z(M,N) \to \frac{Z(M,N)}{Y(M,N)} = T(M,N)$$

to $M \times N$. Further, the map $f: M \times N \to P$ can be naturally extended to the free abelian group $Z(M,N)$ with basis $M \times N$, yielding a homomorphism

$$\bar{f}: Z(M,N) \to P.$$

But since f is a balanced map, $\bar{f}$ vanishes on the subgroup $Y(M,N)$ and hence induces

$$\bar{f}: Z(M,N)/Y(M,N) \to P$$

that is, from $T(M,N)$ to P. It is clear that $\bar{f}\phi = f$. We note $\bar{f}$ is unique. Else, let $g: T(M,N) \to P$ be a homomorphism such that $g\phi = f$. Then $g\phi(m,n) = \bar{f}\phi(m,n)$ for all $m \in M$, $n \in N$. But since $\{\phi(m,n) | m \in M, n \in N\}$ generates $T(M,N)$, it follows $g = \bar{f}$. This proves that $T(M,N) = Z(M,N)/Y(M,N)$ is the tensor product of M and N.

We shall write $M \otimes_R N$ for $T(M,N)$ and $m \otimes n$ for $(m,n) + Y(M,N) = \phi(m,n)$, $m \in M$, $n \in N$. Since every element of $Y(M,N)$ is mapped into zero under the canonical mapping ϕ, we have

2.2 Theorem. *Let M and N be right and left R-modules, respectively. Then*

(i) $(m_1 + m_2) \otimes n = m_1 \otimes n + m_2 \otimes n,$
(ii) $m \otimes (n_1 + n_2) = m \otimes n_1 + m \otimes n_2,$
(iii) $mr \otimes n = m \otimes rn,$
for all $m, m_1, m_2 \in M$, $n, n_1, n_2 \in N$, and $r \in R$.

Next we prove

2.3 Theorem. *Let M and N be right and left R-modules, respectively.*

(a) *Let $\lambda \in \mathbf{Z}$, $m \in M$, and $n \in N$. Then $\lambda m \otimes n = \lambda(m \otimes n) = m \otimes \lambda n$. In particular, $0 \otimes n = 0 = m \otimes 0$ and $-m \otimes n = -(m \otimes n) = m \otimes (-n)$.*
(b) *Every element of $M \otimes_R N$ is of the form $\sum_{\text{finite}} m_i \otimes n_i$, $m_i \in M$, $n_i \in N$.*

Proof. (a) If n is kept fixed, the mapping given by $m \to m \otimes n$ is a group homomorphism of M into $M \otimes_R N$. Thus for $\lambda \in \mathbf{Z}$,

$$\lambda m \otimes n = \lambda(m \otimes n).$$

Similarly

$$m \otimes \lambda n = \lambda(m \otimes n).$$

(b) Since every element of $Z(M,N)$ is of the form

$$\sum_{\text{finite}} \lambda_i(m_i, n_i),\ \lambda_i \in \mathbf{Z},\ m_i \in M,\ n_i \in N,$$

it follows that every element of $M \otimes_R N$ is a finite sum $\sum \lambda_i(m_i \otimes n_i)$. But

$$\lambda_i(m_i \otimes n_i) = \lambda_i m_i \otimes n_i,$$

and so every element of $M \otimes_R N$ is a finite sum of elements of the form $m \otimes n$, $m \in M$, $n \in N$. □

Problems

1. Prove the following:

   (i) $\mathbf{Q} \otimes_{\mathbf{Z}} \mathbf{Z}_8 = 0$.
   (ii) $\mathbf{Z}_6 \otimes_{\mathbf{Z}} \mathbf{Z}_7 = 0$.
   (iii) $\mathbf{Q} \otimes_{\mathbf{Z}} \mathbf{Z} \simeq \mathbf{Q}$, as additive groups.
   (iv) $\mathbf{Z}_m \otimes_{\mathbf{Z}} \mathbf{Z}_n \simeq \mathbf{Z}_d$, as additive groups, where m, n are positive integers and $(m, n) = d$.
   (v) $\mathbf{Z}_2 \otimes_{\mathbf{Z}^2} \mathbf{Z} \simeq \mathbf{Z}_2$.

2. Let A be a torsion abelian group. That is, for each $a \in A$ there exists a positive integer n (depending on a) such that $na = 0$. Show that

   $$A \otimes_{\mathbf{Z}} \mathbf{Q} = 0.$$

3 Module structure of tensor product

If M and N are right and left R-modules, respectively, over an arbitrary ring R, then $M \otimes_R N$ is an abelian group. However, if R is commutative, then $M \otimes_R N$ can be given the structure of an R-module. A natural definition for making $M \otimes_R N$ a right R-module is $(m \otimes n)r = mr \otimes n$, $m \in M$, $n \in N$, and $r \in R$. But it is quite nontrivial (though straightforward) to verify that this is indeed well defined. We proceed to show this, assuming R is a commutative ring.

Let $r \in R$. Consider the homomorphism

$$f: Z(M, N) \to M \otimes_R N$$

of free abelian group $Z(M, N)$ to $M \otimes_R N$ given by $f(m, n) = mr \otimes n$. We note that the elements of the form

$$(m_1 + m_2, n) - (m_1, n) - (m_2, n) \tag{1}$$

and

$$(m, n_1 + n_2) - (m, n_1) - (m, n_2) \tag{2}$$

map, respectively, to

$$(m_1 r + m_2 r) \otimes n - m_1 r \otimes n - m_2 r \otimes n$$
$$= m_1 r \otimes n + m_2 r \otimes n - m_1 r \otimes n - m_2 r \otimes n = 0$$

and

$$mr \otimes (n_1 + n_2) - mr \otimes n_1 - mr \otimes n_2 = 0,$$

for all $m, m_1, m_2 \in M$ and $n, n_1, n_2 \in N$. Furthermore, for all $m \in M$, $n \in N$, and $s \in R$, the elements of the form

$$(ms, n) - (m, sn) \tag{3}$$

map to

$$(msr) \otimes n - mr \otimes sn = (msr) \otimes n - (mrs) \otimes n = 0,$$

since R is commutative.

Thus f vanishes on the subgroup $Y(M, N)$ of $Z(M, N)$ generated by elements of the form (1), (2), and (3). So f induces a homomorphism

$$\bar{f}: Z(M, N)/Y(M, N) = M \otimes_R N \to M \otimes_R N,$$

where $\bar{f}((m, n) + Y(M, N)) = mr \otimes n$. That is, $\bar{f}(m \otimes n) = mr \otimes n$. This proves that given $m \otimes n \in M \otimes_R N$ and $r \in R$, the element $mr \otimes n$ is uniquely defined. Thus $(m \otimes n)r = mr \otimes n$ is well defined, and it is easy to verify that $M \otimes_R N$ becomes a right R-module. Incidentally, $M \otimes_R N$ is a left R-module in a natural way. (Over a commutative ring R a right R-module can be naturally looked upon as a left R-module.)

3.1 Theorem. *Let M be a right R-module. Then $M \otimes_R R \simeq M$ as additive abelian groups. Further, if R is commutative, then $M \otimes_R R \simeq M$ as R-modules.*

Proof. Consider a group homomorphism

$$f: Z(M, R) \to M$$

given by $f(m, r) = mr$. Clearly $Y(M, R) \subset \operatorname{Ker} f$ and so f induces homomorphism $\bar{f}$ from $Z(M, R)/Y(M, R) = M \otimes_R R$ to M given by

$$\bar{f}(m \otimes r) = mr.$$

Also the map $g: M \to M \otimes_R R$ given by $g(m) = m \otimes 1$ is a group homomorphism whose inverse is $\bar{f}$. Thus $\bar{f}$ is an isomorphism as desired. In case R is commutative, then by definition of scalar multiplication of elements of R with those of $M \otimes_R N$

$$\bar{f}((m \otimes r)s) = \bar{f}(ms \otimes r) = (ms)r = m(sr) = m(rs) = (mr)s = \bar{f}(m \otimes r)s.$$

Thus $\bar{f}$ is an R-homomorphism and is indeed an R-isomorphism as shown in the first case. □

3.2 Theorem. *Let M, N, and P be R-modules over a commutative ring R. Then*

(i) $M \otimes_R N \simeq N \otimes_R M$ *as R-modules.*
(ii) $(M \otimes_R N) \otimes_R P \simeq M \otimes_R (N \otimes_R P)$ *as R-modules.*

Proof. First one needs to recall that if M is a right R-module over a commutative ring R, then M is a left R-module by defining $rm = mr$, $r \in R$, $m \in M$. Consider the homomorphism $f: \mathbf{Z}(M, N) \to N \otimes_R M$ given by $f(m, n) = n \otimes m$. Since $Y(M, N) \subset \operatorname{Ker} f$, f induces the homomorphism $\bar{f}: M \otimes_R N \to N \otimes_R M$ given by $\bar{f}(m \otimes n) = n \otimes m$. One can, similarly, obtain a homomorphism from $N \otimes_R M$ to $M \otimes_R N$ which is the inverse of $\bar{f}$, proving that $M \otimes_R N \simeq N \otimes_R M$. The proof of part (ii) is similar. □

Problems

1. Let M be a right R-module and e be an idempotent in R. Show $M \otimes_R Re \simeq Me$ (as groups).
2. Give an example of an R-module M and a left ideal I in R such that $M \otimes_R I \not\simeq MI$.

4 Tensor product of homomorphisms

Let M, M' be right R-modules and N, N' be left R-modules. Let $f: M \to M'$ and $g: N \to N'$ be R-homomorphisms. Then the mapping

$$Z(M, N) \to M' \otimes_R N'$$

in which

$$(m, n) \to f(m) \otimes g(n)$$

is a group homomorphism which vanishes on the subgroup $Y(M, N)$. Thus the above mapping induces a homomorphism from $Z(M, N)/Y(M, N) = M \otimes_R N$ to $M' \otimes_R N'$, where the element $(m, n) + Y(M, N) = m \otimes n$ maps to $f(m) \otimes g(n)$. We denote this homomorphism as $f \otimes g$. Thus

$$(f \otimes g)(m \otimes n) = f(m) \otimes g(n).$$

The mapping $f \otimes g$ is called the tensor product of f and g. When R is commutative, $f \otimes g$, as defined above, becomes an R-homomorphism from the R-module $M \otimes_R N$ to the R-module $M' \otimes_R N'$.

4.1 Theorem. *Let M, M', M'' be right R-modules and N, N', N'' be left R-modules. Suppose $f: M \to M'$, $f': M' \to M''$, $g: N \to N'$, and $g': N' \to N''$ are R-homomorphisms. Then*

$$(f' \otimes g')(f \otimes g) = (ff' \otimes gg').$$

Further, if f and g are isomorphisms of M onto M' and M' onto M'', respecttively, then $f \otimes g$ is also an isomorphism.

Proof. The proof follows from the definition of the tensor product of homomorphisms. □

4.2 Theorem. *Let $(M_i)_{i \in I}$ and $(N_j)_{j \in J}$ be families of right and left R-modules, respectively. Then*

$$\left(\bigoplus_{i \in I} \sum M_i\right) \otimes_R \left(\bigoplus_{j \in J} \sum N_j\right) \simeq \bigoplus_{(i,j) \in I \times J} \sum (M_i \otimes_R N_j)$$

which maps

$$\left(\sum_i x_i\right) \otimes \left(\sum_j y_j\right)$$

onto

$$\sum_{i,j} x_i \otimes y_j.$$

Proof. The map

$$f: \left(\bigoplus_{i \in I} \sum M_i\right) \times \left(\bigoplus_{j \in J} \sum N_j\right) \to \bigoplus_{(i,j) \in I \times J} \sum (M_i \otimes_R N_j)$$

given by

$$f\left(\left(\sum_i x_i, \sum_j y_j\right)\right) = \sum_{i,j} x_i \otimes y_j$$

is balanced and gives rise to a homomorphism

$$\bar{f}: \left(\bigoplus_{i \in I} \sum M_i\right) \otimes_R \left(\bigoplus_{j \in J} \sum N_j\right) \to \bigoplus_{(i,j) \in I \times J} \sum (M_i \otimes_R N_j)$$

such that

$$\bar{f}\left(\left(\sum_i x_i\right)\otimes\left(\sum_j y_j\right)\right)=\sum x_i\otimes y_j.$$

On the other hand, the inclusions

$$\lambda_i: M_i \to \bigoplus_{i\in I}\sum M_i \quad \text{and} \quad \mu_j: N_j \to \bigoplus_{j\in I}\sum N_j$$

give rise to homomorphisms

$$g_{ij}=\lambda_i\otimes\mu_j: M_i\otimes_R N_j\to\left(\sum_{i\in I}\bigoplus M_i\right)\otimes_R\left(\bigoplus_{j\in J}\sum N_j\right).$$

There exists a unique homomorphism

$$\bar{g}: \bigoplus_{(i,j)\in I\times J}\sum(M_i\otimes_R N_j)\to\left(\sum_{i\in I}\bigoplus M_i\right)\otimes_R\left(\bigoplus_{j\in J}\sum N_j\right)$$

which extends all the g_{ij}'s. Clearly $\bar{f}$ and $\bar{g}$ are inverses of each other and hence they are isomorphisms. □

Problems

1. Let F be a free left R-module with a basis $(e_j)_{j\in J}$ say, and let M be a right R-module. Then an element of $M\otimes_R F$ can be uniquely written in the form $\sum_j x_j\otimes e_j$ with $x_j\in M$ and $x_j=0$ for almost all $j\in J$.
2. If F is a free left R-module and $f: M\to M'$ is a monomorphism of right R-modules, then the induced map $f\otimes 1: M\otimes_R F\to M'\otimes_R F$ is also a monomorphism.
3. Let $h: A\to B$ be a homomorphism of rings and let F be a free left A-module with a basis $(e_j)_{j\in J}$. Then if one regards B as a right A-module by defining $ba=bh(a)$, $B\otimes_A F$ is a free left B-module with a basis $(1\otimes e_j)_{j\in J}$.
4. Let R be a commutative ring, and let F' and F be free R-modules with bases $(e'_i)_{i\in I}$ and $(e_j)_{j\in J}$, respectively. Then $F'\otimes_R F$ is a free R-module with $(e'_i\otimes e_j)_{(i,j)\in I\times J}$ as a basis.
5. Let R be a commutative ring with 1, and let F be a free R-module. Then any two bases of F have the same cardinality. (Note that Theorem 5.4, Chapter 14, is a special case of this problem.)
6. Let K be a field, V and W vector spaces over K, V_1 a subspace of V, and W_1 a subspace of W. Show that $(V_1\otimes_K W)\cap(V\otimes_K W_1)=$

$V_1 \otimes_K W_1$, where the vector spaces involved are identified to subspaces of $V \otimes_K W$ by means of obvious canonical maps.

7. Let R be a commutative ring. Show$(\cdot)\otimes_R N$ is a right exact covariant functor from mod-R to mod-R. Show T need not be left exact. [A covariant functor T: mod-$R \to$ mod-R is right (or left) exact if for each short exact sequence $0 \to A \to B \to C \to 0$, $T(A) \to T(B) \to T(C) \to 0$ (or $0 \to T(A) \to T(B) \to T(C)$) is also exact.]
8. Let $M, M' \in$ mod-R and $N, N' \in R$-mod. Suppose $f: M \to M'$ and $g: N \to N'$ be onto R-homomorphisms. Show that $\mathrm{Ker}(f \otimes g)$ is generated by elements of the type $a \otimes n$ and $m \otimes b$ where $a \in \mathrm{Ker} f$, $b \in \mathrm{Ker} g$, $m \in M$, and $n \in N$.

5 Tensor product of algebras

Let K be a commutative ring. If A is a K-algebra, then the mapping $A \times A \to A$ given by $(a, b) \to ab$ is balanced, giving rise to a K-homomorphism

$$m_A: A \otimes_K A \to A$$

such that $m_A(a \otimes a') = aa'$. If B is another K-algebra, we have a sequence of maps

$$(A \otimes_K B) \times (A \otimes_K B) \xrightarrow{\phi} (A \otimes_K B) \otimes_K (A \otimes_K B) \xrightarrow{\tau}$$

$$(A \otimes_K A) \otimes_K (B \otimes_K B) \xrightarrow{m_A \otimes m_B} A \otimes_K B,$$

where the map ϕ is obvious and τ is the unique isomorphism of K-modules such that $\tau((a \otimes b) \otimes (a' \otimes b')) = (a \otimes a') \otimes (b \otimes b')$. The composition of these three maps is a mapping of $(A \otimes_K B) \times (A \otimes_K B)$ into $A \otimes_K B$. This defines a multiplication in the K-module $A \otimes_K B$, namely,

$$\left(\sum_i a_i \otimes b_i\right)\left(\sum_j a_j \otimes b_j\right) = \sum_{i,j} a_i a_j \otimes b_i b_j,$$

which is easily seen to be associative. Also $1 \otimes 1$ acts as identity. Thus $A \otimes_K B$ becomes a K-algebra and is called the tensor product of the K-algebras A and B.

5.1 Theorem. *If A, B, and C are K-algebras, then as K-algebras,*

(i) $K \otimes_K A \simeq A$,
(ii) $A \otimes_K B \simeq B \otimes_K A$, *and*
(iii) $A \otimes_K (B \otimes_K C) \simeq (A \otimes_K B) \otimes_K C$.

Proof. The isomorphism between the K-algebras follows from the definition of the tensor product of algebras by considering canonical maps. □

5.2 Example

Let R be a commutative ring with unity. Let A, B are ideals in R. Then $R/A \otimes_{\mathbf{R}} R/B \simeq R/A + B$ as R-algebras.

Solution. Define a map $\alpha: R/A \times R/B \to R/A + B$ by $\alpha(x + A, y + B) = xy + A + B$. α is balanced and induces the group homomorphism $\bar{\alpha}: R/A \otimes_{\mathbf{R}} R/B \to R/A + B$. The map $\beta: R/A + B \to R/A \otimes_{\mathbf{R}} R/B$ given by $\beta(x + A + B) = (x + A) \otimes (1 + B)$ is well-defined, for if $x = a + b$, $a \in A$, $b \in B$, then $(x + A) \otimes (1 + B) = (b + A) \otimes (1 + B) = (1 + A) \otimes (b + B) = (1 + A) \otimes (0 + B) = 0$. Clearly, $\bar{\alpha}, \beta$ are algebra homomorphisms with $\bar{\alpha}\beta =$ identity on $R/A + B$ and $\beta\bar{\alpha} =$ identity on $R/A \otimes_{\mathbf{R}} R/B$.

Problems

1. Let A be an algebra over a commutative ring K. Show $A \otimes_K M_n(K) \simeq M_n(A)$ as K-algebras.
2. Show $M_m(K) \otimes_K M_n(K) \simeq M_{mn}(K)$ as K-algebras.
3. Show $\mathbf{Q} \otimes_{\mathbf{Z}} \mathbf{Q} \simeq \mathbf{Q}$ as $\mathbf{Z}$-algebras.
4. Let K be a commutative integral domain with 1 and let F be the field of fractions. Show $F \otimes_K F \simeq F$ as K-algebras.
5. Let E be an extension of a field F. Show $F[x] \otimes_F E \simeq E[x]$.

Solutions to odd-numbered problems

Chapter 1, Section 1

1. (iv) $x \in A \cup (B \cap C) \Leftrightarrow x \in A \text{ or } x \in B \cap C$

$$\begin{aligned} &\Leftrightarrow x \in A \text{ or } (x \in B \text{ and } x \in C) \\ &\Leftrightarrow (x \in A \text{ or } x \in B) \text{ and } (x \in A \text{ or } x \in C) \\ &\Leftrightarrow x \in A \cup B \text{ and } x \in A \cup C \\ &\Leftrightarrow x \in (A \cup B) \cap (A \cup C) \end{aligned}$$

3. $|A \cup B| = m + n - k$

5. A = TVs with defective picture tube.
B = Those with defective sound system.
C = Those with defective remote control system.

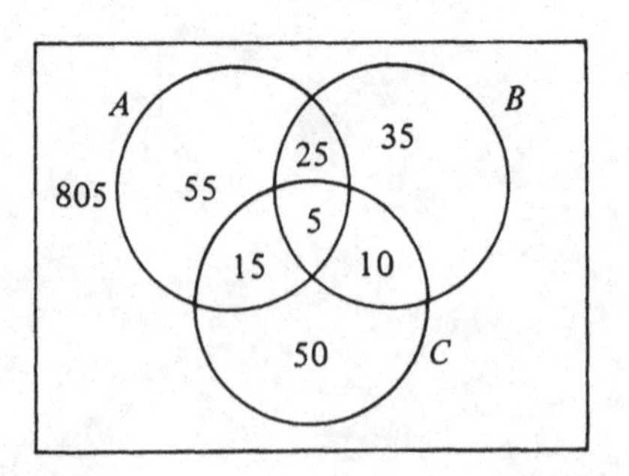

Chapter 1, Section 2

1. $A \times B = \{(a,0) | a \in [0,1]\}$
$\cup \{(a,3) | a \in [0,1]\}$
$\cup \{(a,5) | a \in [0,1]\}$. Hence the graph consists of the three strips as shown.

y
5
3
0 1 x

Because $(0,3) \in A \times B$, but $(0,3) \notin B \times A$, we have $A \times B \neq B \times A$.

3. First part obvious. Second part not necessarily true. Take $A, B \neq \varnothing$, $X = \varnothing$, $Y \not\supseteq B$.
5. Define $x \sim y$ iff $x - y$ is a multiple of 4. $\sim$ is an equivalence relation on **Z** corresponding to the given partition.
7. By (ii) $x\,R\,y$ and $y\,R\,y \Rightarrow y\,R\,x$. So the relation is symmetric. Then by (ii) and the symmetric property, $x\,R\,y$ and $y\,R\,z \Rightarrow z\,R\,x \Rightarrow x\,R\,z$, proving transitivity. The converse is proved similarly.
9. $\sim$ is an equivalence relation, and $\mathbf{Z}/\sim = \{\ldots,-3,-1,1,3,\ldots\} \cup \{\ldots,-4,-2,0,2,4,\ldots\}$.

Chapter, 1, Section 3

1. (a) Let $a_1, a_2 \in A$. Then
$$(gf)(a_1) = (gf)(a_2) \Rightarrow g(f(a_1)) = g(f(a_2)) \Rightarrow f(a_1) = f(a_2) \Rightarrow a_1 = a_2.$$
3. Let $A \xrightarrow{g} B \xrightarrow{f} C$. Suppose fg is injective. Then g must be injective. For
$$g(a_1) = g(a_2) \Rightarrow (fg)(a_1) = (fg)(a_2) \Rightarrow a_1 = a_2.$$
But f need not be injective. For take $A = \mathbf{Z}^+$, the set of positive integers, $B = C = \mathbf{Z}$, and let $f(x) = x^2$ and $g(x) = -x$. Then fg is injective, but f is not. Similarly, if fg is surjective, then f is surjective, but g need not be.
5. By Theorem 3.5, g injective (or f surjective) $\Rightarrow g$ has a left inverse (or f has a right inverse) $\Rightarrow g$ (or f) is invertible, since $gf = I_A$.
7. Let $A = \{a_1,\ldots,a_m\}$, $B = \{b_1,\ldots,b_n\}$, and let $f\colon A \to B$. Suppose $m \leq n$. The possible choices for $f(a_1)$ are n. Having fixed $f(a_1)$, the possible choices for $f(a_2)$ are $n - 1$, and so on. Thus, there are $n(n-1) \cdots (n-m+1)$ possible injective mappings from A to B. When $m > n$, the elements of B are exhausted before we have assigned an image to each element of A (since f is injective, no two distinct elements can have the same image). Therefore, there is no injective mapping from A to B when $m > n$.
9. n^n. Among n^n, $n!$ are injective and, hence, bijective (Theorem 3.3).

Chapter 1, Section 4

1. (a) $(2 \circ 3) \circ 4 = 12 \circ 4 = 576$. $2 \circ (3 \circ 4) = 2 \circ 36 = 144$. So this operation is not associative. Also $2 \circ 3 = 12$, $3 \circ 2 = 18$; hence, $2 \circ 3 \neq 3 \circ 2$. Thus, $\circ$ is not commutative either.
   (b) $x \circ (y \circ z) = x \circ (\min(y,z)) = \min(x,y,z)$. Similarly, $(x \circ y) \circ z = \min(x,y,z)$. Hence, this operation is associative; it is also commutative.
   (c) This is commutative but not associative.
   (d) This is both commutative and associative.

3. $(x \circ y) \circ z = (x \circ a \circ y) \circ z = x \circ a \circ y \circ a \circ z = x \circ (y \circ z)$. Further,

$$x \circ y = x \circ a \circ y = x \circ y \circ a = y \circ x \circ a = y \circ a \circ x = y \circ x.$$

5. First, the total number of binary operations is the number of mappings from a set with n^2 elements to a set with n elements, which is n^{n^2}. Among these the commutative ones are

$$n^{(n^2-n)/2+n} = n^{n(n+1)/2}.$$

Chapter 2, Section 1

1. (a) Let $P(n) = 1^2 + 3^2 + \cdots + (2n-1)^2$. We want to show

$$P(n) = \frac{n(4n^2-1)}{3}.$$

Clearly, $P(n)$ holds for $n = 1$. Assume $P(n)$ holds for $n = k$.

$$\begin{aligned} P(k+1) &= 1^2 + 3^2 + \cdots + (2k-1)^2 + (2k+1)^2 \\ &= P(k) + (2k+1)^2 = \frac{k(4k^2-1)}{3} + (2k+1)^2 \\ &= \frac{(k+1)(4k^2+8k+3)}{3} = \frac{(k+1)(4(k+1)^2-1)}{3}, \end{aligned}$$

so $P(n)$ holds for $n = k+1$. This proves $P(n)$ holds for all n.

(b) Similar.

3. (a) Trivial for $n = 1,2$. Assume for the case $n = k$. Consider

$$\left(\bigcap_{i=1}^{k+1} A_i\right)' = \left(\left(\bigcap_{i=1}^{k} A_i\right) \cap A_{k+1}\right)' = \left(\bigcap_{i=1}^{k} A_i\right)' \cup A'_{k+1}$$

(by the case $n = 2$)

$$= \left(\bigcup_{i=1}^{k} A'_i\right) \cup A'_{k+1} \qquad \text{(by induction hypothesis)}$$

$$= \bigcup_{i=1}^{k+1} A'_i,$$

showing that the result is true for $n = k+1$ and, hence, for all n.

(b) Similar.

5. (a) Let d,d' be l.c.m. of a and b. Then $a|d$, $b|d$ and $a|d'$, $b|d'$.Thus, (ii) gives $d|d'$. Similarly, $d'|d$. Thus, $d = d'$ because d and d' are positive integers.

(b) Suppose $d = p_1^{g_1} \cdots p_k^{g_k}$ and $g_i = \max(e_i, f_i)$. Then $a|d$ and $b|d$. Further, if $a|x$ and $b|x$, then $x = ay = bz$ for some $y,z \in \mathbf{Z}$. Clearly, $p_i^{g_i}$, where $g_i = \max(e_i, f_i)$, must divide x, $1 \le i \le k$. Hence, $d|x$. Thus, $d = [a,b]$.

(c) First observe that $(a,b) = p_1^{h_1} \cdots p_k^{h_k}$, where $h_i = \min(e_i, f_i)$. Using this fact gives $ab = [a,b](a,b)$.

7. Let a/b be a rational solution. Then

$$\frac{a^n}{b^n} + \frac{a_1 a^{n-1}}{b^{n-1}} + \cdots + \frac{a_{n-1}a}{b} + a_n = 0.$$

This implies

$$a_1 a^{n-1} + \cdots + b^{n-2}a_{n-1}a + b^{n-1}a_n = -\frac{a^n}{b}.$$

But then $a^n/b \in \mathbf{Z}$; that is, $b|a$. Hence, a/b is an integer. Further, if α is an integral solution, then

$$\alpha(\alpha^{n-1} + a_1\alpha^{n-2} + \cdots + a_{n-1}) = -a_n \Rightarrow \alpha | a_n.$$

9. $a \equiv 1'$ (mod p), $2a \equiv 2'$ (mod p),...,$(p-1)a \equiv (p-1)$ (mod p), where $\{1', 2',...,(p-1)'\} = \{1, 2,...,(p-1)\}$ and $(a,p) = 1$. Thus,

$$(p-1)!\ a^{p-1} \equiv (p-1)! \pmod p,$$

so $a^{p-1} \equiv 1 \pmod p$. Therefore, $a^p \equiv a \pmod p$.

Chapter 3, Section 5

1. Suppose A is an $n \times n$ upper triangular matrix, so that $a_{ij} = 0$ whenever $i > j$. For every $\sigma \in S_n$, $\sigma \neq e$, there is some i such that $\sigma(i) < i$. Then $a_{i\sigma(i)} = 0$. Hence,

$$\det A = \sum_{\sigma \in S_n} \epsilon_\sigma a_{1\sigma(1)} \cdots a_{n\sigma(n)} = a_{11}a_{22} \cdots a_{nn}.$$

3. Using the definition gives

$$\begin{vmatrix} 1 & a & a^2 \\ 1 & b & b^2 \\ 1 & c & c^2 \end{vmatrix} = bc^2 + ab^2 + a^2c - a^2b - b^2c - ac^2 = (a-b)(b-c)(c-a).$$

5. $\det A = \det A^2 = (\det A)^2$ (Theorem 5.4) $\Rightarrow \det A = 0$ or 1.

Chapter 3, Section 6

1. $$\begin{vmatrix} 1 & a & a^3 \\ 1 & b & b^3 \\ 1 & c & c^3 \end{vmatrix} = \begin{vmatrix} 1 & a & a^3 \\ 0 & b-a & b^3-a^3 \\ 0 & c-a & c^3-a^3 \end{vmatrix}$$ (on subtracting the first row from every other row)

$$= \begin{vmatrix} b-a & b^3-a^3 \\ c-a & c^3-a^3 \end{vmatrix}$$ (on expanding according to the first column)

$$= (b-a)(c-a)\begin{vmatrix} 1 & b^2+a^2+ab \\ 1 & c^2+a^2+ac \end{vmatrix}$$

$$= (b-a)(c-a)(c-b)(a+b+c).$$

3.(a) $$\begin{vmatrix} (b+c)^2 & a^2 & a^2 \\ b^2 & (c+a)^2 & b^2 \\ c^2 & c^2 & (a+b)^2 \end{vmatrix}$$

$$= \begin{vmatrix} (b+c)^2 & a^2-(b+c)^2 & a^2-(b+c)^2 \\ b^2 & (c+a)^2-b^2 & 0 \\ c^2 & 0 & (a+b)^2-c^2 \end{vmatrix}$$ (on subtracting the first column from the other two columns)

$$= (a+b+c)^2\begin{vmatrix} (b+c)^2 & a-b-c & a-b-c \\ b^2 & c+a-b & 0 \\ c^2 & 0 & a+b-c \end{vmatrix}$$ (on taking out the common factor $(a+b+c)$ in the second and the third columns)

$$= (a+b+c)^2\begin{vmatrix} 2bc & -2c & -2b \\ b^2 & c+a-b & 0 \\ c^2 & 0 & a+b-c \end{vmatrix}$$ (on subtracting the second and third rows from the first row)

$$= \frac{(a+b+c)^2}{bc}\begin{vmatrix} 2bc & -2bc & -2bc \\ b^2 & b(c+a-b) & 0 \\ c^2 & 0 & c(a+b-c) \end{vmatrix}$$ (on multiplying the second and third columns by b and c, respectively)

$$= 2(a+b+c)^2\begin{vmatrix} 1 & -1 & -1 \\ b^2 & b(c+a-b) & 0 \\ c^2 & 0 & c(a+b-c) \end{vmatrix}$$ (on taking out the common factor $2bc$ in the first row)

$$= 2(a+b+c)^2\begin{vmatrix} 1 & 0 & 0 \\ b^2 & b(c+a) & b^2 \\ c^2 & c^2 & c(a+b) \end{vmatrix}$$ (on adding the first column to the other two columns)

$$= 2(a+b+c)^2\{b(c+a)c(a+b) - b^2c^2\}$$
$$= 2abc(a+b+c)^3.$$

3.(b) $$\begin{vmatrix} a-b-c & 2a & 2a \\ 2b & b-c-a & 2b \\ 2c & 2c & c-a-b \end{vmatrix}$$

$$= \begin{vmatrix} a+b+c & a+b+c & a+b+c \\ 2b & b-c-a & 2b \\ 2c & 2c & c-a-b \end{vmatrix}$$ (on adding the second and third rows to the first)

$$= (a+b+c)\begin{vmatrix} 1 & 1 & 1 \\ 2b & b-c-a & 2b \\ 2c & 2c & c-a-b \end{vmatrix}$$

$$= (a+b+c)\begin{vmatrix} 1 & 0 & 0 \\ 2b & -b-c-a & 0 \\ 2c & 0 & -c-a-b \end{vmatrix}$$

$$= (a+b+c)^3.$$

Chapter 4, Section 1

1. Because addition in **R** is associative,

   $$(f(x)+g(x))+h(x) = f(x)+(g(x)+h(x))$$

   for all $x \in \mathbf{R}$ and all mappings, $f,g,h \in A$. The mapping given by $x \mapsto 0$ for all $x \in \mathbf{R}$ is clearly the identity in A. Finally, each mapping $f\colon \mathbf{R} \to \mathbf{R}$ has an inverse $f^- \in A$ given by $f^-(x) = -f(x)$ for all $x \in \mathbf{R}$.
3. For all a,b in G,

   $$ab = aeb = a(ab)^2b = aababb = ebae = ba.$$

5. Writing

   $$e = \begin{pmatrix} 1 & 0 \\ 0 & 1 \end{pmatrix},\quad a = \begin{pmatrix} 1 & 0 \\ 0 & -1 \end{pmatrix},\quad b = \begin{pmatrix} -1 & 0 \\ 0 & 1 \end{pmatrix},\quad c = \begin{pmatrix} -1 & 0 \\ 0 & -1 \end{pmatrix},$$

   we easily verify that $a^2 = b^2 = c^2 = e$, $ab = c = ba$, $bc = a = cb$, and $ca = b = ac$. Hence, $G = \{e,a,b,c\}$ is closed under multiplication, e is the identity, and each element is its own inverse. Because matrix multiplication is associative, G is a group.
7. Let $a \in G$. The set $\{a^i | i \in \mathbf{N}\} \subset G$ is finite. Hence, $a^i = a^j$ for some positive integer $i < j$. Choose a positive integer p such that $i < p(j-i)$, and set $e = a^{p(j-i)}$. Then $ea^i = a^i$. Hence,

   $$e^2 = e \cdot a^i \cdot a^{p(j-i)-i} = a^i \cdot a^{p(j-i)-i} = e.$$

9. There exist integers x and y such that $xm + yn = 1$. Hence, for all a,b in G,

$$\begin{aligned} a^m b^n &= (a^m b^n)^{xm+yn} \\ &= a^m(b^n a^m)^{xm+yn-1} b^n \\ &= a^m(b^n a^m)^{xm}(b^n a^m)^{-1}(b^n a^m)^{yn} b^n \\ &= (b^n a^m)^{xm} a^m (b^n a^m)^{-1} b^n (b^n a^m)^{yn} \\ &= b^n a^m. \end{aligned}$$

Thus, every mth power commutes with every nth power. Hence,

$$ab = a^{xm+yn} b^{xm+yn} = (a^x)^m (a^y)^n (b^x)^m (b^y)^n = ba.$$

Chapter 4, Section 2

1. $G = \{e^{2k\pi i/n} | k = 0,1,\ldots,n-1\}$. Let $f\colon \mathbf{Z}/(n) \to G$ be defined by $f(\bar{k}) = e^{2k\pi i/n}$. f is well defined and is an isomorphism.
3. $f(x) = x^2 \Rightarrow (xy)^2 = f(xy) = f(x)f(y) = x^2y^2$. Thus $yx = xy$.
5. Since the order of the image of an element under any homomorphism divides the order of the element, elements of order 2 in S_3 must map to 0 in $\mathbf{Z}_3$.

Chapter 4, Section 3

1. $Ha = H \Rightarrow a = ea \in Ha \Rightarrow a \in H$. Conversely, if $a \in H$, then for every h in H, $ha \in H$ and $h = ha^{-1}a \in Ha$. Hence, $Ha = H$.
3. The mapping $f\colon H \to x^{-1}Hx$ given by $f(a) = x^{-1}ax$ is bijective.
5. Let $H < G$, $a \in G$. For every h in H, $(ah)^{-1} = h^{-1}a^{-1} \in Ha^{-1}$ and $ha^{-1} = (ah^{-1})^{-1}$. Hence, Ha^{-1} consists of inverses of all elements of aH.
7. (a) $|H \cap K|$ is a divisor of $|H|$ as well as of $|K|$. Hence, $|H \cap K| = 1$.
   (b) $|H \cap K|$ divides $|H| = p$. Thus $H \cap K = \{e\}$, or $H \cap K = H$ which implies $H < K$.
9. Suppose $A = aH = bK$ for some subgroups H,K of G. Then $a \in bK$. Hence, $a = bk$ for some $k \in K$. Therefore, $a^{-1}b = k^{-1} \in K$. Hence, $H = a^{-1}(aH) = a^{-1}bK = K$. Again, $A = aH \Rightarrow A = (aHa^{-1})a$. Hence, A is a right coset of the subgroup aHa^{-1}.
11. Suppose $o(ab) = m$. Then $(ba)^m = a^{-1}a(ba)^m = a^{-1}(ab)^m a = e$. Hence, $o(ba)|m$. Similarly, $o(ab)|o(ba)$.
13. Because $o(x^{-1}ax) = o(a)$, it follows that $x^{-1}ax = a$ for every $x \in G$. Hence, $a \in Z(G)$. Further, if $n > 2$, there exists a positive integer m (> 1) relatively prime to n. Then $o(a^m) = o(a)$, which contradicts the hypothesis that a is the only element of order n.
15. Because $x^2 \neq e \Rightarrow x \neq x^{-1}$, it follows that there are an even number of elements x satisfying $x^2 \neq e$. Hence, there are an even number of elements x such that $x^2 = e$. Because e is one of these, there are an odd number of elements of order 2.
17. Let $o(a^m) = k$. Then $a^{mk} = e$. Hence, $r|mk$. But $(r,m) = 1$. Hence, $r|k$. On the other hand, $(a^m)^r = (a^r)^m = e$. Hence, $k|r$.
19. Let $o(a) = m$ and $o(g(a)) = k$. Then

$$(g(a))^m = g(a) \cdots g(a) = g(a^m) = g(e) = e'.$$

Hence, $k|m$. Further, if g is injective, then $g(a^k) = (g(a))^k = e'$ implies $a^k = e$. Therefore. $m|k$.

21. Clearly, $e \in C(S)$. If $x,y \in C(S)$, then

$$xy^{-1}s = xy^{-1}syy^{-1} = xy^{-1}ysy^{-1} = xsy^{-1} = sxy^{-1}$$

for all $s \in S$. Hence, $xy^{-1} \in C(S)$. $C(G)$ is the center of G.

Chapter 4, Section 4

1. The roots of $x^n = 1$ are $\cos\frac{2k\pi}{n} + i\sin\frac{2k\pi}{n}$, where $k = 0, 1, \ldots, n-1$. These roots form a group generated by $\omega = \cos\frac{2\pi}{n} + i\sin\frac{2\pi}{n}$ (see Problem 1, Section 2).
3. $G = (\mathbf{Z},+)$. If $a \in \mathbf{Z}$ generates $(\mathbf{Z},+)$, then $1 = ma$ for some $m \in \mathbf{Z}$, so $a = \pm 1$.
5. (Only if) Let $a \in G$, $a \neq e$. Then $G = [a]$. Now apply Theorem 4.4.
7. Since S_3 is not abelian, there is no nonzero homomorphism from S_3 onto a cyclic group of order 6; nor onto a cyclic group of order 3 (see Problem 5, Section 2). There is only one nonzero homomorphism from $S_3 = \{e, a, a^2, b, ab, a^2b\}$ onto $\mathbf{Z}_2$, given by $a \to 0$ and $b \to 1$.
9. $\mathbf{Z}_2 \times \mathbf{Z}_2$ is isomorphic to the Klein four-group, which is not cyclic.

Chapter 4, Section 5

1. The only proper subgroups of S_3 are the cyclic subgroups generated by $\sigma = (1\ 2\ 3)$, $\tau_1 = (2\ 3)$, $\tau_2 = (1\ 3)$, and $\tau_3 = (1\ 2)$.
3. Using the notation in the text, we get $\sigma^i\tau = \tau_{1+i}$ and $\tau\sigma^i = \tau_{n+1-i}$. Hence, σ^i,τ commute if and only if $i = 0$ or $2i = n$. If n is even, $Z = \{e, \sigma^{n/2}\}$.

Chapter 5, Section 1

1. For all $a \in Z(G)$, $x \in G$,

   $xax^{-1} = axx^{-1} = a \in Z(G)$.
3. For all $n \in N$, $m \in M$, $x \in G$,

   $x(nm)x^{-1} = (xnx^{-1})(xmx^{-1}) \in NM$.
5. The cyclic subgroup $[b]$ generated by $b = \begin{pmatrix} 1 & 2 & 3 \\ 2 & 1 & 3 \end{pmatrix}$ is of order 2, and $a^{-1}ba \notin [b]$, where $a = \begin{pmatrix} 1 & 2 & 3 \\ 2 & 3 & 1 \end{pmatrix}$.
7. For all $h \in H$, $x \in G$,

   $xhx^{-1} = (xh)^2h^{-1}(x^{-1})^2 \in H$.

   Hence, $H \triangleleft G$. Moreover, for every coset $xH \in G/H$, $(xH)(xH) = x^2H = H$. Hence, G/H is abelian.
9. N is a subgroup of index 2 in S_3 and, hence, normal. $S_3/N = \{N, \tau N\}$, where $\tau N = \{(2\,3), (3\ 1), (1\ 2)\}$.
11. $\det(B) = 1 \Rightarrow \det(A^{-1}BA) = \det(A^{-1})\det(B)\det(A) = \det B = 1$.
13. Clearly, $O(a/b + \mathbf{Z})$ divides b.

15. $G/Z(G)$ is cyclic, hence G is abelian (see Example 1.5(h)), which contradicts $|G/Z(G)| = 37$.
17. In the notation of Theorem 5.2, Chapter 4, choose

    $A = [\sigma]$, $B = [\tau]$.

19. The set of all normal subgroups of G is a lattice under partial ordering by inclusion, with $H \vee K =$ subgroup generated by H and $K = HK$ and $H \wedge K = H \cap K$. Since H, K are normal subgroups, HK is also normal. Furthermore, if $H \subseteq L$, then $H(K \cap L) = HK \cap L$. The lattice of all subgroups of G will not, in general, be modular. Take $G = D_4 = \{e, a, a^2, a^3, b, ab, a^2b, a^3b\}$, the dihedral group, $L = [a]$, $H = [a^2]$, and $K = [b]$. Then $H(K \cap L) \neq HK \cap L$.

Chapter 5, Section 2

1. $(\sigma(a))^n = \sigma(a) \cdots \sigma(a) = \sigma(a^n)$. Because σ is injective, $\sigma(a^n) = e' = \sigma(e)$ holds if and only if $a^n = e$.
3. If G is generated by a, consider $\phi: G \to \mathbf{Z}_4$, given by $a^i \mapsto r(i) =$ remainder of i modulo 4.
5. Consider $\phi: D_n \to \mathbf{Z}_2$ given by $\sigma^i \mapsto 0$, $\sigma^i\tau \mapsto 1$ for all i.

Chapter 5, Section 3

1. $\text{Aut}(\mathbf{Z}) \simeq \mathbf{Z}_2$.
3. $\text{Aut}(\mathbf{K}) \simeq S_3$.
5. Verify that the indicated homomorphisms are onto.
7. By invoking Theorem 2.1, Chapter 8, the given abelian group $A \simeq \mathbf{Z}_{n_1} \times \mathbf{Z}_{n_2} \times K$ where $n_1 | n_2$ and K is some suitable abelian group. Let $\text{Hom}(A, A)$ denote the set of all homomorphisms of any abelian group A. It can be directly vertified that if $A \simeq \mathbf{Z}_{n_1} \times \mathbf{Z}_{n_2} \times K$, then as additive groups

$$\text{Hom}(A, A) \simeq \begin{bmatrix} \text{Hom}(\mathbf{Z}_{n_1}, \mathbf{Z}_{n_1}) & \text{Hom}(\mathbf{Z}_{n_2}, \mathbf{Z}_{n_1}) & \text{Hom}(K, \mathbf{Z}_{n_1}) \\ \text{Hom}(\mathbf{Z}_{n_1}, \mathbf{Z}_{n_2}) & \text{Hom}(\mathbf{Z}_{n_2}, \mathbf{Z}_{n_2}) & \text{Hom}(K, \mathbf{Z}_{n_2}) \\ \text{Hom}(\mathbf{Z}_{n_1}, K) & \text{Hom}(\mathbf{Z}_{n_2}, K) & \text{Hom}(K, K) \end{bmatrix} = M(\text{say}),$$

acting on A, whose elements are viewed as $\begin{bmatrix} a_1 \\ a_2 \\ a_3 \end{bmatrix}, a_1 \in \mathbf{Z}_{n_1}, a_2 \in \mathbf{Z}_{n_2}, a_3 \in K$. Indeed

$\text{Aut}(A) \simeq \text{Aut}(M)$. Choosing $\sigma \in \text{Aut}(\mathbf{Z}_{n_1})$, $\theta \in \text{Hom}(\mathbf{Z}_{n_2}, \mathbf{Z}_{n_1})$ suitably one can produce two automorphisms $\alpha, \beta \in \text{Aut}(A)$ given by

$$\alpha = \begin{bmatrix} \sigma & \theta & 0 \\ 0 & 1 & 0 \\ 0 & 0 & 1 \end{bmatrix} \qquad \beta = \begin{bmatrix} 1 & \theta & 0 \\ 0 & 1 & 0 \\ 0 & 0 & 1 \end{bmatrix}$$

such that $\alpha\beta \neq \beta\alpha$.

Chapter 5, Section 4

1. $x \in Z(G) \Leftrightarrow |C(x)| = 1$.
3. $\{e\}$, $\{(2\ 3), (3\ 1), (1\ 2)\}$, $\{(1\ 2\ 3), (1\ 3\ 2)\}$.

5. By Theorem 4.4 there is a homomorphism $\phi: G \to S_n$ with $\operatorname{Ker}\phi = N = \bigcap_{x\in G} xHx^{-1} \subset H$. Now $N \lhd G$ and $G/N \simeq \operatorname{Im}\phi \subset S_n$. Hence, G/N is of finite order.
7. Let $x \in G$ and $o(x) \| |N|$. $[x]N/N < G/N$, so $|[x]N/N|$ divides $|G/N|$. But then $(|[x]N/N|, |N|) = 1$. From Example 3.11(e) in Chapter 4, $(|[x]|/|N \cap [x]|, |N|) = 1$. But since $o(x) = |[x]|$ divides $|N|$, $o(x)$ must divide $|N \cap [x]|$. This yields $[x] = N \cap [x]$, so $x \in N$.
9. If b is a conjugate of a, then

$$b = xax^{-1} = (xax^{-1}a^{-1})a \in G'a.$$

Hence, $C(a) \subset G'a$.
11. G acts on N under conjugation. Since $N \not\subset Z(G)$, there exists an element $n \in N$ such that $1 < |C(n)| < 3$, since $N = \bigcup C(n)$. Thus $|C(n)| = 2$ and hence $[G:G_n] = 2$, where G_n is the stabilizer ($=$normalizer) of n.
13. Writing $S_3 = \{[a, b] | a = (1\,2\,3),\ b = (1\,2)\}$ we find $|X_e| = 8$, $|X_a| = 2$, $X_{a^2} = 2$, $X_b = 4$, $X_{ab} = 4$, and $X_{a^2b} = 4$. Thus by the Burnside Theorem the number of orbits $= \frac{1}{6}(24) = 4$. The four orbits are precisely

```
⎧  2            2            ·2   ⎫   ⎧  2            2            2     ⎫
⎨ /              \                 ⎬ , ⎨ /            / \            \    ⎬ ,
⎩1     ·3 ,  1·    3 ,  1———3      ⎭   ⎩1———3 ,    1     3 ,  1———3       ⎭

⎧  2   ⎫         ⎧   ·2    ⎫
⎨ / \  ⎬ , and   ⎨         ⎬ .
⎩1———3 ⎭         ⎩1·    ·3 ⎭
```

15. Let X be the set of all possible necklaces. As in Example (c), $|X| = 2^6$ and $\mathbf{Z}_6$ acts on X. We compute X_g for all $g \in \mathbf{Z}_6$. Writing $\mathbf{Z}_6 = \{1, 2, 3, 4, 5, 6 = 0\}$ and beads $a_1, a_2, \ldots, a_6$, the action of 2 will transform the beads (in this order) $a_1a_2a_3a_4a_5a_6$ to $a_3a_4a_5a_6a_1a_2$. In order that these are the same necklaces, the beads a_1, a_3, a_5 must be of one color and a_2, a_4, a_6 must also be of one color. Thus the number of choices $= 2\cdot 2 = 4$. So $|X_2| = 4$. Similarly the action of 3 yields that beads a_1, a_4 must be of the same color, beads a_2, a_5 must be of the same color, and beads a_3, a_6 must also be of the same color; and this yields $X_3 = 8$. Similarly $X_1 = 2$, $X_4 = 4$, $X_5 = 2$, and $X_6 = 2^6$. Thus by the Burnside Theorem, the number of different necklaces $= \frac{1}{6}(84) = 14$.
17. Because the top and the bottom of the neckties are interchangeable, the neckties

1
2
⋮
n

and

n
$n-1$
⋮
1

are the same. Let X be the set of all ordered sets $(a_1, a_2, \ldots, a_n)$ of n colored strips and G be the subgroup of S_n generated by the element $\alpha = \begin{pmatrix} 1 & 2 & \cdots n \\ n & n-1 & \cdots 1 \end{pmatrix}$ of order 2. Then X is a G-set and the number of distinct neckties is the number of orbits. Let $n = 2m$; then $\alpha = (1\ n)(2\ n-1)\cdots(m\ m+1)$. The ordered set $(a_1, \ldots, a_n)$ is fixed by α if and only if $a_1 = a_n$, $a_2 = a_{n-1}, \ldots, a_m = a_{m+1}$. Thus $X_\alpha = k^m = k^{n/2}$. Thus the number of distinct neckties $= \frac{1}{2}(k^n + k^{n/2})$. A similar argument yields the desired result for the case when n is odd.

19. $N \lhd G$ is clear. The orbit of $\begin{pmatrix} 1 \\ 0 \end{pmatrix}$ consists of $\begin{pmatrix} a & b \\ 0 & a^{-1} \end{pmatrix}\begin{pmatrix} 1 \\ 0 \end{pmatrix}$, $a = 1,2$, or 4. Thus the orbit of $\left(\begin{pmatrix} 1 \\ 0 \end{pmatrix}, \begin{pmatrix} 3 \\ 0 \end{pmatrix}\right)$ is $\left\{\left(\begin{pmatrix} a_1 \\ 0 \end{pmatrix}, \begin{pmatrix} 3a_2 \\ 0 \end{pmatrix}\right) \middle| a_1, a_2 = 1,2, \text{ or } 4\right\}$.

Chapter 6, Section 1

1. The quaternion group Q has three composition series of the form $\{e\} \subset \{e,-e\} \subset \{e,-e,x,-x\} \subset Q$, where x is

$$\begin{bmatrix} \sqrt{-1} & 0 \\ 0 & -\sqrt{-1} \end{bmatrix}, \begin{bmatrix} 0 & 1 \\ -1 & 0 \end{bmatrix}, \text{ or } \begin{bmatrix} 0 & \sqrt{-1} \\ \sqrt{-1} & 0 \end{bmatrix}.$$

3. $\{e\} \subset \{e,a\} \subset \{e,a,b,ab\} = G$.
5. $\{0\} \subset [15] \subset [5] \subset G$, $\{0\} \subset [15] \subset [3] \subset G$,
   $\{0\} \subset [10] \subset [5] \subset G$, $\{0\} \subset [10] \subset [2] \subset G$,
   $\{0\} \subset [6] \subset [3] \subset G$, $\{0\} \subset [6] \subset [2] \subset G$,
   where $G = \mathbf{Z}/(30)$, and $[m]$ denotes the cyclic subgroup generated by m.
7. Let $N_0 = N \supset N_1 \supset N_2 \supset \cdots \supset \{e\}$ be a composition series of N and $G_0/N = G/N \supset G_1/N \supset \cdots N/N = \{eN\}$ be a composition series of G/N. Then $G \supset G_1 \supset \cdots \supset N \supset N_1 \supset \cdots \supset \{e\}$ is a desired series.
9. If any abelian group A has a finite composition series $A_0 = A \supset A_1 \supset A_2 \supset \cdots \supset \{e\} = A_{r+1}$, then since A_i/A_{i+1} is a simple abelian group, $|A_i/A_{i+1}| = p_i$ for some prime p_i. This gives $|A| = p_0p_1 \cdots p_r$, where p_i are primes. Thus, $\mathbf{Z}$ has no finite composition series.
   Clearly $\mathbf{Z} \supset 2\mathbf{Z} \supset 2^2\mathbf{Z} \supset 2^3\mathbf{Z} \supset \cdots$ and $\mathbf{Z} \supset 3\mathbf{Z} \supset 3^2\mathbf{Z} \supset 3^3\mathbf{Z} \supset \cdots$ are infinite composition series and the Jordan–Hölder theorem fails.

Chapter 6, Section 2

1. $G = G_0 = \left\{ \begin{bmatrix} 1 & a & b \\ 0 & 1 & c \\ 0 & 0 & 1 \end{bmatrix} \middle| \, a, b, c \in \mathbf{R} \right\}, G_1 = \left\{ \begin{bmatrix} 1 & 0 & b \\ 0 & 1 & 0 \\ 0 & 0 & 1 \end{bmatrix} \middle| \, b \in \mathbf{R} \right\}$, and $G_2 = \begin{bmatrix} 1 & 0 & 0 \\ 0 & 1 & 0 \\ 0 & 0 & 1 \end{bmatrix} = \{\mathbf{e}\}$.
   Then $G_1 \triangle G$, and G/G_1 and G_1/G_2 are abelian. Thus G is solvable.
3. Let $G = A \times B$. Clearly $G/(\{e\} \times B) \simeq A$. Since A and B are solvable, it follows that G is solvable.

Chapter 6, Section 3

1. S_n has a trivial center, if $n \geq 3$, and hence is not nilpotent. For if $Z(G) = \{e\}$, then $Z_i(G) = \{e\}$ for all i.
3. S_3 has a normal subgroup $N = \{e,(1\ 2\ 3),(1\ 3\ 2)\}$, where N and S_3/N are both abelian and, hence, nilpotent. But S_3 is not nilpotent.
5. Assume the result is false. Let G be a group of minimal order such that G is nilpotent, $|G| = p^\beta s$, $(p,s) = 1$, and $p \nmid |Z(G)|$. Because $|Z(G)| \neq 1$, $|G/Z(G)| < |G|$. Also, $G/Z(G)$ is a nilpotent group, and p divides $|G/Z(G)|$ because $p \nmid |Z(G)|$. Thus, by the minimality assumption p divides $Z(G/Z(G)) = Z_1(G)/Z(G)$. Let $\bar{x} = xZ(G) \in Z_1(G)/Z(G)$ be of order p. Clearly, $p | o(x)$. By replacing x by a suitable power of itself, we may assume that x itself is of order p. Now $y = x^{-1}g^{-1}xg \in Z(G)$ for all $g \in G$. Because $x \notin Z(G)$, there exists $g \in G$ such that $x^{-1}g^{-1}xg \neq e$. We note that p does not divide $o(x^{-1}g^{-1}xg)$, because p does not divide $|Z(G)|$. Further, $x^{-1}g^{-1}xg \in Z(G)$ implies $o(x \cdot x^{-1}g^{-1}xg) = o(x)o(x^{-1}g^{-1}xg)$ [Example 3.11(d), Chapter 4]. Thus, $o(g^{-1}xg) = o(x)o(x^{-1}g^{-1}xg)$. But $o(x) = o(g^{-1}xg)$. Hence, $o(x^{-1}g^{-1}xg) = 1$; that is, $x^{-1}g^{-1}xg = e$, a contradiction.

Chapter 7, Section 1

1. If an index i occurs in α only, then $\alpha\beta(i) = \alpha(i) = \beta\alpha(i)$. If i occurs in neither α nor β, then $\alpha\beta(i) = i = \beta\alpha(i)$.
3. For any cycle γ, $o(\gamma) =$ length of γ. Let $\sigma = \gamma_1 \cdots \gamma_k$ be a cycle decomposition, with $o(\gamma_i) = m_i$, $i = 1,\dots,k$. Let $o(\sigma) = m$, and l.c.m.$(m_1,\dots,m_k) = l$. Then

   $$e = \sigma^m = \gamma_1^m \cdots \gamma_k^m \Rightarrow \gamma_i^m = e \Rightarrow m_i | m, \qquad i = 1,\dots,k.$$

   Hence, $l|m$. On the other hand, $\sigma^l = \gamma_1^l \cdots \gamma_k^l = e$. Hence, $m|l$.

Chapter 7, Section 3

1. Any transposition $(ij) \in S_n$ can be decomposed as $(ij) = (in)(jn)(in)$. Because every permutation is a product of transpositions, it follows that S_n is generated by the set $(1\,n)$, $(2n),\dots,(n-1\,n)$. Further, it is easily verified that for any i, $1 \le i \le n-1$,

   $$(in) = (1\,2 \cdots n-1)^{-[n-1-i]}(n-1\;n)(1\,2 \cdots n-1)^{n-1-i}.$$

   Hence, S_n is generated by $(n-1\;n)$ and $(1\,2 \cdots n-1)$.
3. Let H be a group of order $4n+2$. Then $H < S_{4n+2}$ (Theorem 5.1, Chapter 4) by identifying each $a \in H$ with the corresponding permutation f_a given by left multiplication by a. We claim that H contains an odd permutation; that is, $H \not\subset A_{4n+2}$. Now by Cauchy's theorem $\exists a \in H$ such that $o(a) = 2$. Regarding H as an $[a]$-set, we obtain $H = \bigcup_{x \in H}\{x,ax\}$ as a disjoint union of orbits. Thus, $a = (x_1\; ax_1) \cdots (x_k\; ax_k)$, a product of $k = 2n+1$ disjoint transpositions. Therefore, H contains an odd permutation, as claimed. To complete the solution, we note that $HA_{4n+2} \supsetneq A_{4n+2}$, so $S_{4n+2} = HA_{4n+2}$. This implies

   $$\frac{S_{4n+2}}{A_{4n+2}} = \frac{HA_{4n+2}}{A_{4n+2}} = \frac{H}{A_{4n+2} \cap H},$$

   proving that $[H : A_{4n+2} \cap H] = 2$.
5. $\dfrac{7 \times 6}{2} \cdot \dfrac{5 \times 4 \times 3}{3} = 420$. Thus the number of elements in the conjugate class of $x = (ab)(cde) =$ 420. Now $|C(x)| = \dfrac{|S_7|}{|N_7(x)|}$ yields $|N_7(x)| = 12$.
7. The following are six possible positions by rigid motions with vertex 1 fixed:

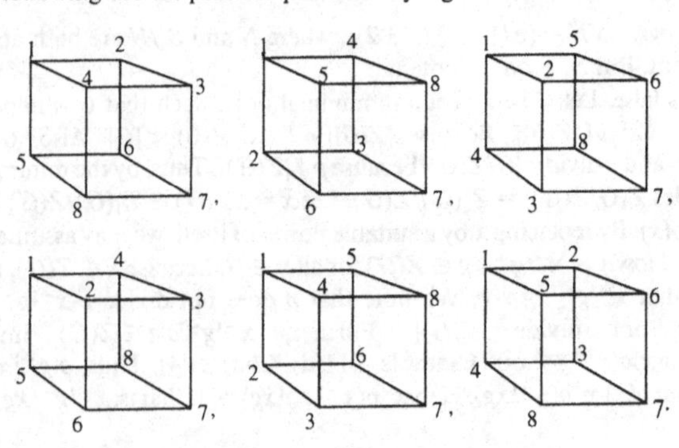

$$a = \begin{pmatrix} 1 & 2 & 3 & 4 & 5 & 6 & 7 & 8 \\ 1 & 4 & 8 & 5 & 2 & 3 & 7 & 6 \end{pmatrix}, \quad a^3 = e,$$

$$b = \begin{pmatrix} 1 & 2 & 3 & 4 & 5 & 6 & 7 & 8 \\ 1 & 4 & 3 & 2 & 5 & 8 & 7 & 6 \end{pmatrix}, \quad b^2 = e,$$

$ab = ba^2$.

This proves $H \simeq S_3$. Clearly, $|G| = 8 \cdot 6 = 48$.

9. By Cayley's Theorem $G < S_n$ for some $n > 0$. Define a mapping $f: S_n \to S_{n+2}$ given by $f(\sigma) = \sigma$, if σ is an even permutation, and $f(\sigma) = \sigma\tau$, otherwise, where $\tau = (n+1 \quad n+2)$ is a transposition. f is a monomorphism. Thus, S_n is embeddable in A_{n+2}.

Chapter 8, Section 1

1. Clearly, H and K are subgroups of $\mathbf{Z}/(10)$ such that $H \cap K = \{\bar{0}\}$ and $H + K = \{\overline{0},\overline{2},\overline{4},\overline{6},\overline{8},\overline{5},\overline{7},\overline{9},\overline{11},\overline{13}\} = \{\overline{0},\overline{1},\overline{2},\overline{3},\overline{4},\overline{5},\overline{6},\ldots,\overline{9}\} = \mathbf{Z}/(10)$.
3. Suppose, if possible, $\mathbf{Z}/(8) = H \oplus K$, where H and K are nontrivial subgroups of $\mathbf{Z}/(8)$. Then one of the subgroups must be of order 2, and the other subgroup must be of order 4. Suppose $|H| = 2$ and $|K| = 4$. The only subgroup of order 2 in $\mathbf{Z}/(8)$ is $\{\bar{0},\bar{4}\}$. Thus, $H = \{\bar{0},\bar{4}\}$. Now $\bar{2} \in \mathbf{Z}/(8)$ implies $\bar{2} = a + b$, $a \in H$, $b \in K$. If $a = \bar{0}$, then $\bar{2} \in K$, so $\bar{4} = \bar{2} + \bar{2} \in K$, a contradiction because $H \cap K = \bar{0}$. If $a = \bar{4}$, then $b = -\bar{2} \in K$, and again we get a contradiction. Hence, $\mathbf{Z}/(8)$ cannot be decomposed as a direct sum of two nontrivial subgroups.

Chapter 8, Section 5

1. The mapping $\sigma: G \to K$ defined by $\sigma(a,b) = b$, $(a,b) \in H \times K = G$ is a homomorphism of G onto K. Ker $\sigma = H \times \{e\}$. Thus, $G/H \times \{e\} \simeq K$.
3. Suppose, if possible, $\mathbf{Q} = H \oplus K$. Let $0 \neq a/b \in H$, $0 \neq c/d \in K$. Then $ac = (bc)(a/b) \in H$ and $ac = (ad)(c/d) \in K$. Hence, $H \cap K \neq \{0\}$, a contradiction. Thus, $\mathbf{Q}$ cannot be decomposed as a direct sum of two nontrivial subgroups.
5. By Theorem 3.2 and by hypothesis $A = A_1 \oplus \cdots \oplus A_m$, where $|A_i| = p_i$. Hence, each A_i is a cyclic group of order p_i, $1 \leq i \leq m$. But then A is cyclic of order $p_1 \cdots p_m$.
7. By Theorem 3.1, $A = A_1 \oplus \cdots \oplus A_k$ is a direct sum of cyclic groups A_i, where $|A_i| = m_i$ and $m_1 | m_2 | \cdots | m_k$. If $k = 1$, A is cyclic, so $k > 1$. Let $p | m_1$. Then $p | m_2$. Thus, by Cauchy's theorem A_1 and A_2 contain subgroups B_1 and B_2, respectively, each of order p. Then $B = B_1 \oplus B_2$ is a subgroup of type (p,p) of A.
9. Use the first Sylow theorem and the result that a subgroup of index 2 is normal.
11. $200 = 2^3 \cdot 5^2$. The number of Sylow 5-subgroups is $1 + 5k$ and $1 + 5k | 200$. This implies $k = 0$. Hence, there is a unique Sylow 5-subgroup that must be normal.
13. There are $1 + 7k$ Sylow 7-subgroups. Also $(1 + 7k) | 42$ implies $k = 0$. Hence, there is a unique Sylow 7-subgroup.
15. Let $|G| = p_1^{e_1} \cdots p_k^{e_k}$. Let $H_1,\ldots,H_k$ be Sylow p_i-subgroups of G. Because $H_i \lhd G$, each H_i is the unique Sylow p_i-subgroup of G. Clearly, $H_i \cap \Pi_{j=1, j\neq i}^{n} H_j = \{e\}$. Thus, by Theorem 1.1, G is the direct product of $H_1,\ldots,H_k$.
17. If $|G/Z(G)| = 77 = 11 \cdot 7$, then by Example 4.7(f), $G/Z(G)$ is cyclic. Hence, G is abelian, a contradiction.

19. We know $|Z(G)|>1$. If $|G/Z(G)|=1, p$, then G is abelian, a contradiction. Thus, $|G/Z(G)|=p^2$, so $|Z(G)|=p$. Further,

    $|G/Z(G)|=p^2 \Rightarrow G/Z(G)$ is abelian $\Rightarrow G' \subset Z(G) \Rightarrow G'=Z(G)$.

21. Let $H \supset N \supset NP$, and let $x \in N(H)$. Now P and $x^{-1}Px$ are Sylow p-subgroups in H. Thus, there exists $y \in H$ such that $P=y^{-1}x^{-1}Pxy$. This implies $xy \in N(P) \subset H$, so $x \in H$.

23. First we show that if H is a proper subgroup of nilpotent group G, then $N(H) \neq H$. Let n be maximal such that $Z_n \subsetneq H$. Choose $x \in Z_{n+1}$, $x \notin H$. Let $h \in H$. Then $x^{-1}hx = h(h^{-1}x^{-1}hx) \in HZ_n = H$, so $x \in N(H)$. This proves our assertion. Now, let P be a Sylow p-subgroup. Suppose P is not normal in G; that is, $N(P) \neq G$. Then P is a proper subgroup of G, so by the assertion in the first sentence, $N(P) \neq P$. Using the same fact again for the proper subgroup $N(P)$, we have $N(N(P)) \neq N(P)$, a contradiction (Problem 21). This proves that every Sylow p-subgroup is normal. The "only if" part is completed by Problem 15. The converse follows from the fact that a finite p-group is nilpotent, and a finite direct product of nilpotent groups is nilpotent.

25. $P < N(P)$, so $H \cap P < H \cap N(P)$. Also, $(H \cap N(P))P < G$. This implies $(H \cap N(P))P = P$, because P is Sylow p-subgroup. Then

$$\{e\} = \frac{(H \cap N(P))P}{P} \simeq \frac{H \cap N(P)}{H \cap N(P) \cap P} = \frac{H \cap N(P)}{H \cap P}$$
$$\Rightarrow H \cap N(P) = H \cap P.$$

27. Let L be a Sylow p-subgroup of K. Then $L \subset M$, where M is a Sylow p-subgroup of G. Thus, $L \subset M = x^{-1}Px$ for some $x \in G$. But $x = hk$, $h \in H$, $k \in K$. So $L \subset k^{-1}h^{-1}Phk \Rightarrow kLk^{-1} \subset h^{-1}Ph \Rightarrow kLk^{-1} \subset h^{-1}Ph \cap K$. But then $kLk^{-1} = (h^{-1}Ph) \cap K$.

Chapter 9, Section 4

1. (a)

$$(fg)(x) = f(x)g(x) = g(x)f(x) = (gf)(x) \Rightarrow fg = gf.$$
$$(f1)x = f(x)1 = f(x) \Rightarrow f1 = f.$$

   (b) Let

$$f(x) = \begin{cases} 0, & 0 \le x \le \frac{1}{2}, \\ x - \frac{1}{2}, & \frac{1}{2} \le x \le 1, \end{cases}$$
$$g(x) = \begin{cases} x - \frac{1}{2}, & 0 \le x \le \frac{1}{2}, \\ 0, & \frac{1}{2} \le x \le 1. \end{cases}$$

   Then, $f, g \in S$, $f \neq 0$, $g \neq 0$, but $fg = 0$.

   (c) Let $f = f^2 \in S$ and let $a \in [0,1]$. Then either $f(a)=0$ or $f(a)=1$. Suppose $f \neq 0$ and $f \neq 1$. Let $A = \{x \in [0,1] | f(x)=0\}$, and let $B = \{x \in [0,1] | f(x) = 1\}$. Then $A \cup B = [0,1]$, $A \neq \varnothing$, $B \neq \varnothing$, $A \neq [0,1]$, and $B \neq [0,1]$. From real analysis, A and B are closed; also, they are bounded. Because $A \cap B = \varnothing$, we may assume that l.u.b. $A \neq 1$. Let $p = $ l.u.b.A. Then, clearly, f is not continuous at p, a contradiction. Thus, either $f = 0$ or $f = 1$.

   (d) Let $f_1, f_2 \in T$. Then $(f_1 - f_2)(a) = f_1(a) - f_2(a) = 0$. Thus, $f_1 - f_2 \in T$. Similarly, $f_1 f_2 \in T$. Therefore, T is a subring. The last statement is clear.

3. (a) (i) Let $a+b\sqrt{-1}, c+d\sqrt{-1} \in A$. Then
$$(a+b\sqrt{-1})-(c+d\sqrt{-1})=(a-c)+(b-d)\sqrt{-1} \in A.$$
Also,
$$(a+b\sqrt{-1})(c+d\sqrt{-1})=(ac-bd)+(ad+bc)\sqrt{-1} \in A.$$
Hence, A is a subring.

(ii) Let $a+b\sqrt{-3}, c+d\sqrt{-3} \in B$. Then
$$(a+b\sqrt{-3})-(c+d\sqrt{-3})=(a-c)+(b-d)\sqrt{-3}.$$
Case I. If $a,b,c,d \in \mathbf{Z}$, then $(a-c),(b-d) \in \mathbf{Z}$.
Case II. If $a,b \in \mathbf{Z}$ and c,d are halves of odd integers, then both $a-c$ and $b-d$ are halves of odd integers. Proceeding like this, we get in all possible cases $(a+b\sqrt{-3})-(c+d\sqrt{-3}) \in B$. Next,
$$(a+b\sqrt{-3})(c+d\sqrt{-3})=(ac-3bd)+(ad+bc)\sqrt{-3}.$$
If $a,b,c,d \in \mathbf{Z}$, then $(a+b\sqrt{-3})(c+d\sqrt{-3}) \in B$. Suppose $a,b \in \mathbf{Z}$ and c,d are halves of odd integers. Then $ac-3bd$, $ad+bc \in \mathbf{Z}$, if a and b are both even or both odd integers. But $ac-3bd$ and $ad+bc$ are halves of odd integers if a is odd, b is even, or vice versa. Proceeding in this manner, we get in each case that $(a+b\sqrt{-3})(c+d\sqrt{-3}) \in B$. Hence, B is a subring of $\mathbf{C}$.

(b) Because $eae-ebe=e(a-b)e \in eRe$ and $(eae)(ebe)=e(aeb)e \in eRe$, eRe is a subring. It is clear that e is a unity.

5. (a) (i) $\{\overline{0},\overline{1}\}$ are idempotents. $\{\overline{0},\overline{2}\}$ are nilpotents. $\{\overline{1},\overline{3}\}$ are invertible elements.
(ii) $\{\overline{0},\overline{1},\overline{5},\overline{16}\}$ are idempotents. $\{\overline{0},\overline{10}\}$ are nilpotents. $\{\overline{1},\overline{3},\overline{7},\overline{9},\overline{11},\overline{13},\overline{17},\overline{19}\}$ are invertible elements.

(b) Let $a,b \in (U(R),\cdot)$. Then $(ab)^{-1}=b^{-1}a^{-1}$, so $ab \in U(R)$. Because $(R,\cdot)$ is a semigroup, $(U(R),\cdot)$ becomes a group.

(c) $\overline{x} \in \mathbf{Z}/(n)$ invertible $\Rightarrow \exists \overline{y} \in \mathbf{Z}/(n)$ such that $\overline{x}\overline{y}=\overline{1} \Rightarrow 1-xy=nk$ for some integer $k \Rightarrow 1=xy+nk \Rightarrow (x,n)=1$. Conversely, $(x,n)=1 \Rightarrow \exists a,b \in \mathbf{Z}$ such that $1=xa+nb \Rightarrow \overline{1}=\overline{x}\overline{a}+\overline{n}\overline{b}=\overline{x}\overline{a} \Rightarrow \overline{x}$ is invertible. Let $U(R)=\{\overline{x} \in \mathbf{Z}/(n) \mid (x,n)=1\}$. Then $(U(R),\cdot)$ is a multiplicative group of order $\phi(n)$. Thus $\overline{x}^{\phi(n)}=\overline{1}$; that is, $x^{\phi(n)} \equiv 1 \pmod{n}$.

7. Let $a^m=0$, $b^n=0$. Let $k=\max(m,n)$. Then
$$(a+b)^{2k}=\sum_{i=0}^{2k}\binom{2k}{i}a^{2k-i}b^i=0.$$
Next, if $R=F_2$, the 2×2 matrix ring over a field, and if $a=\left(\begin{smallmatrix}0&1\\0&0\end{smallmatrix}\right)$, $b=\left(\begin{smallmatrix}0&0\\1&0\end{smallmatrix}\right)$, then $a^2=0=b^2$, but $a+b$ is not nilpotent; in fact, $a+b$ is invertible.

9. Let R be an integral domain with characteristic $n \neq 0$. Suppose $n=pm$, $p,m<n$. Then there exists $a \in R$ such that $pa \neq 0$. Let $x \in R$. Then $(pa)(mx)=(pm)(ax)=n(ax)=0$ implies $mx=0$ for all $x \in R$, a contradiction, because $m<n$. Hence, n must be prime.

11. Let $R=\{0,a_1,\ldots,a_n\}$ be an integral domain. Let $a_i \in R$. Then $a_ia_1,\ldots,a_ia_n$ are all nonzero and are distinct elements of R. Thus, given $a_k \in R$, there exists $a_j \in R$ such that $a_ia_j=a_k$; that is, the equation $ax=b$ has a solution for all nonzero $a \in R$ and for all nonzero $b \in R$. Also, if $b=0$, then, clearly, $x=0$ is a solution. Hence, by Example 4.4(d), R is a division ring.

13. $aua=a \Rightarrow auab=ab \Rightarrow au=1 \Rightarrow ua=1$
$$\Rightarrow uab=b \Rightarrow u=b \Rightarrow ba=1.$$

15. Let $0 \neq a \in R$ and $b \in R$. Then there exists $c \in R$ such that $ac=e$. This implies $acb=eb=b$. But then $ax=b$ has a solution. Hence, R is a division ring [see Example 4.4(d)].

17. Let $(ab)^m = 0$. Then $(ba)^{m+1} = b(ab)^m a = 0$.
Let $ab + c - abc = 0$. Then it is straightforward to verify that

$$ba + (-ba + bca) - ba(-ba + bca) = 0.$$

19. Let $a + c - ac = 0$. Then

$$ba + bc - bac = 0 \Rightarrow b + bc - bc = 0 \Rightarrow b = 0.$$

21. Let $e + b - eb = 0$. Then $e^2 + eb - e^2b = 0 \Rightarrow e^2 = 0$; that is, $e = 0$.
23. Let $e \in R$ be such that e is not r.q.r. Clearly, $e \neq 0$. If e^2 is r.q.r., then e is also r.q.r. (Problem 20). Thus, e^2 is not r.q.r. and, hence, $e = e^2$. Let $a \in R$. Suppose $ea \neq a$. Then $a - ea + e \neq e$ and so is r.q.r. Thus, $\exists b \in R$ such that $a - ea + e + b - (a - ea + e)b = 0$. Multiplying on the left by e and using the fact that $e = e^2$, we obtain $e = 0$, a contradiction. Therefore, $a = ea$ for all $a \in R$. Further, if $ae \neq a$, then $ae - a + e \neq e$; so there exists $c \in R$ such that $(ae - a + e) + c - (ae - a + e)c = 0$. Because e is a left identity as already shown, we obtain $ae - a + e + c - ac + ac - c = 0$; in other words, $ae - a + e = 0$. This yields, by multiplying by e on the right, $e = 0$, a contradiction. Hence, e is the identity of R.
Now, let $0 \neq a \in R$. Then $e - a \neq e$, so there exists $b \in R$ such that $e - a + b - (e - a)b = 0$; that is, $e - a + ab = 0$. This implies $a(e - b) = e$. Hence, each nonzero element has a right inverse. Therefore, R is a division ring [you may use Example 4.4(d)].
25. Let $x = (a_1,0,...,0)$, $y = (0,a_2,0,...,0) \in R_1 \oplus \cdots \oplus R_n$, where $0 \neq a_1 \in R_1$, $0 \neq a_2 \in R_2$. Then $xy = 0$, but $x \neq 0$ and $y \neq 0$. If $a^m = 0$, then $a_i^m = 0$, $1 \leq i \leq n$. Conversely, if $a_i^{m_i} = 0$, $1 \leq i \leq n$, then by choosing $m = \max(m_1,...,m_n)$, it follows that $a^m = 0$.

Chapter 10, Section 1

1. Let $0 \neq a \in R$. Then aR is a nonzero ideal, so $aR = R$, which proves that $ab = 1$ for some $b \in R$.
3. Let $0 \neq a \in R$. Then aR is a nonzero right ideal, so $aR = R$; that is, every nonzero element has a right inverse, proving that R is a division ring.
5. As remarked just after the definition of a PID, any ideal in $F[x]$ is of the form $(f(x))$, $f(x) \in F[x]$.
7. First we show that an element $f(x) = \Sigma_{i=0}^{\infty} a_i x^i$ is invertible in $F[[x]]$ iff $a_0 \neq 0$. If $f(x)$ is invertible with inverse $g(x) = \Sigma_{i=0}^{\infty} b_i x^i$, then, clearly, $a_0 b_0 = 1$, so $a_0 \neq 0$. Conversely, if $a_0 \neq 0$, we show that there exists $g(x) = \Sigma_{i=0}^{\infty} b_i x^i$ that is the inverse of $f(x)$. Tentatively, we assume $f(x)g(x) = 1$ and obtain the following equations:

$$\begin{aligned} a_0 b_0 &= 1, \\ a_0 b_1 + a_1 b_0 &= 0, \\ a_0 b_2 + a_1 b_1 + a_2 b_0 &= 0, \\ &\vdots \\ a_0 b_n + a_1 b_{n-1} + \ldots + a_n b_0 &= 0, \\ &\ldots \end{aligned}$$

Because $a_0 \neq 0$, this system of equations determines b_0, b_1, b_2, and so on. Having determined $b_0, b_1, \ldots, b_{n-1}$, we can find b_n from the last equation. Thus, by induction on n, each b_i can be determined. This proves $f(x)$ is invertible with $g(x)$ as its inverse.

Let A be a nonzero ideal, and let $S = \{m \geq 0 | a_m x^m + a_{m+1}x^{m+1} + \cdots \in A$, and $a_m \neq 0\}$. Obviously $S \neq \varnothing$. Let m be the smallest positive integer in S. Then $x^m(a_m + a_{m+1}x + \cdots) \in A \Rightarrow x^m \in A$. Thus, $A = (x^m)$.

Chapter 10, Section 2

1. Ideals in $\mathbf{Z}/(n)$ are of the form $(a) + (n)/(n)$, $a \in \mathbf{Z}$ (Theorem 2.6). Because $(a) + (n) = (d)$, where $(a,n) = d$, it follows that the ideals in $\mathbf{Z}/(n)$ are $(d)/(n)$, where $d|n$.
3. Let f: $\mathbf{Z}/(m) \rightarrow \mathbf{Z}/(n)$ be a homomorphism, where $f(\bar{1}) = \bar{1}$. Then $\bar{0} = f(\bar{0}) = f(m\bar{1}) = m\bar{1} = \bar{m} \in \mathbf{Z}/(n) \Rightarrow n|m$. For the converse, define $f(a + (m)) = a + (n)$. f is well defined because $n|m$.
5. Let $x, y \in r(S)$, $r \in R$. Then $S(x - y) = Sx - Sy = 0$, $S(xr) = (Sx)r = 0$. Thus, $r(S)$ is a right ideal. Similarly, $l(S)$ is a left ideal.
7. Let $\sigma: F \rightarrow R$ be a homomorphism. Because $\text{Ker}\,\sigma$ is an ideal, $\text{Ker}\,\sigma = (0)$ or F. Hence, σ is 1-1 or 0.
9. Let $x, y \in A$, and let B be any nonzero right ideal. Then $r(x) \cap r(y) \cap B \neq 0$ gives that there exists $0 \neq b \in B$ such that $xb = 0 = yb$; that is, $(x - y)b = 0$. Thus, $r(x - y) \cap B \neq 0$, so $x - y \in A$. Next, let $a \in A$. If $aB = 0$, then $xaB = 0$. This implies $r(xa) \cap B \neq 0$. Therefore, $xa \in A$. But if $aB \neq 0$, then $r(x) \cap aB \neq 0$ implies that there exists $0 \neq b \in B$ such that $xab = 0$; that is, $r(xa) \cap B \neq 0$. Hence, in any case for all $x \in A$, $a \in R$, $xa \in A$. We now show $ax \in A$. Clearly, $r(x) \subset r(ax)$. Because $r(x) \cap B \neq 0$, we obtain $r(ax) \cap B \neq 0$. Hence, $ax \in A$, proving that A is an ideal in R.
11. Suppose $A \not\subset B$ and $A \not\subset C$. Then there exists $a_1 \in A$, $a_1 \notin B$ and $a_2 \in A$, $a_2 \notin C$. Now consider $a_1 + a_2$, which is in A and, hence, in $B \cup C$. Thus, $a_1 + a_2 \in B$ or $a_1 + a_2 \in C$. Suppose $a_1 + a_2 \in B$. Note that $a_2 \in A$, $a_2 \notin C$ implies $a_2 \in B$. Thus, $a_1 + a_2 \in B$ implies $a_1 \in B$, a contradiction. Thus, $a_1 + a_2 \notin B$. Similarly, $a_1 + a_2 \notin C$. But this is impossible. Therefore, our assumption that $A \not\subset B$ and $A \not\subset C$ is false.
13. Let $(x,0),(y,0) \in R_1^*$ and $(a,b) \in R$. Then $(x,0) - (y,0) = (x - y, 0) \in R_1^*$, $(x,0)(a,b) = (xa,0) \in R_1^*$, and $(a,b)(x,0) = (ax,0) \in R_1^*$. Hence, R_1^* is an ideal in R. Similarly, R_2^* is an ideal. Clearly, the mapping $x \mapsto (x,0)$ is an isomorphism (as rings) of R_1 onto R_1^*.
15. The mapping σ: $(\bar{a}_{ij}) \rightarrow (a_{ij}) + A_n$ of $(R/A)_n$ into R_n/A_n is clearly well defined. For
$$\begin{aligned}(\bar{a}_{ij}) = (\bar{b}_{ij}) &\Leftrightarrow \bar{a}_{ij} = \bar{b}_{ij} \Leftrightarrow a_{ij} - b_{ij} \in A_n \\ &\Leftrightarrow (a_{ij}) + A_n = (b_{ij}) + A_n.\end{aligned}$$
Also, σ is an onto ring homomorphism. Hence, σ is an isomorphism.
17. $$\begin{aligned}(fg)(xy) = (f(g(xy)) &= f(g(y)g(x)) \\ &= f(g(x))f(g(y)) = (fg)(x)(fg)(y).\end{aligned}$$
Hence, fg is a homomorphism (or isomorphism).
19. (a) $\bar{0} \neq x + (p^2) \in \mathbf{Z}/(p^2)$ is invertible
$$\begin{aligned}&\Leftrightarrow 1 - xy \in (p^2) \text{ for some } \bar{0} \neq y + (p^2) \in \mathbf{Z}/(p^2) \\ &\Leftrightarrow 1 - xy \in p^2\mathbf{Z} \\ &\Leftrightarrow (x,p) = 1 \Leftrightarrow p \nmid x.\end{aligned}$$
Clearly, $(p)/(p^2)$, the set of noninvertible elements, forms an ideal.
(b) Note that $a_0 + a_1x + a_2x^2 + \cdots \in F[[x]]$ is invertible iff $a_0 \neq 0$. The set (x) of noninvertible elements is clearly an ideal (see Problem 7, Section 1, Chapter 10).

Chapter 10, Section 3

1. Let $a \in A$. Then $a = a + 0 \in A + A$; that is, $A \subset A + A$. Let $a + b \in A + A$, $a,b \in A$. Then $a + b \in A$; that is, $A + A \subset A$. Hence, $A + A = A$.
3. The mapping

$$x + \bigcap_{i=1}^{n} A_i \mapsto (x + A_1, \ldots, x + A_n)$$

is an isomorphism of $R/\bigcap_{i=1}^{n} A_i$ onto $R/A_1 \times \cdots \times R/A_n$ [see Example 3.4(a)(ii)]. So if $(x_1 + A_1, \ldots, x_n + A_n) \in R/A_1 \times \cdots \times R/A_n$, then there exists $x + \bigcap_{i=1}^{n} A_i \in R/\bigcap_{i=1}^{n} A_i$ such that $(x + A_1, \ldots, x + A_n) = (x_1 + A_1, \ldots, x_n + A_n)$; that is, $x - x_i \in A_i$, $1 \leq i \leq n$.

5. (a) $f_1((a_1,a_2) + (b_1,b_2)) = f_1(a_1 + b_1, a_2 + b_2)$
$= a_1 + b_1 = f_1(a_1,a_2) + f_1(b_1,b_2)$.
$f_1((a_1,a_2)(b_1,b_2)) = f_1(a_1b_1, a_2b_2) = a_1b_1 = f_1(a_1,a_2)f_1(b_1,b_2)$.
Thus, f_1 is a homomorphism. Clearly, f_1 is onto.

(b) Let A be a right ideal of the ring $R = R_1 \oplus R_2$. Let $a = (a_1,a_2) \in A$. Then $f_1(a) = a_1$ and $f_2(a) = a_2$. So $a \in f_1(A) \oplus f_2(A)$; that is, $A \subset f_1(A) \oplus f_2(A)$. Next, let $x_1 \in f_1(A)$. Then $\exists x_2 \in R$ such that $(x_1,x_2) \in A$. This implies $(x_1,x_2)(1,0) \in A$; that is, $(x_1,0) \in A$. Similarly, if $y_1 \in f_2(A)$, then $(0,y_1) \in A$. So $(x_1,y_1) = (x_1,0) + (0,y_1) \in A$. This proves $f_1(A) \oplus f_2(A) \subset A$. Hence, $A = f_1(A) \oplus f_2(A)$.

(c) Follows from (b).

7. Define a mapping σ: $R \times S \rightarrow R/A \times S/B$ by $\sigma(x,y) = (x + A, y + B)$. Then σ is an onto ring homomorphism and

Ker $\sigma = \{(x,y) | x \in A, y \in B\} = A \times B$.

Hence, by the fundamental theorem of homomorphisms

$$\frac{R \times S}{A \times B} \simeq \frac{R}{A} \times \frac{S}{B}.$$

Chapter 10, Section 4

1. By Corollary 1.3 in Chapter 10, F_m has no nontrivial ideals. Hence, any ideal in F_m is (0) or F_m. So if $AB \subset (0)$, where A and B are ideals in F_m, then we must have $A = (0)$ or $B = (0)$. (Both A and B cannot be equal to F_m, since $AB \subset (0)$.)
3. $x^4 + 4 = (x^2 + 2x + 2)(x^2 - 2x + 2)$. Thus, $(x^2 + 2x + 2)(x^2 - 2x + 2) \in (x^4 + 4)$. But $x^2 + 2x + 2 \notin (x^4 + 4)$ and $x^2 - 2x + 2 \notin (x^4 + 4)$. Hence, by Theorem 4.6, $(x^4 + 4)$ is not a prime ideal in $\mathbf{Q}[x]$.
5. (a) $\Rightarrow$ (b) Let $aRb \subseteq P$. Let

$$A = aR + Ra + RaR + \mathbf{Z}a, \qquad B = bR + Rb + RbR + \mathbf{Z}b$$

be the two-sided ideals generated by a and b, respectively. Then

$$AB = (aR + Ra + RaR + \mathbf{Z}a)(bR + Rb + RbR + \mathbf{Z}b) \subseteq P,$$

because $aRb \subseteq P$. Hence, $A \subseteq P$ or $B \subseteq P$; that is, $a \in P$ or $b \in P$.

(b) $\Rightarrow$ (c) Let $AB \subseteq P$, where A and B are right ideals of R. Let $A_1 = RA + A$, $A_2 = RB + B$, where

$$RA = \left\{ \sum_{\text{finite sum}} r_i a_i | r_i \in R, a_i \in A \right\}.$$

RB has a similar meaning. Then A_1 and A_2 are ideals in R, and

$$A_1A_2 = (RA + A)(RB + B) = RARB + RAB + ARB + AB \subseteq P,$$

because $AB \subseteq P$. By (b) $A_1 \subseteq P$ or $A_2 \subseteq P$. Hence, $A \subseteq P$ or $B \subseteq P$.

(c) $\Rightarrow$ (d) Let $AB \subseteq P$, where A and B are left ideals in R. Let $A_1 = A + AR$, $A_2 = B + BR$. As before, $A_1A_2 \subseteq P$, and A_1 and A_2 are ideals (hence right ideals). Thus, by (c) $A_1 \subseteq P$ or $A_2 \subseteq P$. This yields $A \subseteq P$ or $B \subseteq P$.

(d) $\Rightarrow$ (a) Trivial.

7. Let $x \in \bigcap_{i\in\Lambda} P_i$. Suppose x is not nilpotent. Consider the multiplicative semigroup $S = \{x, x^2, x^3, \ldots\}$. Then $0 \notin S$. Let

$$\mathscr{C} = \{I \mid I \text{ is an ideal in } R \text{ with } I \cap S = \varnothing\}.$$

$\mathscr{C}$ is a po set. By Zorn's lemma $\mathscr{C}$ has a maximal member, say P. We claim P is prime. Let A and B be ideals in R such that $AB \subseteq P$. If possible, let $A \not\subseteq P$ and $B \not\subseteq P$. By maximality of P, $x^m \in A + P$ and $x^k \in B + P$ for some positive integers m and k. Then $(A + P)(B + P) = AB + AP + PB + P^2 \subset P$ implies $x^{m+k} \in P$, a contradiction. Hence, P is prime. But then $x \in P$, again a contradiction. This proves x must be a nil element.

Chapter 10, Section 5

1. Let $A^m = 0$, $B^n = 0$. Let $k = \max(m,n)$. Then we have

$$(A + B)^{2k} = \sum A^{e_1}B^{f_1} \cdots A^{e_t}B^{f_t},$$

where e_i and f_i are nonnegative integers, $\Sigma_i(e_i + f_i) = 2k$. Thus, each term is a product of at least k factors all belonging to A or B. Because $A^k = 0 = B^k$, it follows from this equation that $(A + B)^{2k} = 0$ and that $A + B$ is a nilpotent ideal.

3. Let $a = (a_{ij}) \in F_n$ be an element of a nil right ideal, say A. Now write $a = \Sigma a_{ij}e_{ij}$, where e_{ij} are matrix units. Then $(\Sigma a_{ij}e_{ij})(e_{k1}) \in A$; that is, $\Sigma_{i=1}^{n} a_{ik}e_{i1} \in A$; writing in matrix notation, we get

$$\begin{bmatrix} a_{1k} & 0 & \cdots & 0 \\ a_{2k} & 0 & \cdots & 0 \\ \cdot & \cdot & & \\ \cdot & \cdot & & \\ \cdot & \cdot & & \\ a_{nk} & 0 & \cdots & 0 \end{bmatrix} \in A.$$

Then for any positive integer m, we have

$$\begin{bmatrix} a_{1k} & 0 & \cdots & 0 \\ a_{2k} & 0 & \cdots & 0 \\ \cdot & \cdot & & \\ \cdot & \cdot & & \\ \cdot & \cdot & & \\ a_{nk} & 0 & \cdots & 0 \end{bmatrix}^m = \begin{bmatrix} a_{1k}^m & 0 & \cdots & 0 \\ a_{2k}a_{1k}^{m-1} & 0 & \cdots & 0 \\ \cdot & \cdot & & \\ \cdot & \cdot & & \\ \cdot & \cdot & & \\ a_{nk}a_{1k}^{m-1} & 0 & \cdots & 0 \end{bmatrix}$$

Because A is a nil right ideal, $a_{1k}^m = 0 = a_{2k}a_{1k}^{m-1} = \cdots = a_{nk}a_{1k}^{m-1}$ for some positive integer m. This implies $a_{1k} = 0$. Because k is arbitrary, we get that the first row of the matrix a is zero. Similarly, by multiplying $\Sigma a_{ij}e_{ij}$ by e_{k2} on the right we obtain that the second row of the matrix a is zero. Proceeding like this we obtain $a = 0$.

Chapter 10, Section 6

The proof is exactly similar to that of Theorem 6.1.

Chapter 11, Section 1

1. Let $(a,b) = d$ and $(ca,cb) = e$. Then $d|a \Rightarrow cd|ca$. Similarly, $cd|cb$. Thus, $cd|(ca,cb) \Rightarrow cdx = e$ for some $x \in R$. Now $ca = ey$ for some $y \in R$. Thus $ca = ey = cdxy$. Hence, $a = dxy$; that is, $dx|a$. Similarly, $dx|b$, so $dx|d$. This implies x is a unit; hence, $(ca,cb) = e = cdx = c(a,b)$.
3. $b|ac \Rightarrow bd = ac$ for some $d \in R$. Now $c = c(a,b) = (ca.cb) = (bd,bc) = b(c,d) \Rightarrow b|c$.
5. See Example 1.2(d), where it is shown that $2 + \sqrt{-5}$ is irreducible. Also, $(2 + \sqrt{-5})|3 \cdot 3$, but $(2 + \sqrt{-5}) \nmid 3$. Hence, $2 + \sqrt{-5}$ is not prime.
7. Indeed, both $1 + \sqrt{-3}$ and $1 - \sqrt{-3}$ are irreducible. For if $1 + \sqrt{-3} = (a + b\sqrt{-3})(c + d\sqrt{-3})$, then $4 = (a^2 + 3b^2)(c^2 + 3d^2)$. This implies $a = \pm 1, b = 0$ (or $c = \pm 1, d = 0$).
9. Consider $\dfrac{10 + 11i}{8 + i} = \dfrac{(10 + 11i)(8 - i)}{(8 + i)(8 - i)} = \dfrac{91 + 78i}{65} = (1 + i) + \left(\dfrac{2}{5} + \dfrac{1}{5}i\right)$. Then by multiplying both sides with $8 + i$, $10 + 11i = (1 + i)(8 + i) + (3 + 2i)$. Next, divide $8 + i$ by the remainder $3 + 2i$ and obtain $\dfrac{8 + i}{3 + 2i} = 2 - i$. This implies $8 + i = (2 - i)(3 + 2i)$ and so the remainder is zero, showing $3 + 2i$ is the g.c.d.
11. Considering $\left(\dfrac{a + b\sqrt{-3}}{2}\right)\left(\dfrac{c + d\sqrt{-3}}{2}\right) = 1$, we obtain $(a^2 + 3b^2)(c^2 + 3d^2) = 16$. Now, neither $a^2 + 3b^2$ nor $c^2 + 3d^2$ can be equal to 1, 2, or 16. Thus $a^2 + 3b^2 = 4$. This yields $(a, b) = (\pm 2, 0)$, or $(\pm 1, \pm 1)$. This gives that units of $\mathbf{Q}[\sqrt{-3}]$ are ± 1, $\dfrac{\pm 1 \pm \sqrt{-3}}{2}$.

Chapter 11, Section 3

1. (a) Write $b = bq + r$, $\phi(r) < \phi(b)$. Suppose $r \neq 0$. Then $b(1 - q) = r$ implies $b|r$; so $\phi(b) \leq \phi(r)$, a contradiction. Thus, $r = 0$, so $\phi(0) < \phi(b)$.
   (b) We have $a|b$ and $b|a$; so $\phi(a) \leq \phi(b)$ and $\phi(b) \leq \phi(a)$; that is, $\phi(a) = \phi(b)$.
   (c) Write $a = bq + r$, $\phi(r) < \phi(b)$. Now $a|b \Rightarrow ax = b$. Thus,
   $$\begin{aligned} a = bq + r = axq + r &\Rightarrow a(1 - xq) = r \Rightarrow a|r \\ &\Rightarrow \phi(a) \leq \phi(r) < \phi(b) = \phi(a) \end{aligned}$$
   unless $r = 0$. Then $b|a$. Therefore, a and b are associates.
3. Let $0 \neq a \in R$ be noninvertible. Then the ideal $xR[x] + aR[x]$ is not a principal ideal. For let
   $$xR[x] + aR[x] = f(x)R[x]. \tag{1}$$
   Then $a \in f(x)R[x]$ implies $a = f(x)g(x)$. This gives $f(x) = \beta$, $g(x) = \gamma$, where $\beta,\gamma \in R$. Also,
   $$x \in f(x)R[x] \Rightarrow x = f(x)h(x) = \beta h(x) = \beta(bx) \Rightarrow 1 = \beta b, \qquad b \in R.$$

Thus, β is invertible, so $f(x)R[x] = \beta R[x] = R[x]$. Then from (1),

$$1 = xu(x) + av(x), \qquad u(x),v(x) \in R[x]. \tag{2}$$

(2) gives $u(x) = 0$, and $v(x) = c$, $c \in R$. But then $1 = ac$, so a is invertible, a contradiction. Hence, $R[x]$ is not a PID.

5. Apply Theorem 3.2 of Chapter 11 and Theorem 4.7 of Chapter 10.
7. Indeed every ideal contains a product of prime ideals. Otherwise, consider a family F of all ideals which are not products of prime ideals. We claim F has a maximal member. Otherwise, we can produce an infinite properly ascending chain

$$A_1 \subset A_2 \subset A_3 \subset \cdots$$

of ideals $A_i \in F$. But if we consider $B = \cup A_i$, B is an ideal, say bR. Let $b \in A_k$. Then $A_k = A_{k+1} = \cdots$, a contradiction. Thus $F = \varnothing$. In particular, $0 = P_1 P_2 \cdots P_n$, where P_i are prime ideals. The uniqueness follows easily.
9. Since $b|a$ implies $\phi(b) \leq \phi(a)$, it follows that if $ac = 1$ then $\phi(a) = \phi(1)$. Conversely, write $1 = aq + r$. If $r \neq 0$, then $\phi(r) < \phi(a) = \phi(1) \leq \phi(r)$, a contradiction. Thus $r = 0$.
11. $\sqrt{P} = aR$. Write $a = p_1 \cdots p_k$, where p_i are prime elements. Since $\sqrt{P}$ is a prime ideal, for some $p_i \in \sqrt{P}$. It is easy to check $\sqrt{P} = p_i R$, and $P = p_i^e R, e > 0$.

Chapter 11, Section 4

1. Let $R = F[x]$. Then R is a commutative integral domain that is not a field. Then $R[y] = F[x,y]$ is not a PID (see Problem 3, Section 3).
3. First suppose $f(x)$ is a zero divisor in $R[x]$. Let $K = \{g(x) \in R[x] | g(x)f(x) = 0\}$. We want to show $K \cap R \neq (0)$. If $f(x) \in R$, then it is clear that $K \cap R \neq 0$. So assume the degree of $f(x) > 0$. If possible, let $K \cap R = (0)$. This implies that if $0 \neq c \in R$, then $cf(x) \neq 0$. Let $g(x) \in K$ be of minimum degree, say m. Write

$$g(x) = b_0 + b_1 x + \cdots + b_m x^m.$$

Then $b_m \neq 0$. Also, by assumption, $m > 0$. Let

$$f(x) = a_0 + a_1 x + \cdots + a_n x^n, \qquad a_n \neq 0.$$

Because $b_m f(x) \neq 0$, $a_i g(x) \neq 0$ for some $i = 0,1,...,n$. Choose p to be the largest positive integer such that $a_p g(x) \neq 0$; that is, $a_{p+1} g(x) = 0 = a_{p+2} g(x) = \cdots = a_n g(x)$. Then

$$\begin{aligned} 0 = f(x)g(x) &= (a_0 + a_1 x + \cdots + a_p x^p + a_{p+1} x^{p+1} + \cdots + a_n x^n) g(x) \\ &= (a_0 + a_1 x + \cdots + a_p x^p) g(x) \end{aligned}$$

implies $a_p b_m = 0$. Thus, the degree of $a_p g(x) < m$. Moreover, $(a_p g(x)) f(x) = 0$. Hence, $a_p g(x) \in K$, degree $a_p g(x) < m$, a contradiction. Thus, $K \cap R \neq 0$, so there exists $b \in R$, $bf(x) = 0$. The converse is clear.

Chapter 12, Section 1

1. Define $h: R_S \to R'$ by $h(a/s) = g(a)(g(s))^{-1}$. Then $hf(a) = h(a/1) = g(a)$, so $hf = g$.
3. Let A be an ideal in R_S. Let $A^c = \{a \in R | a/s \in A \text{ for some } s \in S\}$. Then A^c is an ideal in R. Next, let A and B be ideals in R_S such that $A \subsetneq B$. We show that $A^c \subsetneq B^c$. This will

prove an infinite properly ascending chain of ideals in R_S gives rise to an infinite properly ascending chain of ideals in R. This proves the required result. Let $A \subsetneq B$. Clearly, $A^c \subset B^c$. If possible, let $A^c = B^c$. Choose $b/s \in B$, $b/s \notin A$. Then $b \in B^c = A^c$. Thus, there exists $s_1 \in S$ such that $b/s_1 \in A$. This implies $(b/s_1) \cdot (s_1/s) \in A$; that is, $b/s \in A$, a contradiction. Hence, $A \subsetneq B \Rightarrow A^c \subsetneq B^c$, as desired.

5. Note that $\mathbf{R}(x) \neq \mathbf{R}(x,y)$.

Chapter 12, Section 2

1. We first claim that R cannot possess an infinite properly ascending chain of right ideals. For, let $a_1R \subset a_2R \subset a_3R \subset \cdots$ be a properly ascending chain of right ideals. Let $A = \bigcup_{i=1}^{\infty} a_iR$. Then A is a right ideal, so $A = cR$ for some $c \in R$. Now $c \in A \Rightarrow c \in a_iR$ for some i. Then $cR \subset a_iR \subset A = cR$. Thus, $A = a_iR$. This yields $a_iR = a_{i+1}R = \ldots$, a contradiction. Further, if possible, let I, J be nonzero right ideals such that $I \cap J = (0)$. Let $0 \neq a \in J$. Then for any positive integer m, $\Sigma_{k=1}^{m} a^kI$ is a direct sum of right ideals. This yields an infinite properly ascending chain of right ideals,

$$aI \subset aI \oplus a^2I \subset aI \oplus a^2I \oplus a^3I \subset \cdots,$$

a contradiction. Hence $I \cap J \neq (0)$, which shows that R is a right Ore domain.

3. Let R be an integral domain satisfying a standard identity

$$S_n(x_1,\ldots,x_n) = \sum \pm x_{i_1}x_{i_2}\ldots x_{i_n},$$

where the summation runs over every permutation $(i_1,\ldots,i_n)$ of $(1,\ldots,n)$, and the sign is positive or negative according to whether $(i_1,\ldots,i_n)$ is an even or an odd permutation.

We first show that any direct sum $\oplus\Sigma_{i=1}^{k} A_i$ of right ideals A_i cannot contain more than $n-1$ terms. If possible, let $A_i \oplus \cdots \oplus A_n$ be a direct sum of n right ideals. Let $0 \neq a_i \in A_i$, $1 \le i \le n$. Then $S_n(a_1,\ldots,a_n) = 0$. Rearranging the terms in $S_n(a_1,\ldots,a_n)$, we have

$$a_1S_{n-1}(a_2,\ldots,a_n) - a_2S_{n-1}(a_3,\ldots,a_n,a_1) + a_3S_{n-1}(a_4,\ldots,a_n,a_1,a_2) - \cdots = 0.$$

This implies that $a_1S_{n-1}(a_2,\ldots,a_n) = 0$, $a_2S_{n-1}(a_3,\ldots,a_1) = 0$, and so on, because $A_1 \oplus \cdots \oplus A_n$ is a direct sum. But then $S_{n-1}(a_2,\ldots,a_n) = 0$ for all $0 \neq a_i \in A_i$, $2 \le i \le n$. Continuing likewise, we obtain $a_n = 0$, a contradiction.

Now, we show that if A,B are right ideals, then $A \cap B \neq (0)$. If possible, let $A \cap B = 0$. Let $0 \neq c \in B$. Then $\Sigma_{i=1}^{n} c^iA$ is a direct sum of n right ideals, a contradiction to what we proved before. Hence, $A \cap B \neq (0)$, proving that R is a right Ore domain. Similarly, we can show that R is a left Ore domain.

Chapter 14, Section 1

1. $(R[x],+)$ is an additive abelian group. Let $r \in R, f(x) = a_0 + a_1x + \cdots + a_nx^n \in R[x]$. Define $rf(x) = ra_0 + (ra_1)x + \cdots + (ra_n)x^n$. Then, clearly,

$$\begin{aligned}(r_1+r_2)f(x) &= r_1f(x) + r_2f(x),\\ r(f_1(x)+f_2(x)) &= rf_1(x) + rf_2(x),\\ (r_1r_2)f(x) &= r_1(r_2f(x)),\\ 1f(x) &= f(x) \quad \text{if } 1 \in R.\end{aligned}$$

Hence, $R[x]$ is a left R-module.

3. M is a $\mathbf{Z}$-module by defining

$$am = \begin{cases} \overbrace{m + \cdots + m}^{a \text{ times}}, & a > 0, \\ \overbrace{-m + \cdots + -m}^{-a \text{ times}}, & a < 0, \\ 0, & a = 0, \end{cases}$$

where $a \in \mathbf{Z}$, $m \in M$. Suppose M is also a $\mathbf{Z}$-module by a mapping $f: \mathbf{Z} \times M \to M$. Then by axiom (iv) for a module, $f(1,m) = m$. Also, by axiom (ii), if $a \in \mathbf{Z}$, $a > 0$, then

$$f(a,m) = f(\overbrace{1 + \cdots + 1}^{a \text{ times}}, m) = \overbrace{f(1,m) + \cdots + f(1,m)}^{a \text{ times}}$$
$$= af(1,m) = am.$$

Next, using axiom (ii) again, we have

$$f(-1,m) = f(-1+0,m) = f(-1,m) + f(0,m).$$

Hence, $f(0,m) = 0$. Further,

$$0 = f(0,m) = f(1 + (-1),m) = f(1,m) + f(-1,m),$$

so $f(-1,m) = -f(1,m) = -m$. Then $f(-a,m) = -am$, $a > 0$. Hence, for all $a \in \mathbf{Z}$, $f(a,m) = am$. This proves that there is only one way of making M a $\mathbf{Z}$-module.

Chapter 14, Section 2

1. (a) $W = \{(0,\lambda_2,\lambda_3) | \lambda_2,\lambda_3 \in \mathbf{R}\}$ is clearly an additive subgroup of $(\mathbf{R}^3,+)$, and if $\alpha \in \mathbf{R}$, $(0,\lambda_2,\lambda_3) \in W$, then $\alpha(0,\lambda_2,\lambda_3) = (0,\alpha\lambda_2,\alpha\lambda_3) \in W$. Hence, W is a subspace of $\mathbf{R}^3$.
   (b) $W = \{(\lambda_1,\lambda_1,\lambda_3) | \lambda_1,\lambda_3 \in \mathbf{R}\}$ is clearly a subspace.
   (c) $W = \{(\lambda_1,\lambda_2,\lambda_3) | \lambda_1,\lambda_2,\lambda_3 \in \mathbf{R}, \lambda_1\lambda_2 = 0\}$. Now $(1,0,1), (0,1,1) \in W$. But $(1,0,1) - (0,1,1) = (1,-1,0) \notin W$, because $1,-1 \neq 0$. Thus, W is not a subspace.
   (d) $W = \{(2\lambda_2 - 1,\lambda_2,\lambda_3) | \lambda_2,\lambda_3 \in \mathbf{R}\}$. Now $(1,1,1) \in W$ and $(3,2,1) \in W$. But $(3,2,1) - (1,1,1) = (2,1,0) \notin W$. So W is not a subspace.
   (e) $W = \{(\lambda_1,\lambda_2,\lambda_3) \in \mathbf{R}^3 | \lambda_1 + \lambda_2 \geq 0\}$. Now $(1,2,3) \in W$, but $-(1,2,3) = (-1,-2,-3) \notin W$. Thus, W is not a subspace.
3. (a) $W = \{f: \mathbf{R} \to \mathbf{R} | f(1) = 0\}$. Let $f,g \in W$, $a \in \mathbf{R}$. $(f-g)(1) = f(1) - g(1) = 0$, $(af)(1) = a(f(1)) = 0$. Thus, $f-g$, $af \in W$. Hence, W is a subspace.
   (b) $W = \{f: \mathbf{R} \to \mathbf{R} | f(0) = 1\}$. Let $f,g \in W$. Then $(f-g)(0) = f(0) - g(0) = 1 - 1 = 0$. Hence, $f-g \notin W$. Thus, W is not a subspace.
   (c) $W = \{f: \mathbf{R} \to \mathbf{R} | f(3) = 2f(2)\}$. Let $f,g \in W$, $a \in \mathbf{R}$. Then

   $$(f-g)(3) = f(3) - g(3) = 2(f(2) - g(2)) = 2(f-g)(2),$$

   and

   $$(af)(3) = a(f(3)) = a(2f(2)) = 2(af)(2).$$

   Thus, $f-g$, $af \in W$. Hence, W is a subspace.
   (d) $W = \{f: \mathbf{R} \to \mathbf{R} | f(3) \geq 0\}$. Now $(-f)(3) = -f(3) \leq 0$. Hence, $-f \notin W$. Thus, W is not a subspace.

(e) $W = \{f: \mathbf{R} \to \mathbf{R} \mid \lim_{t\to 1} f(t)$ exists$\}$. Clearly, if $f,g \in W$, then $f - g \in \mathbf{W}$, and if $a \in \mathbf{R}$, then $af \in W$. Hence, W is a subspace.

5. Let $A = \{x \in R \mid xM = 0\}$. Suppose $x,y \in A$ and $r \in R$. Then $(x-y)M = xM - yM = 0$, $(xr)M \subset xM = 0$, and $(rx)M = r(xM) = 0$. Hence, $x - y, xr, rx \in A$. This proves that A is an ideal in R.

7. Let $(a,b,c) \in V$. Then

$$(a,b,c) = (a-b)x_1 + (b-c)x_2 + cx_3 \in Rx_1 + Rx_2 + Rx_3.$$

Thus, $V = Rx_1 + Rx_2 + Rx_3$. Now if $0 = \alpha x_1 + \beta x_2 + \gamma x_3$, $\alpha,\beta,\gamma \in R$, then $\alpha + \beta + \gamma = 0$, $\beta + \gamma = 0$, $\gamma = 0$. Thus, $\alpha = 0 = \beta = \gamma$. Hence, $Rx_1 + Rx_2 + Rx_3$ is a direct sum.

9. The ring R has eight elements:

$$a_1 = 0 = \begin{pmatrix} 0 & 0 \\ 0 & 0 \end{pmatrix}, \quad a_2 = \begin{pmatrix} 1 & 0 \\ 0 & 0 \end{pmatrix}, \quad a_3 = \begin{pmatrix} 0 & 1 \\ 0 & 0 \end{pmatrix}, \quad a_4 = \begin{pmatrix} 0 & 0 \\ 0 & 1 \end{pmatrix},$$

$$a_5 = \begin{pmatrix} 1 & 1 \\ 0 & 0 \end{pmatrix}, \quad a_6 = \begin{pmatrix} 0 & 1 \\ 0 & 1 \end{pmatrix}, \quad a_7 = \begin{pmatrix} 1 & 1 \\ 0 & 1 \end{pmatrix}, \quad a_8 = \begin{pmatrix} 1 & 0 \\ 0 & 1 \end{pmatrix}.$$

Here a_1,a_2,a_4,a_5,a_6,a_8 are idempotents. Because a left ideal A is a direct summand if and only if $A = Re$ for some idempotent $e \in A$, it follows that

$$Ra_1 = \left\{\begin{pmatrix} 0 & 0 \\ 0 & 0 \end{pmatrix}\right\}, \qquad Ra_2 = \left\{\begin{pmatrix} 0 & 0 \\ 0 & 0 \end{pmatrix}, \begin{pmatrix} 1 & 0 \\ 0 & 0 \end{pmatrix}\right\},$$

$$Ra_4 = \left\{\begin{pmatrix} 0 & 0 \\ 0 & 0 \end{pmatrix}, \begin{pmatrix} 0 & 0 \\ 0 & 1 \end{pmatrix}, \begin{pmatrix} 0 & 1 \\ 0 & 0 \end{pmatrix}, \begin{pmatrix} 0 & 1 \\ 0 & 1 \end{pmatrix}\right\},$$

$$Ra_5 = \left\{\begin{pmatrix} 0 & 0 \\ 0 & 0 \end{pmatrix}, \begin{pmatrix} 1 & 1 \\ 0 & 0 \end{pmatrix}\right\},$$

and $Ra_8 = R$ are direct summands of R as a left R-module (note that the left ideal generated by a_6 does not yield another direct summand, because $Ra_6 = Ra_4$).

11. Let $a,b \in R$. Then

$$ae + bf = ae + b(f - fe) + bfe = (a + bf)e + b(f - fe).$$

Thus, $Re + Rf \subset Re + R(f - fe)$. Similarly, $Re + R(f - fe) \subset Re + Rf$. Hence, $Re + R(f - fe) = Re + Rf$. Next, let $a \in Re \cap R(f - fe)$. Then $a = xe = y(f - fe)$. This implies $xe^2 = y(f - fe)e = 0$; that is, $xe = 0$. Hence, $Re \cap R(f - fe) = (0)$, which proves that $Re \oplus R(f - fe) = Re + Rf$.

13. Let $I = Rx_1 + \cdots + Rx_n$ and $I = I^2$. Then $I = Ix_1 + \cdots + Ix_n$. So there exist $a_{ij} \in I$ such that

$$x_i = a_{i1}x_1 + \cdots + a_{in}x_n, \qquad 1 \le i \le n;$$

so

$$\begin{aligned} (1 - a_{11})x_1 + \cdots + a_{1n}x_n &= 0, \\ &\vdots \\ a_{n1}x_1 + \cdots + (1 - a_{nn})x_n &= 0. \end{aligned}$$

Because R is a commutative ring, by Cramer's rule for solving a system of linear equations, we have

$$\begin{vmatrix} 1 - a_{11} & \cdots & a_{1n} \\ \vdots & & \vdots \\ a_{n1} & \cdots & 1 - a_{nn} \end{vmatrix} x_i = 0, \qquad 1 \le i \le n.$$

Hence, $(1-z)x_i = 0$ for some $z \in I$. This yields that $(1-z)I = 0$, so $I = zI$, where $z = z^2$ and this proves that I is a direct summand.

Let $R = \mathbf{R}^{\mathbf{R}}$ be the ring of functions from $\mathbf{R}$ to $\mathbf{R}$. Let $I = \{f \in R \mid f(a) = 0$ for all but countably many $a \in \mathbf{R}\}$. Then I is an ideal of R that is not finitely generated; $I^2 = I$, but I is not a direct summand of R.

Chapter 14, Section 3

1. (a) Let $x, y \in \operatorname{Ker} f$, $a \in R$. Then $f(x-y) = f(x) - f(y) = 0$, $f(ax) = af(x) = 0$. So $x - y, ax \in \operatorname{Ker} f$, and, hence, $\operatorname{Ker} f$ is an R-submodule of M.
   (b) Let $f(x), f(y) \in \operatorname{Im} f$, $a \in R$. Then $f(x) - f(y) = f(x-y) \in \operatorname{Im} f$, and $af(x) = f(ax) \in \operatorname{Im} f$. Hence, $\operatorname{Im} f$ is an R-submodule of M.
3. Let $q \in \mathbf{Q}$. Define a mapping $q^*: \mathbf{Q} \to \mathbf{Q}$ defined by $q^*(x) = qx$. Then

   $$q^*(x+y) = q(x+y) = qx + qy = q^*(x) + q^*(y)$$

   where $x, y \in \mathbf{Q}$. Also, for $a \in \mathbf{Z}$,

   $$q^*(ax) = q(ax) = a(qx) = aq^*(x).$$

   Thus, q^* is a $\mathbf{Z}$-homomorphism of $\mathbf{Q}$ to $\mathbf{Q}$. Next, let us define a mapping f: $\mathbf{Q} \to \operatorname{Hom}_{\mathbf{Z}}(\mathbf{Q},\mathbf{Q})$ by $f(q) = q^*$. Let $q_1, q_2, x \in \mathbf{Q}$. Then

   $$(q_1q_2)^*(x) = (q_1q_2)x = q_1(q_2x) = q_1(q_2^*(x)) = q_1^*(q_2^*(x)) = (q_1^*q_2^*)(x).$$

   Also,

   $$(q_1 + q_2)^*(x) = (q_1 + q_2)x = q_1x + q_2x = q_1^*(x) + q_2^*(x).$$

   Hence, f is a ring homomorphism. f is also injective, for $q^* = 0$ implies $q^*(x) = 0$ for all $x \in \mathbf{Q}$. In particular, $q^*(1) = 0$; that is, $q = 0$. Finally, we show that f is surjective. Let $\sigma \in \operatorname{Hom}_{\mathbf{Z}}(\mathbf{Q},\mathbf{Q})$, and let $\sigma(1) = q$. Let $a/b \in \mathbf{Q}$. Then

   $$q = \sigma(1) = \sigma\left(b \cdot \frac{1}{b}\right) = b\sigma\left(\frac{1}{b}\right) \quad \text{implies} \quad \frac{q}{b} = \sigma\left(\frac{1}{b}\right).$$

   Further,

   $$\sigma\left(\frac{a}{b}\right) = a\sigma\left(\frac{1}{b}\right) = \frac{aq}{b} = q^*\left(\frac{a}{b}\right).$$

   Hence, $\sigma = q^* = f(q)$; that is, f is surjective. This proves $\mathbf{Q} \simeq \operatorname{Hom}_{\mathbf{Z}}(\mathbf{Q},\mathbf{Q})$.
5. Define a mapping f: $F^n \to V$ by $f(a_1, \dots, a_n) = a_1x_1 + \cdots + a_nx_n$. f is clearly an F-homomorphism. f is also injective, for $a_1x_1 + \cdots + a_nx_n = 0$ implies, by hypothesis, that $(a_1, \dots, a_n) = 0$. Also, since V is generated by $x_1, \dots, x_n$, f is surjective. Thus, $F^n \simeq V$.
7. Let $\phi: R \to R/I$ be an isomorphism. Let $x, u \in R$ be such that $\phi(x) = 1 + I$ and $\phi(1) = u + I$. Then for all $a \in I$, $\phi(ax) = a\phi(x) = a + I = 0$. Because ϕ is injective, $ax = 0$. Thus, $Ix = 0$. Also, $1 + I = \phi(x) = x\phi(1) = xu + I$. Then $1 - xu \in I$. Because $Ix = 0$, $(1 - xu)x = 0$; that is, $x = xux$. Let $f = xu$. Then f is an idempotent and $1 - f \in I$. Also, $If = 0$, so $I = R(1 - f) = Re$, $e = 1 - f$.

Chapter 14, Section 4

3. $\mathbf{Z}/(p_1p_2) \simeq \mathbf{Z}/(p_1) \times \mathbf{Z}/(p_2)$. If A is a $\mathbf{Z}$-submodule of $\mathbf{Z}/(p_1)$, then A is also a $\mathbf{Z}/(p_1)$-submodule of $\mathbf{Z}/(p_1)$, because $A(p_1) = (0)$. But then $A = (0)$ or $A = \mathbf{Z}/(p_1)$, because $\mathbf{Z}/(p_1)$ as $\mathbf{Z}/(p_1)$-module has no nontrivial submodules. Thus, $\mathbf{Z}/(p_1)$ and, similarly,

$\mathbf{Z}/(p_2)$ are simple $\mathbf{Z}$-modules, which proves that $\mathbf{Z}/(p_1 p_2)$ is a completely reducible $\mathbf{Z}$-module.

5. $R/M \simeq \begin{pmatrix} F & 0 \\ 0 & 0 \end{pmatrix} \oplus \begin{pmatrix} 0 & 0 \\ 0 & F \end{pmatrix}$, and each $\begin{pmatrix} F & 0 \\ 0 & 0 \end{pmatrix}$, $\begin{pmatrix} 0 & 0 \\ 0 & F \end{pmatrix}$ is a simple R-(as well as) R/M-module.

Chapter 14, Section 5

1. If xe is any element in a basis (if it exists) of Re, then $(1-e)xe = x(1-e)e = 0$ implies $1-e=0$, a contradiction.
3. Follows from the definition of basis of a free module.
5. ${}_{\mathbf{Z}}\mathbf{Q}$ free implies $\mathbf{Q} \simeq \mathbf{Z} \oplus K$ for some $\mathbf{Z}$-module $K \neq (0)$.

   $$\mathbf{Q} \simeq \mathrm{Hom}_{\mathbf{Z}}(\mathbf{Q},\mathbf{Q}) \simeq \mathrm{Hom}_{\mathbf{Z}}(\mathbf{Z} \oplus K, \mathbf{Z} \oplus K),$$

   but the latter (which is isomorphic to $\mathbf{Q}$) has nontrivial idempotents, a contradiction.
7. $Rx \simeq R$ under natural mapping, if $x \neq 0$.
9. Take in Problem 4, $M = A$, $N = B = F$, and $f =$ identity map. Then there exists a homomorphism $h: B \to A$ such that $gh = I$. It can be shown now $A = \mathrm{Ker}\, g \oplus \mathrm{Im}\, h$.

Chapter 14, Section 6

1. (a) $D(1) = 0 + 0x + 0x^2 + 0x^3$,
   $D(x) = 1 + 0x + 0x^2 + 0x^3$,
   $D(x^2) = 0 + 2x + 0x^2 + 0x^3$,
   $D(x^3) = 0 + 0x + 3x^2 + 0x^3$.

   The matrix of D with respect to $\mathscr{B}_1$ is

   $$\begin{pmatrix} 0 & 1 & 0 & 0 \\ 0 & 0 & 2 & 0 \\ 0 & 0 & 0 & 3 \\ 0 & 0 & 0 & 0 \end{pmatrix}.$$

   (b) $D(1) = 0$, $D(x+1) = 1$, $D(x+1)^2 = 2(x+1)$, $D(x+1)^3 = 3(x+1)^2$. Thus, the matrix of D with respect to $\mathscr{B}_2$ is

   $$\begin{pmatrix} 0 & 1 & 0 & 0 \\ 0 & 0 & 2 & 0 \\ 0 & 0 & 0 & 3 \\ 0 & 0 & 0 & 0 \end{pmatrix}.$$

   (c) $1 = 1 + 0(x+1) + 0(x+1)^2 + 0(x+1)^3$,
   $x = -1 + 1(x+1) + 0(x+1)^2 + 0(x+1)^3$,
   $x^2 = 1 - 2(x+1) + 1(x+1)^2 + 0(x+1)^3$,
   $x^3 = -1 + 3(x+1) - 3(x+1)^2 + 1(x+1)^3$.

   Thus, the matrix of transformation of basis $\mathscr{B}_1$ to basis $\mathscr{B}_2$ is

   $$\begin{pmatrix} 1 & -1 & 1 & -1 \\ 0 & 1 & -2 & 3 \\ 0 & 0 & 1 & -3 \\ 0 & 0 & 0 & 1 \end{pmatrix}.$$

3. (a) $\begin{bmatrix} 2 & 0 \\ 2 & 3 \end{bmatrix}$.

(b) $\begin{bmatrix} \cos\alpha & \sin\alpha \\ -\sin\alpha & \cos\alpha \end{bmatrix}$.

(c) $\begin{bmatrix} 2+\sin^2\alpha+2\sin\alpha\cos\alpha & \cos\alpha\sin\alpha-2\sin^2\alpha \\ 2\cos^2\alpha+\sin\alpha\cos\alpha & 2+\cos^2\alpha-2\sin\alpha\cos\alpha \end{bmatrix}$.

Chapter 14, Section 7

1. (a) The matrix of ϕ with respect to standard bases is

$$A = \begin{bmatrix} 2 & -1 & 3 & 1 \\ 1 & -8 & 6 & 8 \\ 1 & 2 & 0 & -2 \end{bmatrix}.$$

If R_1, R_2, R_3 denote the first, second, third rows, respectively, then by considering $\alpha R_1 + \beta R_2 + \gamma R_3 = 0$, we obtain that $-2R_1 + R_2 + 3R_3 = 0$. So R_1, R_2, and R_3 are linearly dependent. But $\alpha R_1 + \beta R_2 = 0$ implies $\alpha = 0 = \beta$. Thus, row rank $A = 2$; hence, rank $\phi = 2$.

(b) Proceeding as in (a), rank $\phi = 3$.

3. (a) Let $V = \{x \mid Bx = 0\}$, $W = \{x \mid ABx = 0\}$. Then $\dim V = n - \operatorname{rank} B = q$, say; and $\dim W = n - \operatorname{rank}(AB) = r$, say. Clearly $V \subset W$. Choose a basis $x_1, \ldots, x_q$ of V and extend (if necessary) to a basis $x_1, \ldots, x_q, x_{q+1}, \ldots, x_r$ of W. Then $Bx_{q+1}, \ldots, Bx_r$ is a basis of $U = \{Bx \mid ABx = 0,\ x \in F^n\}$. Thus $\dim U = r - q = \operatorname{rank}(B) - \operatorname{rank}(AB)$. By the rank nullity theorem, $\dim U \le n - \operatorname{rank}(A)$. This proves the desired inequality.

(b) Follows by problems 2(a) and 3(a) on choosing $B = I - A$.

5. Now $A^TAx = 0 \Rightarrow x^TA^TAx = 0 \Rightarrow (Ax)^T(Ax) = 0 \Rightarrow Ax = 0$. Thus nullity$(A^TA)$ = nullity(A) and so by the rank nullity theorem, $\operatorname{rank}(A^TA) = \operatorname{rank} A$.

Chapter 15, Section 1

1. Let $f(x) = x^3 + 3x + 2$, and 0,1,2,3,4,5,6 be the elements of $\mathbf{Z}/(7)$. By Proposition 1.3, $f(x)$ is reducible over $\mathbf{Z}/(7)$ iff $f(x)$ has a root in $\mathbf{Z}/(7)$. It is easily checked that none of the elements 0,1,2,3,4,5,6 are roots of $f(x)$. So $f(x)$ is irreducible over $\mathbf{Z}/(7)$.

3. Apply Theorem 1.7.

5. (a) The only irreducible polynomial of degree 2 is $x^2 + x + 1$.

(b) Irreducible polynomials of degree 3 are of the form $f(x) = x^3 + bx^2 + cx + 1$, where $b, c \in \mathbf{Z}/(2)$. So (b,c) can be $(1,0), (1,1), (0,0), (0,1)$. For $f(x)$ to have no root in $\mathbf{Z}/(2)$, we must have $(b,c) = (1,0)$ and $(0,1)$. So the irreducible polynomials of degree 3 are $x^3 + x^2 + 1$ and $x^3 + x + 1$.

(c) Polynomials of degree 4 with nonzero constant term are of the form

$$f(x) = x^4 + bx^3 + cx^2 + dx + 1.$$

If $f(x)$ is reducible over $\mathbf{Z}/(2)$, then $f(x)$ must be one and only one of the following:

$$f(x) = (x^2+x+1)(x^2+x+1) = x^4+x^2+1; \quad (b,c,d) = (0,1,0);$$
$$f(x) = (x^3+x^2+1)(x+1) = x^4+x^2+x+1; \quad (b,c,d) = (0,1,1);$$
$$f(x) = (x^3+x+1)(x+1) = x^4+x^3+x^2+1; \quad (b,c,d) = (1,1,0);$$

$f(x) = (x^2 + x + 1)(x + 1)^2 = x^4 + x^3 + x + 1$; $(b,c,d) = (1,0,1)$;
$f(x) = (x + 1)^4 = x^4 + 1$.

So the irreducible biquadratics are those corresponding to $(b,c,d) = (0,0,1),(1,1,1)$, and $(1,0,0)$; so they are $x^4 + x + 1$, $x^4 + x^3 + x^2 + x + 1$, and $x^4 + x^3 + 1$.

Chapter 15, Section 2

1. Let 0,1,2 be the elements of $\mathbf{Z}/(3)$. It is easily seen that none of these are roots of $x^2 - x - 1$. Hence, $x^2 - x - 1$ is irreducible over $\mathbf{Z}/(3)$.
   Consider the set $K = \{a + bu | a,b \in \mathbf{Z}/(3)\}$, where $u^2 - u - 1 = 0$. Then K is a field containing $\mathbf{Z}/(3)$ that contains one root and, hence, both the roots of $f(x)$.
3. The required extension is the field
   $K = \{a + bu + cu^2 + du^3 | a,b,c,d \in \mathbf{Q}\}$,
   where $u^4 - 2 = 0$. We can write $u = \sqrt[4]{2}$.
5. We invoke the Kronecker theorem (Corollary 2.4) to get successive fields $K_1 \subset K_2 \subset \cdots \subset K_m$ such that K_1 contains a root of $f_1(x)$, K_2 contains a root of $f_2(x)$, and so on.

Chapter 15, Section 3

1. $F \subset K \subset E$. Now α is algebraic over $K \Rightarrow \alpha$ is a root of an irreducible polynomial, say
   $f(x) = a_0 + a_1 + \cdots + a_m x^m, \quad a_i \in K, i = 0,\ldots,m.$
   Consider the field $L = F(a_0,\ldots,a_m)$. Because K is algebraic over F, each a_i is algebraic over F; so L is a finite extension (and so algebraic) of F (Theorem 3.6). Now α is algebraic over $L \Rightarrow L(\alpha)$ is a finite extension of $L \Rightarrow [L(\alpha):L] < \infty$, so, by Theorem 2.1,
   $[L(\alpha):F] = [L(\alpha):L][L:F] < \infty.$
   This implies each element of $L(\alpha)$ is algebraic over F, and, in particular, α is algebraic over F.
3. (a) Let $u = \sqrt{2} + 5$. Then $(u - 5)^2 = 2$; that is, $u^2 - 10u + 23 = 0$; so the minimal polynomial of $\sqrt{2} + 5$ over $\mathbf{Q}$ is $x^2 - 10x + 23$.
   (b) $u = 3\sqrt{2} + 5 \Rightarrow (u - 5)^2 = 18$; that is, $u^2 - 10u + 7 = 0$; so the minimal polynomial is $x^2 - 10x + 7$.
   (c) $u = \sqrt{-1 + \sqrt{2}} \Rightarrow (u^2 + 1) = \sqrt{2} \Rightarrow u^4 + 2u^2 - 1 = 0$; so the minimal polynomial is $x^4 + 2x^2 - 1$.
   (d) $u = \sqrt{2} - 3\sqrt{3} \Rightarrow u^2 = 2 + 27 - 6\sqrt{6} \Rightarrow (u^2 - 29)^2 = 36 \cdot 6 \Rightarrow u^4 - 58u^2 + 625 = 0$. So the minimal polynomial is $x^4 - 58x^2 + 625$.
5. $F \subset E$, $a,b \in E$. Now $[F(a,b):F] = [F(a,b):F(a)] \cdot [F(a):F]$, which shows that m divides $[F(a,b):F]$. Similarly, n divides $[F(a,b):F]$. Because $(m,n) = 1$, $mn | [F(a,b):F]$. But $[F(a,b):F] = [F(a,b):F(a)] \cdot [F(a):F]$ shows that $[F(a,b):F] \le mn$. Hence, $[F(a,b):F] = mn$.
7. Let $[F(a):F] = n$ and $[F(a^2):F] = m$. Obviously $m \le n$. Let $\phi(x) = a_0 + a_1 x + \cdots + x^m$ be the minimal polynomial of a^2 over F. Then $a_0 + a_1 a^2 + \cdots + a_m a^{2m} = 0 \Rightarrow a$ satisfies the polynomial $a_0 + a_1 x^2 + \cdots + a_m x^{2m} \in F[x]$. So $n | 2m$. Because n is odd, $n | m$. But $m \le n$. Hence, $m = n$.

9. We have a tower of fields

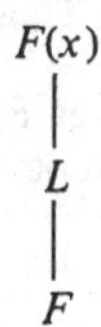

Let $y = p(x)/q(x) \in L, p(x), 0 \neq q(x) \in F[x], y \notin F$. Then $p(x) - yq(x) = 0$. This shows that x is algebraic over L. Thus, $F(x)$ is algebraic over L.

11. $\left[\omega - \cos\frac{2\pi}{n}\right]^2 = -\sin^2\frac{2\pi}{n}$. So $\omega^2 - 2\omega\cos\frac{2\pi}{n} + \cos^2\frac{2\pi}{n} = -1 + \cos^2\frac{2\pi}{n}$, yielding $\omega^2 - 2\omega\cos\frac{2\pi}{n} + 1 = 0$. Thus ω satisfies an irreducible polynomial of degree 2 over $\mathbf{Q}(u)$.

Chapter 16, Section 1

1. (a) Let $1 \neq \omega$ be a cube root of 1. Then the splitting field of $x^3 - 1$ over $\mathbf{Q}$ is K. $K = \mathbf{Q}(\omega)$, where $\omega = (-1 + \sqrt{3}i)/2$; that is, $K = \mathbf{Q}(\sqrt{3}i)$.
   (b) $x^4 + 1$ is irreducible over $\mathbf{Q}$, because $(x+1)^4 + 1 = x^4 + 4x^3 + 6x^2 + 4x + 2$ is irreducible over $\mathbf{Q}$ [take $p = 2$ in the Eisenstein criterion]. Also, $\omega = \cos(\pi/4) + i\sin(\pi/4) = (1/\sqrt{2})(1 + i)$ is a root of $x^4 + 1$, and the other roots are

$$\omega^3 = \cos\frac{3\pi}{4} + i\sin\frac{3\pi}{4} = \frac{1}{\sqrt{2}}(-1+i),$$

$$\omega^5 = \frac{1}{\sqrt{2}}(-1-i),$$

$$\omega^7 = \frac{1}{\sqrt{2}}(1-i),$$

   so the splitting field is $\mathbf{Q}(\sqrt{2}, i)$.
   (c) The splitting field is $\mathbf{Q}(\sqrt{3}i)$.
   (d) The splitting field is $\mathbf{Q}(\sqrt{2}, \sqrt[3]{3}, \omega)$, where $\omega = (-1+\sqrt{3}i)/2$. In (a), $[K:\mathbf{Q}] = 2$; in (b) $[K:\mathbf{Q}] = 4$; in (c) $[K:\mathbf{Q}] = 2$; in (d) $[K:\mathbf{Q}] = 12$.

3. $f(x) = x^3 + ax + b$ must be irreducible over $\mathbf{Q}$, otherwise the degree of the splitting field of $f(x)$ over $\mathbf{Q}$ is 1 or 2. So let $f(x)$ be irreducible over $\mathbf{Q}$, and let $f(x) = (x - \alpha_1)(x - \alpha_2)(x - \alpha_3)$. Then the splitting field of $f(x)$ over $\mathbf{Q}$ is $K = \mathbf{Q}(\alpha_1, \alpha_2, \alpha_3)$. From the equation

$$\alpha_1 + \alpha_2 + \alpha_3 = 0, \qquad \alpha_1\alpha_2 + \alpha_2\alpha_3 + \alpha_3\alpha_1 = a, \qquad \alpha_1\alpha_2\alpha_3 = -b, \tag{1}$$

we calculate

$$D = [(\alpha_1 - \alpha_2)(\alpha_2 - \alpha_3)(\alpha_3 - \alpha_1)]^2 = -4a^3 - 27b^2,$$

which is called the discriminant of $x^3 + ax + b$. If $\sqrt{D} \in \mathbf{Q}$, then the splitting field of $f(x)$ is $K = \mathbf{Q}(\alpha)$, where α is any one of $\alpha_1, \alpha_2, \alpha_3$. For suppose $\alpha = \alpha_3$. Then

$$(\alpha_1 - \alpha_2)(\alpha_1 - \alpha_3)(\alpha_2 - \alpha_3) \in \mathbf{Q} \Rightarrow (\alpha_1 - \alpha_2)[\alpha_1\alpha_2 - \alpha(\alpha_1 + \alpha_2) + \alpha^2] \in \mathbf{Q};$$

that is, $\alpha_1 - \alpha_2 \in \mathbf{Q}(\alpha)$, from (1). Also by (1), $\alpha_1 + \alpha_2 \in \mathbf{Q}(\alpha)$, so $\alpha_1, \alpha_2, \alpha_3 \in \mathbf{Q}(\alpha)$;

that is, $\mathbf{Q}(\alpha)$ is the splitting field, and $[\mathbf{Q}(\alpha):\mathbf{Q}]$ = degree of the irreducible polynomial $f(x) = 3$. If $\sqrt{D} \notin \mathbf{Q}$, then the splitting field K is $\mathbf{Q}(\sqrt{D},\alpha)$ and $[K:\mathbf{Q}] = 6$. Thus, $D = -4a^3 - 27b^2$ must be a square in $\mathbf{Q}$ in order that $[K:\mathbf{Q}] = 3$.

5. We use the following result: If u satisfies a polynomial of degree m over F, then $[F(u):F] \leq m$. Let $f(x) \in F[x]$ be of degree n over F. Let $\alpha_1,\alpha_2,\ldots,\alpha_n$ be the roots of $f(x)$ in E. Then

$$[F(\alpha_1):F] \leq n,$$
$$[F(\alpha_1,\alpha_2):F(\alpha_1)] \leq (n-1),$$
$$[F(\alpha_1,\ldots,\alpha_n):F(\alpha_1,\ldots,\alpha_{n-1})] \leq 2.$$

So

$$[F(\alpha_1,\alpha_2,\ldots,\alpha_n):F] = [F(\alpha_1,\ldots,\alpha_n):F(\alpha_1,\ldots,\alpha_{n-1})]\,[F(\alpha_1,\ldots,\alpha_{n-1}):F(\alpha_1,\ldots,\alpha_{n-2})] \cdots [F(\alpha_1):F] < 2\cdot 3\cdots n = n!.$$

7. Let $\alpha \in E$ be a root of $p(x)$. Let β be another root of $p(x)$ in some extension of F. Then since $F[x]/(p(x)) \simeq F(\theta)$, where θ is any root of $p(x)$, it follows that $F(\alpha) \simeq F(\beta)$. We can regard $f(x)$ as a polynomial over $F(\alpha)$ as well as over $F(\beta)$. Then E is also a splitting field of $f(x)$ regarded as a polynomial over $F(\alpha)$. Let K be a splitting field regarded as a polynomial over $F(\beta)$. Then by the uniqueness of splitting fields (up to isomorphism), we have $E \simeq K$. Further, because $f(x)$ as a polynomial over F splits in E, it follows that splitting field of $f(x)$ over $F(\beta)$ must be contained in $E(\beta)$; that is, $K \subset E(\beta)$. However, K contains all the roots of $f(x)$, including β. Thus, $K \supset E(\beta)$. Therefore, $K = E(\beta)$. Further,

$$[E:F] = [E:F(\alpha)][F(\alpha):F] = [K:F(\beta)][F(\beta):F] = [K:F] = [E(\beta):F].$$

Therefore, $E = E(\beta)$, so $\beta \in E$.

9. $f(x)$ is irreducible by Eisenstein's criterion. Choose any two roots, say, $\alpha = \sqrt{1+\sqrt{3}}$, $\beta = \sqrt{1-\sqrt{3}}$. Since the minimal polynomial of α and β is the same, $\mathbf{Q}(\alpha) \simeq \mathbf{Q}(\beta)$.

Chapter 16, Section 2

1. (a) The minimal polynomial of $\sqrt{-2}$ over $\mathbf{Q}$ is $x^2 + 2$, its roots in $\mathbf{C}$ are $\sqrt{-2}, -\sqrt{-2}$, and both $\in \mathbf{Q}(\sqrt{-2})$. Hence, $\mathbf{Q}(\sqrt{-2})$ is a normal extension of $\mathbf{Q}$.
   (b) The minimal polynomial of $5\sqrt{7}$ is $x^2 - 175$, whose roots are $5\sqrt{7}$ and $-5\sqrt{7}$, and they lie in $\mathbf{Q}(5\sqrt{7})$. So $\mathbf{Q}(5\sqrt{7})$ is normal over $\mathbf{Q}$.
   (c) By a similar argument, $\mathbf{Q}(\sqrt{-1})$ is normal over $\mathbf{Q}$.
   (d) If x is not algebraic over $\mathbf{Q}$, then $\mathbf{Q}(x)$ is not an algebraic extension of $\mathbf{Q}$. So $\mathbf{Q}(x)$ is not a normal extension of $\mathbf{Q}$.

3. E is a normal extension of $F \Rightarrow E$ is a splitting field of a family of polynomials over $F \Rightarrow E$ is a splitting field of a family of polynomials over K, because $F \subset K$, and any polynomial over F can also be considered as a polynomial over K. Hence, E is normal over K.

   Take $K = \mathbf{Q}(\sqrt[3]{2})$ and $E = \mathbf{Q}(\sqrt[3]{2},\omega)$. Then $\mathbf{Q} \subset \mathbf{Q}(\sqrt[3]{2}) \subset \mathbf{Q}(\sqrt[3]{2},\omega)$, where $\omega = (-1+\sqrt{3}i)/2$. Clearly, $\mathbf{Q}(\sqrt[3]{2})$ is not normal over $\mathbf{Q}$, but $\mathbf{Q}(\sqrt[3]{2},\omega)$ is normal over $\mathbf{Q}(\sqrt[3]{2})$.

5. Let E be a finite extension of a finite field F, and $[E:F] = n$. Then E is also a finite field, and E^*, the multiplicative group of E, is cyclic and generated by u, say, such that $u^{n-1} = 1$. Thus, u is a root of $x^n - x \in F[x]$. All the other roots of $x^n - x$ are 0 and powers of u, and so are in E. Hence, E is a splitting field of $x^n - x \in F[x]$, and, therefore, E is normal extension of F.

7. $x^3 - x - 1$ is irreducible and has only one real root (consider graph of $y = x^3 - x - 1$ or refer to Chapter 16, Section 1, Problems 3 and 4). Thus $\mathbf{Q}(\alpha)$ is not a normal extension because its minimal polynomial $x^3 - x - 1$ does not split into linear factors over $\mathbf{Q}$.
9. The smallest normal extension of $\mathbf{Q}(2^{1/4}, 3^{1/4})$ is the splitting field of $(x^4 - 2)(x^4 - 3)$, namely, $\mathbf{Q}(2^{1/4}, 3^{1/4}, i)$.
11. Follows from the fact that any field E is normal over F iff each embedding $\sigma: E \to \bar{F}$ that keeps elements of F fixed sends E to E.

Chapter 16, Section 3

1. Let $f(x) = a_0 + a_1x + a_2x^2 + \cdots$ and $g(x) = b_0 + b_1x + b_2x^2 + \cdots$. Then

$$f(x) + g(x) = (a_0 + b_0) + (a_1 + b_1)x + \cdots + (a_i + b_i)x^i + \cdots$$
$$= \sum_{i=0}^{r} (a_i + b_i)x^i.$$

By definition of the derivative,

$$(f(x) + g(x))' = \sum_{i=1}^{r} i(a_i + b_i)x^{i-1} = \sum_{i=1}^{r} ia_ix^{i-1} + \sum_{i=1}^{r} ib_ix^{i-1}$$
$$= f'(x) + g'(x).$$

3. Suppose $f(x) \in F[x]$ has a root α of multiplicity n. Then

$$f(x) = (x - \alpha)^n\phi(x), \qquad \phi(\alpha) \neq 0.$$

If $\phi(x) \in F$, then $f^{(k)}(\alpha) = 0$, $k = 1,\ldots,n - 1$, and $f^{(n)}(\alpha) = n! \neq 0$. Suppose $\phi(x)$ is of degree ≥ 1. Then

$$f^{(1)}(x) = n(x - \alpha)^{n-1}\phi(x) + (x - \alpha)^n\phi'(x) = (x - \alpha)^{n-1}\phi_1(x), \qquad \phi_1(\alpha) \neq 0$$
$$f^{(2)}(x) = (x - \alpha)^{n-2}\phi_2(x), \qquad \phi_2(\alpha) \neq 0,$$
$$\vdots$$
$$f^{(n-1)}(x) = (x - \alpha)\phi_{n-1}(x), \qquad \phi_1(\alpha) \neq 0,$$
$$f^{(n)}(x) = \phi_{n-1}(x) + (x - \alpha)\phi'_{n-1}(x).$$

Thus, $f^{(n-1)}(\alpha) = 0$ and $f^{(n)}(\alpha) = \phi_{n-1}(\alpha) \neq 0$. The converse is obtained by retracing the above steps.

Chapter 16, Section 4

1. The mapping $\phi: x \mapsto x^p$ is an endomorphism of the ring F, since $(x + y)^p = x^p + y^p$ and $(xy)^p = x^py^p$. ϕ is also 1-1, since Ker ϕ, an ideal of F, must be (0). Because $|F|$ is finite, ϕ is 1-1 and onto. That is, ϕ is an automorphism of F. Thus, if $a \in F$, $\exists b \in F$ such that $b^p = a$. The unique b is called the pth root of a and written $\sqrt[p]{a}$.
3. By Theorem 4.3, $GF(2^3)$, the Galois field with eight elements, is the splitting field of $x^8 - x$ over $GF(2)$. Also, the multiplicative group of $GF(2^3)$ is cyclic, of order 7, and generated by any element $\delta \neq 1$. So $GF(2^3) = \{0, \delta, \ldots, \delta^7\}$, where $1 + \delta + \cdots + \delta^6 = 0$, $\delta^8 = \delta$.

$$GF(13) = \text{the set of integers modulo } 13 = (0, 1, 2, \ldots, 12),$$

which is generated by 2.

5. The field F of four elements is the set $\{0,1,\alpha,1+\alpha\}$, where $1+\alpha+\alpha^2=0$. An irreducible polynomial of degree 2 over F is $x^2+\alpha x+1$. An irreducible polynomial of degree 3 over F is $x^3+x^2+\alpha x+\alpha+1$. An irreducible polynomial of degree 4 over F is $x^4+\alpha x^3+\alpha x^2+(\alpha+1)x+1$.
7. Let $|F|=p^n$. First assume that $p\neq 2$. For any fixed $a\in F$, let $A=\{a-x^2|x\in F\}$ and $B=\{y^2|y\in F\}$. Note that $|A|=(p^n-1)/2+1=(p^n+1)/2$. Also, $|B|=(p^n+1)/2$. This implies that $A\cap B\neq\emptyset$. Hence, a can be written as a sum of squares. Next, we assume that $p=2$. Then the mapping $\phi(x)=x^2$ of F into F is bijective. So if $a\in F$, then there exists $x\in F$ such that $a=x^2$.
9. If $f(x)=x^p-x-1$, then $f'(x)\neq 0$. So the roots of $f(x)$ are distinct. Further, if α is a root, then $\alpha+1$ is also a root. Thus the p roots may be written as $\alpha,\alpha+1,\ldots,\alpha+(p-1)$. If $\alpha\in\mathbf{Z}_p$, then all roots lie in $\mathbf{Z}_p$ and so 0 must be a root of $f(x)$, which is not true. This shows $\mathbf{Z}_p(\alpha)$ $(\neq\mathbf{Z}_p)$ is a splitting field of $f(x)$ over $\mathbf{Z}_p$ and $[\mathbf{Z}_p(\alpha):\mathbf{Z}_p]=p$, proving $f(x)$ is irreducible over $\mathbf{Z}_p$.

Chapter 16, Section 5

1. $\sqrt{3}-\sqrt{2}=1/(\sqrt{3}+\sqrt{2})\in\mathbf{Q}(\sqrt{3}+\sqrt{2})$, so $\sqrt{3},\sqrt{2}\in\mathbf{Q}(\sqrt{3}+\sqrt{2})$; that is, $\mathbf{Q}(\sqrt{3},\sqrt{2})\subset\mathbf{Q}(\sqrt{3}+\sqrt{2})$. Trivially, $\mathbf{Q}(\sqrt{3}+\sqrt{2})\subset\mathbf{Q}(\sqrt{3},\sqrt{2})$. Hence, $\mathbf{Q}(\sqrt{2},\sqrt{3})=\mathbf{Q}(\sqrt{2}+\sqrt{3})$.
3. Put $\theta=\sqrt{2}+\omega$. Then $\theta^2=2+2\sqrt{2}\omega-(\omega+1)\in\mathbf{Q}(\theta)$. This implies $\omega(2\sqrt{2}-1)\in\mathbf{Q}(\theta)$. So $(2\sqrt{2}-1)^3\in\mathbf{Q}(\theta)$; that is, $8\cdot 2\sqrt{2}-3\cdot 4\cdot 2+3\cdot 2\sqrt{2}-1\in\mathbf{Q}(\theta)$; that is, $\sqrt{2}\in\mathbf{Q}(\theta)$. This together with $\omega(2\sqrt{2}-1)\in\mathbf{Q}(\theta)$ gives $\omega\in\mathbf{Q}(\theta)$. Hence, $\mathbf{Q}(\sqrt{2}+\omega)=\mathbf{Q}(\sqrt{2},\omega)$.
5. Let F be a field with p^m elements and E be an extension of F having p^n elements. Then $E=F(\alpha)$ where $\alpha\in E$ and so $\alpha^{p^n}-\alpha=0$. This implies α is a separable element, and hence $F(\alpha)$ is a separable extension of F.
7. x^p-x-1 or any of its factors cannot have repeated roots.

Chapter 17, Section 1

1. (a) $S_1=\{1,\sigma_2\}$, $\sigma_2:\sqrt[3]{2}\to\sqrt[3]{2}\omega^2$, $\omega\to\omega^2$. The basis of E over $\mathbf{Q}$ is $(1,\sqrt[3]{2},\sqrt[3]{4},\omega,\omega\sqrt[3]{2},\omega\sqrt[3]{4})$. Let

$$\alpha=x_1+x_2\sqrt[3]{2}+x_3\sqrt[3]{4}+x_4\omega+x_5\omega\sqrt[3]{2}+x_6\omega\sqrt[3]{4}\in\mathbf{Q}(\sqrt[3]{2},\omega),\qquad x_i\in\mathbf{Q},\ i=1,2,\ldots,6.$$

Then $\sigma_2(\alpha)=\alpha$ gives

$$x_1+x_2(\sqrt[3]{2}\omega^2)+x_3(\sqrt[3]{2}\omega^2\cdot\sqrt[3]{2}\omega^2)+x_4\omega^2+x_5\omega^2\sqrt[3]{2}\omega^2+x_6\omega^2(\sqrt[3]{2}\omega^2\cdot\sqrt[3]{2}\omega^2)$$
$$=x_1+x_2\sqrt[3]{2}+x_3\sqrt[3]{4}+x_4\omega+x_5\omega\sqrt[3]{2}+x_6\omega\sqrt[3]{4};$$

that is,

$$-x_4+\sqrt[3]{2}(-x_2-x_2)+\sqrt[3]{4}(x_6-x_3)+\omega(-x_4-x_4)+\omega\sqrt[3]{2}(-x_2)+\omega\sqrt[3]{4}(-x_6+x_3)=0,$$

which gives

$$x_4=0=x_2,\qquad x_3=x_6.$$

So

$$\alpha=x_1+x_3\sqrt[3]{4}+x_3\omega\sqrt[3]{4}+x_5\omega\sqrt[3]{2}.$$

Thus,

$\alpha = x_1 + x_3\sqrt[3]{4}(-\omega^2) + x_5\omega\sqrt[3]{2} \in \mathbf{Q}(\omega\sqrt[3]{2})$.

Conversely, if $\alpha \in \mathbf{Q}(\omega\sqrt[3]{2})$, $\sigma_2(\alpha) = \alpha$. Hence, $E_{S_1} = \mathbf{Q}(\omega^2\sqrt[3]{4})$.

(b) As in (a) we can show that if

$\alpha = x_1 + x_2\sqrt[3]{2} + x_3\sqrt[3]{4} + x_4\omega + x_5\omega\sqrt[3]{2} + x_6\omega\sqrt[3]{4} \in E_{S_2}$,

then $x_3 = 0 = x_4$, and $x_2 = x_5$. Therefore,

$\alpha = x_1 + x_2\sqrt[3]{2}(1+\omega) + x_6\omega\sqrt[3]{4} = x_1 - x_2\omega^2\sqrt[3]{2} + x_6\omega\sqrt[3]{4} \in \mathbf{Q}(\omega^2\sqrt[3]{2})$.

Also, $\mathbf{Q}(\omega^2\sqrt[3]{2}) \subset E_{S_2}$. Thus, $E_{S_2} = \mathbf{Q}(\omega^2\sqrt[3]{2})$.

(c) With α as in (a), $\sigma_4(\alpha) = \alpha$ gives $x_4 = x_5 = x_6 = 0$. Therefore, $E_{S_3} = \mathbf{Q}(\sqrt[3]{2})$.

(d) As in (a), we can show that if

$\alpha = x_1 + x_2\sqrt[3]{2} + x_3\sqrt[3]{4} + x_4\omega + x_5\omega\sqrt[3]{2} + x_6\omega\sqrt[3]{4} \in E_{S_4}$,

then $x_2 = 0 = x_3 = x_5 = x_6$; so $\alpha = x_1 + x_4\omega \in \mathbf{Q}(\omega)$. Hence, $E_{S_4} = \mathbf{Q}(\omega)$.

Chapter 17, Section 2

1. (a) $G(K/\mathbf{Q}) = \{1,\sigma_1,\sigma_2,\sigma_3\}$, where

$\sigma_1: \sqrt{3} \to -\sqrt{3}, \sqrt{5} \to \sqrt{5}$,
$\sigma_2: \sqrt{3} \to -\sqrt{3}, \sqrt{5} \to -\sqrt{5}$,
$\sigma_3: \sqrt{3} \to \sqrt{3}, \sqrt{5} \to -\sqrt{5}$.

(b) $G(\mathbf{Q}(\alpha)/\mathbf{Q}) = \{1,\sigma\}$, where $\sigma: \alpha \to \alpha^2$.

(c) $x^4 - 3x^2 + 4 = 0$. So

$$x^2 = \frac{3 \pm \sqrt{-7}}{2} = \frac{3 \pm 2\sqrt{-7/4}}{2} = \frac{(\sqrt{7/2} \pm \sqrt{-1/2})^2}{2};$$

hence, $x = \pm(\sqrt{7} \pm i)/2$. Thus, the roots of $x^4 - 3x^2 + 4$ are

$$\frac{\sqrt{7}+i}{2}, \quad \frac{-\sqrt{7}-i}{2}, \quad \frac{\sqrt{7}-i}{2}, \quad \frac{-\sqrt{7}+i}{2}.$$

The splitting field is $E = \mathbf{Q}(\sqrt{7}, i)$, and $[E:\mathbf{Q}] = 4$. The Galois group obtained exactly before is the set of automorphisms $\{1,\sigma_1,\sigma_2,\sigma_3\}$, where

$\sigma_1: \sqrt{7} \to -\sqrt{7}, \quad i \to i$,
$\sigma_2: \sqrt{7} \to -\sqrt{7}, \quad i \to -i$,
$\sigma_3: \sqrt{7} \to \sqrt{7}, \quad i \to -i$.

3. Write down a composition series of the Galois group $G(\mathbf{Q}(u)/\mathbf{Q})$ by considering Sylow 2-subgroups.

Chapter 18, Section 2

1. Let ω be a primitive nth root of unity in F. Then $1, \omega, \omega^2, \ldots, \omega^{n-1}$ are the n distinct roots of $f(x) = x^n - 1 \in F[x]$. So $f(x)$ does not possess multiple roots in F, and so $f'(x) = nx^{n-1} \neq 0$. This implies that either char $F = 0$ or char $F \nmid n$.

3. Let char $F = p$. Suppose $x^p - b \in F[x]$ is reducible over F. Let α be a root of $x^p - b$ in its splitting field E. Then $x^p - b = (x - \alpha)^p$. Thus, by hypothesis, some factor $(x - \alpha)^r$ of $(x - \alpha)^p$ belongs to $F[x]$. This implies $-r\alpha \in F$, yielding $\alpha \in F$. Therefore, $E = F$.
   Now let char $F \neq p$. If α is a root of $x^p - b$ in its splitting field E, then $\alpha, \alpha\omega, \ldots, \alpha\omega^{p-1}$ are the p roots of $x^p - b$, where ω is a primitive pth root of unity in E. Suppose $f(x) \in F[x]$ is a factor of $x^p - b$, where degree $f(x) = r > 1$. The product of the roots of $f(x)$ is of the form $a = \alpha^r\omega_1$, $\omega_1^p = 1$. Thus, $a = \alpha^r\omega_1 \in F$, so $a^p = \alpha^{rp} = b^r, 0 < r < p$. Choose integers s,t such that $rs + pt = 1$. It then follows that $b = b^{rs}b^{pt} = a^{ps}b^{pt} = (a^sb^t)^p = \beta^p$, say, where $\beta \in F$. Therefore, β is a root of $x^p - b$; so $E = F(\omega)$.

Chapter 18, Section 3

1. (a) $F(x) = x^5 - 9x + 3$ has a real root in the interval $(0,1)$ and another real root in $(1,2)$. (Use the intermediate value theorem of analysis.) By Descartes's rule of signs the number of positive roots ≤ 2, and the number of negative roots $= 1$. So $f(x)$ has exactly two nonreal roots. By Theorem 3.6 the Galois group of $f(x)$ is isomorphic to S_5. Hence, by Theorem 3.2, $f(x)$ is not solvable by radicals.
   (b) Let $p(x) = 2x^5 - 5x^4 + 5$. If we put $y = 2x$, then $p(x)$ transforms to $f(y) = \frac{1}{16}(y^5 - 5y^4 + 80)$. Let $g(y) = y^5 - 5y^4 + 80$. Then the roots of $f(y)$ or those of $g(y)$ are twice the roots of $p(x)$. So the solvability of $p(x)$ by radicals is equivalent to the solvability by radicals of $g(y)$. The purpose of this transformation was to enable us to be able to apply Theorem 3.6, which holds for monic irreducible polynomials. It is clear, by Eisenstein's criterion, that $g(y)$ is irreducible over **Q**.
   Further, by Descartes's rule of signs, the number of positive (negative) real roots is ≤ 2 (1). Therefore, the total number of real roots ≤ 3. But by the intermediate value theorem, $g(y)$ has three roots, one in each of the intervals $(-2,-1)$, $(2,3)$, and $(4,5)$. Hence, there are exactly three real and two nonreal roots of $g(y)$ (and so of $p(x)$). Thus, by Theorem 3.6 the Galois group of $p(x)$ is S_5. This proves $p(x)$ is not solvable by radicals.
   (c) $f(x) = x^5 - 8x + 6$. As explained before, there are at most three real roots. There is one real root in $(0,1)$ and one in $(1,2)$. Hence, there are exactly two nonreal roots of $f(x)$ in **C**, which shows that $f(x)$ is not solvable by radicals over **Q**.
   (d) $f(x) = x^5 - 4x + 2$. By Descartes's rule of signs the number of real roots ≤ 3. There is one real root in the interval $(0,1)$ and one in $(1,2)$. Hence there are exactly two nonreal roots of $f(x)$ in **C**, which shows that $f(x)$ is not solvable by radicals over **Q**.

Chapter 18, Section 4

(a) $x_1^3 + x_2^3 + x_3^3 = s_1^3 - 3s_1s_2 + 3s_3$, where $s_1 = x_1 + x_2 + x_3$, $s_2 = x_1x_2 + x_2x_3 + x_3x_1$, $s_3 = x_1x_2x_3$.
(b) Let $s = x_1^2x_2^2 + x_2^2x_3^2 + x_3^2x_1^2$. Then $s_2^2 = (x_1x_2 + x_2x_3 + x_3x_1)^2 = s + 2s_3s_1$; that is, $s = s_2^2 - 2s_3s_1$.
(c)

$$(x_1^2 + x_2^2)(x_2^2 + x_3^2)(x_3^2 + x_1^2) = \prod_{i=1}^{3}(x_1^2 + x_2^2 + x_3^2 - x_i^2)$$

$$= \prod_{i=1}^{3}(t - x_i^2) \qquad (\text{where } t = x_1^2 + x_2^2 + x_3^2)$$

$$= t^3 - t^2(x_1^2 + x_2^2 + x_3^2) + t(x_1^2x_2^2 + x_2^2x_3^2 + x_3^2x_1^2) - x_1^2x_2^2x_3^2$$
$$= t^3 - t^3 + t(s_2^2 - 2s_1s_3) - s_3^2$$
$$= (s_1^2 - 2s_2)(s_2^2 - 2s_1s_3) - s_3^2.$$

(d)

$$(x_1 + x_2)(x_2 + x_3)(x_3 + x_1) = \prod_{i=1}^{3}(s_1 - x_i)^3$$
$$= s_1^3 - s_1^2(x_1 + x_2 + x_3) + s_1(x_1x_2 + x_1x_3 + x_2x_3) - x_1x_2x_3$$
$$= s_1^3 - s_1^3 + s_1s_2 - s_3 = s_1s_2 - s_3.$$

Chapter 18, Section 5

1. Let $\alpha =$ the angle $2\pi/15$; that is, α is 24°. But $\sin 24° = \frac{1}{8}(\sqrt{15} + \sqrt{3} - \sqrt{10 - 2\sqrt{5}})$, which is a constructible number by Theorem 5.2. Hence, by Remark 5.9, the angle $2\pi/15$ is constructible; that is, it can be trisected by ruler and compass.
3. $\cos 18° = \sqrt{10 + 2\sqrt{5}}/4$; so by Theorem 5.2, cos 18° is a constructible number. Hence, by Remark 5.9, angle 18° is constructible; that is, angle 54° can be trisected by ruler and compass.
5. (i), (iii), and (v).

Chapter 19, Section 1

1. Apply Theorem 1.1 and Schur's lemma.

Chapter 19, Section 2

1. We use induction on n and Theorem 2.6. Assume the result is true for $n = m$. Then

$$\frac{R_1 \oplus \cdots \oplus R_m \oplus R_{m+1}}{(0) \oplus \cdots \oplus (0) \oplus R_{m+1}} \simeq R_1 \oplus \cdots \oplus R_m;$$

so $R_1 \oplus \cdots \oplus R_m$ and R_{m+1} noetherian imply, by Theorem 2.6, that $R_1 \oplus \cdots \oplus R_{m+1}$ is noetherian.
3. Follows from Theorem 2.3.
5. See Problem 3 in Section 1 of Chapter 12.
7. Assume $Rc \cap A = (0)$. Let $0 = a_1c + a_2c^2 + \cdots + a_mc^m$, $a_i \in A$. Then $a_1 + a_2c + \cdots + a_mc^{m-1} = 0$; so $-a_1 = a_2c + \cdots + a_mc^{m-1} \in A \cap Rc = (0)$. Thus, $a_1 = 0$. Continuing like this, we obtain $a_i = 0$ for all $i = 1,\dots,m$. Hence, $\Sigma_{i=1}^{\infty} Ac^i$ is a direct sum of left ideals Ac^i. This gives an infinite properly ascending chain of left ideals: $Ac \subset Ac \oplus Ac^2 \subset Ac \oplus Ac^2 \oplus Ac^3 \subset \cdots$, a contradiction. Thus, $A \cap Rc \neq (0)$.
9. Any descending chain $A_1 \supset A_2 \supset A_3 \supset \cdots$ of left ideals of the ring R is also a descending chain of subspaces of the vector space R over D. Because $[R:D] < \infty$, there must exist a positive integer n such that $A_n = A_{n+1} = \cdots$. Hence, R is left artinian.
11. If possible, let $\oplus\Sigma_{i=1}^{\infty} A_i$ be the direct sum of left ideals in R. Then $\oplus\Sigma_{i=1}^{\infty} A_i \supset \oplus\Sigma_{i=2}^{\infty} A_i \supset \oplus\Sigma_{i=3}^{\infty} A_i \supset \cdots$ is an infinite properly descending chain of left ideals, a contradiction.

Chapter 19, Section 3

1. By the Wedderburn–Artin theorem $R = R_1 \oplus \cdots \oplus R_n$, where each R_k is a matrix ring over a division ring. If I is an ideal in R, then $I = I_1 \oplus \cdots \oplus I_n$, where I_k is an ideal in R_k. This implies $I_k = (0)$ or R_k, since R_k is a simple ring. Hence, $R_k/I_k \simeq R_k$ or (0). After renumbering if necessary, we may assume $R/I \simeq R_1 \times \cdots \times R_s$, where each R_i is a nonzero matrix ring over a division ring. Clearly, the R_i's do not possess nonzero nilpotent ideals. Hence, R/I cannot possess nonzero nilpotent ideals.
3. Let A be a nil ideal. Because R is artinian, A is nilpotent. Thus, $N \subset U$. Next, let B be a nilpotent left ideal and suppose $B^m = 0$. Then $(B + BR)^m \subseteq B^mR + B^m$ (by induction). Hence, $B + BR$ is a nilpotent ideal. Thus, $B + BR \subset V$ or $B \subset V$. This implies $U \subset V$. Finally, if C is a nilpotent right ideal, then, as before, $C + RC$ is a nilpotent ideal. This implies $C + RC$ is a nil ideal, so $C \subset N$. This yields $V \subset N$. Hence, $N \subset U \subset V \subset N$. Therefore, $N = U = V$.
4. *Hint:* $R = \{ax + by + c \mid a,b,c \in \mathbf{Z}\}$, subject to $x^2 = 1$, $xy = y$, $y^2 = 2y$, and R is not artinian.
5. Note that R is artinian with no nonzero nilpotent ideals; so by the Wedderburn–Artin theorem, R = direct sum of matrix rings over division rings D_i. That each division ring D_i is F follows exactly as in the proof of the Maschke theorem.

Chapter 20, Section 3

1. (a)

$$\begin{bmatrix} 1 & 0 & 0 \\ 0 & 1 & 0 \\ 0 & 0 & 10 \end{bmatrix}$$

is the Smith normal form of the given matrix, which may be obtained by performing $R_3 - R_1, R_2 + 3R_1, C_2 + 3C_1, C_2 + 2C_3, R_2 + 2R_3, R_3 - 2R_2, C_2 - 5C_1$, and so on. Thus, rank = 3.

(b)

$$\begin{bmatrix} 1 & 0 & 0 \\ 0 & 1 & 0 \\ 0 & 0 & (x+1)^2(x+3) \end{bmatrix}$$

is the Smith normal form of the given matrix. Rank = 3.

3. (a) Clearly, the two elements are linearly dependent over $\mathbf{Z}$. So rank = 1.
(b) By performing elementary operations on

$$A = \begin{bmatrix} 2 & 3 & 1 & 4 \\ 1 & 2 & 3 & 0 \\ 1 & 1 & 1 & 4 \end{bmatrix}$$

we can reduce A to

$$\begin{bmatrix} 1 & 0 & 0 & 0 \\ 0 & 1 & 0 & 0 \\ 0 & 0 & 3 & 0 \end{bmatrix}.$$

Thus, the rank of the subgroup is 3.
(c) As explained in (b), we find rank = 3.

Chapter 21, Section 3

(a) Because [1 1] is equivalent to [1 0], the abelian group is isomorphic to $\mathbf{Z} \times \mathbf{Z}/1\mathbf{Z} \simeq \mathbf{Z}$.

(b) Let

$$A = \begin{bmatrix} 3 & -2 & 0 \\ 1 & 0 & 1 \\ -1 & 3 & 2 \end{bmatrix}.$$

By performing elementary operations, A can be transformed to

$$\begin{bmatrix} 1 & 0 & 0 \\ 0 & 1 & 0 \\ 0 & 0 & 3 \end{bmatrix}.$$

Thus, the abelian group is $\mathbf{Z}/1\mathbf{Z} \times \mathbf{Z}/1\mathbf{Z} \times \mathbf{Z}/3\mathbf{Z} \simeq \mathbf{Z}/3\mathbf{Z}$.

(c) Proceeding as in the preceding problem, we find that the abelian group is $\mathbf{Z}/2\mathbf{Z}$.

Chapter 21, Section 4

(a) Let A be the given matrix. The invariant factors of $A - xI$ are 1, 1, $(x-4)(x-1)^2 = x^3 - 6x^2 + 9x - 4$. Thus, the rational canonical form of A is

$$\begin{bmatrix} 0 & 0 & 4 \\ 1 & 0 & -9 \\ 0 & 1 & 6 \end{bmatrix}.$$

(b) If A is the given matrix, then the invariant factors of $A - xI$ are 1, 1, $x^3 - 9x^2 + 14x - 8$. Thus, the rational canonical form of A is

$$\begin{bmatrix} 0 & 0 & 8 \\ 1 & 0 & -14 \\ 0 & 1 & 9 \end{bmatrix}.$$

(c) The rational canonical form of the given matrix is

$$\begin{bmatrix} 0 & 0 & 0 & 0 \\ 1 & 0 & 0 & 1 \\ 0 & 1 & 0 & -3 \\ 0 & 0 & 1 & 3 \end{bmatrix}.$$

Chapter 21, Section 5

1. (a)

$$A = \begin{bmatrix} 0 & 4 & 2 \\ -3 & 8 & 3 \\ 4 & -8 & -2 \end{bmatrix}.$$

On $A - xI$, we perform (i) $R_2 + R_3$, (ii) $R_1 + xR_2$, (iii) $R_3 - 4R_2$, (iv) $C_2 + xC_1$, (v) $C_3 + (x-1)C_1$, (vi) $R_1 + \frac{1}{4}(x+2)R_3$, (vii) $C_3 - \frac{3}{4}C_2$, (viii) $R_1 \leftrightarrow R_2$, (ix) $R_2 \leftrightarrow R_3$,

(x) $\frac{1}{4}R_2$, and (xi) $-4R_3$, successively, and obtain

$$\begin{bmatrix} 1 & 0 & 0 \\ 0 & (x-2) & 0 \\ 0 & 0 & (x-2)^2 \end{bmatrix}.$$

The invariant factors of $A - xI$ are $1,(x-2),(x-2)^2$, and the elementary divisors are $(x-2)$, $(x-2)^2$. Thus, the Jordan canonical form is

$$\begin{bmatrix} 2 & & \\ & 2 & 0 \\ & 1 & 2 \end{bmatrix}.$$

(b) Let A be the given matrix. The invariant factors of $A - xI$ are $1,1,1,(x-4),(x-5)^3$. Thus the elementary divisors are $(x-4)$, $(x-5)^3$; so the Jordan canonical form is

$$\begin{bmatrix} 4 & & & \\ & 5 & 0 & 0 \\ & 1 & 5 & 0 \\ & 0 & 1 & 5 \end{bmatrix}.$$

Chapter 22, Section 2

1. (i) $\frac{a}{b} \otimes \bar{c} = \frac{a}{8b} \otimes 8\bar{c} = \bar{0}$.
   (ii) Choose integers a, b such that $6a + 7b = 1$. Then
   $$\bar{x} \otimes \bar{y} = \bar{x}(6a + 7b) \otimes \bar{y} = \bar{x}6a \otimes \bar{y} + \bar{x}7b \otimes \bar{y} = \bar{0} + \bar{x}b \otimes 7\bar{y} = \bar{0}.$$
   (iii) Consider the map from $\mathbf{Q} \times \mathbf{Z}$ to $\mathbf{Q}$ given by $(q, a) \mapsto qa$ which is clearly balanced. Thus there exists a homomorphism from $\mathbf{Q} \otimes_{\mathbf{Z}} \mathbf{Z}$ to $\mathbf{Q}$ that is invertible.
   (iv) As usual consider the balanced map $\mathbf{Z}_m \times \mathbf{Z}_n \to \mathbf{Z}_d$ where $(\bar{a}, \bar{b}) \mapsto \overline{ab}$. This is well defined, since d is the g.c.d. of m and n and induces a homomorphism $\alpha: \mathbf{Z}_m \otimes_{\mathbf{Z}} \mathbf{Z}_n \to \mathbf{Z}_d$. We can define a suitable homomorphism $\beta: \mathbf{Z}_d \to \mathbf{Z}_m \otimes_{\mathbf{Z}} \mathbf{Z}_n$ which is the inverse of α.
   (v) Use canonical maps (e.g., see Theorem 3.1).

Chapter 22, Section 3

1. Consider the balanced map from $M \times Re$ to Me given by $(m, ae) \to mae$. This induces a homomorphism from $M \otimes Re$ to Me which is invertible.

Chapter 22, Section 4

1. Any element $\sum_i m_i \otimes (\sum_j r_{ij} e_j)$ of $\mathbf{M} \otimes_{\mathbf{R}} \mathbf{F}$ can be rewritten as $\sum_j (\sum_i m_i r_{ij}) \otimes e_j = \sum_j x_j \otimes e_j$, $x_j \in M$. Now $\mathbf{M} \otimes_{\mathbf{R}} \mathbf{F} \simeq \oplus \sum_{j \in J} M_j$, $M_j = M$, under $\sum x_j \otimes e_j \to \sum x_j$. Thus $\sum x_j \otimes e_j = 0$ iff each $x_j = 0$, proving uniqueness of representation of elements of $\mathbf{M} \otimes_{\mathbf{R}} \mathbf{F}$, as desired.
3. $F \simeq \oplus \sum_{j \in J} A_j$, $A_j = A$. $B \otimes_A F \simeq B \otimes_A \sum_j A_j \simeq \sum_j B \otimes_A A_j \simeq \oplus \sum B$, direct sum of $|J|$ copies of B. That $(1 \otimes e_j)_{j \in J}$ is a basis of $B \otimes_A F$ is clear (see Problem 1).
5. In Problem 3, choose $A = R$, $B = R/M$, where M is a maximal ideal. R/M is a field. Since $B \otimes_A F$ is a vector space over the field B, any two bases have the same cardinality. The same is thus true for the free module F (see Problem 3).

7. Let $0 \to A \xrightarrow{f} B \xrightarrow{g} C \to 0$ be an exact sequence of right R-modules. Consider the induced sequence $A \otimes_R N \xrightarrow{f^*} B \otimes_R N \xrightarrow{g^*} C \otimes_R N$, where $f^* = f \otimes I_N$, $g^* = g \otimes I_N$. Since g is onto, it immediately follows g^* is onto. Also, $g^* f^* = (g \otimes I_N)(f \otimes I_N) = gf \otimes I_N = 0$, since $gf = 0$. Thus, $\operatorname{Im} f^* \subset \operatorname{Ker} g^*$. We proceed to show $\operatorname{Im} f^* = \ker g^*$, and thus proving the desired result.

   Consider a mapping from $C \times N \to B \otimes_{\mathbf{R}} N/\operatorname{Im} f^*$ given by $(c, n) \to b \otimes n + \operatorname{Im} f^*$, where $g(b) = c$. Since $\operatorname{Im} f^* \subset \operatorname{Ker} g^*$, the mapping is a well-defined balanced map and induces a homomorphism $\alpha: C \otimes_{\mathbf{R}} N \to B \otimes_{\mathbf{R}} N/\operatorname{Im} f^*$. Furthermore, g^* induces a homomorphism $\beta: B \otimes_{\mathbf{R}} N/\operatorname{Im} f^* \to C \otimes_{\mathbf{R}} N$, whose kernel is $\operatorname{Ker} g^*/\operatorname{Im} f^*$. Observe that $\alpha\beta =$ identity, and so β is $1-1$. This gives $\operatorname{Im} f^* \supset \operatorname{Ker} g^*$ and so $\operatorname{Im} f^* = \operatorname{Ker} g^*$.

   $(\cdot) \otimes N$ need not be a left exact functor. For $0 \to \mathbf{Z} \to \mathbf{Q} \to \mathbf{Q}/\mathbf{Z} \to 0$ is exact. However, $0 \to \mathbf{Z}_2 \otimes_{\mathbf{Z}} \mathbf{Z} \to \mathbf{Z}_2 \otimes_{\mathbf{Z}} \mathbf{Q}$ is not exact since $\mathbf{Z}_2 \otimes_{\mathbf{Z}} \mathbf{Z} \cong \mathbf{Z}_2$ and $\mathbf{Z}_2 \otimes_{\mathbf{Z}} \mathbf{Q} = 0$.

Chapter 22, Section 5

1. $B = M_n(K)$ has a basis $\{e_{ij} \mid 1 \le i, j \le n\}$ consisting of matrix units, as an algebra over K. A typical element of $A \otimes_{\mathbf{K}} B$ is $\sum_{i,j} a_{ij} \otimes e_{ij}$, $a_{ij} \in A$. Define canonical maps to show that $A \otimes_{\mathbf{K}} M_n(K) \simeq M_n(A)$.

3. Let $a, b, c, d \in \mathbf{Z}$, $b \ne 0$, $d \ne 0$. The map ϕ given by $(a/b, c/d) \to \dfrac{ac}{bd}$ is balanced and induces an algebra homomorphism $\phi: \mathbf{Q} \otimes_{\mathbf{Z}} \mathbf{Q} \to \mathbf{Q}$, which has an inverse ψ given by $\psi(a/b) = a/b \otimes 1$, for $\phi\psi(a/b) = a/b$ and $\psi\phi(a/b \otimes c/d) = \psi(ac/bd) = ac/bd \otimes 1 = \dfrac{a}{bd} \otimes c = \dfrac{a}{bd} \otimes \dfrac{cd}{d} = \dfrac{a}{b} \otimes \dfrac{c}{d}$.

5. Let $(e_i)_{i \in \Lambda}$ be a basis of E over F. Thus $p(x) = \sum_{i=1}^m x^i a_i \in E[x]$ can be rewritten as $p(x) = \sum_{i=1}^m \sum_{j \in \Lambda} x^i \alpha_{ij} e_j$, $\alpha_{ij} \in F$, and where all but finitely many α_{ij}, $j \in \Lambda$, are zero. This gives $p(x) = \sum_{j \in \Lambda} p_j(x) e_j$, $p_j(x) \in F[x]$.

   Now the map $(f(x), a) \to f(x)a$ of $F[x] \times E \to E[x]$ is balanced and induces a homomorphism $F[x] \otimes_{\mathbf{F}} E \to E[x]$, and its inverse is given by $p(x) \to \sum_j p_j(x) \otimes e_j$.

Selected bibliography

1. Artin, M., *Algebra*, Prentice-Hall, Englewood Cliffs, N.J., 1991.
2. Bourbaki, N., *Algèbra*, vol. 1, Hermann, Paris, 1970.
3. Cohn, P. M., *Algebra*, vols. 1 and 2, John Wiley, New York, 1974, 1977.
4. Halmos, P. R., *Naive Set Theory*, Van Nostrand, New York, 1960.
5. Hardy, G. H. and Wright, E. M., *An Introduction to the Theory of Numbers*, Clarendon Press, Oxford, 1945.
6. Isaacs, I. M., *Algebra*, Brooks/Cole Publishing Co., a Division of Wardsworth Inc., Pacific Grove, 1994.
7. Jacobson, N., *Basic Algebra*. I, II, W. H. Freeman, San Francisco, 1974, 1980.
8. Kaplansky, I., *Fields and Rings*, University Press, Chicago, 1969.
9. Lang, S., *Algebra*, Addison-Wesley, Reading, Mass., 1965.
10. McCoy, N. H., *Theory of Rings*, Macmillan, New York, 1964.
11. Rotman, J. J., *The Theory of Groups: An Introduction*, Allyn and Bacon, Boston, 1965.
12. Zariski, O. and Samuel, P., *Commutative Algebra*. I, Van Nostrand, New York, 1960.

Index

Printed in the United States
By Bookmasters